Education

Presentations at the 2009 AIChE Annual Meeting

Nashville, Tennessee, USA
8-13 November 2009

ISBN: 978-1-61567-928-7

Printed from e-media with permission by:

Curran Associates, Inc.
57 Morehouse Lane
Red Hook, NY 12571
www.proceedings.com

Some format issues inherent in the e-media version may also appear in this print version.

Printed by Curran Associates, Inc. (2010)

AIChE
3 Park Avenue
New York, NY 10016-5991

Phone: (203) 702-7660
Fax: (203) 775-5177

www.aiche.org

TABLE OF CONTENTS

Metal Organic Framework Based Mixed Matrix Membranes for Gas Separations

Ryan T. Adams, Chemical & Biomolecular Engineering, Georgia Institute of Technology, Atlanta, GA

Cantwell G. Carson, Chemical & Biomolecular Engineering, Georgia Institute of Technology, Atlanta, GA

Willilam J. Koros, Chemical & Biomolecular Engineering, Georgia Institute of Technology, Atlanta, GA

Rina Tannenbaum, Materials Science & Engineering, Georgia Institute of Technology, Atlanta, GA

Traditional mixed matrix membranes (MMMs) are comprised of zeolite particles dispersed in a polymer matrix. Zeolites exhibit high penetrant sorption capacities and improved penetrant size-based selectivities compared to polymers due to large micropore volumes and molecular sieving pore windows. However, zeolite-based MMMs have a number of drawbacks. These include 1) the difficulty of preparing defect-free crystals, 2) the limited number of zeolite structures and compositions, 3) the unfavorable aspect ratios of the crystallites, and 4) costly surface chemistries are often used to improve adhesion to polymer matrices, and such modifications can be detrimental to zeolite transport properties.

Metal organic frameworks (MOFs) are a relatively new class of microporous materials that are an attractive alternative to zeolites as a MMM dispersed phase. Many MOFs can be synthesized easily and quickly at low cost, potential MOF compositions and structures-including high aspect ratio MOFs-are effectively infinite, and the organic linkages provide a useful platform to improve adhesion to polymer matrices and to tune the pore window sizes.

In the present work, a MOF of copper terephthalate (CuTPA) was dispersed in poly(vinyl acetate) to create MMMs. Pure He, H2, CO2, O2, Ar, N2, and CH4 permeation experiments were performed. The permeabilities of all penetrants were enhanced due to the large, 1-dimensional pore window of CuTPA (~ 1 nm). Some selectivity enhancements were observed despite the pore window of CuTPA being much larger than the largest penetrant tested. The permeation results and potential for future advances with MOF based MMMs are discussed.

See more of this Session: Poster Session: Meet the Faculty Candidate
See more of this Group/Topical: Education

Investigating Dynamic Penetrant Transport in Glassy Polymers for High-Tech Applications

Adam K. Ekenseair, Department of Chemical Engineering, The University of Texas at Austin, Austin, TX

Nicholas A. Peppas, Department of Chemical Engineering, The University of Texas at Austin, Austin, TX

Glassy polymers are increasingly used in the construction and fabrication of high-tech devices. These devices are often used in applications, such as microfluidics, membrane separations, food packaging, artificial tissues, bones, and organs, and controlled drug delivery, where the polymers come into contact with fluids capable of penetrating the polymer network. To examine the behavior of such systems both on the macroscopic and molecular level, it is important to study the penetrant diffusion process. In such transport, the macromolecular chains rearrange toward new conformations, where the rate of relaxation depends on the penetrant concentration. The relative rates of penetrant diffusion and macromolecular chain relaxation determine the nature of the transport process and lead to a wide variety of penetrant transport phenomena, such as Fickian, anomalous (non-Fickian), and Case II sorption behavior.

While previous models account for anomalous behavior, there is still a disconnect between theory and experiment, as data must be fit to the model with previously determined independent parameters. With trends leading to smaller device scales and increasingly complex polymer structures, there is a need for a quantitative understanding of the manner in which a polymer's network structure alters both the rate and the mode of penetrant transport. To this end, poly(methyl methacrylate) (PMMA) was synthesized in an iniferter-mediated radical polymerization process to produce a variety of well-characterized polymer network structures. The effects of basic network parameters, including the degree of crosslinking, polymer mesh size, and the crosslink size, on the transport process were studied. The effects of sub-Tg annealing/aging, temperature, and the presence of un-reacted monomer were also investigated.

Utilizing both gravimetric sorption data and in-situ ultra-high-resolution X-ray computed tomography studies, the transport dynamics of methanol and other solvents in PMMA discs were investigated. Controlling the relative timescale of the relaxational process by altering the polymer network structure is shown to directly influence the Case II front propagation velocity and the nature of the observed transport behavior.

See more of this Session: Poster Session: Meet the Faculty Candidate
See more of this Group/Topical: Education

Constraining Ozonolysis Product Distributions with Temperature Programmed Reaction Spectroscopy and Computational Chemistry

Scott A. Epstein, Department of Chemical Engineering, Carnegie Mellon University, Pittsburgh, PA

Neil M. Donahue, Department of Chemical Engineering, Carnegie Mellon University, Pittsburgh, PA

Reactions between biogenic alkenes and ozone are a significant source of atmospheric particulate matter (PM), contribute to the aging of PM, and in the gas phase, produce hydroxyl radicals-another important atmospheric oxidant. All ozonolysis reactions proceed through an initial highly unstable 1,2,3-trioxolane, the primary ozonide (POZ). Depending on the structure of the parent alkene, decomposition of the POZ can proceed through two different reaction pathways, sometimes leading to drastically different subsequent chemistry. We strive to constrain product distributions by examining the POZ decomposition branching using laboratory experiments and computational chemistry models.

In order to isolate the highly unstable POZ, an apparatus was constructed to perform temperature programmed reaction spectroscopy (TPRS) on a liquid nitrogen cooled surface with Fourier transform infrared spectroscopy. We will show results from experiments with SOA precursors and their surrogates-cyclohexene, 1-methyl-cyclohexene as a surrogate for α-pinene, and methylene-cyclohexene as a surrogate for β-pinene. With this instrument, we can determine POZ decomposition rate constants and characterize decomposition products for these three alkenes. We will also show that Density Functional Theory (DFT) calculations are a reliable way to predict POZ decomposition branching and extend these calculations to include several other atmospherically relevant terpenes.

See more of this Session: Poster Session: Meet the Faculty Candidate
See more of this Group/Topical: Education

Designed Syntheses of Micro and Meso Porous Materials as Catalysts for Petrochemistry, Fine Chemistry and Conversion of Biomass

Wei Fan, Department of Chemical Engineering and Materials Science, University of Minnesota, Minneapolis, MN

Crude oil as the major feedstock for the petrochemical industry is the dominant energy source driving the world economy, but known oil reserves will cover demand for no more than 50 years at the current rate of consumption. This situation requests more efficient strategies for converting crude oil into fuel, petrochemical products and fine chemical products. Moreover, with decreasing crude oil reserves, enhanced demand for fuels worldwide, increased climate concerns about the use of fossil-based energy carriers, and political commitment, the focus has recently turned towards improved utilization of renewable energy resources. Among the renewable energy sources, biomass is an abundant and carbon-neutral renewable energy resource for the production of biofuels and valuable chemicals. Heterogeneous catalysts are primarily used for catalytic fuel-upgrading in refineries and also for the production of various petrochemicals. Their superior performance can often be attributed to the existence of a well-defined system of micropore with uniform shape and size. However, in many cases, the sole presence of micropore can also be a major limitation because mass transport to and from the active sites located within the micropore is slow, which is often limiting the performance of industrial catalysts. I aim to design and prepare a new class of solid catalysts with tunable hierarchical pore structure, as well as controllable surface properties, with the consideration of the applications in conversion of crude oil, synthesis of fine chemicals and conversion of biomass. With the presence of ordered microporosity and mesoporosity, the catalysts are expected to exhibit the molecular sieving properties of zeolites without the mass transfer limitations, and high accessibility for large components from biomass. Furthermore, tunable pore structures and controllable surface properties allow us to rationally design the catalysts for various chemical reactions.

See more of this Session: Poster Session: Meet the Faculty Candidate
See more of this Group/Topical: Education

Atom-by-Atom Metrology of Materials

Domingo Ferrer, Microelectronics Research Center, University of Texas at Austin, Austin, TX

Sanjay K. Banerjee, Microelectronics Research Center, The University of Texas at Austin, Austin, TX

Metallic and semiconductor nanostructures possess an array of unique magnetic, electronic and photonic behavior that often differs from their bulk counterparts. The overarching goal of the research proposed herein is elucidating the synthesis-structure-performance relations of atomic-scaled materials by advanced transmission electron microscopy (TEM) techniques. The availability of various imaging, diffraction and spectroscopic modes within a single instrument makes of TEM a powerful tool for characterizing functional nanostructures. These efforts enable us to understand the surface chemistry and electronic structure of diverse atomic-scale systems such as nanocatalysts, transistors or functional nanostructures as therapeutic devices. TEM metrology analysis is used to determine feature size/functions and physical interface states of atomically-designed materials.

One of the core-strengths of our research program is the production of wide variety of semiconductor and metal nanocrystals (and nanowires) by liquid-phase colloidal synthesis in aqueous or non-hydrolytic medium. The colloidal synthetic approach is an especially powerful tool for convenient and reproducible synthesis of nanostructures. An example of our current work on the synthesis of shape-controlled nanocrystals corresponds to the fabrication of sphere- and cubic-shaped PbTe and PbSe NCs. These NCs were synthesized using a one-pot approach and the noncoordinating solvent, 1-octadecene (ODE) as capping agent. This one-pot technique does not need cooling of the precursors in a glovebox and is advantageously reproducible. Materials displaying rock salt structures such as PbTe and PbSe show a size-dependent sphere to cube transition, which occurs at high-temperature regimes. The growth of {100} facets leads to a cubic morphology and lowers total surface energy. Therefore, the sphere to cube transition tends to occur at smaller sizes for PbTe NCs compared to those of PbSe NCs. The different surface energies that exists among {111} and {100} faces is greater for PbTe than for PbSe, favoring a faster growth in the <111> direction for PbTe and a predominant {100} face for nanocrystals with reduced sizes. The ultimate goal of this research involves engineering the work function of the floating gate of non-volatile memory devices with semiconductor nanocrystals of different shapes and sizes.

Axial heterostructures, namely core-shell semiconductor nanowires offer unique features for diverse applications including field-effect transistors, sensors, detectors, light-emitting diodes, and solar cells. Our research efforts placed on the synthesis of core-shell nanowires involved using a combination of silane (SiH4) and germane (GeH4) gases by a ultrahigh vacuum chemical vapor deposition (UHV-CVD) method for successful fabrication of a Si1-xGex shell around the Ge nanowires at wafer temperatures of approximately 400 °C. Another example of recent work on metrology of nanostructures-based electronic devices is the cross sectional view of dual-gated graphene field effect transistor (FET). Graphene is a single layer of carbon atoms, densely packed in a honeycomb two-dimensional (2D) lattice. The waviness nature of the graphene acting as channel in this device indicates the presence of mechanical deformations, which could have also been introduced during the fabrication process of the device. The corrugation of graphene in FET devices could dramatically affect its transport properties.

The efforts of this research program demonstrate how electron microscopy can be used to answer several critical issues in chemistry, biology and materials science aimed at revolutionizing the current understanding of nanostructures applications.

See more of this Session: Poster Session: Meet the Faculty Candidate
See more of this Group/Topical: Education

Solvent Engineering for Biorefining Catalysis and Separations

Jackson W. Ford, Center for Environmentally Beneficial Catalysis and Department of Chemical & Petroleum Engineering, University of Kansas, Lawrence, KS

Bala Subramaniam, Center for Environmentally Beneficial Catalysis and Department of Chemical & Petroleum Engineering, University of Kansas, Lawrence, KS

Raghunath V. Chaudhari, Center for Environmentally Beneficial Catalysis and Department of Chemical & Petroleum Engineering, University of Kansas, Lawrence, KS

Charles A. Eckert, Chemical and Biomoelcular Engineering, Georgia Institute of Technology, Atlanta, GA

Efficient biomass utilization to produce fuels and chemicals is one of the key challenges facing chemical engineers today. At the core of this problem is the need to embrace renewable feedstocks while leveraging our existing petrochemical infrastructure. Energy-efficient unit operations, novel processes, and advanced separation and catalytic reaction engineering concepts must be developed for the biorefinery of the future. Alternative solvents including supercritical fluids and switchable solvents have tremendous potential in these areas. My research is aimed at exploiting the synergies between catalysis and solvent engineering to develop unique processes for the biorefinery.

For example, biomass pyrolysis oil (bio-oil) has the potential to dramatically impact the transportation fuel sector of the energy market, but new approaches must be used to efficiently upgrade bio-oil to a refinery-ready feedstock. Switchable solvents may be able to selectively extract polar compounds from bio-oil, separating the bio-oil into more homogeneous fractions for upgrading to hydrocarbon fuels or value-added chemical intermediates. For chemical processes, supercritical solvents can enhance hydrogen availability such as in the hydrogenation of triglycerides, enabling milder reaction conditions and reducing both capital and raw materials costs. Specific fundamental and practical aspects of these challenging problems will be highlighted.

See more of this Session: Poster Session: Meet the Faculty Candidate
See more of this Group/Topical: Education

Rational Design of Materials for Control of Chemical Processes

Rachel B. Getman, Department of Chemical and Biological Engineering, Northwestern University, Evanston, IL

One of the grand challenges in chemical engineering is to find new ways to manipulate chemical phenomena. A prominent application of this goal is controlling the rates of chemical reactions, commonly achieved with catalysts. Catalysts are typically applied on a large scale, but they function at the molecular level. Effective large-scale catalyst design requires an intimate understanding of molecular-level catalytic behavior. Simulation provides a means for understanding, but a dominant challenge is bridging the gap between microscopic models and macroscopic behavior. In this work we show how careful molecular simulation and thermodynamic and kinetic modeling were used in collaboration with experimental observation to elucidate the key features in Pt-catalyzed NO oxidation, an important process in NOx remediation that occurs in strongly oxidizing conditions. The results pave the way for catalyst design and make implications for other catalyzed oxidations. Future work in these areas will be discussed.

See more of this Session: Poster Session: Meet the Faculty Candidate
See more of this Group/Topical: Education

Integrated Nanotechnology for Molecular Biophysics

Edgar D. Goluch, Kavli Institute of Nanoscience, Section Molecular Biophysics, Delft University of Technology, 2628 CJ Delft, Netherlands

Observation of individual biomolecules allows for characterization of real-time dynamics, molecular subpopulations, and heterogeneous biological behavior. This information, in turn, provides a molecular-based understanding of biological mechanisms and may elucidate the underlying causes of disease states. With these ideas in mind, the goal of my proposed research is to develop and employ biophysical tools for the investigation of biological phenomena at the molecular level. Specifically, I plan to exploit the advantages that nanofluidic structures provide when combined with various detection modalities, such as, surface plasmon resonance, surface enhanced Raman spectroscopy, and infra-red imaging.

During my doctoral studies at the University of Illinois, I developed microfluidic systems for biochemical and medical diagnostics. In collaboration with the Sligar Group, I employed a microfluidic device to pattern nanoscale lipid bilayers, known as Nanodiscs, and investigate their interactions with proteins in a multiplexed manner. I also designed and tested, with the Mirkin Group at Northwestern University, a disposable device for ultra-sensitive detection of protein cancer markers. My postdoctoral research involves the development of nanofluidic devices for single molecule analysis. I am employing these devices to study the dynamics of enzymatic activity via electrochemical detection.

See more of this Session: Poster Session: Meet the Faculty Candidate
See more of this Group/Topical: Education

Porous Materials: Design and Modeling Challenges

Bradley R. Gordon, Department of Chemical and Biomolecular Engineering, University of Maryland, College Park, MD

Porous materials are characterized as having a "sponge-like" structure and encapsulate a range of materials including catalysts, membranes, and skin. In particular these materials have the potential to be incredibly useful for drug delivery and energy-related applications as they have large surface areas and can be used to store compounds. My current work focuses on the controlling the transport in these materials using surface modification techniques as well developing an understanding of the phenomenological behavior using numerical approaches. In particular, my research aims to create smart porous materials that have altered transport characteristics in response to external stimuli. In addition, my current work focuses on the digital design of porous materials using modeling techniques in an effort to create novel materials with intriguing properties.

See more of this Session: Poster Session: Meet the Faculty Candidate
See more of this Group/Topical: Education

Enzymatic Synthesis in Deep Eutectic Solvents

Johnathan T. Gorke, Chemical Engineering and Materials Science, University of Minnesota, Minneapolis, MN

Friedrich Srienc, Chemical Engineering and Materials Science / Biotechnology Institute, University of Minnesota, Saint Paul, MN

Romas J. Kazlauskas, Biochemistry, Molecular Biology and Biophysics, University of Minnesota, Saint Paul, MN

Enzymes are desirable catalysts to study because they give high selectivity with minimal byproducts, and industrial enzymatic processes are becoming increasingly common. Often, non-aqueous conditions are required for desired reactions, particularly with industrial hydrolases. These processes commonly use volatile organic solvents (VOS), which are often flammable, toxic, and environmentally harmful. Ionic liquids (ILs) are molten salts that are recyclable, non-volatile, and non-flammable alternatives to VOS that initially showed promise in enzymatic reactions. ILs offer advantages in enzymatic reactions such as high enzyme stability, ease of product separation, and recyclability; but their high cost is prohibitive, and the lack of sustainable materials is undesirable. However, current ILs are mainly prepared from petrochemically-derived, toxic materials such as imidazole, fluorinated amides, and halogenated alkanes. Further, ILs require high purity, are not available on large scales, and are typically orders of magnitude more expensive than VOS, limiting their widespread use. Deep eutectic solvents (DES's) are physical mixtures of ammonium or metal salts and hydrogen bond donors, such as 1:2 choline chloride (a naturally occurring vitamin used in chicken feed): urea (fertilizer or animal waste product) or 1:2 choline chloride:glycerol (biodiesel byproduct) are promising replacements for both VOS and ILs. In addition to offering the recyclability, low volatility, and low flammability of ILs, DES's are composed of non-toxic and inexpensive materials including vitamins, amides, sugars, and alcohols that have costs and production scales comparable to VOS. Many of these components are natural products or can be made from renewable materials, meaning that DES's are a sustainable alternative to common ILs and to many petroleum-based organic solvents. Further, DES's are trivial to synthesize compared to most ILs, Figure 1. We have found that DES's are effective replacements for a number of hydrolase-catalyzed reactions, including transesterification, epoxide hydrolysis, and ring-opening polymerization. This efficacy was not obvious, however. The use of DES's in enzymatic reactions was initially questionable due to the presence of strong denaturants such as chloride anions strong hydrogen bond donors that inhibit enzyme activity in ILs. However, we hypothesized that intrasolvent hydrogen bonding between DES components would be preferred over hydrogen bonding between solvent and enzyme, a key step in denaturation. We have found conversions, selectivity, and activities in DES's comparable to those found under similar conditions in VOS. For example, we found nearly identical polymerization activity of *Candida antarctica* lipase B in the production of polycaprolactone from ε-caprolactone in 1:2 choline chloride:urea compared to in toluene. We also found a 20-fold conversion enhancement of epoxide hydrolase-catalyzed hydrolysis of styrene oxide to styrene glycol (an important reaction in pharmaceutical production) in 25% 1:2 choline chloride:glycerol compared to in water or water/VOS mixtures.

See more of this Session: Poster Session: Meet the Faculty Candidate
See more of this Group/Topical: Education

High Capacity Silicon Anode Composite for Li-Ion Battery and Novel Anion Conductive Membrane for Alkaline Fuel Cell

Juchen Guo, Department of Chemical and Biomolecular Engineering, University of Maryland, College Park, MD

Chunsheng Wang, Department of Chemical & Biomolecular Engineering, University of Maryland, College Park, MD

Yanting Luo, Chemical and Biomolecular Engineering, University of Maryland, College Park, MD

Two parallel projects will be presented in this poster. The first one is the development of high energy high power anode composite based on silicon for Lithium-ion battery. Novel polymer scaffold-Si electrodes were prepared using slurry spray technique. This type of electrodes showed high energy density, power density and superb cycle stability, because the porous polymer scaffold can accommodate more volume changes of Si and enhance the lithium ion transport in the electrodes, as well as being able to improve the electrochemical reaction kinetics. The electrochemical performance can be further improved by carbonization of the polymer scaffold to form a porous carbon matrix in which silicon particles are uniformly dispersed.

Second project is the development of an anion (hydroxyl group) conductive membrane based on copolymer made of three monomers, methyl methacrylate, butyl acrylate and vinylbenzyl chloride. The preliminary results demonstrated that the random copolymer from these three monomers has promising anion conductivity with great physical strength even at fully water saturated condition. The comb-structure copolymer based on these monomers is currently under investigation. The conductivity and fuel cell performance of this anion conducting membrane will be presented.

See more of this Session: Poster Session: Meet the Faculty Candidate
See more of this Group/Topical: Education

Fibronectin Domain Engineering: Improved Agents for Biotechnology and Tumor Targeting

Benjamin J. Hackel, Chemical Engineering, Massachusetts Institute of Technology, Cambridge, MA

K. Dane Wittrup, Chemical Engineering, Massachusetts Institute of Technology, Cambridge, MA

Molecular recognition, the ability to provide high affinity, high specificity binding to a target of interest, is of critical importance to biology and biotechnology. Binding reagents enable selective detection and purification for laboratory research and biotechnology applications as well as potent delivery of therapeutic or diagnostic agents for clinical applications. The natural solution to molecular recognition, the antibody, has historically dominated binding reagent selection because of its broad binding capacity and the ability to generate binders by immunization. However, antibodies are large, complex, multi-domain proteins of moderate stability; thus alternative protein scaffolds can provide a multitude of advantages. In particular, the fibronectin type III domain provides facile cysteine- and amine-based chemical conjugation; small size for in vivo imaging and proteomics, a cysteine-free fold for intracellular use; single-domain architecture for protein fusions; production in bacterial culture; and improved stability and solubility.

Yet, native fibronectin binds a single target and must be engineered to provide broad binding capacity akin to antibodies. To robustly engineer high affinity binders we have developed a platform in which a combinatorial protein library is screened for binders and lead clones are improved by directed evolution. Multiple studies of synthetic combinatorial library design elucidated the critical elements including loop length diversity, stability bias, and a tailored amino acid repertoire; these elements were combined to produce a vastly improved synthetic library. A magnetic bead technique was developed to exploit avidity for selection of low affinity binders prior to affinity maturation. To improve the efficiency and efficacy of the search of combinatorial sequence space during clone evolution, a novel dual-mutagenesis method was developed and validated. Techniques and insights generated during development of the fibronectin platform are directly applicable to protein engineering in general.

This platform has enabled us to generate high affinity binders to myriad targets including immunoglobulin G for biotechnology applications and the tumor targets carcinoembryonic antigen, CD276, and epidermal growth factor receptor. Of multiple tumor targeting directions, particular focus was placed on downregulation of epidermal growth factor receptor by novel bispecific fibronectin constructs. Unique heterobivalent constructs were developed that robustly and swiftly decrease surface receptor levels resulting in reduced proliferation and migration of cells thereby presenting potential for therapeutic use.

See more of this Session: Poster Session: Meet the Faculty Candidate
See more of this Group/Topical: Education

Quantifying Physical and Chemical Defects in Zeolites and Zeolite Membranes

Karl D. Hammond, Chemical Engineering, University of Massachusetts, Amherst, MA

Zeolites are some of the most intriguing materials ever discovered. Their crystal structures contain pores the size of small molecules, giving them a "molecular sieving" property and making them ideal materials for use as adsorbents. Zeolites with over 190 different framework connectivities have been discovered, and applications involving zeolites exist in such diverse fields as landscaping, water treatment, mixture separations, catalysis, and even nuclear waste remediation. Particular interest has surfaced in recent years to develop zeolites and similar materials that do two things: (1) Serve as separation membranes with high throughput and high selectivity, and (2) Catalyze reactions that require strong bases in a selective fashion. The microporous nature of zeolites is what provides the selectivity: some molecules are large enough to fit into the zeolite pores, while others are not. In membranes, high selectivity via such molecular sieving requires that all paths through the membrane pass through a zeolite pore. Predicting the properties of zeolite membranes therefore requires knowledge of the presence and influence of *physical defects* in the membranes, which facilitate transport via non-zeolitic porosity. In the case of zeolites as strongly basic catalysts, it is *chemical* defects that are of import: strong base sites do not exist in zeolites, and must be created. Such sites can be created by a post-synthetic treatment which substitutes framework oxygen for nitrogen, but these treatments can also create physical defects and even destroy the microporous structure of the zeolite. This work addresses various challenges associated with the characterization of zeolite membranes and nitrogen-substituted zeolites, using techniques such as X-ray diffraction, NMR spectroscopy, vibrational spectroscopy, and high-resolution physical adsorption to quantify defects and structural changes in zeolites and zeolite membranes. Experimental work is combined with theoretical methods to yield unique insight into these intriguing materials.

See more of this Session: Poster Session: Meet the Faculty Candidate
See more of this Group/Topical: Education

Development of One-Component Reversible Ionic Liquids for Energy Applications

Ryan Hart, School of Chemical & Biomolecular Engineering, Georgia Institute of Technology, Atlanta, GA

Vittoria Blasucci, School of Chemical and Biomolecular Engineering, Georgia Institute of Technology, Atlanta, GA

Pamela Pollet, School of Chemistry and Biochemistry, Georgia Institute of Technology, Atlanta, GA

Charles Eckert, Chemical & Biomolecular Engineering, Georgia Institute of Technology, Atlanta, GA

Charles L. Liotta, School of Chemistry and Biochemistry, Georgia Institute of Technology, Atlanta, GA

We have recently reported the synthesis and characterization of one-component reversible ionic liquids, and presented here is the direct application of the novel solvent systems to carbon capture and extraction of crude oil from tar sands/oil shale. We look to develop optimal solvent systems for the given application by examining how the molecular structure of our reversible ionic liquid solvent systems affects their physical and thermodynamic properties, namely the viscosity, equilibrium constant (K), enthalpy of reaction (ΔHrxn), and Gibbs free energy (ΔGrxn). Addition of carbon dioxide (CO2) to a liquid tri-substituted silylpropyl amine causes the conversion of the molecular liquid to a room-temperature ionic liquid, which is easily reversed by stripping with an inert gas or the addition of heat. This molecular-to-ionic liquid conversion results in a drastic change in the chemical and physical properties of the solvent, advantageous for the development of sustainable processes. An example of this switch is seen where non-polar alkanes are soluble in the molecular liquid but insoluble in the ionic liquid, leading to the development of a process for tar sands/oil shale extraction.

We have systematically examined the effect of altering the molecular structure on K by using ATR-FTIR at CO2 pressures up to 70 bar and temperatures up to 150°C. Knowledge of K as a function of temperature and pressure allows for determination of ΔHrxn and ΔGrxn at different processing conditions, critical values for the optimization of solvent systems that can be economically viable for commercial applications. Such reversible ionic liquids have been shown to combine a homogeneous reaction with heterogeneous separation, for facile product purification and catalyst recycle. Another application that we are investigating is the recovery of CO2 from large point-source emitters. Our reversible ionic liquids combine the notable CO2 carrying capacities of conventional ionic liquids with chemical sorption through reaction, and can be optimized for a given feed stream by use of structure-property relationships.

See more of this Session: Poster Session: Meet the Faculty Candidate
See more of this Group/Topical: Education

Back to the Future - a New Generation of Chemical Reaction Engineering Challenges

Ryan L. Hartman, Department of Chemical Engineering, Massachusetts Institute of Technology, Cambridge, MA

Klavs F. Jensen, Massachusetts Institute of Technology, Cambridge, MA

Onsite conversion of biomass, dissolution reactions in small-scale flows, and membrane science all include emerging challenges that chemical reaction engineering fundamentals can tackle. Furthermore, these research topics show potential for impacting science and education over the next decade. Onsite conversion of biomass from second and third generation feedstock to chemical products would help address transportation and storage issues that reduce efficiency. Modular, meso-scale processing units could potentially overcome this challenge, but require understanding chemical mechanisms, engineering catalysts, and creating novel unit operations. Dissolution reactions remain an important class of chemical transformations that impact our lives. Drug delivery, feedstock conversion (e.g., solid waste and forest residues), and synthetic chemistry applications can take advantage of accurate kinetics and mechanisms. The use of small-scale flow systems to elucidate this knowledge is limited by a lack of strategies for handling solid particulates, which presents tremendous opportunity for new insight. Finally, membrane science is on the brink of a revolution by potentially impacting energy, environmental (including water sustainability), and biological problems that demand robust and more efficient materials. Next generation membranes include novel hybrid materials that meet application requirements yet are efficiently manufactured. Nevertheless, new science is needed to continue moving this field forward. These emerging challenges warrant a multifaceted approach, which includes interdisciplinary collaboration, partnerships with industry and academia, and integration of modern problems into our curriculum.

Current research work at the Massachusetts Institute of Technology concerns application of microchemical systems for discovery and development. Over the past decade, microchemical systems have evolved from simple devices for basic chemistry (e.g., microreactors) to more complex systems for multi-step synthesis. Early studies using microreactors revealed the many advantages of chemical synthesis in small-scale flows such as enhanced heat and mass transfer characteristics, safer synthesis of dangerous compounds, isolation of air and moisture sensitive chemistry, and reduction of hazardous waste. Optimized reaction conditions and rapid experimentation also add value to the technology by shortening product development life cycles. These reasons and others have provoked recent interest in continuous-flow synthetic chemistry using microchemical systems. Nevertheless, small-scale continuous-flow chemistry is not without challenges. Efficient separations and handling solids are two challenges facing discovery and development with microchemical systems. These topics will be presented and discussed.

See more of this Session: Poster Session: Meet the Faculty Candidate
See more of this Group/Topical: Education

Transport in Polymer Electrolyte Membranes for Energy Applications

Daniel T. Hallinan Jr., Department of Chemical Engineering, University of California, Berkeley, CA

Yossef A. Elabd, Chemical and Biological Engineering, Drexel University, Philadelphia, PA

Polymer electrolyte membranes (PEMs) hold potential to improve performance in fuel cells and batteries, electrochemical devices that can generate electricity efficiently. In particular, hydrogen fuel cells are promising for clean, renewable transportation. Direct methanol fuel cells (DMFCs) and lithium ion batteries can operate at room temperate and have high energy density, making them ideal for portable power applications. The electrolyte is an integral part of any electrochemical cell because it serves to conduct ions while separating the electrodes. Useful hydrogen fuel cell performance requires high water sorption and dynamics (diffusion) to maintain sufficient ion conductivity in Nafion (most frequently used PEM to date), which conflicts with the goal of high temperature and low humidity operation. DMFC power output suffers from methanol crossover (when methanol permeates across Nafion from the anode to the cathode). Therefore, understanding water and methanol diffusion in Nafion is critical. Although many investigators have reproduced water sorption isotherms in Nafion, reported diffusion coefficients varied considerably. Also, many investigators have focused on reducing methanol diffusion in PEM development for the DMFC, despite no clear evidence for that approach.

Water and methanol diffusion in Nafion was studied using time-resolved Fourier transform infrared - attenuated total reflectance (FTIR-ATR) spectroscopy. Critical assessment of water transport phenomena in Nafion identified vapor-phase mass transfer resistance and two non-Fickian regimes, explaining some of the variation in reported diffusion coefficients and improving the fundamental understanding of the transport mechanisms. Methanol crossover (flux) was explicitly shown to increase with increasing methanol concentration. More importantly, the increase was found to be more strongly dependent on methanol sorption or partitioning (chemical interaction) rather than methanol diffusion. The implication is that an effective PEM for the DMFC must be chemically incompatible with methanol.

See more of this Session: Poster Session: Meet the Faculty Candidate
See more of this Group/Topical: Education

Electrochemical Synthesis of One-Dimensional Nanostructures

Carlos M. Hangarter, Chemical and Environmental Engineering, University of California, Riverside, Riverside, CA

One-dimensional (1-D) nanostructures, such as nanowires (NWs) and nanotubes (NTs), are attractive materials for electronics because of their potential for device miniaturization and unique properties which arise from size dependent quantum confinement effects. These building blocks of nanotechnology can be further complexed with axial and radial interfaces, for segmented, multilayered and core/shell structures. However, to fully exploit these higher level configurations, nanomaterials must be engineered for specific properties and interfacial characteristics that give rise to enhanced functionalities for communications, data storage, logic operations, sensing, and energy conversion. This ardent task requires a fundamental understanding of the synthesis approach on the structure-property relationship as well as size dependent effects. Yet the true values of these physical properties are not always accurately measured at this level, as many can only be assessed by interrogation of individual nanostructures. Additionally, the commercialization of nanowire-based devices has been severely stunted by stunted by controllable integration, including contact and alignment, into existing technologies.

During my time at UCR I have employed various electrochemical techniques to synthesize a barrage of nanostructures for sensing, spintronic and optoelectronic applications. The bulk of my work focused on conjugated polymers and the development of hybrid organic/inorganic nanoconstructs, but also extended to nanopeapods or chains nanoparticles. To describe fundamental characteristics of these materials we have developed methodologies for isolating single nanowires and characterizing their true magneto and electron transport properties to distinguish interfacial and size effects. Rational assembly of such nanoentities is however, a continuing challenge in the field. To this end, the inherent magnetic and electric properties of these materials were utilized for aligning single nanowires and electrical contact has also been explored with annealing, lithography, electron beam lithography (e-beam), and focused ion beam (FIB).

See more of this Session: Poster Session: Meet the Faculty Candidate
See more of this Group/Topical: Education

Molecular Recognition in Bioprocessing, Biosensing and Disease Researchmolecular Recognition in Bioprocessing, Biosensing and Disease Research

Caryn L. Heldt, Rensselaer Polytechnic Institute, Troy, NY

Many different biological processes involve the interaction of proteins with other proteins or small chemical molecules, including enzymatic reactions, viral entry into cells, and cell-to-cell signaling. These molecular recognition events are the bases for life as we know it, yet they are not well understood. The chemical engineer is ideally suited to study a plethora of these protein interactions and apply them to many different challenges. Some of these challenges include improving the sustainability of biological purification processes by creating new affinity resins for more efficient purification and virus removal steps. Molecular recognition could also be used to create surfaces that could function as a biosensor for environmental and human toxins. The continued study of protein-protein interactions could also reveal the mechanism of cell destruction in many diseases, leading to improved treatments and possibly cures to these diseases. All of these challenges can be met by studying the biophysical interaction of proteins with other entities at a molecular level. This can be accomplished with a focus on thermodynamics and kinetics, both of which are traditional chemical engineering areas. I have experience in different aspects of molecular recognition. I have worked on the discovery of small peptide ligands that bind to porcine parvovirus (PPV) a small nonenveloped virus. The peptides were discovered from the screening of a solid-phase combinatorial library. Many trimeric peptides were discovered that completely remove PPV when it is spiked into saline. One trimer, WRW, has been found that can clear PPV from the first three column volumes from a solution of 7.5% human blood plasma spiked with PPV. These peptide columns were compared to a weak ion exchange resin that was able to clear considerably less PPV from the same spiked solutions. This peptide ligand has the ability to specifically bind to the target virus. This molecular recognition can be used for removal, detection, and concentration of virus solutions. I would like to continue my work by discovering new and interesting molecular recognition surfaces and probing their specificity. I have also worked on the aggregation of amyloid proteins and their implications to human diseases. The aggregation of amyloid proteins can be manipulated to change the rate of aggregation formation, the morphology of the aggregate formed, and the toxicity of the aggregates. By adding different osmolytes to solutions of the amyloid peptide Aβ, we have been able to discover different morphological changes in the aggregates formed with little change to the rate of aggregate formation. This will continued to be probed to examine the toxicity and stability of these aggregates. All of this research involves the molecular recognition of proteins and peptides. I plan to continue my research career in this area to improve bioprocessing, create biosensors, and to probe the mechanisms of human diseases.

See more of this Session: Poster Session: Meet the Faculty Candidate
See more of this Group/Topical: Education

Bridging the Gap: Multiscale Approaches to Integrated Colloids and Nanomaterials

Matthew E. Helgeson, Chemical Engineering, MIT, Cambridge, MA

Norman J. Wagner, Chemical Engineering, University of Delaware, Newark, DE

Eric W. Kaler, Stony Brook University, Stony Brook, NY

Patrick S. Doyle, Chemical Engineering, MIT, Cambridge, MA

Next-generation soft materials involve the integration of molecular and colloidal components, such as self-assembled surfactants, polymers, and nanoparticles, whose function and behavior are governed by increasingly complex interactions. As such, they require a detailed understanding of the hierarchical nature of self-assembly, whereby molecular interactions, mesoscale structure, and macroscopic properties are intimately connected. Although these materials hold potential for breakthroughs in the fields of medicine, energy, and sustainability, control and scaleup of their formulation and processing poses a significant technological challenge. My research aims to develop a robust understanding of mixed colloidal systems for the synthesis of functional colloids and nanomaterials. An experimental toolbox that combines techniques such as microscopy, scattering, and rheology that bridge the micro- and macroscales provides a foundation for understanding and modeling colloidal assemblies and their response to applied fields and external stimuli. Furthermore, the combination of bottom-up and top-down approaches to material synthesis leads to versatile platforms for the production of functional soft materials. This presentation summarizes results from my doctoral (Department of Chemical Engineering, University of Delaware) and postdoctoral (Department of Chemical Engineering, Massachusetts Institute of Technology) research.

Controlling the structure and dynamics of wormlike micelles using colloidal particles (and vice versa)

Wormlike micelles (WLMs) have become widely used in a number of products and processes where they come in contact with colloidal species, yet relatively little is understood regarding the interactions between WLMs and colloids and resulting changes in macroscopic properties. This work demonstrates that the addition of model nanoparticles to WLMs allows for unique tunability of the structure and rheology of the resulting fluid. Interactions at the surfactant-nanoparticle interface can be tuned to promote the fusion of micelles to an adsorbed surfactant layer to form micelle-nanoparticle junctions. These junctions act as physical cross-links between micelles and impart significant viscosity and viscoelasticity to otherwise Newtonian micellar solutions. This rheological enhancement is understood by the formation of a so-called "double network", where the viscoelasticity can be tuned through two energetic scales, the micellar end cap energy and micelle-nanoparticle junction energy, which are readily manipulated by adjusting solution conditions.

Conversely, the formation of micelle-nanoparticle junctions gives rise to new colloidal interactions between nanoparticles which can be tuned through the self-assembled micellar structure, leading to thermoreversible phase separation of the colloid. A statistical mechanical model has been developed to describe these interactions in terms of experimentally measurable properties. This model allows for a priori predictions of the interaction potential between colloids dispersed in WLMs, and shows quantitative agreement with independent scattering measurements and phase behavior.

Shear-induced phase transitions in surfactant liquids

Because the structure of complex fluids typically depends on a delicate balance of forces, applied flow fields such as simple shear can result in significant changes in microstructure and phase behavior. Shear banding is a particularly prominent flow instability in complex fluids, including WLMs. Here, model shear banding

WLMs near an equilibrium isotropic-nematic (I-N) transition are studied using a combination of spatially-resolved rheological and scattering measurements to determine the underlying mechanism of banding and its relation to equilibrium surfactant phase behavior. The results reveal that shear banding is due to a first order, shear-induced I-N transition upon a critical degree of micellar flow-alignment. Constitutive modeling allows for combination of rheological and microstructural data to produce a non-equilibrium phase diagram which explains both the microstructural underpinnings as well as the macroscopic flow kinematics and stability during shear banding.

One-step platforms for the synthesis of polymeric colloids and nanomaterials

The assembly of polymer and composite materials is typically performed by one of two methods: bottom-up approaches using molecular self-assembly suffer from low yields and slow kinetics, whereas top-down approaches lack precise control of final material properties. Here, two methods are investigated that combine both approaches resulting in versatile platforms for the synthesis of soft materials. The first is electrospinning, which has gained significant attention for the production of polymeric particles and nanofibers. Fundamental studies and modeling of electrospinning jets yield simplified theories for jet kinematics and rheology, ultimately resulting in correlations for the final fiber morphology in terms of measurable fluid properties and controlled process parameters. These results are used to guide the synthesis of novel nanomaterials using electrospinning, including polymer-nanoparticle composite fibers for functional membranes and textiles, and reversibly-swelling surfactant polyelectrolyte coacervate nanoparticles for encapsulation and release.

Stop flow lithography (SFL) also provides a facile route to functional colloidal materials with arbitrary shape, whereby the conditions and substituents of photopolymerization are finely controlled using microfluidic devices. As such, SFL provides an inexpensive, customizable platform for the continuous manufacture of pharmaceuticals with direct incorporation of APIs and other biomaterials. Of particular interest to this work is the synthesis of functional hydrogels with hierarchical topologies for the creation of "patchy" and multi-compartment particles for sophisticated degredation and release profiles of encapsulated materials.

See more of this Session: Poster Session: Meet the Faculty Candidate
See more of this Group/Topical: Education

Developing the Link Between Atomic Level Structure and Macroscopic Properties

Beverly Brooks Hinojosa, Department of Chemical Engineering, University of Florida, Gainesville, FL

Continuous advancements in computational power and efficiency have significantly expanded the system sizes that may be simulated using high accuracy quantum mechanical calculations. I have used density function theory (DFT) to provide further insight into two areas of interest in modern day chemical engineering, electronics and surface chemistry. In particular, I have focused on understanding the role atomic structure plays within the pyrochlore systems ($A_2B_2O_7$), which have received significant attention for use in high-permittivity dielectrics, capacitors, and high-frequency filter applications. The highly flexible structure allows for incorporation of a wide selection of A and B cations. There is a need to understand the effect of local atomic structure and dynamics on the macroscopic properties to enable rational design within the large range of possible chemical compositions. Through DFT, I have explained the development of atomic displacements within bismuth titanate ($Bi_2B^{4+}{}_2O_7$) and examined the role of chemical substitutions on the local geometric and electronic structure within more technologically relevant bismuth pyrochlores such as $Bi_{3/2}M^{2+}Nb_{3/2}O_7$ and $Bi_{3/2}M^{2+}Ta_{3/2}O_7$ with M^{2+} = Zn, Mg. I have connected my DFT results with available experimental data to provide further explanation of the role atomic substitutions and structure on the material properties.

My other area of interest is surface chemistry. I have collaborated with Dr. Jason Weaver's experimental group in the investigation of adsorption of simple molecules on polar and non-polar surfaces of $SrTiO_3$. Specifically, I have examined the role of surface termination and coverage on the favorability of molecular and dissociative adsorption of NH_3 and H_2O. These two molecules serve as simple analogs of a broad range of alcohol and amine reactions that can occur on the oxide surface. The goal of the computational study is to map the effect of local surface structure on the adsorption and reactive behavior observed in ultrahigh vacuum surface science experiments. These two areas of interest have provided an opportunity to simultaneously understand the underlying principles of condensed matter physics and surface science and to collaborate directly with experimentalists within these areas.

See more of this Session: Poster Session: Meet the Faculty Candidate
See more of this Group/Topical: Education

Quantitative Engineering Approach to Particle and Particulate Systems Design

Defne Kayrak-Talay, School of Chemical Engineering, Purdue University, West Lafayette, IN

James D. Litster, School of Chemical Engineering; Department of Industrial and Physical Pharmacy, Purdue University, West Lafayette, IN

Particles and particulate systems are very important to a wide range of industries including pharmaceutical, food, detergents and consumer goods, specialty chemicals manufacturing, minerals and metals processing. Particle/particulate system formation processes need to be well understood at different levels in order to control these processes. Models which combine process parameters and feed material properties to product characteristics should be developed and used to increase process efficiency and product quality.

See more of this Session: Poster Session: Meet the Faculty Candidate
See more of this Group/Topical: Education

Complex Flows of Complex Materials at the Micro- and Nano-Scale

Aditya S. Khair, Chemical Engineering, University of California at Santa Barbara, Santa Barbara, CA

Advancements in cutting-edge fields such as micro- and nano-fluidics, self assembly, cellular mechanics, and complex fluids, require a thorough knowledge of fluid mechanics and transport processes --- a traditional pillar of chemical engineering science. In this poster, I present selected vignettes from my research on fluid dynamics and transport at the micro- and nano-scale. More specifically, I will discuss the following: (i) a variety of problems concerning electrically driven, or electrokinetic, flows, including electro-osmosis over surfaces bearing nano-scale inhomogeneities [1], electrophoresis of "slippery" colloids [2], steric effects on electrophoresis, and voltage-gated nanochannels for sample preconcentration (ii) Active microrheology of complex microstructured fluids, as a means of inferring nonlinear rheological properties, such as micro-viscosity [3] and normal stress differences. Finally, I will outline directions for future research in these and other areas.

References:

[1] A. S. Khair and T. M. Squires, "Surprising consequences of ion conservation in electro-osmosis over a surface charge discontinuity," J. Fluid Mech. 615, 323-334 (2008).

[2] A. S. Khair and T. M. Squires, "The influence of hydrodynamic slip on the electrophoretic mobility of a spherical colloidal particle," Phys. Fluids 21, 042001 (2009).

[3] A. S. Khair and J. F. Brady "Single particle motion in colloidal dispersions: a simple model for active and nonlinear microrheology," J. Fluid Mech. 557, 73 (2006).

For further information please visit: http://engineering.ucsb.edu/~akhair/

See more of this Session: Poster Session: Meet the Faculty Candidate
See more of this Group/Topical: Education

Application of Nano-Basis Technique Onto FEW (Food, Energy, Water) Research Via Membrane Technology

Myung-man Kim, Chemical and Biomolecular Engineering, University of California, Los Angeles, Los Angeles, CA

Developments of polymeric membranes, primarily over the last two decades, have advanced the use of ultrafiltration (UF), nanofiltration (NF) and reverse osmosis (RO) in water treatment, industrial separation processes, and pollution prevention applications. Most challengeable problems in membrane processes are membrane fouling and scaling. These problems not only result in a decreased membrane permeate flux but also protein adhesion and mineral salt scale formation that may permanently alter the physical features of the surface and lead to irreparable membrane damage. In order to develop robust and efficient large-scale membrane separation processes membranes, it is crucial to understand and quantify membrane surface impact on fouling, permeate flux and selectivity.

Although numerous methods of scale and fouling detection have been proposed, it is only recently that real-time early detection of the onset of scale formation has become possible. Direct visual observations and detection of mineral scale on RO/NF membranes under high pressure have been made possible with an ex-situ scale observation detector (EXSOD) along with digital image analysis. Efficient on-line image analysis software was developed assisted with neural networks algorithms to enhance image analysis by providing image family groups to increase the accuracy of single crystal analysis, surface area covered by scale and shape and thus crystal type identification. Direct information on surface nucleation by mineral salt crystals and the rate of single crystal growth was determined over a range of operating conditions and different antiscalants, generating, for the first time, direct fundamental data on the kinetics of surface mineral salt crystallization on RO/NF membranes. These measurements, along with a comprehensive numerical concentration polarization model, enabled evaluation of the direct relationship between the observed flux decline and the surface area covered by mineral scale.

A promising approach to increasing membrane performance, while mitigating fouling, is the structuring of membrane surfaces at the nano-size levels. A novel atmospheric pressure plasma-induced graft polymerization method was developed to enable the generation of a high surface density of active surface sites for subsequent graft polymerization using a suitable monomer. Surface graft polymerization was then carried out to form a dense layer of grafted polymer chains that are covalently and terminally bound to the surface. The chemical and physical features of the resulting grafted polymer film may be tuned by altering the monomer chemistry as well as the reaction conditions to achieve unique architectures for effective advanced materials in membrane separations. Filtration tests of fouling-resistant RO/NF membrane are discussed with protein and polysaccharide solution. Finally, the technical feasibility of nanostructured polymeric membrane for sustainable energy will be discussed.

Discharge of engineered nanomaterials (ENMs) into freshwater and marine ecosystems is expected to increase with accelerating use of nanoparticles in manufacturing. Both biogenic and anthropogenic nanoparticles that are in the atmosphere may find their way to the aquatic environment via dry and wet deposition as well as via runoff. Studies of the potential environmental and human health impacts of ENMs are in their infancy despite and information on their removal from water systems is scarce. The literature is divided on the effectiveness of conventional methods for the removal of nanoparticles from wastewater as well as water sources intended for potable use. Given current knowledge of polyelectrolyte's coagulation of nanoparticles into aggregates or flocs, it is postulated that coagulation can be an effective method that can be coupled with membrane filtration for the removal of nanoparticles from aqueous systems with membrane clogging. With a series of laboratory studies, the potential for polyelectrolyte-enhanced ultrafiltration removal of nanoparticles was investigated. This study will evaluate the environmental impact of nanoparticles to address the application of theory and research in pollution remediation, as well as treatment of wastewater streams that may contain nanoparticles.

See more of this Session: Poster Session: Meet the Faculty Candidate
See more of this Group/Topical: Education

Photochemical and Thermally Adaptable Networks

Christopher J. Kloxin, Chemical and Biological Engineering, University of Colorado, Boulder, CO

Brian Adzima, Chemical and Biological Engineering, University of Colorado, Boulder, CO

Timothy F. Scott, Center for Bioengineering, Department of Mechanical Engineering, University of Colorado, Boulder, CO

Chistopher N. Bowman, Department of Chemical and Biological Engineering, University of Colorado, Boulder, CO

Traditionally, thermosets are synthesized using irreversible covalent bonds, and as a consequence cannot be melted, molded, or dissolved. Here, we have explored a new class of materials, known as 'adaptable networks', which incorporate reversible covalent linkages throughout the network backbone to facilitate controllable bond rearrangement and material reshaping. These linkages are activated either photochemically or thermally as in, for example, the photo-induced addition-fragmentation of the mid-chain allyl sulfide functionality or the thermo-reversible cycloaddition of furan and maleimide, respectively. We have explored both post- and during-polymerization stress reduction utilizing the mid-chain allyl sulfide functionality within the network chemistry. The novel stress relieving properties of fully cured networks containing allyl sulfides within their backbone were demonstrated by the reshaping and even actuation of the material. Furthermore, we have studied the ability of the allyl sulfide functionality to undergo reversible addition-fragmentation chain transfer during photo-polymerization of thiol-ene monomers, demonstrating significant stress relaxation via bond rearrangement in the latter stages of photo-polymerization (i.e., post gel-point). Thermally reversible Diels-Alder networks allow an avenue for complete reverse gelation, and are thus fully recyclable and can be 'healed' upon material cracking or failure. We have employed rheometry and Fourier transform infrared (FTIR) spectroscopy to characterize the mechanical properties, reaction rate, and thermodynamic equilibrium over a broad temperature range. The gel-point temperature, as determined by rheometry using the Winter-Chambon criterion, corresponded to a gel-point conversion of 71%, as determined by FTIR, which is consistent with the Flory-Stockmayer equation. Furthermore, the material exhibited a low frequency relaxation, which compares well with the characteristic time-scale of bond rearrangement determined from the FTIR kinetic studies.

See more of this Session: Poster Session: Meet the Faculty Candidate
See more of this Group/Topical: Education

Hydrogels for Probing the Influence of Microenvironment Structure and Chemistry On Cell Function

April M. Kloxin, Chemical and Biological Engineering and the Howard Hughes Medical Institute, University of Colorado, Boulder, CO

Kristi S. Anseth, Chemical and Biological Engineering and the Howard Hughes Medical Institute, University of Colorado, Boulder, CO

Cell-cell and cell-material interactions play an important role in directing cellular processes such as cytoskeletal organization, migration, and differentiation. To probe the influence of these interactions on cell function, polymer-based, three-dimensional cell culture platforms have emerged, which offer control of the cellular microenvironment for examining its influence on cell function in vitro. In my research program, biomaterials and hydrogels that offer control of the physical and chemical structures within the microenvironment will be developed. For example, photolabile moieties have been incorporated within the gel backbone to allow real-time, in situ tuning of the gel structure and chemistry in the presence of cells. These gel systems will enable the experimentalist to precisely test the effect of an individual cue on cell function in both space and time.

See more of this Session: Poster Session: Meet the Faculty Candidate
See more of this Group/Topical: Education

Interfacial Forces in Nanoparticle and Biological Systems at the Micro- and Nano-Scales

Kai Kristiansen, Department of Chemical Engineering, University of California at Santa Barbara, Santa Barbara, CA

The overall theme of my research is to understand the molecular and nanoparticle behavior at the interface of surfaces. As the size of particles reaches characteristic length scales, novel physical properties (compare to bulk properties) start to emerge [1]. The carbon nanotubes are one candidate and I will present a study of aqueous film of carbon nanotubes on mica surfaces, and discuss the frictional properties and structure formation of this system.

Another interesting interfacial system is bio-membrane. We have focused on the myelin membrane. The myelin sheath is a multilamellar membrane surrounding the axon of neurons in the nervous system. This sheath acts as an electrical insulator, which allows for faster and more efficient conduction of nerve impulses than unmyelinated nerves. Myelin dysfunctions vary from deterioration of signal transduction to demyelinating diseases such as multiple sclerosis (MS). We want to understand the molecular synergistic interaction responsible for a healthy myelin membrane structure. Using a Surface Forces Apparatus we have studied the interaction forces and adhesion between supported myelin lipid bilayers mediated by myelin basic protein [2].

In the last part of this poster I want to present the development and implementation of new devices for interfacial science and how to integrate them into a Surface Forces Apparatus [3].

[1] Min Y J, Akbulut M, Kristiansen K, Golan Y, and Israelachvili J; "The role of interparticle and external forces in nanoparticle assembly". Nature Materials 7 (2008) 527-538.

[2] Min Y, Kristiansen K, Boggs J M, Husted C, Zasadzinski J A, and Israelachvili J; "Interaction forces and adhesion of supported myelin lipid bilayers modulated by myelin basic protein". Proceeding of the National Academy of Science 106 (2009) 3156-3159.

[3] Kristiansen K, McGuiggan P M, Carver G, Meinhart C, and Israelachvili J; "3D Force and Displacement Sensor for AFM and SFA measurements". Langmuir 24 (2008) 1541-1549.

See more of this Session: Poster Session: Meet the Faculty Candidate
See more of this Group/Topical: Education

Functional Nanoassemblies for Electrochemical Energy Conversion Devices

Avni A. Argun, Chemical Engineering, Massachusetts Institute of Technology, Cambridge, MA

The increasing global focus on alternative energy sources has led to a renewed interest in electrochemical devices such as batteries, dye-sensitized solar cells, and fuel cells. All electrochemical energy originates from the transfer of electrons and ions at interfaces, and it is crucial to study these nanoscale interfaces via unique integration of electrochemistry and polymer engineering. The alternating adsorption of oppositely charged molecular species, known as the electrostatic layer-by-layer technique, is a simple and elegant method of constructing highly tailored ultrathin polymer films and organic-inorganic hybrid nanocomposites. We have utilized this method to engineer a number of functional thin film systems, including the use of conjugated polymers for electrochromic displays, the formation of thin films as proton exchange membranes in fuel cells, and the use of these multilayers in other electrochemical energy devices.

See more of this Session: Poster Session: Meet the Faculty Candidate
See more of this Group/Topical: Education

Soft Materials for High Performance Structural Materials, Bio-Implants, and Energy Storage Materials

Santanu Kundu, Polymer Characterization and Measurement Group, Polymer Division, National Institute of Standards and Technology (NIST), Gaithersburg, MD

My research interest is investigating the processing/fabrication-structure-property relationships across multiple length scales for various technologically relevant soft materials. Common examples of soft materials include gels/hydrogels, polymer melts, biological tissues, etc [1, 2]. The motivation of my research will be applying these soft materials in various challenging applications ranging from energy storage to new generation smart materials.

Presently, I am a postdoctoral researcher working in the polymer characterization and measurement group at the National Institute of Standards and Technology (NIST), where I am developing microfluidics/lab-on-chip techniques to obtain polymeric materials from sustainable sources. Before joining NIST, I was a post-doctoral research associate at the Polymer Science and Engineering Department of University of Massachusetts-Amherst. I was involved in investigating surface, near surface and mechanical characterization of gel/hydrogel/elastomeric materials. The mechanical properties of gels present qualitatively contradictory behavior; they are commonly soft but also notoriously brittle. We investigated the elasticity and fracture behavior of swollen polymer networks using a simple experimental method to induce cavitation within polyacrylamide hydrogel, a common material used in many biological applications. A transition from reversible cavitation to irreversible fracture was observed with the increase of polymer volume fraction [3]. Adapting scaling theories, it is shown quantitatively that the transition from reversible cavitation to irreversible fracture depends on the polymer volume fraction and an initial defect length scale. The use of cavitation experiments permits characterization of network properties across length scales ranging from μm to mm. We anticipate that these results significantly enhance the understanding of mechanical properties of soft materials, both synthetic and biological.

For some applications it is important to know the adhesion and frictional behavior of hydrogels and one such example is contact lenses. In an industry sponsored project, I have investigated the adhesion and mechanical properties of commercially available contact lenses (hydrogels) by developing a custom built setup. The set-up is also equipped with a liquid cell, which enabled us to measure adhesion forces for various gels/hydrogels/elastomers in different environmental conditions. Adhesion properties of polymeric materials can be altered by incorporating topological patterns on the surfaces. However, conventional patterning methods, including photo- and imprint lithography, are difficult to apply to non- planar surfaces. During my postdoctoral research at Amherst, I developed a technology based on surface wrinkling technique to incorporate novel structures on the non-planar contact lens surfaces [4, 5]. Incorporation of such structure resulted in significant performance improvement (clinical trial will be performed).

In my Ph.D. research I have studied the flow and three dimensional microstructure of a liquid crystalline carbonaceous material (mesophase pitch) for different flow conditions, such as steady shear flow, dynamic flow, and processing flow using polarizing optical microscopy and X-ray diffraction [6-10]. The evolution of microstructure in different flow conditions was uniquely studied in three orthogonal planes. The systematic understanding of flow and its effect on microstructure helped us to predict/model the complex flow behavior and microstructural evolution of this material, which in turn will help to design carbon materials, such as carbon fibers, carbon-carbon composites, and fuel cell separators in more efficient ways. My dissertation, 'Investigation of flow and microstructure in rheometric and processing flow conditions for liquid crystalline pitch', has been awarded the Best Dissertation in Carbon Science, 2007 for 'outstanding scientific achievement' by the Elsevier-Carbon Journal.

The research program that I plan to develop will largely build on my postdoctoral and doctoral research and specifically, I would like to develop a research program in the following areas:

• Mechanics of soft materials

• Hierarchical carbon materials for energy storage applications

• Large scale development of carbon thin films for electronics materials

In addition to research, I am passionate about teaching and am inspired to educate and to mentor future scientists and engineers to meet the challenges of today's ever-changing technologies. My teaching philosophy has been developed through my experience as a lab and teaching assistant during my graduate school days, as a mentor during my postdoctoral work, and as a supervisor during my industrial tenure. I can teach undergraduate and graduate level courses on transport phenomenon, engineering mechanics, engineering mathematics, chemical reaction engineering, polymer processing, and soft materials mechanics. I also intend to develop a new interdisciplinary course on fabrication and characterization of nanostructured materials for engineering applications, viscoelasticity of polymers and biopolymers: theory and simulation, Soft Materials Mechanics.

References:

1. Ladet S, Laurent David L and Alain Domard A. Multi-membrane hydrogels. Nature 452, 76-79 (2008).

2. Dzenis Y. Structural Nanocomposites. Science 319, 419-420 (2008).

3. Kundu S, Crosby AJ. Cavitation rheology and fracture behavior of polyacrylamide gels, submitted; Kundu S, Crosby AJ. Cavitation rheology and fracture mechanics of polyacrylamide hydrogels. American Physical Society -March meeting, Pittsburgh, USA, 2009.

4. Kundu S, Crosby AJ, Sharma R. Adhesion Behavior of Non-planar Wrinkled Surfaces, in preparation; Kundu S, Crosby AJ, Sharma R. Adhesion Behavior of Non-planar Wrinkled Surfaces. American Physical Society -March meeting, Pittsburgh, USA, 2009.

5. Breid D, Kundu S, Denic V, Crosby AJ. Curvature effect on the wrinkle morphology, in preparation.

6. Kundu S, Grecov D, Rey AD, Ogale AA. Shear flow induced microstructure of a synthetic mesophase pitch, Journal of Rheology 53(1):85-113(2009).

7. Kundu S, Ogale AA. Rheostructural studies on a synthetic mesophase pitch during transient shear flow, Carbon 44(11): 2224-2235 (2006).

8. Kundu S, Ogale AA. Microstructural effects on the dynamic rheology of a discotic mesophase pitch, Rheologica Acta 46(9):1211-1222 (2007).

9. Kundu S, Naskar AK, Ogale AA, Anderson D, Arnold JR. Observations on a low-angle x-ray diffraction peak for AR-HP mesophase pitch, Carbon 46(8):1166-1169 (2008).

10. Kundu S, Ogale AA. Microstructure development of a synthetic mesophase pitch during processing flow, in preparation.

See more of this Session: Poster Session: Meet the Faculty Candidate
See more of this Group/Topical: Education

DNA Therapeutics - New Challenges for Separation of Plasmid Isoforms

David R. Latulippe, Chemical Engineering, Pennsylvania State University, University Park, PA

Andrew L. Zydney, Chemical Engineering, The Pennsylvania State University, University Park, PA

Recent progress in the development of gene therapies and DNA-based vaccines has created a need for new separations technologies suitable for the large scale production and purification of plasmid DNA. Plasmids exist in one of three topological morphologies or isoforms: supercoiled, open-circular, and linear. These double-stranded DNA molecules are all naturally produced in typical plasmid manufacturing processes. The supercoiled isoform has the greatest transfection efficiency and is thus the FDA recommended morphology for therapeutic applications; the other isoforms are considered process impurities. The objective of our research has been two-fold: to develop a quantitative understanding of the behavior of the different plasmid isoforms in membrane ultrafiltration (UF) and size exclusion chromatography (SEC), and to use appropriate theoretical frameworks to analyze the experimental results in terms of the underlying bio-physical properties of the different isoforms.

UF experiments were conducted over a range of filtrate flux, membrane pore size, and solution properties. Plasmid transmission was a strong function of the filtrate flux with very high transmission obtained at the higher filtration rates even for plasmids that were more than an order of magnitude larger than the effective membrane pore size. Plasmid transmission was nearly independent of the plasmid size (number of base pairs), but varied with solution ionic strength and membrane pore size. The data were analyzed using a model for flexible polymers that accounts for the elongation of the plasmid in the converging flow-field above the membrane pores. At a given filtrate flux, transmission of the linear isoform was greater than that for the supercoiled plasmid, with the open-circular plasmid being almost completely retained out to very high filtrate flux. These differences were directly related to the underlying differences in the flexibility of the three isoforms and were successfully exploited to develop a novel 'flexibility-based' UF process that was able to achieve high-resolution purification of the different plasmid isoforms.

In contrast to the behavior seen in UF, plasmid separation in SEC was governed primarily by the plasmid size. Thus, the linear isoform showed the least retention while the supercoiled isoform had the greatest retention (for plasmids with the same number of base pairs). These trends were confirmed by static light scattering measurements of the radii of gyration. The resolution between the different isoforms was dramatically reduced for the larger plasmids, resulting in low product yields and quality under these conditions. The experimental data were analyzed using theoretical models for partitioning behavior of flexible linear and cyclic polymers. The results provide a framework for analysis and design of SEC systems for plasmid separations.

The results obtained in these studies clearly demonstrate the critical relationship between the underlying physical properties and the separation characteristics of the plasmid DNA isoforms. Ongoing advances in our understanding of the biophysics of plasmid DNA could provide exciting opportunities for the development of new / improved separations technologies that specifically exploit the unique physical properties of plasmid isoforms.

See more of this Session: Poster Session: Meet the Faculty Candidate
See more of this Group/Topical: Education

Rational Design of Advanced Materials: Combining Modeling, Synthesis, Processing, and Characterization to Generate Superior Materials

Richard A. Lawson, School of Chemical & Biomolecular Engineering, Georgia Institute of Technology, Atlanta, GA

Laren M. Tolbert, Chemistry & Biochemistry, Georgia Institute of Technology, Atlanta, GA

Clifford L. Henderson, Chemical & Biomolecular Engineering, Georgia Institute of Technology, Atlanta, GA

In the late 1800's, Thomas Edison tried thousands of different materials to find a suitable filament material for his incandescent light bulb. Over a hundred years later, this "Edisonian" approach is still one of the dominant methods used in the development of new materials. While this method may eventually yield improved materials, it is far too costly and slow for many advanced applications such as fuel cell membranes, organic electronics, organic photovoltaics, photoresists, biomaterials, and many other important purposes. Modern materials require modern, interdisciplinary approaches to their design and development. It is no longer sufficient to make thousands of different materials to identify a few superior materials. Instead, it would be much preferred to utilize predictive models to guide material design and selection and utilize a more limited experimental data set to calibrate and refine these model predictions. This is a daunting task even at the level of simply predicting the bulk properties of a material, but unfortunately in many emerging applications (e.g. organic electronics) the manner in which the material is processed and used can strongly affect its physiochemical properties. Thus, coupling in this information about processing effects will also be important in many advanced material applications. Applying these fundamentals to the synthesis of new materials combined with intelligent synthetic schemes quickly yield new and better materials.

Our work has piloted the use of such a holistic rational materials design scheme in the area patterning materials for microelectronics processing. More specifically, we have used a combination of atomistic and mesoscale modeling to guide the design of novel molecular glass photoresist materials that can be used for high resolution patterning in semiconductor fabrication. Conventional state-of-the-art photoresist materials are composed multi-component blends of polymers with small molecule photoacid generators, dissolution inhibitors, and base quenchers. While this design approach has worked for the patterning of multiple generations of smaller and smaller transistors, it cannot be used for future generations because this design is unable to print features small enough with sufficient smoothness (line edge roughness) and sensitivity. This is especially true since the patterned feature sizes have truly approached the scale of individual polymers. Custom written kinetic Monte Carlo mesoscale simulations of the lithographic process have allowed us to investigate multiple different materials design strategies and processing conditions that lead to greatly improved material performance and helped guide the design of better materials. These simulations not only accurately model current performance; they have successfully predicted the improved performance of new materials. We have also developed thermodynamic models of molecular glass materials that successfully predict the glass transition temperature and accurately predict the full range of dissolution properties of these materials. These models allow the reduction of thousands of candidate materials and conditions to tens of materials. Even if excellent materials are designed, they are only useful if they can be made and made economically. To that end, we have created synthetic strategies that have allowed us to synthesize several generations of these materials, each with improved performance. After modeling and synthesizing new materials, we have improved their performance even further thorough the use of fundamental reaction engineering. By modifying the kinetics of cross-linking of one negative tone resist, the resolution was improved from 35 nm down to 25 nm. This rational design approach has yielded truly superior performing materials.

See more of this Session: Poster Session: Meet the Faculty Candidate
See more of this Group/Topical: Education

Development of Novel Water-Soluble Gold Nanoparticles with Atypical Fluorescence and Study On Its Mechanism

Chang-won Lee, Bioengineering, University of Utah, Salt Lake City, UT

I have received BS and MS degrees in Chemical Engineering department, Seoul National University and have worked on enzyme entrapment in silica micro-particles and covalent immobilization on the surface of hydrophilic chitosan micro-particles. Before the graduation of the MS program, I have worked in Pacific Northwest National Laboratory as a graduate fellow to work in single enzyme nanoparticle development and characterization. I am currently a PhD candidate at the Bioengineering department, University of Utah, expected to graduate by Fall 2009, working on the development and characterization of water-soluble gold nanoparticle with enhanced emission and studying the mechanism of the atypical fluorescence due to electron transfer from surface bound molecules. The summary of my current PhD work is given below.

Water-soluble gold nanoparticles with atypical fluorescence have been developed with mercaptooctanoic acid as a capping agent under reduction condition with sodium borohydride. The product was purified with ethanol trituration for further application or characterization. TEM analysis of the nanoparticle showed its diameter to be 2.2 ± 0.6 nm. The gold nanoparticle solution shows bright red emission with the excitation with UV.

In comparison to the widely used quantum dots which usually have broad excitation spectra and overlapped region with emission spectra, it showed distinct excitation and emission spectra (excitation spectra peak @ 280 nm and emission spectra peak @ 610 nm) with 200 nm gap between them with large Stokes shift.

It shows the possibility of two band gap modalities with responsible for excitation and emission and electron transfer between them. The loss during the electron transfer is considered to be the reason of distinct excitation and emission peak location with large Stokes shift. In this hypothetical diagram, excitation occurs in the left hand side band gap from surface covering mercaptooctanoic acid (MOA) for 280 nm excitation and transferred to the surface of the gold nanoparticles and the emission process occurs in the right hand side with relatively smaller band gap responsible for 610 nm emission.

This mercaptooctanoic acid stabilized gold nanoparticle also showed strong relationship with pH on its emission. The fluorescence intensity weakened at lower pH while it remained high at higher pH. The dependency of the fluorescence on the pH change also supports the proposed mechanism of excitation on surface capping mercaptooctanoic acid molecules. The charge state of carboxylic acid groups of mercaptooctanoic acid is believed to be responsible for this phenomenon and the hypothesis is proposed.

This electron energy density distribution change eventually results higher energy level distribution of ground state for the charged carboxylic acid group comparing to neutral carboxylic groups.

In addition to the pH effect, computation simulation study was also performed for the change of the band gap of mercaptooctanoic acid with and without Au-S bond formation. Gaussian03 program was used with b3pw91/lanl2dz functional/basis set combination. The result showed similar trend of band gap change as it is experimentally confirmed by absorption band edge measurement. Detailed result and its hypothesis will be discussed in the presentation.

As a future research plan, I'd like to develop different types of optically active nanoparticles by understanding its intrinsic fluorescent mechanisms and apply these novel fluorescence systems to many applications including bio-imaging agents and nano-pH sensors. My work philosophy is collaboration and maximize its synergy effect. There are always many branching projects coming from a single specified work and it is hard for one person to execute professionally in all the branching projects efficiently. By collaborating with other professionals on the project together, it would be more efficient to achieve the break-through results and get professionalized on that field during the collaboration. I understand that providing specialties from my side is also essential key factor for the successful collaborations. In my case, I am specialized in enzyme/protein engineering, organic synthesis and bio-nanotechnology based on previous experiences during my education and professional life.

See more of this Session: Poster Session: Meet the Faculty Candidate
See more of this Group/Topical: Education

Multiscale-Simulation and Experimental Studies of the Interactions of Membranes with Peptides, Synthetic Polymers and Nanoparticles

Hwankyu Lee, Lab. of Computational biology, National Heard, Lung and Blood Institute, National Institutes of Health, Rockville, MD

All-atom and coarse-grained (CG) molecular dynamics (MD) simulations were performed to investigate (1) peptide-peptide, peptide-monolayer, and peptide-micelle interactions; (2) effects of the dendrimer surface, size, and concentration on pore formation in lipid bilayers; and (3) hydrodynamics and shape anisotropy of polymers as they pass through the pores of membrane proteins. Circular dichroism spectroscopy and single-channel current recording experiments were also performed to complement simulation results. Future research directions are presented.

(1) Peptide interactions with lipid monolayers and micelles.

SP-B_{1-25} is a fragment of 79-amino acid lung surfactant protein B, an important component for preventing respiratory distress syndrome. Multiple copies of the peptide in palmitic acid monolayer were simulated with all atoms, showing that the final peptide conformation favorably compared with experiments. The peptides are anchored by hydrogen bond interactions between the cationic residues Arg-12 and Arg-17 and the hydrogen bond acceptors of the ionized monolayer headgroup, and the peptide tilt angle is modulated by the interactions of Tyr-7 and Gln-19 with the monolayer headgroup. These results can be applied to the rational design of synthetic lung surfactant peptides. [1]

Coiled-coil peptides [2,3] and the BAR peptide-SDS micelle [4] were simulated to investigate models of membrane curvature by BAR domains. Both simulation and experiment indicate that the N-terminal region (helix-0) is disordered, and that the peptide curves to adopt the micelle shape. This is consistent with notion that helix increases the peptide-membrane binding affinity.

(2) Nanoparticle induced-pore formation in the lipid bilayer: Effects of the nanoparticle size, surface charge, and concentration.

Polyamidoamine (PAMAM) dendrimers, which consist of regularly branched monomeric building blocks and surface terminal groups, are good candidate nanoparticles for use as anti-tumor therapeutics and drug delivery. To investigate the effect on pore formation of the dendrimer properties (size, surface charge, and concentration) and solution conditions (temperature and salt concentration), we simulated PAMAM dendrimers in DMPC bilayers with explicit water using the CG model. When initially clustered together near the bilayer, neutral acetylated dendrimers aggregate, whereas cationic un-acetylated dendrimers disperse, in agreement with experiments. Bilayers interacting with un-acetylated dendrimers of higher concentration are significantly deformed and show pore formation on the positively curved portions, while acetylated dendrimers are unable to form pores. Larger un-acetylated dendrimers bring more water molecules into the pores than do smaller ones. At higher salt concentration (~500mM NaCl) or lower temperature (277 K), un-acetylated dendrimers do not insert into the bilayer. These results highlight the applicability of CG simulations, and help explain why charged dendrimers deform membranes and form pores at high concentration. [5,6,7,8,9]

(3) Hydrodynamic radius and shape anisotropy of polymers as they pass through the pores of membrane proteins

Polyethylene oxide (PEO) and polyethylene glycol (PEG) are used as probes of pore sizes of membrane channels, although many molecular details are unclear. PEO and PEG were simulated to examine anisotropy of the shape and translational diffusion tensors of these polymers as a step toward understanding their interaction with membrane pores. The calculated persistence length, the exponent u relating the radius of gyration and molecular weight ($R_g \mu M_w^{\nu}$), and hydrodynamic radii are in quantitative agreement with experiment values. The dimension of the middle length for each of the polymers nearly equals the hydrodynamic radius obtained from diffusion measurements in solution. This implies that a polymer diffuses with its long axis parallel to the membrane channel without substantial distortion. This quantitative study also helps development of coarse-grained models of PEG and PEO for ongoing investigation of the effects of the molecular flexibility (soft linear vs. hard sphere) in membrane channels at the microsecond time scale. Single-channel current recording experiments of PEG and dendrimers are also being performed to complement simulation results. [10,11]

References

[1] **Lee H**, Kandasamy SK, and Larson RG, Molecular dynamics simulations of the anchoring and tilting of the lung-surfactant peptide SP-$B_{1\text{-}25}$ in palmitic acid monolayers. *Biophysical J.*, **2005**, 89:3807-3821

[2] Sayer JA, Otto EA, O'toole JF, Nurnberg G, Kennedy MA, Becker C, Hennies HC, Helou J, Attanasio M, Fausett BV, Utsch B, Khanna H, Liu Y, **Lee H**, Larson RG, and Hildebrandt F et al., The centrosomal protein nephrocystin-6 is mutated in Joubert syndrome and activates transcription factor ATF4. *Nature Genetics*, **2006**, 38:674-681

[3] **Lee H**, and Larson RG, Prediction of the stability of coiled coils using molecular dynamics simulations. *Molecular Simulation*, **2007**, 33:463-473

[4] Low C, Weininger U, **Lee H**, Schweimer K, Pastor RW, Balbach J. Structure and dynamics of Helix-0 of the N-BAR domain in lipid micelles and bilayers. *Biophysical J.*, **2008**, 95:4315-4323

[5] **Lee H**, Baker JR, and Larson RG, Molecular dynamics studies of the size, shape, and internal structure of 0% and 90%-acetylated G5 PAMAM dendrimers in water and methanol. *J. Physical Chemistry B.*, **2006**, 110:4014-4019

[6] **Lee H**, and Larson RG, Molecular dynamics simulations of PAMAM dendrimer-induced pore formation in DPPC bilayers using a coarse grained model. *J. Physical Chemistry B.*, **2006**, 110:18204-18211

[7] **Lee H**, and Larson RG. Coarse-grained molecular dynamics studies of the concentration and size dependence of fifth- and seventh-generation PAMAM dendrimers on pore formation in DMPC bilayer. *J. Physical Chemistry B.*, **2008**, 112:7778-7784

[8] **Lee H**, and Larson RG, Lipid bilayer curvature and pore formation induced by charged linear polymers and dendrimers: the effect of molecular shape. *J. Physical Chemistry B.*, **2008**, 112:12279-12285

[9] **Lee H**, and Larson RG, Multiscale modeling of dendrimers and their interactions with bilayers and polyelectrolytes. *Molecules*, **2009**, 14:423-438

[10] **Lee H**, Venable RM, MacKerell AD, and Pastor RW. Molecular dynamics studies of polyethylene oxide and polyethylene glycol: Hydrodynamic radius and shape anisotropy. *Biophysical J.*, **2008**, 95:1590-1599

[11] **Lee H**, de Vries AH, and Pastor RW. A coarse-grained model of polyethylene oxide: conformation and hydrodynamics. *Submitted.*

See more of this Session: Poster Session: Meet the Faculty Candidate
See more of this Group/Topical: Education

Bioactive Materials for Controlled Release and Tissue Engineering Applications

Chien-Chi Lin, Department of Chemical and Biological Engineering, University of Colorado at Boulder, Howard Hughes Medical Institute, Boulder, CO

Kristi S. Anseth, Chemical & Biological Engineering, Howard Hughes Medical Institute and University of Colorado Boulder, Boulder, CO

Poly(ethylene glycol) (PEG) hydrogels possess many desirable properties making them an excellent material choice for numerous biomedical applications, including controlled delivery, tissue engineering, and regenerative medicine [1, 2]. The high water content of the PEG hydrogel environment preserves the bioactivity of protein therapeutics and facilitates rapid nutrient diffusion to support the survival of encapsulated cells. Further, the crosslinked, semi-permeable PEG network prevents direct contact of the encapsulated cells with immune cells and antibodies, and hence creates an immuno-isolation barrier. However, the use of PEG hydrogels also has disadvantages for protein and cell delivery. Firstly, due to the high water content of PEG hydrogels, it is a challenging task to effectively control molecular diffusion rates through the gel network without sacrificing the preferential hydrophilicity of the gels [3-5]. Further, the lack of biorecognition sites on the inert PEG polymer backbone hinders the survival and function of encapsulated anchorage-dependent cells.

Toward this end, we are developing bioactive peptide-functionalized PEG hydrogels for growth factors and cell delivery. Specifically, crosslinkable peptides exhibiting monovalent or multivalent affinity for basic fibroblast growth factor (bFGF) were incorporated in PEG hydrogels [5]. The architecture and presentation of the peptide motifs play an important role in regulating growth factor delivery. Utilizing the concept of affinity binding, we are also developing cytokine-antagonizing and chemokine-sequestering hydrogels to enhance the survival, function, and differentiation of encapsulated cells through controlling local inflammation [6]. Rat pheochromocytoma cells (PC12), human mesenchymal stem cells (hMSCs), and pancreatic beta-cells encapsulated in the cytokine-scavenging PEG hydrogels are less prone to pro-inflammatory cytokine TNF alpha-induced apoptosis. Finally, PEG hydrogels co-polymerized with bioactive peptides that signal pancreatic beta-cells directly are fabricated via facile thiol-acrylate photopolymerization to improve the survival and function of the encapsulated islets for the treatment of type 1 diabetes [7].

References:

1. C. -C. Lin and A. T. Metters, "Hydrogels for controlled release formulations - Network design and mathematical modeling." Advanced Drug Delivery Reviews. 2006(58) 1379-1408.

2. C. -C. Lin and K. S. Anseth, "PEG hydrogels for the controlled release of biomolecules in regenerative medicine." Pharmaceutical Research. 2009(26) 631-43.

3. C. -C. Lin and A. T. Metters, "Metal-chelating affinity hydrogels for sustained protein release." Journal of Biomedical Materials Research: Part A. 2007(83) 954-964.

4. C. -C. Lin and A. T. Metters, "Bifunctional monolithic affinity hydrogels for dual-protein delivery." Biomacromolecules. 2008(9) 789-795.

5. C. -C. Lin and K. S. Anseth, "Controlling affinity binding in peptide functionalized poly(ethylene glycol) hydrogels." Advanced Functional Materials. 2009 (In press).

6. C. -C. Lin, A. T. Metters, and K. S. Anseth, "Bio-interactive PEG-peptide hydrogels to modulate local inflammation induced by the pro-inflammatory cytokine TNFα." (Submitted).

7. C. -C. Lin and K. S. Anseth, "Glucagon-like peptide-1 functionalized PEG hydrogels promote survival and

function of encapsulated pancreatic β-cells" (Submitted).

See more of this Session: Poster Session: Meet the Faculty Candidate
See more of this Group/Topical: Education

Structure and Dynamics of Water near Mica Surface

Ateeque Malani, Department of Chemical Engineering, Indian Institute of Science, Bangalore, India

K. G. Ayappa, Department of Chemical Engineering, Indian Institute of Science, Bangalore, India

Mica is representative of a naturally occurring hydrophilic solid substrate. Due to its atomically smooth structure mica has been widely used in a variety of experiments which seek to probe the structure and dynamics of water near mica surface. Understanding the structure and dynamics of molecularly thin films is an active field of research and has wide applications in areas ranging from biology to geology.

Adsorption isotherms of water on muscovite mica are obtained using grand canonical Monte Carlo simulations over a wide range of relative vapor pressures, p/p0 at 298 K. Three distinct stages are observed in the adsorption isotherm. A sharp rise in the water coverage occurs for $0 < p/p_0 < 0.1$, followed by a relatively slow increase in the coverage for $0.1 < p/p_0 < 0.7$. Above $p/p_0 = 0.7$, a second increase in the coverage occurs due to the adsorption of water with bulklike features. The derived film thickness and isotherm shape for the simple point charge (SPC) water model is in excellent agreement with recent experiments of Balmer et al. [*Langmuir* 2008, 24, 1566]. A novel observation is the significant redistribution of water between adsorbed layers as the water film develops. This redistribution is most pronounced for $0.1 < p/p_0 < 0.7$, where water is depleted from the inner layers and film growth is initiated on the outer layer. Isosteric heats of adsorption were found to be in good agreement with the differential microcalorimetric data of Rakhmatkariev (*Clays Clay Miner.* 2006, 54, 402), over the entire range of pressures investigated.

The dynamics of interfacial water near mica surface is evaluated using molecular dynamics simulations. The analysis shows that water in the contact layer has a high residence time near the surface. The potassium ions have a strong influence on the structure and dynamics of interfacial water. The presence of potassium ions, gives rise to a non-monotonic decoupling behaviour in the translational and orientational motions toward the mica surface. It was observed that orientational motions are much more suppressed when compared with translational motions. Despite slowing down of water relaxation near the surface, dynamics of water in the hydration shell of potassium ion is comparable to that observed in bulk water. The analysis also shows the presence of orientational jumps in the contact layer of mica surface. Recently these molecular jumps have been observed in molecular dynamics simulations of bulk water.[1]

References

1. Laage, D. and Hynes, J. T., A molecular jump mechanism of water reorientation. *Science* **311**, 832-835, 2006.

See more of this Session: Poster Session: Meet the Faculty Candidate
See more of this Group/Topical: Education

Materials Engineering: Applications to Biomedical Implants and Magnesium Corrosion

Holly J. Martin, Center for Advanced Vehicular Systems (CAVS), Mississippi State University, Mississippi State, MS

Graduate/Post-Doctoral Advisor: Dr. Kirk H. Schulz - President, Kansas State University

Post-Doctoral Advisor: Dr. Mark F. Horstemeyer - Center for Advanced Vehicular Systems (CAVS), Mississippi State University

Corrosive effects lead to the destruction and failure of many different materials in many different situations, including implants placed in the human body and metal used in engine cradles. One thing that all of these materials have in common is a necessity to engineer a material that can withstand these corrosive effects. My graduate and post-doctoral research has provided me with opportunities to examine methods to prevent corrosion in two different manners. In addition to exploring corrosion using two different methodologies, I have also had the opportunity to collaborate with other researchers, mentor graduate students, teach two different classes, and write proposals based on my research.

My initial work as a graduate student allowed me the chance to work on a way to improve the strength of adhesion of a coating on titanium to prevent corrosion caused by the human body and improve osseointegration. We began by chemically analyzing the surface of titanium as the coating was built in a three-step process, involving aminopropyltriethoxysilane, or a two-step process, involving triethoxsilylbutyraldehyde. We demonstrated that the silanes were unchanged and bound to the titanium surface when toluene was used as the solvent. Following the chemical analysis, mechanical testing was performed, which demonstrated that the films were strongly bound to the titanium surface. However, as expected, the toluene used to deposit the silanes resulted in slow growth of bone cells, making toluene an inappropriate solvent for use in biomedical coatings. My graduate work, and my post-doctoral work, in this area provided me with one method to prevent corrosion, which was to creating a coating that protects the metal beneath it.

A second way to prevent corrosion is what I am currently working on as a post-doctoral researcher in the Center for Advanced Vehicular Systems (CAVS). In order to improve a material, one must understand how it corrodes. We are currently working to determine the corrosion of two types of magnesium and how the application method of a NaCl aqueous solution affects corrosion. Pitting is the initial mechanism of corrosion and can ultimately lead to failure of material in which pitting occurs. Understanding how pit characteristics, such as depth, area, volume, and number density, change due to exposure ultimately will allow us to model this behavior. By modeling corrosive effects, the numerous magnesium alloys currently in production can be analyzed to determine which is most resistive to salt corrosion, allowing the lightweight material to be used in the automotive industry. This post-doctoral work has introduced me to a second method to prevent corrosion, which is to develop a material that can withstand corrosive effects by adding elements to improve the alloy.

I have had the opportunity to collaborate with many different people, departments, and universities, as both a graduate student and as a post-doctoral researcher. As the laboratory manager of the XPS equipment, I ran and analyzed samples for professors in Chemical Engineering, Plant and Soil Sciences, and Electrical and Computer Engineering. I also had the opportunity to work with people from the University of Texas Health Science Center and Georgia Tech. Because of my experience working with professors and other departments, I believe that I can develop strong collaborations between myself and other professors, including those in Mechanical Engineering, Biology, and Biological Engineering.

As a graduate student, I had the opportunity to assist with several classes and teach two classes by myself. The first class I taught was an entirely new laboratory for an established class, focusing on process design simulation. As part of my responsibilities, I created all of the materials needed, including the syllabus, Powerpoint presentations covering individual unit operations, homework, and a final project combining the individual unit operations into a working flowsheet. I also was responsible for grading both the homeworks and the final projects. I taught this class for the Fall and Spring semesters and was evaluated by the students, receiving 4.7/5.0 for both semesters. I also taught Unit Operations Laboratory I and II in the Spring semester. I was responsible for teaching the class, including assigning the groups to the experiments each week, for grading the laboratory reports, and for assigning final grades. As with the class I developed, I was evaluated by the students, receiving a 4.5/5.0 average.

As an assistant professor, I plan to continue collaborating with professors in both my research area, biomedical implants and biomaterials, and in other research areas and departments. I truly enjoy teaching and look forward to teaching both established undergraduate and graduate classes and classes I create based on my research interests.

See more of this Session: Poster Session: Meet the Faculty Candidate
See more of this Group/Topical: Education

Advanced Nanosystems for Energy, Environmental and Biomedical Applications

Ken Martirosyan, Chemical and Biomolecular Engineering, University of Houston, Houston, TX

The main topic of this presentation will focus on development of nanostructured particulate systems and fabrication of advanced devices for energy storage, environmental protection, national security and health care. I will present novel nanoenergetic systems that have the potential to enable a more concentrated energy release and potentially can be used for various military applications such as an actuation parts, igniter, propulsion unit, gas-generators as well as an active part for high power electromagnetic pulse generators. Most previous research on metastable intermolecular composites or nanoenergetic materials focused on the impact of the particle size on the amplitude and velocity of the temperature front and the ignition features of the binary thermite reacting systems. Only few studies were concerned with the gas pressure evolution. We have recently shown that among common thermite reactions, the Al/Bi2O3 mixture generated the highest pressure pulse and may be used as a nanoenergetic gas generator (NGG) and has the potential of being used in next generation of weapons and explosives.

I will describe a novel cost-effective and energy efficient production of nanostructured complex oxides that we referred to as Carbon Combustion Synthesis of Oxides (CCSO). In this process, the reactive oxidation of carbon/graphite nanoparticles generates a steep thermal wave (temperature gradient of up to 500 "aC/cm) that propagates through the solid reactant mixture (oxides, carbonates or nitrates) converting it to the desired products. The high rate of gas release enables synthesis of highly porous complex oxides having a particle size in the range of 50-800 nm. The experimental results of fabrication of various systems such as hard and soft magnetic materials, superconductors, multiferroics, bulk ceramic resistors, capacitors, photocatalysts with p-n junction, MRI contrast agents and cancer hyperthermia will be presented. Key factors that affected to the device characteristics (magnetization, conductivity, magnetic resonance relaxivity and other) will be discussed. The process may be scaled up for continuous large-scale production of complex oxides nanoparticles. Development of this emerging technologies warrant a multifaceted approach, which includes interdisciplinary collaboration, partnerships with industry and academia, and integration of modern problems into our curriculum.

See more of this Session: Poster Session: Meet the Faculty Candidate
See more of this Group/Topical: Education

Understanding Brain Adaptation Processes During Alcohol Withdrawal

Mary K. McDonald, Chemical Engineering, University of Delaware, Newark, DE

Alcohol abuse affects at least 14 million Americans, and nearly 35,000 die each year due to chronic alcohol dependence and alcohol withdrawal. At the clinical level, individuals with chronic alcohol dependence experience a variety of physical and emotional consequences upon cessation of alcohol intake that may include severe anxiety, hallucinations, elevated blood pressure, increased heart rate, seizure, tremens, cardiac arrhythmias, or in some cases, sudden death. This collection of symptoms during alcohol withdrawal suggests a role for two interconnected brain regions known to regulate homeostasis and emotion: the nucleus tractus solitarius (NTS) and the central nucleus of the amygdala (CeA), according to mechanisms that are currently not well understood. As a first step towards understanding these mechanisms in the brain, the goal of this work is to identify and characterize the molecular processes of adaptation and regulation in alcohol withdrawal using a combination of experimental and computational techniques. Specifically, we have: (1) utilized emerging high-fidelity, high-throughput techniques to collect dynamic gene expression profiles in the NTS and CeA; (2) developed novel techniques to analyze these data sets, and finally, (3) determined the primary control mechanism for compensation following alcohol blockage of a key brain receptor using a computational model. The results provide a unique systems level understanding of dynamic adaptation processes in the brain during alcohol withdrawal, which may lead to strategies for prevention of the lethal consequences of withdrawal and thereby enable the successful recovery of alcohol dependent individuals.

Advisors: Babatunde A. Ogunnaike[1] and James S. Schwaber[2] ([1]Department of Chemical Engineering, University of Delaware, Newark, DE; [2]Department of Pathology, Anatomy, and Cell Biology, Jefferson Medical College, Thomas Jefferson University, Philadelphia, PA)

See more of this Session: Poster Session: Meet the Faculty Candidate
See more of this Group/Topical: Education

Modeling and Simulation of Polymeric and Biophysical Systems

Aruna Mohan, Chemical Engineering, University of California, Santa Barbara, Santa Barbara, CA

There is much interest in understanding the dynamical behavior of biopolymers such as DNA, proteins and cellular networks for many medical and biotechnological applications. Computer simulations combined with theoretical modeling provide a powerful technique for the study of these systems. As part of my graduate thesis, I studied the size-dependent separation of DNA chains induced by collisions with obstacles in microfluidic obstacle arrays. This research developed and utilized theoretical models of chain dynamics in conjunction with Brownian dynamics simulations of single polymer chains to elucidate the behavior of DNA in obstacle array-based microfluidic separation devices. These studies enable the optimized design and operation of DNA separation devices.

My postdoctoral research involved the development of theoretical stochastic models of polymer translocation through nanometer-size pores. Specifically, this work considered the effect of pore-polymer interactions and pore geometry on the time of polymer transport through the pore. These studies aid the understanding of biological processes such as the transport of proteins through cellular membranes, and facilitate the design of nanopores for technological applications such as DNA sequencing.

More recently, my postdoctoral research has resulted in the development of a reduced-order computational model for the molecular dynamics simulation of entangled polymer systems. This novel model can accommodate both semiflexible polymer systems, such as networks of biopolymers, as well as flexible polymer melts, and consequently, may be applied to study a wide range of polymer systems.

See more of this Session: Poster Session: Meet the Faculty Candidate
See more of this Group/Topical: Education

Synthetic Biology: An Evolving Field to Create New Biological Components and Systems

Tae Seok Moon, Chemical Engineering, Massachusetts Institute of Technology, Cambridge, MA

Kristala Jones Prather, Chemical Engineering, Massachusetts Institute of Technology, Cambridge, MA

Synthetic organic chemists approach a synthesis scheme for a compound, based on the structure of the target compound. This retrosynthetic design is a powerful tool, giving organic chemists a lot of flexibility: a wide range of target products and a large set of possible synthetic steps. Despite this flexibility, chemical synthesis is not efficient when the target compounds are complex and have many functional groups. To synthesize these challenging compounds, biochemists and metabolic engineers have recruited enzymes and have shown the potential of these biocatalysts. However, most biosynthetic approaches have been made by utilizing "forward biosynthesis," a substrate-focused design of biological pathways, limiting the range of the target products. My research focuses on combining these two approaches: retrosynthesis and biocatalysis. Using tools of "Synthetic Biology," an evolving field involving the creation of new biological components and systems, I aim to create novel biosynthetic pathways for highly functionalized compounds. In addition, I seek to design and characterize interchangeable parts to build synthetic devices and systems that help to understand and to harness biological systems.

See more of this Session: Poster Session: Meet the Faculty Candidate
See more of this Group/Topical: Education

Synergy of Protein and Genome Engineering

Nikhil U. Nair, Chemical & Biomolecular Engineering, University of Illinois at Urbana-Champaign, Urbana, IL

Huimin Zhao, Chemical & Biomolecular Engineering, University of Illinois at Urbana-Champaign, Urbana, IL

The focus of my work has been to use protein and metabolic engineering to produce industrially important chemicals and fuels from biomass. At the same time, I have developed genome engineering tools to facilitate studies in metabolic engineering, synthetic biology, and systems biology.

1) Selective catalysis for xylitol production: We have engineered a highly substrate specific D-xylose reductase (XR) that retains very little of its natural activity toward the epimeric substrate L-arabinose. Using structure-function analysis and directed evolution, we evolved this mutant with >16-fold preference for xylose compared to arabinose. Whole cell biocatalysis studies demonstrate that our mutant produces very little byproduct (arabinitol) from a mixture of hemicellulosic sugars, while producing xylitol with near-wild-type productivity.

2) A tool for E. coli genome modification: As one of the most important platform organism for biocatalysis, there is always room for better methods for E. coli genome manipulation. However, the eccentricities of its recombination machinery have necessitated the use of heterologous (phage) proteins for quick, effective genome modification. We have developed a tool for easily modifying the E. coli genome that does not rely on any heterologous proteins. PCR products are directly used as mutagenic cassettes and recombinants with desired deletion or point mutation can be obtained with high frequency.

3) Biobutanol from S. cerevisiae: Butanol has been identified as an alternative to ethanol due to several superior properties as a liquid biofuel. Heterologous production in yeast was challenging until we performed metabolic engineering on the expression strain and directed evolution on one of the butanol biosynthetic enzymes for soluble expression. The synergy of the two efforts finally yielded a strain capable of producing butanol. Our efforts demonstrate that yeast is a viable platform organism to produce butanol from glucose.

4) A tool for S. cerevisiae genome modification: S. cerevisiae is a model organism for industrial biocatalysis. We developed a tool for quick, precise modification of the yeast genome. It is a powerful, yet simple, tool for point mutagenesis, insertions, and deletion studies alike. Named MIRAGE (mutagenic inverted repeat assisted genome engineering), this method uses yeast's highly efficient homologous recombination machinery to integrate a mutagenic cassette containing an inverted repeat (IR) into the chromosome. Due to its inherent instability in the chromosome, the IR catalyzes its own excision leaving behind only the desired mutation without any scars at very high frequency.

See more of this Session: Poster Session: Meet the Faculty Candidate
See more of this Group/Topical: Education

Inertial Microfluidics for Diagnostics and Therapeutics

John Oakey, Center for Engineering In Medicine, Massachusetts General Hospital, Charlestown, MA

Mehmet Toner, Center for Engineering In Medicine, Massachusetts General Hospital, Charlestown, MA

Microfluidic platforms have attracted much interest for their ability to improve the efficiency, reliability and access of diagnostic and therapeutic tools in a variety of health care settings. A primary limitation of these devices, particularly microscale flow cytometry platforms, is their relatively low throughput, relative to their conventional counterparts. Inertial microfluidics offers a solution to this bottleneck. Inherently high-throughput, inertial microfluidics exploits hydrodynamic forces resulting from interactions between particles and surfaces. Lift forces, shear gradient forces and Dean drag forces are precisely controlled in microfluidic channels by geometric design to produce well-defined, predictable behavior. While inertial focusing devices are operationally simple, they exhibit rich, complex physics that may be adapted for a wide variety of applications. My work has addressed the physics of inertial migration and particle focusing in microfluidic channels, design aspects of inertial focusing devices and applications for inertial focusing technology in medicine and biomedical research. My poster will provide an overview of each of these topics. My work in these areas is detailed in the following publications:

Inertial focusing:

Controlled Microfluidic Encapsulation of Particles and Cells in Poly(Ethylene Glycol) Microparticles, Oakey, J., Edd, J., Toner, M., In Preparation.

Inertial Particle Focusing and Ordering in Microfluidic Channels, Oakey, J., Lee, E., Di Carlo, D., Toner, M., In Preparation.

Microscale Optics and Flow Cytometry:

Fiber Focused diode bar trapping for microfluidic flow manipulation, Applegate, R.A., Jr., Squier, J., Vestad, T., Oakey, J., Marr, D.W.M., Applied Physics Letters, 92, 013904, 2008.

Optically Integrated Microfluidic Systems for Cellular Characterization and Manipulation, Applegate, R.A., Jr., Schafer, D.N., Amir, W., Squier, J., Vestad, T., Oakey, J., Marr, D.W.M., J. Opt A: Pure Appl. Opt, 9, S122, 2007.

Optical Waveguides via Viscosity-Mismatched Microfluidic Flows, M. Brown, T. Vestad, J. Oakey, D.W.M. Marr, Applied Physics Letters, 88, 134106, 2006.

Microfluidic Sorting System Based on Optical Waveguide Integration and Diode Laser Bar Trapping, Applegate, R.W., Jr., Squier, J., Vestad, T., Oakey, J., Marr, D.W.M., Bado, P., Dugan, M.A., Said, A., Lab on a Chip, 6, 422-426, 2006.

Optical Trapping, Manipulation and Sorting of Cells and Colloids in Microfluidic Systems With Diode Laser Bars, Applegate, R.W., Jr., Squier, J., Vestad, T., Oakey, J., Marr, D.W.M., Optics Express, 12, 4390-4398, 2004.

Laminar Flow-Based Separations at the Microscale, Oakey, J.S., Marr, D.W.M., Biotechology Progress, 18, 1439-1442l, 2002.

Microfluidic Control Using Colloidal Devices, Terray, A., Oakey, J.S., Marr, D.W.M., Science, 296, 1841-1844, 2002.

See more of this Session: Poster Session: Meet the Faculty Candidate
See more of this Group/Topical: Education

Engineering the Yeast Saccharomyces Cerevisiae for Protein Production and Bioenergy Applications

Michelle A. O'Malley, Department of Biology, Massachusetts Institute of Technology, Cambridge, MA

Chris A. Kaiser, Department of Biology, Massachusetts Institute of Technology, Cambridge, MA

Anne Skaja Robinson, Department of Chemical Engineering, University of Delaware, Newark, DE

Molecular engineering has enabled the economical production of several valuable products with the aid of microbial systems, from vaccines and therapeutics to biofuels. Yeast systems, such as *S. cerevisiae*, are ideal for protein production and other industrial applications because they are easy to grow and scale-up, inexpensive to work with, and straightforward to genetically manipulate. Since yeast have a eukaryotic secretory pathway, they also are equipped with the cellular machinery to facilitate protein folding and critical post-translational processing steps that often are necessary for the production of authentic proteins.

My doctoral research with Anne Robinson (Dept. of Chemical Engineering, University of Delaware) involved the expression, purification, and biophysical characterization of human G-protein coupled receptors (GPCRs) from *S. cerevisiae*. GPCRs are membrane proteins that mediate responses to extracellular stimuli, represent over 50% of all drug targets, and have been directly linked to several diseases including HIV infection, cancer, and diabetes. However, the development of improved, structure-based therapeutics to target these proteins requires milligram amounts of properly folded, purified receptors, which are generally not achievable from native tissues. To structurally characterize this important class of proteins, we over-expressed GPCRs in *S. cerevisae* to circumvent the problem of low natural abundance, and systematically evaluated their trafficking, folding, and activity. Although all GPCRs were successfully expressed at the mg/L scale, we discovered that most receptors trigger cellular stress responses when expressed in yeast and exhibit compromised activity. We found that these observations were closely linked to improper processing of the N-terminal leader sequence, which suggests that translocation is a critical bottleneck for receptor over-expression in yeast. Of all the receptors analyzed, only the human adenosine A_2a receptor (hA_2aR) was found in large quantities at the plasma membrane, where it was able to bind to adenosine analogs. The hA_2aR was purified to homogeneity in milligram amounts and reconstituted with mixed surfactant micelles, which allowed for extensive structural characterization using biophysical techniques.

Building on my expertise with *S. cerevisiae*, my postdoctoral research within Chris Kaiser's lab (Dept. of Biology, MIT) is focused on engineering *S. cerevisiae* as a tool for ethanol production from diverse cellulose-rich feedstocks. Our approach involves the characterization of fungal cellulosomes, large multisubunit complexes of carbohydrate hydrolyzing enzymes and cellulose binding domains, from anaerobic fungi that are native to the rumen of large herbivorous animals. The multicomponent cellulosome complexes produced by these fungi are superior to similar complexes that have been widely studied from bacteria since they are much more efficient at degrading cellulose. As the components of the fungal cellulosome have not been completely identified, we are using powerful genomic tools to identify genes within anaerobic fungi that are upregulated when grown in the presence of different cellulosic feedstocks. We are currently working towards expressing and reconstituting components of the anaerobic fungal cellulosome within *S. cerevisiae* to facilitate consolidated bioprocessing of cellulose to ethanol. Furthermore, we seek to understand the structural basis for efficient cellulose degredation, and optimize the cellulosome through genetic engineering to maximize *S. cerevisiae*'s ability to degrade complex forms of cellulose.

See more of this Session: Poster Session: Meet the Faculty Candidate
See more of this Group/Topical: Education

Describing DNA at the Nanometer Length-Scale

Vanessa Ortiz, Chemical and Biological Engineering, University of Wisconsin-Madison, Madison, WI

Although any DNA segment can form nucleosomes in vitro, the relative affinity of histones for different DNA sequences may differ by as much as 1000-fold (> 4 kcal/mol in free-energy). Recent experimental studies of nucleosome positioning have established a link between the affinity of a DNA segment for the histone protein complex and its sequence. Crystal structures of nuclesomes show no direct contacts between DNA bases and the histones; therefore, any influence of sequence on nucleosome positioning should be indirect. It has been suggested suggested that certain sequence patterns could influence nucleosome positioning by facilitating the tight bending required in DNA to wrap around the histones. Sequence could facilitate this bending either by creating permanent kinks in the segment (modification of "bendedness") or by reducing the amount of energy required to form a bend (modification of "bendability").

Experimental efforts to study the effects of sequence on DNA conformation have provided some understanding of the problem, but the complexity of the experimental systems has hindered attempts to distinguish bendedness from bendability in the experimental data. On a theoretical level, the study of sequence effects on DNA conformation requires the ability to simulate large systems during long timescales and in a diverse set of conditions (different sequences, temperatures, ionic strength, etc.). Even with current technological advances, it still becomes computationally expensive to meet these conditions with the use of atomistic molecular dynamics. Attempts to provide a theoretical underpinning using coarser models have been limited, largely because models capable of describing the mechanical properties of dsDNA beyond the elastic limit and its ability to de-hybridize locally on a sequence-dependent manner were not available until recently.

My research has focused on developing a mesoscale DNA model that can reproduce experimentally observed sequence-dependent flexibility. The model has been parameterized to reproduce the experimentally observed melting temperature of DNA, and it can describe the mechanism of DNA hybridization. As an application for this model I have employed umbrella sampling molecular dynamics simulations to examine the mechanical properties of DNA as a function of sequence. Because the model allows for de-hybridization, it becomes possible to examine the effects of sequence on "pure" bending and on the overall stability of the DNA double-helical structure when the molecule is subject to severe bending constraints. Omitting the histones from the simulated systems allows discerning whether the sharp bends observed in crystal structures of nucleosomes are a cause of or an effect from the formation of the complex. Results demonstrate that when bent, DNA has a sequence-dependent preference for hybridization and orientation, even in the absence of a protein. Analysis of the results in the context of previous experimental observations suggests a mechanism for nucleosome positioning in which sequence can determine the affinity of a DNA segment for the histones by altering the direction (via anistropic bending) and lifetimes (via non-native hybridization) of the bends it adopts.

See more of this Session: Poster Session: Meet the Faculty Candidate
See more of this Group/Topical: Education

Metabolic Engineering of Terpenoid Production in Microbes: Challenges and Opportunities

Ajikumar Parayil, Chemical Engineering, Massachusetts Institute of Technology, Cambridge, MA

Gregory Stephanopoulos, Chemical Engineering, Massachusetts Institute of Technology, Cambridge, MA

Recent developments in synthetic biology and metabolic engineering offer new possibilities for the large scale production of complex natural products through the engineering of biosynthetic pathways in surrogate microbial hosts. However, challenges associated with this task include (1) the expression of many foreign, large, or multidomain enzymes that may complicate processes and (2) the accumulation of toxic foreign metabolites in the new host. With my chemical and biochemical expertise, my research interest in this area is to reconstruct complex biosynthetic pathways through engineering of natural product molecular machinery in microbial hosts. My ultimate research goals will be on pathway engineering of bioactive molecules and developing microbial cell factories for overproduction of biochemical and biofuels from renewable resources.

The anticancer drug Taxol is a unique and challenging example with many interesting biochemical features. The complete biosynthetic pathway consists of 19 enzymes in the native organism, a formidable challenge to transplant to the microbial host. Additionally, production in the native organism, as well as chemical synthesis is uneconomical and unscalable for the production of Taxol. Thus engineering the biosynthesis of Taxol is a challenging option to accelerate the development of complex biosynthetic pathway engineering in microbes for large scale sustainable production of Taxol and analogous terpenoid molecules. It requires the design of a microbial cellular environment, capable of supplying precursors needed for Taxol biosynthesis and the functional reconstitution of the heterologous Taxol biosynthetic enzymes. In our research we used metabolic engineering and synthetic biology tools to modulate the upstream endogenous non-mevalonate pathway in E. coli with downstream heterologous Taxol biosynthetic pathways, resulting in a strain capable of overproducing early Taxol precursors. Additionally, our multivariate modular approach to the pathway engineering unraveled the regulation of the non-mevalonate isoprenoid pathway in bacteria. Thus the current study has opened up the new possibilities high level microbial production rich and diverse chemical structures in terpenoid molecules. As engineered pathways are extended to include very long pathways, such as the complete 19 enzyme Taxol pathway, new cell engineering, pathway engineering and protein engineering, tools will need to be developed to stably express the pathway and synthesize the product.

See more of this Session: Poster Session: Meet the Faculty Candidate
See more of this Group/Topical: Education

Fluctuations of Water near Extended Hydrophobic and Hydrophilic Surfaces

Amish J. Patel, Chemistry, University of California, Berkeley, Berkeley, CA

David Chandler, Chemistry, University of California, Berkeley, Berkeley, CA

The primary goal of this work is to study water density fluctuations in probe volumes of different shapes and sizes, both in the bulk as well as near various hydrophobic and hydrophilic surfaces. We introduce a formalism involving a coarse-grained density that allows us to readily combine umbrella sampling with available molecular dynamics simulation packages. It has been established, both theoretically and using simulations, that the presence of a purely repulsive extended hydrophobic surface induces drying in its vicinity, resulting in the formation of a vapor-liquid interface. Recent reports in the literature have called attention to an apparent discrepancy between the theory and simulations on addition of weak dispersive attractions between the extended hydrophobic surface and water, which results in the disappearance of drying. However, an examination of the fluctuations of water near the attractive surface reveals that the probability of density depletion near the surface is significantly larger than that in bulk, akin to density fluctuations at the water-vapor interface. Hence, there is no discrepancy between theory and simulation and the absence of drying is simply the result of the interface been drawn to the hydrophobic surface in the presence of attractions. As the attraction between the hydrophobic surface and water increases, we expect the adhesion of the interface to the surface to get stronger, resulting in the probability of density depletion getting smaller. Near a model hydrophobic surface with attractions that are typical of alkanes, water density depletion is easier than in the bulk for up to 3 molecular diameters. In contrast, fluctuations near a model hydrophilic surface are very similar to that in the bulk.

See more of this Session: Poster Session: Meet the Faculty Candidate
See more of this Group/Topical: Education

Molecular Handles: Nanoscale Chemical Engineering with Optoelectronic Tweezers

Peter Pauzauskie, Physical and Life Sciences Directorate, Chemical Sciences Division, Lawrence Livermore National Laboratory, Livermore, CA

Joe H. Satcher Jr., Chemical Sciences Division, Lawrence Livermore National Laboratory, Livermore, CA

The design and fabrication of future nanoscale devices for applications in biotechnology, communications, information processing, and energy conversion will require massive parallelism, low cost, as well as the ability to address a range of both inorganic and organic nanostructures with many different sizes, shapes, and chemistries. In addition to massive parallelism, an ideal fabrication platform will also have the ability to address single structures at molecular length scales. Employing chemical engineering fundamentals such as momentum, heat, and mass transfer will be crucial in realizing future nanoscale design goals. For instance, due to the extremely low Reynolds numbers at sub-micron dimensions, the forces employed in solution-phase processing must overcome viscous drag while withstanding significant randomizing effects from Brownian mass transfer.

Optoelectronic tweezers (OET) have emerged in recent years [1] as a low-cost, massively parallel means for the optical assembly of both inorganic and biological building blocks using 100,000x less optical power than conventional laser tweezers [2]. Recent work with OET [3] has demonstrated that the separation of semiconducting and metallic nanowires is possible based on the large intrinsic differences in conductivity and dielectric constant between silicon and silver nanowires. A similar approach can be extended to multi-walled carbon nanotubes (CNTs) [4] and fluorescently-labeled single-walled CNTs, promising a flexible new approach to nanoscale separations with far-reaching implications for future engineering with near-molecular scale materials.

Furthermore, heat transfer and temperature distributions are significant constraints in addressing biological nanostructures. Current work with the ellipsoidal pathogenic bacterium 'Y. pestis' (bubonic plague) requires a thorough understanding of heat generation and transfer within the OET device. Thermal measurements with a thermographic infrared camera correlate well with finite-element simulations of heat transfer in OET chambers, suggesting that OET is a powerful new platform for future chemical engineering with inorganic and biological nanostructures.

[1] Chiou, P.Y. et al., Nature (2005)

[2] Pauzauskie, P.J. et al. Nature Materials (2006); Nakayama, Y.*; Pauzauskie, P.J.*; et al., Nature (2007)

[3] Jamshidi, A.*; Pauzauskie, P.J.*; et al., Nature Photonics (2008)

[4] Pauzauskie, P.J.*; Jamshidi, A.*; et al., Applied Physics Letters (in press, 2009)

* - equal contribution

See more of this Session: Poster Session: Meet the Faculty Candidate
See more of this Group/Topical: Education

Microfluidic Platforms for the Study of in Meso Membrane Protein Crystallization

S.L. Perry, Department of Chemical & Biomolecular Engineering, University of Illinois at Urbana-Champaign, Urbana, IL

C.F. Zukoski, Department of Chemical & Biomolecular Engineering, University of Illinois at Urbana-Champaign, Urbana, IL

P.J.A. Kenis, Department of Chemical & Biomolecular Engineering, University of Illinois at Urbana-Champaign, Urbana, IL

Microfluidic technology originally gained prominence because of the ability to work at very small volumes with precise control. Research in this field has advanced to the point where an idea of a complete "lab on a chip" is a realistic goal, even to the extent of performing clinical and diagnostic tests. Microfluidic approaches also enable the efficient handling of a large number of individual trials simultaneously. In the field of crystallization and structural biology this kind of high throughput screening capability is critical to identify conditions at which biological macromolecules will crystallize so their structure can be determined.

Membrane proteins play a critical role in a variety of biological processes including signaling and material or energy transduction. As such they are prime targets for pharmacological treatments. Knowledge of their structure and function would be of tremendous fundamental and medical benefit. Despite their importance, the precise structure of a disproportionate number of membrane proteins is still unknown. One of the key bottlenecks in this process is creating and maintaining membrane protein functionality, not only during protein expression and purification, but also during crystallization studies for structural characterization. To counter this difficulty the *in meso* (lipidic cubic phase, LCP) method was developed, allowing membrane proteins to remain in a membranous environment during crystallization [1]. This method has had marked successes with membrane proteins such as bacteriorhodopsin that were not successfully crystallized by traditional methods. More recently, the crystallization and structure determination of two human G protein-coupled receptors has been reported [2,3]. Despite its successes, the biggest challenge of this method is the preparation of the highly viscous lipid mesophase used as the crystallization medium, at sub-microliter volumes to enable screening of many potential crystallization conditions.

In my graduate research I have developed a method for preparing *in meso* crystallization trials at the 20 nL level using multilayer microfluidic technology in polydimethylsiloxane (PDMS). Within these chips, complex patterns of fluid flow for mixing can be achieved by pneumatic actuation of integrated valves and pumps. A novel multi-chamber design enables the mixing of highly viscous lipids with an inviscid aqueous protein solution to prepare the lipid mesophase needed for *in meso* crystallization trials. Arrays of these mixing and crystallization elements enable the screening of a wide range of crystallization conditions. We demonstrated feasibility with the successful on-chip crystallization of the previously characterized membrane proteins bacteriorhodopsin [4] and photosynthetic reaction center. Presently we are extending application of these microfluidic *in meso* crystallization chips to novel proteins for which no structure is known.

A secondary challenge in the structure determination of proteins is the harvesting and mounting of fragile and potentially tiny crystals, a task that is exacerbated when trying to harvest a crystal from a tiny microfluidic compartment. The ability to perform *in situ* X-ray analysis of crystals grown on-chip would circumvent these issues. To accomplish this, we created an X-ray transparent, PDMS / polyimide hybrid device architecture that retains the ability to perform complex fluid handling while allowing for on-chip X-ray analysis. Testing of these all-integrated on-chip crystallization and *in situ* X-ray analysis platforms is in progress.

With these microfluidic platforms now in hand, I am also pursuing studies of the phase behavior of novel lipid/water systems for use with *in meso* crystallization trials as well as mechanistic studies of how *in meso* crystallization occurs. All of this work together has the potential to develop the scientific understanding of *in meso* protein crystallization to the point where some level of intelligent design could be incorporated into future crystallization trials, leading to improved rates of success.

References:

[1] Landau, E. M.; Rosenbusch, J. P., *P Natl Acad Sci USA* **1996**, *93*, 14532-14535.

[2] Cherezov, V.; Rosenbaum, D. M.; Hanson, M. A.; Rasmussen, S. G. F.; Thian, F. S.; Kobilka, T. S.; Choi, H. J.; Kuhn, P.; Weis, W. I.; Kobilka, B. K.; Stevens, R. C., *Science* **2007**, *318*, 1258-1265.

[3] Jaakola, V. P.; Griffith, M. T.; Hanson, M. A.; Cherezov, V.; Chien, E. Y. T.; Lane, J. R.; IJzerman, A. P.; Stevens, R. C., *Science* **2008**, *322*, 1211-1217.

[4] Perry, S. L.; Roberts, G. W.; Tice, J. D.; Gennis, R. B.; Kenis, P. J. A., *Cryst Growth Des* **2009** (in press, DOI: 10.1021/cg900289d).

See more of this Session: Poster Session: Meet the Faculty Candidate
See more of this Group/Topical: Education

Biomaterial Platforms to Investigate the Fundamentals of Cell Migration

Shelly R. Peyton, Biological Engineering, MIT, Cambridge, MA

Douglas Lauffenburger, Biological Engineering, MIT, Cambridge, MA

Linda Griffith, Biological Engineering, MIT, Cambridge, MA

Increasing evidence suggests that the extracellular matrix (ECM) plays a critical role in regulating the behavior of cells in a multitude of tissue systems. The ECM contains a number of biochemical and biophysical properties, some of which have already been shown to direct cell migration in vitro, namely the stiffness of the ECM (durotaxis), the ECM's adhesivity (haptotaxis), and its presentation of growth factors (chemotaxis). Since these parameters are often lumped in naturally-derived materials, such as collagen and Matrigel, it is difficult to determine which of these cues (or which combination of cues) may be most important in regulating cell behavior in these systems. For that reason, many are now attempting to use synthetic biomaterials, which have the ability to independently tune these biochemical and biophysical properties, to study the affect of these cues on cell behavior.

Here, I will discuss a variety of biomaterial systems that maybe be used to study the effects of ECM adhesivity, stiffness, growth factor ability, porosity, and degradability in a systematically controlled environment. These systems can be adapted to study cell migration in a variety of tissue systems, though I will be discussing their utility mostly with primary smooth muscle cells and adult mesenchymal stem cells.

See more of this Session: Poster Session: Meet the Faculty Candidate
See more of this Group/Topical: Education

Vesicle Transformations: Unilamellar to Multilamellar Structures

Arun Ramachandran, Chemical Engineering, University of California at Santa Barbara, Santa Barbara, CA

Vesicle Transformations: The Whys and Hows of Multilamellar Structures

Vesicles are an extremely important class of materials, being the foundation for a wide variety of existing industrial products, as well as serving as vehicles for drug delivery, imaging and separations. However, in spite of their ubiquitous applications and appearance in many products, much of the underlying product development has been empirical, and the basic principles that are necessary for improved product performance (e.g. longer shelf life), and for the development of whole new classes of materials and applications, are largely unexplored.

Typically, for most vesicle-based products, there are at least two desirable properties:

(a) There is a target pour viscosity for the product

(b) The vesicles should not gravitationally phase separate from the suspending fluid over the shelf life of the product.

These two properties are very intimately connected to the size, shape and lamellarity distributions of the vesicles. It is recognized in the research community that these distributions depend on the following factors:

(i) The constituent lipids of the bilayer, and the weight fraction of the lipid in the solution

(ii) Any additives (such as salts, short chain alcohols etc.)

(iii) The exact procedure by which the lipids are mixed with aqueous phase, including when the additives are added.

Unfortunately, systematic investigations and reliable models connecting the above variables to the final size and lamellarity distributions are lacking, and the consequences of different surfactant combinations on vesicle formation and properties are understood only at a broad overview level. In addition, there is no guarantee that once vesicles are prepared, their size and lamellarity distributions remain fixed. Even if products are prepared with their initial distributions tailored to satisfy conditions (a) and (b), they may eventually evolve towards distributions that may compromise these requirements. Therefore, the task of systematically achieving favorable vesicle suspensions can be broken down into understanding two major questions:

(A) How do factors (i) to (iii) control whether a certain formulation and preparation procedure will yield unilamellar or multilamellar structures of certain sizes and shapes?

(B) How and over what time scales do these structures evolve from being, say, unilamellar to multilamellar after formation?

The questions posed above represent an extremely large number of investigations. We have chosen to begin with a small subset, namely the exploration of question (B) with the class of vesicles composed completely of positively charged bilayers. Even though such surfactants are frequently used in fabric softeners, shampoos and conditioners, the number of studies on the vesicles constituted from such surfactants is surprisingly limited as compared to vesicles composed of neutral lipids or even negatively charged lipids. Our recent observations show that these vesicle suspensions can undergo a transition over a several day period from an initial state of mainly unilamellar vesicles to a state of smaller bilamellar and multilamellar vesicles, without any added salt or other forcing, such as changes of temperature. Our hypothesis is that this transition could be driven by repulsive forces between excessively "crowded" charged vesicles. We also propose a mechanism driving this transformation - unilamellar vesicles deflate, curl up and fuse at the tips to form bilamellar vesicles (see figure 1 below). We were able to verify this mechanism by fabricating vesicles with a mixture of single and double tailed cationic lipids, which did not exhibit these transformations, presumably because the vesicles prepared

We are currently attempting to delineate the mechanism for the formation of multilamellar vesicles from unilamellar ones. Since a mechanism involving curling and folding of the vesicle [and in fact, answering both questions (A) and (B)] would invariably depend on the properties of the bilayer membrane, one of the key components of our studies is the measurement of the membrane properties for vesicles made from lipid mixtures, which we are currently performing using micropipette aspiration of giant vesicles prepared from these mixtures. We have recently performed measurements on giant vesicles prepared from a C18-double tailed cationic lipid, which is, to the best of our knowledge, the first measurement performed for vesicles composed fully of charged lipids. We also intend to develop a theoretical description of vesicle folding and other processes, both from a molecular and a continuum perspective.

The research outlined here is extremely interesting from a fundamental point of view. For example, the transformation observed in our and prior experiments is reminiscent of the process of *endocytosis*, whereby a cell ingests some volume of extracellular fluid by initiating an invagination on the cell wall, followed by pinch-off. Further, we anticipate that this work will enable a number of important technological applications:

(i) The phenomenon may be utilized as a procedure to encapsulate drugs or other materials within a bilamellar (or multilamellar) vesicle. Bilamellar vesicles are expected to have longer lifetimes *in vivo* than unilamellar vesicles, and this is a key property for drug delivery formulations.

(ii) Evaporation of the suspending fluid (or other ways of increasing the volume fraction of vesicles) may lead to vesicle transformations, from unilamellar to multilamellar, resulting in a reduction in the average size and vesicle volume fraction, which may be unintended and may have unfavorable consequences.

(iii) Alternatively, this phenomenon may be utilized as a way of manipulating the vesicle structure without the need to introduce salt or other "additive" sources of osmotic pressure that can reduce the volume of vesicles and potentially drive similar transitions.

It should be emphasized that the changes in density and the volumetric concentration of vesicles caused by a transition from unilamellar to multilamellar structures will also play an important role in the development of gravitationally-induced phase separation (sedimentation or creaming), which is a critical issue for many of the suspension-based commercial products.

The results from our experiments and future directions will be discussed in greater detail in the poster.

SHAPE * MERGEFORMAT

Deflated unilamellar vesicle

Curled up stomatocyte
Bilamellar Vesicle

Figure 1. Proposed mechanism for bilamellar vesicle formation.

See more of this Session: Poster Session: Meet the Faculty Candidate
See more of this Group/Topical: Education

Wiring Enzymes and Cells On Biomimetic Nanoscale Supports for Biological Devices

Ramaraja P. Ramasamy, Airbase Sciences Branch RXQL, U.S. Air Force Research Laboratory, Tyndall Air Force Base, FL

Bio-based electrochemical devices such as biofuel cells, biosensors, MEMS actuation devices, bioelectronics, and micro-power systems integrate biotechnology, material science and electrochemistry. While amperometric biosensor is an established technology now, a closely similar biological fuel cell is still a budding area of research. Biofuel cells are a subset of fuel cells that employ biocatalysts. Biofuel cells offer a viable way for direct electricity production from organic compounds, biomass and biomass waste. Microbial biofuel cells employ living cells to catalyze the oxidation of the fuel, whereas enzymatic biofuel cells use enzymes and redox proteins for this purpose. The current advantage of microbial biofuel cells is that they typically have long lifetimes (up to five years) and are capable of completely oxidizing simple sugars to carbon dioxide. They are limited, however by low power densities. By contrast, enzymatic biofuel cells typically possess orders of magnitude higher power densities, but can only partially oxidize the fuel and have limited lifetimes owing to the fragile nature of the enzyme. Major advancement is needed in increasing enzyme lifetime, enhancing enzyme-electrode electron transfer, improving the kinetics of the bio-electrochemical reactions before this technology can be successfully commercialized. Promising near term applications of biological fuel cells include power sources for remote military equipments, portable electronic devices, implantable medical devices and biomass waste treatment.

The electrodes in these biological devices require an effective electrical communication with the biological redox active center. Effective electrical coupling of the biological redox center and the electrode material requires a fundamental understanding of the underlying electron transport processes as well as the material properties of the nanocomposite structures in which the biocatalysts are entrapped. A novel and effective strategy is to directly wire the electrons between the enzymes and electrodes using biomimetic supports that serve as conductive hybrid nanocomposites. This benefits in a direct electron transfer (DET) between the enzyme and electrode provides great potential for miniaturization, stability and high power output.

See more of this Session: Poster Session: Meet the Faculty Candidate
See more of this Group/Topical: Education

Designing Nanoparticles with Self-Assembly for Applications in Medicine, Environment and Energy

Hitesh G. Bagaria, Dept of Chemical and Biomolecular Engineering, Rice University, Houston, TX

Nanoparticles - organic, inorganic and organic/inorganic composites - have proven potential for many applications in medicine, environment and energy. However, significant challenges still remain in realizing these applications. In most cases the solutions are sophisticated and there is a need to develop simpler, more practical methods. One of the ways to achieve the desirable complex assemblies is by making use of non-covalent forces or self-assembly. The development of self-assembly based nanoparticles is the motivation of my research.

See more of this Session: Poster Session: Meet the Faculty Candidate
See more of this Group/Topical: Education

Molecular Simulations of Complex Polymeric Materials and Nanocomposites

Robert A. Riggleman, Department of Chemical Engineering, University of California, Santa Barbara, Santa Barbara, CA

The use of computational tools to obtain the properties of polymeric materials becomes increasingly challenging as these materials become increasingly complex. Such complexities that are often present in polymer materials include long-range electrostatic interactions in polyelectrolyte materials, or the inclusion of nanoparticles for mechanical reinforcement. Problems as complex as these can require a diverse set of computational tools to understand the array of phenomena observed in such systems, and in this poster I will describe some of the topics I have studied and the methods I have learned in my graduate and postdoctoral education.

Using a combination of Monte Carlo and molecular dynamics simulations, part of my research has focused on understanding how aging and deformation changes the properties of polymer nanocomposite materials, which is essential to understand as nanocomposite materials are finding increased use in critical applications where they will be subject to deformation. Recent experiments and simulations have shown that the dynamics in glassy solids can be enhanced by several orders of magnitude. Using molecular dynamics simulations on a model nanocomposite material, I have studied how the presence of nanoparticles changes the dynamic enhancement. I have shown that the changes due to the presence of the particles arise from the nanoparticles' ability to stiffen the polymer. Additionally, by examining the entanglement network in a polymer glass and in the nanocomposite, we have found that nanoparticles have the ability to increase the entanglement density in the polymer, leading to increases in the plateau modulus, a key design parameter in many applications.

More recently I have worked with field-theoretic simulations, which have the advantage of capturing the behaviors of polymeric materials on length scales much larger than those accessible by particle-based simulations. A problem commonly faced in field-theoretic simulations requires the simulation of two phases in equilibrium with each other, and difficulties arise when performing such a simulation for two primary reasons. First, equilibration of such a system is computationally demanding. Additionally, it is necessary to perform a simulation large enough so that the interface between the two equilibrium phases does not affect the measured properties, and large simulations exacerbate the problem of slow equilibration. To circumvent this issue, I have helped develop a new method to perform field theoretic simulations in the so-called Gibbs ensemble, where two simulations are performed in conjunction with each other. By allowing the two simulations to exchange both particles and volume with each other, they remain in equilibrium with each other, enabling the efficient simulation of two phases in equilibrium with each other. This method is extremely power in that it will enable efficient simulations of complex phase diagrams using field theory, and I will present several examples on my poster.

See more of this Session: Poster Session: Meet the Faculty Candidate
See more of this Group/Topical: Education

Design of Novel Polymeric Materials

Megan L. Robertson, Chemistry, University of Minnesota, Minneapolis, MN

Polymers are used in a plethora of applications: packaging, medical devices, membranes, fuel cells, adhesives, rubber elastomers, fibers, coatings, building materials, automotive materials, and etc. These applications require control over various polymer properties such as mechanical properties, chemical compatibility, transport (or prevention of transport) of small molecules, adhesion, conductivity, biocompatibility, and many others. A single polymer is unlikely to have the properties required for a desired application. Rather, combining more than one polymer is a powerful way to take advantage of the diversity of properties that polymers have. Various strategies can be employed to create polymer mixtures: binary blending, synthesis of copolymers, and designing polymeric surfactants to compatibilize immiscible polymers.

I have worked previously in two areas of interest related to the design of polymeric materials. First, during the course of my graduate work at the University of California, Berkeley, advised by Prof. Nitash P. Balsara in Chemical Engineering, I utilized block copolymer surfactants to create microstructured mixtures of immiscible polymers. The block copolymers exhibited both repulsive and attractive interactions with the immiscible polymers, which were manipulated in order to control the structure of the resulting material. Thermodynamically stable microstructured materials were prepared with as little as 1 vol. % of the block copolymer surfactant added to the mixture.[1-3]

Second, my postdoctoral research focuses on the area of renewable materials under the guidance of Prof. Marc A. Hillmyer at the University of Minnesota, Department of Chemistry. Presently, the majority of polymers are synthesized from petroleum-derived feedstocks. The world supply of petroleum is finite, and in the future it will be necessary to turn to sustainable alternative resources for plastics raw materials. We have chosen to prepare composites prepared from two renewable materials: polylactide, a biodegradable polymer derived from starch sources such as corn, and soybean oil, an abundant and inexpensive triglyceride oil. Polylactide is one of most commonly studied polymers derived from a renewable resource. However, the brittleness of the polymer is one of the main impediments to its use in many traditional petroleum-based polymer applications. The incorporation of soybean oil into polylactide allows for the reduction of its brittleness and thus broadens the potential applications of this material.

1. Ruegg, M. L.; Reynolds, B. J.; Lin, M. Y.; Lohse, D. J.; Balsara, N. P. "Minimizing the Concentration of Diblock Copolymer Needed to Organize Blends of Weakly Segregated Polymers by Tuning Attractive and Repulsive Interactions", Macromolecules, **2007**, *40*, 1207-1217.

2. Ruegg, M. L.; Reynolds, B. J.; Lin, M. Y.; Lohse, D. J.; Krishnamoorti, R.; Balsara, N. P. "Effect of Pressure on a Multicomponent A/B/A-C Polymer Blend with Attractive and Repulsive Interactions", Macromolecules, **2007**, *40*, 355-365.

3. Ruegg, M. L.; Reynolds, B. J.; Lin, M. Y.; Lohse, D. J.; Balsara, N. P. "Microphase and Macrophase Separation in Multicomponent A/B/A-C Polymer Blends with Attractive and Repulsive Interactions", Macromolecules **2006**, *39*, 1125-1134.

See more of this Session: Poster Session: Meet the Faculty Candidate
See more of this Group/Topical: Education

Engineering Efficacious Controlled Release Therapeutics

Sam N. Rothstein, Department of Chemical Engineering, University of Pittsburgh, Pittsburgh, PA

Steven R. Little, Chemical Engineering, University of Pittsburgh, Pittsburgh, PA

The engineering design methodology can streamline problems in fields as disparate as unit operations and immunotherapy. Specifically, the fundamental mathematics used to design, for example, a packed bed reactor also readily apply to a single injection vaccine and numerous other controlled release therapeutics. However, model-driven design is rarely used to develop medical treatments, like the latter. Changing this paradigm in both classroom and laboratory has been and will continue to be a focus of my career.

Most recently, I have combined *in silico*, *in vitro*, and *in vivo* methods to supplant the heuristic design of controlled release therapeutics with a rational, model-driven methodology. The backbone of this transformation is a newly developed predictive model connecting the fundamental design parameters to a formulation's drug delivery kinetics. Proof-in-principal for this design strategy comes from the development of two biodegradable microparticle formulations with dramatically different dosing kinetics and therapeutic payloads. Further, the efficacy of each formulation was confirmed with disease-specific, animal models. Together these data establish this mathematical approach as a platform for the design and development of controlled release therapeutics.

See more of this Session: Poster Session: Meet the Faculty Candidate
See more of this Group/Topical: Education

Probing Living Systems by Nanoparticles

Gang Ruan, Department of Chemical and Biomolecular Engineering, The Ohio State University, Columbus, OH

Jessica O. Winter, Chemical and Biomolecular Engineering, The Ohio State University, Columbus, OH

R. Sooryakumar, Department of Physics, The Ohio State University, Columbus, OH

Tony Brown, Department of Neuroscience, The Ohio State University, Columbus, OH

Sanjay Rajagopalan, Division of Cardiovascular Medicine, The Ohio State University, Columbus, OH

Michael Paulaitis, Department of Chemical and Biomolecular Engineering, The Ohio State University, Columbus, OH

Sishen Feng, Department of Chemical and Biomolecular Engineering, National University of Singapore, Singapore

Shuming Nie, Biomedical Enigneering, Georgia Institute of Technology and Emory University, Atlanta, GA

Nanotechnology is revolutionizing the ways how physicians prevent, diagnose and treat diseases. The sizes of nanomaterials are close to those of biomolecules, making it possible to probe or intervene in the processes occurring in living systems at molecular level using the numerous special optical, mechanical, electrical, magnetic and thermal properties of nanomaterials. We synthesized nanoparticles of semiconductors, iron oxides, polymers and lipids, and assembled novel nanocomposites based on the combinations of the individual types of nanoparticles. We used semiconductor nanocrystals (quantum dots) to image and track the intracellular transport of Tat peptide, the peptide sequence used by HIV virus to infect cells, with a focus on the downstream events after the Tat-quantum dot conjugates were internalized by the cells. We developed a technique to deliver and image single quantum dots in living cell cytoplasm. Using this technique, we were able to track the motion of single dynein motor proteins in living cells, generating new insights into how the transport system of living cells works.

Further, we combined quantum dot-mediated fluorescent imaging with magnetic force for probing single molecule biophysics. We incorporated quantum dots and iron oxide nanoparticles into nanocontainers of sizes 10-100 nm, forming fluorescent-magnetic nanocomposites. Biomolecules (e.g. peptides, proteins and DNAs) were conjugated with the nanocomposites. Single bioconjugated nanoparticles (nanocomposites) were then, for the first time, simultaneously manipulated and imaged with magnetic tweezers attached to a fluorescent microscope.

On the other hand, we explored the applications of nanocomposites in clinical imaging, diagnostics and therapeutics. Fluorescent-magnetic nanocomposites (combining quantum dots and iron oxide nanoparticles) were used for combined MRI and fluorescent cancer imaging. MRI can reveal the location of tumors in deep tissues, while fluorescent imaging can offer convenience in visualizing the tumors during surgery. Fluorescent-magnetic nanocomposites were also used to detect circulating tumor cells and T-cells. Therapeutic drugs and SiRNAs were further incorporated into these multifunctional nanoparticles using polymeric nanocontainers. In addition, polymer-based and lipid-based systems were combined to form novel nanocomposites for improved therapeutic effects.

My (the presenting author's) independent research program will have the following themes: (1) developing nanomaterial-based platform technologies for clinical medicine, and (2) seeking fundamental biomedical discoveries using nanotechnologies. Furthermore, I have strong interests in teaching. I have delivered several guest lectures in both undergraduate and graduate courses, with positive feedback. I have also mentored undergraduate students' research, successfully leading to publications.

See more of this Session: Poster Session: Meet the Faculty Candidate
See more of this Group/Topical: Education

Compartmentalizing the Application: Microfluidic Chips for Combinatorial Screening

Benjamin R. Schudel, Department of Chemical & Biomolecular Engineering, University of Illinois at Urbana-Champaign, Urbana, IL

The emerging field of microfluidic devices has led to a variety of creative options that surpass operation at the macroscale. The low Reynolds numbers and high surface-to-volume ratios typically found in microfluidic devices can be exploited for excellent control over mass transport and heat transfer. These phenomena have lead to creative and varied devices relying on a rich amalgamation of multiple disciplines from optics and physics to biology and chemistry. At the Kenis Microsystems Group, we have focused on crafting experimental chambers into microfluidic devices capable of combinatorial synthesis, crystallization studies, and fuel cell optimization.

Specifically, my work has focused on the combinatorial aspect of microfluidic design, building chips for microscale combinatorial library synthesis of sub-nanoliter volumes. Working collaboratively with biophysics and electrical engineering groups, we have incorporated screening capabilities within our microfluidic combinatorial chips by using new valving structures utilizing Actuate-to-Open (AtO) valves. For determining viable protein inhibition candidates, we included photonic crystal biosensors to determine product feasibility. We are also using molecular beacons for individually resolved fluorescence response of viral targets for point-of-care clinical applications. Given my current experience with microfluidic routing, my goal is to expand upon the complex designs I have crafted and look at other applications in multistep synthesis, biophysical studies and integrated optical readout.

See more of this Session: Poster Session: Meet the Faculty Candidate
See more of this Group/Topical: Education

Biomass Feedstock Production and Provision for Bioenergy Sector: A System Level Optimization Approach

Yogendra Shastri, Department of Agricultural and Biological Engineering, University of Illinois Urbana-Champaign, Energy Biosciences Institute, Urbana, IL

The success of biomass based sector depends critically on an efficient, cost-effective and sustainable biomass feedstock production system supporting the biorefinery. Spatially distributed collection of the low energy density feedstock demands a highly efficient system to ensure cost competitiveness. Further challenges arise as seasonal availability of energy crops must support year-round demand to operate refinery on a continuous basis. Consequently, an integrated system level analysis is necessary to coordinate various feedstock production related tasks. Such an analysis should incorporate not only planning level (long term) but also management and operational level (short term) aspects. This work presents the research conducted in developing a feedstock production optimization model as a step in developing such a framework. The breadth level optimization model incorporates different tasks such as harvesting, packing, storage, biomass handling and transportation that are essential for feedstock production. The objective is to determine the optimal configuration of the feedstock production system on a regional basis. The decision variables include the design/planning level decisions such as equipment selection and transportation mode selection, as well as the management level decisions such as daily farm management and transportation fleet scheduling. This leads to the formulation of a mixed integer linear programming (MILP) problem. An economic objective function is formulated and established techniques from mathematical programming are used to solve the computationally challenging problem. Other performance indicators such as energy consumption and greenhouse gas emissions are also tracked for comparison. The model has been applied for the case of switchgrass production as the energy crop in southern Illinois. The results show the farm size distribution has a major impact on the delivered cost, farm management activities as well as machine selection decisions. On-farm covered storage is observed to be the most preferred storage option. It is observed that integrated farm management as well as optimized harvesting schedule leads to significant cost savings, while improvements in packing and transportation technologies lead to greatest benefits to the system. The system level optimization model is then extended to explore the effect of optimizing individual stake-holder benefits. The overall goal of the model development work is to provide a decision support platform for the individual stake-holders such a farmers, storage elevators, transportation company, biorefinery as well as policy makers.

See more of this Session: Poster Session: Meet the Faculty Candidate
See more of this Group/Topical: Education

Toward Developing Multi-Functional Microfluidic Platforms to Array, Label, and Detect Microbeads, Lipobeads, and Liposomes

Shahab Shojaei-Zadeh, Benjamin Levich Institute, New York, NY

Bead based microarrays are promising platforms for implementing the high throughput, multiplexed assaying of the binding interactions of biomolecules (for example the binding of antigens and antibodies, or the conjugation of membrane receptors with small molecule ligands). In these arrays, each bead contains a particular probe molecule on its surface, and a code to identify this probe. Particles with different probes are mixed and subsequently bound onto a surface in a regular array. The array is then incubated with a target, and the binding of the target to particular probe molecules is identified (usually by fluorescently labeling the targets and scanning to find luminescent beads). The bead code is then read to identify the probe and complete the multiplexed assay. The promise of this platform lies in the fact that by decreasing the size of the beads, ulra-mininaturized platforms capable of an increased number of binding assays can be constructed.

Bead arraying is essential to the formation of these bead based microarrays, and most research has focused on using the covalent binding of functional groups on the bead surface to functionalized sites on the platform surface. Arraying paradigms which sequester particles without reaction have the distinct advantage that they avoid chemical reaction conjugation.

Microfluidic technology is used to fabricate monodisperse polymeric microbeads, embedded with semiconductor nanocrystals (quantum dots or QDs), to form a spectral bar code. Unique labels are obtained by incorporating, into the beads, QDs of a prescribed set of colors (emitting wavelengths), and for each color, a prescribed concentration. The quantitative incorporation of the nanocrystals into the beads is necessary for constructing a reproducible label, and is most effectively undertaken by a suspension polymerization in which monomer droplets, loaded with the QDs in the prescribed composition, are first formed in a continuous phase, and then the droplets are polymerized. In this method, liquid monomer, with dispersed QDs of a prescribed composition, is formed into droplets in a continuous phase immiscible with the monomer phase in a hydrodynamic focusing flow. The polymerization is then initiated in the dispersed phase of this suspension, and the monomer droplets are converted into particles. The QDs present in the polymerizing droplets are effectively "locked into" the beads since they are insoluble in the continuous phase, and each particle has the same composition since droplets are formed from the same monomer phase.

See more of this Session: Poster Session: Meet the Faculty Candidate
See more of this Group/Topical: Education

LCST-Driven Stimuli Responsive Polymers: Analytical Models

David S. Simmons, Chemical Engineering, University of Texas, Austin, Austin, TX

Isaac C. Sanchez, Chemical Engineering, University of Texas, Austin, Austin, TX

Stimuli responsive polymers are of great interest in a variety of fields ranging from biomedical engineering to robotics. The key property of these 'smart' materials is their ability to respond to small changes in environmental conditions such as temperature, pressure, pH and ionic strength. One physical mechanism commonly driving this behavior is the lower critical solution temperature (LCST) transition. In polymer-solvent solutions above the overlap concentration, this transition takes the form of a phase separation into polymer-rich and polymer-depleted phases. In highly dilute solution, on the other hand, the corresponding transition is the collapse of isolated chains from an expanded coil state to a collapsed globule state. Several theoretical models are offered for these transitions. Systems addressed include uncharged aqueous and non-aqueous polymer systems, aqueous polymers in the presence of free salt, and lightly charged polyelectrolyte systems.

See more of this Session: Poster Session: Meet the Faculty Candidate
See more of this Group/Topical: Education

Catalytic Strategies for the Conversion of Alternative Feedstocks to Fuels and Chemicals Via Selective C-C Bond Formation and Oxygen Removal

Dante A. Simonetti, Department of Chemical Engineering, University of California at Berkeley, Berkeley, CA

Current economic, environmental, and geopolitical issues may detrimentally affect the future of sustainable production of liquid transportation fuels from petroleum. One possible approach to help ameliorate these effects is to synthesize gasoline, diesel, and jet fuel-range hydrocarbon compounds from alternative feedstocks such as coal, natural gas, and biomass. Catalytic processing of these feedstocks requires high selectivity to the desired hydrocarbons in a limited number of steps to be cost-competitive with petroleum refining, and herein, we present catalytic systems for the conversion of functional molecules and H_2/CO streams which can be derived from natural gas, coal, and/or biomass into hydrocarbons which comprise current gasoline, diesel, and jet fuel.

One approach consists of a two-step process involving initial removal of most of the oxygen from biomass-derived carbohydrates on Pt-Re/C catalysts at modest temperatures (~520 K) and pressures (5-20 bar). In this step, sorbitol and glucose are converted to a mixture of hydrophobic molecules consisting of monofunctional hydrocarbon intermediates such as alcohols, ketones, carboxylic acids, and heterocyclic compounds with 4-6 carbon atoms. The H_2 necessary for this initial oxygen removal is produced *in situ* via reforming of a portion of the carbohydrate to H_2/CO_x gas streams. These monofunctional hydrocarbons retain the necessary functionality such that they may be used directly in the chemical industry as chemical intermediates or solvents. Alternatively, these molecules may be upgraded to a variety of hydrocarbons with the appropriate molecular weight, energy content, and combustion properties for liquid fuel applications via selective C-C bond formation combined with removal of the remaining oxygen. Strategies for this subsequent C-C bond formation include aromatization/alkylation on acid catalysts to produce gasoline range hydrocarbons, aldol-condensation of ketones and secondary alcohols on a bi-functional metal-base catalyst to produce diesel and jet fuels, and ketonization of organic acids to ketones on a mixed metal oxide followed by aldol-condensation to produce diesel and jet fuels.

An additional approach for the conversion of alternative feedstocks to fuels is processing of light oxygenates derived from biomass, coal, and natural. These light oxygenates can be derived from biomass conversion processes similar to step one of the previous approach or from initial conversion of biomass/coal/natural gas to H_2/CO followed by synthesis to light oxygenates. The upgrading of light oxygenates occurs via co-homologation reactions with alkane byproducts from petroleum refining on solid acid catalysts (such as zeolites) at low temperatures (~473 K) and pressures (total pressure of 1 bar). Specifically, low value light alkanes, such as isobutane, can be upgraded to gasoline range alkanes via co-homologation with dimethyl ether in the presence of adamantane. Selective C-C bond formation occurs between alkene species derived from the alkane co-feed and methylating species derived from dimethyl ether. High selectivity to 2,2,3-trimethylbutane (research octane number = 112) is achieved as a result of a chain growth pathway involving preferential methylation of growing chains leading to structures that cannot readily change chain length by cracking or isomerization. Adamantane acts a dehydrogenation/hydrogenation co-catalyst that facilitates the conversion of the alkane co-feed to alkenes which serve as primary chain growth centers.

The strategies presented here represent advancements toward the production of fuels and chemicals from biomass, coal, and natural gas by achieving high selectivity to various types of hydrocarbon fuels in a limited number of flow reactors.

See more of this Session: Poster Session: Meet the Faculty Candidate
See more of this Group/Topical: Education

Bottom-up Self-Assembly of Novel Materials and Nanoscale Surface Patterning

Chetana Singh, Chemical Engineering, University of Michigan, Ann Arbor, MI

Sharon C. Glotzer, Chemical Engineering, University of Michigan, Ann Arbor, MI

Some well known methods for obtaining nanopatterned surfaces through self-assembly are block copolymer phase separation, and directed and templated self-assemblies. We explore phase-separation in self-assembled monolayers, or SAMs, as a possible method for obtaining patterned surfaces. This method exploits immiscibility between two types of molecules combined with a long range repulsion that prevents the two components from bulk phase separating. The source of the long range repulsion is entropic in nature and leads to preference of longer or bulkier molecules to stack next to shorter (or less bulky) molecules despite being immiscible with them [1]. The immiscibility drives phase separation as in block copolymers but is different in that the molecules are grafted at one end on the surface and immiscibility is between two different types of molecules rather than between two sections of the same molecule.

We have performed simulation studies of different compositions of grafted molecular mixtures with varying immiscibilities on flat [1,2], spherical [1] and cylindrical [3,4] surfaces. We have obtained nanoscale domains that are shaped as stripes, spherical patches, elongated patches and combinations of these. In some cases we also find complete bulk phase separation or no phase separation. We have predicted that the shapes and feature sizes of these patterns can be changed by tuning parameters such as immiscibility, composition, surface curvature etc. Some of these predictions have recently been confirmed by experimental studies. We demonstrate how surface stress in the form of nanoscale curvature aids in the formation of stripes as opposed to patches [1-4].

This method for obtaining nanopatterned surfaces can provide a powerful tool for design and fabrication of new nano building blocks [4-5]. Furthermore, these patterns also form on flat substrates [2], with applications in the fields of catalysis, electronics, ceramics and biosensing.

1. C Singh, PK Ghorai, MA Horsch, AM Jackson, RG Larson, F Stellacci and SC Glotzer, "Entropy-Mediated Patterning of Surfactant-Coated Nanoparticles and Surfaces", Physical Review Letters 99, 226106 (2007).

2. C Singh, K Kuwabara, F Stellacci and SC Glotzer, "Nanoscale Striped vs. Patchy Phase Separation in Mixed Self-Assembled Monolayers", preprint.

3. C Singh, AM Jackson, F Stellacci and SC Glotzer, "Exploiting Substrate Stress to Modify Nanoscale Phase-Separated Patterns", preprint.

4. C Singh, Y Hu, F Stellacci and SC Glotzer, "Striped Phase Separation on Nanowires and Nanotubes", preprint.

5. AM Jackson, JW Myerson, F Stellacci, "Spontaneous assembly of subnanometre-ordered domains in the ligand shell of monolayer-protected nanoparticles", Nature Materials, 3, 330-336 (2004).

6. RP Carney, GA DeVries, C Dubois, H Kim, JY Kim, C Singh, PK Ghorai, JB Tracy, RL Stiles, RW Murray, SC Glotzer and F Stellacci, "Size Limitations for the Formation of Ordered Striped Nanoparticles", Journal of American Chemical Society, (Communication), 130(3), 798-799 (2008).

See more of this Session: Poster Session: Meet the Faculty Candidate
See more of this Group/Topical: Education

Towards a Microrheological Description of Granular Segregation

Ashley G. Smart, Chemical Engineering, California Institute of Technology, Pasadena, CA

Microrheology has emerged as a useful paradigm for understanding the dynamics of granular segregation: recent experiments suggest that microviscosities in granular flows obey an Einstein relationship, with observed segregation fluxes related to an effective temperature, T_{eff}. Here, we explore the granular microrheology paradigm both theoretically and computationally, considering the motion of a single forced probe in a dilute, rapidly sheared granular bath, and by extension, fluxes in bi-disperse granular flows. In this dilute limit, T_{eff} scales as the "granular temperature", T_k, and particle mobility is characterized by a competition between forcing and collision-induced diffusion - a viscous plateau in the passive forcing regime gives way to order Pe^2 force-thickening and a sharp increase in the relative probe temperature in the active regime. Aided by particle dynamics simulations, we explore the behavior of the ratio T_{eff}/T_k as a function of particle fill ratio, volume fraction, and forcing. We touch briefly upon the formidable challenge of extending this work to more dense granular flows.

See more of this Session: Poster Session: Meet the Faculty Candidate
See more of this Group/Topical: Education

Multi-Scale Chemical Product Design

Charles C. Solvason, Department of Chemical Engineering, Auburn University, Auburn University, AL

Mario R. Eden, Department of Chemical Engineering, Auburn University, Auburn University, AL

The National Research Council (NRC) has recently recognized the importance of developing integrated formulated product design tools. In response, it created the Committee on Integrated Computational Materials Engineering (CICME) which developed and published a road map for what it termed a "Grand Challenge" [1]. In the report, the committee noted that in order to alleviate strain put on U.S. manufacturers from the swiftly changing and increasingly global marketplace, integrated design closely coupling computational models with manufacturing processes would be required. The term 'integrated' recognizes that the properties of products are controlled by a multitude of separate and often competing mechanisms that operate over a wide range of length and time scales. It is the linkage of the scales that remains the 'Grand Challenge' [1]. In response to this call from CICME, the research community has shifted focus from developing only the physical form, function, and aesthetics of assembled products to the design of chemically formulated products [2]. Chemically formulated products are products designed at the molecular level to deliver a specific desired attribute that may exist at multiples scales such as the quantum-, nano-, meso-, macro-, and mega-scales. Examples of formulated products include pharmaceuticals [3], proteins [4], organic semi-conductors [5], nano-structured materials [6] and many more product types.

Using today's technology, designing these optimized molecules for a specific end-use has never more attainable. Recent achievements owe much of their successes to increased computational speeds and the advent of novel algorithms that improved the transfer of information between the formulated scales [6]. One type of promising new algorithms utilize the reverse problem formulation which has been shown to significantly reduce the complexity and required computational time of product design problems. For example, Eden et. al. [7] demonstrated the ability of the reverse problem formulation to decouple product design problems from constitutive equations so that molecules can be designed prior to verification at the various scales, effectively reducing the computational complexity. Furthermore, Eljack et al. [8] extended the reverse problem formulation to include information from processes operating at the macro-scale. Additionally, Satyanarayana et al. [9] recently demonstrated the use of the reverse problem formulation combined with grid technology for polymer design with long range order approaching the meso-scale. While each of these applications of the reverse problem formulation has incorporated elements of multi-scale product design, the integration of the reverse problem formulation into the complete multi-scale framework has yet to be achieved.

The main objective for this research is to extend the reverse problem formulation algorithm to include aspects of each of the length scales, thereby creating a framework where product synthesis, design, and optimization can be achieved with significantly reduced computational time. The research plan to achieve the objective entails the following: (1) the development of a centralized framework, (2) the development of an algorithm that chooses the appropriate properties to bridge each of the length scales, (3) the successful implementation of a property clustering algorithm to load the property descriptions into the framework. Independent validation steps will also be built in to the method and verified using published experimental results because of the "profound importance of experimental results to calibrate and validate computational methods and fill gaps in the theoretical understanding" [1]. Due to the size of this project, proof-of-concept case studies will initially focus on experimentally derived parameters used in acetaminophen tablet design. A more robust case study exploring simulated parameters in polymer nano-composites may also be discussed, provided the expected results are achieved prior to the conference.

References

[1] Committee on Integrated Computational Materials Engineering, National Research Council (2008). Integrated Computational Materials Engineering: A Transformational Discipline for Improved

Competitiveness and National Security. The National Academies Press, USA.

[2] Hill, M. (2004) Product and Process Design for Structured Products. AIChE Journal, 50(8), pp. 1656-1661.

[3] Solvason C.C., Chemmangattuvalappil N.G., Eden M.R. (2009) Decomposition Techniques for Multi-Scale Structured Product Design: Molecular Synthesis. Computer Aided Chemical Engineering (In Press).

[4] Floudas C.A., Fung H.K., McAllister S.R., Monningmann, M., Rajgaria R. (2006) Advances in protein structure prediction and de novo protein design: A review. Chemical Engineering Science, 61(3), pp.966-988.

[5] Farhang A.R and Deeter T. (1999) Theoretical characterization of phenylene-based oligomers, polymers, and dendrimers. Synthetic Metals, 100(1), pp. 141-162.

[6] Fermeglia M. and Pricl S. (2009) Multiscale molecular modeling in nanostructured material design and process system engineering. Computers & Chemical Engineering (In Press).

[7] Eden M.R., Jorgensen S.B., Gani R., and El-Halwagi M.M. (2003) Reverse problem formulation based techniques for process and product synthesis and design. Computer Aided Chemical Engineering, 15(1), pp. 451-456.

[8] Eljack F.T., Eden M.R., Vasiliki K., Kazantzi Q., and El-Hawagi M.M. (2007) Simultaneous process and molecular design - A property based approach. AIChE Journal, 53(5), pp. 1232-1239.

[9] Satyanarayana K.C., Abildskov J., and Gani, R. (2009) Computer-aided polymer design using group contribution plus property models. Computers & Chemical Engineering, 33, pp. 1004-1013.

See more of this Session: Poster Session: Meet the Faculty Candidate
See more of this Group/Topical: Education

Using XRD to Correlate MFI Zeolite Expansion and Membrane Permeation Data

Stephanie G. Sorenson, Department of Chemical and Biological Engineering, University of Colorado, Boulder, CO

E. Andrew Payzant, Oak Ridge National Laboratories, Oak Rige, TN

Joseph R. Smyth, Geological Sciences, University of Colorado, Boulder, CO

Richard D. Noble, Department of Chemical and Biological Engineering, University of Colorado, Boulder, CO

John L. Falconer, Department of Chemical and Biological Engineering, University of Colorado, Boulder, CO

X-ray diffraction measurements showed that silicalite-1 and B-ZSM-5 zeolite crystals expand when n-alkanes, isobutane, ethanol, and SF_6 adsorb at saturation and loadings less than saturation conditions at room temperature. The n-alkane expansion at saturation correlated with the maximum number of carbon atoms in the structure. The higher number of carbon atoms per unit cell, the more the crystals expanded. XRD showed i-butane and low SF_6 loadings contract the MFI crystals. Permporosimetry showed that increases and decreases in helium flow through membranes trended well with the expansion due to adsorbate loading as measured by XRD.

The changes due to adsorbate loading significantly influence the permeation properties of zeolite membranes. When the MFI crystals expand, the size of inter-crystalline defects shrink enough to drop the flux of molecules thru the defects by orders of magnitude. Pervaporation was used to measure flux because a liquid feed creates the highest loading in the defects and thus is the most severe test of the zeolite membrane properties. Isooctane was used as a measure of flux thru defects because the kinetic diameter of isooctane is larger than the MFI pore diameter and therefore only transports through membrane defects. When 1% n-alkane was added to liquid isooctane feed of a B-ZSM-5 membrane, the isooctane flux dropped orders of magnitude. The percentage of the original isooctane flux that remained after n-alkane addition correlated with the XRD volume expansion.

See more of this Session: Poster Session: Meet the Faculty Candidate
See more of this Group/Topical: Education

Generation of Renewable Fuels and Chemicals From Fatty Acids Via Supercritical Fluid Processing

Darrell L. Sparks, Chemical Engineering, Mississippi State University, Mississippi State University, MS

Identifying renewable sources for fuels and chemicals is currently a paramount global research initiative. Lipids derived from renewable feedstocks such as vegetable oils, animal fats, and microorganisms are a promising alternative to petroleum. For example, lipids derived from soybean oil are used to produce biodiesel, which is currently the dominant fuel alternative to petroleum diesel. In order to generate fuels and chemicals from lipids, typically the lipids are first extracted from lipid-containing biomass. Traditionally, this unit operation is accomplished by extraction with an organic solvent such as hexane. However, hexane is a volatile, flammable material that can leave behind a toxic residue. Hence, if the extracted lipids are to be further processed into food-grade products, extra processing costs can be incurred to ensure that any remaining hexane is at a safe level. Extraction of lipids with supercritical carbon dioxide is an appealing alternative because carbon dioxide is inexpensive, readily available, non-toxic, and non-flammable. Also, supercritical carbon dioxide is considered a tunable extraction medium because changes in temperature or density of the solvent can greatly alter its solvating effectiveness.

Supercritical carbon dioxide can also be an effective reaction medium. Many lipids are composed of unsaturated fatty acids, and once lipids have been extracted from a biomass matrix, the unsaturated fatty acids can be oxidized in a controlled manner (as opposed to autoxidation) to produce a variety of valuable chemical intermediates. The oxidation products are useful in a spectrum of products such as pharmaceuticals and polymers. By conducting the reaction in a supercritical fluid, several advantages can be realized. For example, the reaction rate and conversion can be improved by increasing pressure. Also, if the solubilities of the reaction products in the supercritical fluid are different, then a supercritical fluid separation can be used to fractionate the products. For oxidation reactions, the use of carbon dioxide as the supercritical fluid is very appealing because it is already completely oxidized; therefore it will not break down during the reaction.

This poster will emphasize the author's graduate work under the guidance of Dr. Rafael Hernandez at Mississippi State University. In the first phase of the author's research, supercritical carbon dioxide was evaluated as an alternative lipid extraction media to hexane. The first lipid source considered was rice bran, a component of rice kernels removed during whitening and polishing. Extraction of rice bran lipids with supercritical carbon dioxide provided similar oil yields as hexane on a gravimetric basis; however, use of supercritical carbon dioxide resulted in a much higher solvent efficiency [1]. The other lipid source was municipal wastewater sewage sludge [2]. Once again, supercritical carbon dioxide was comparable to hexane in terms of oil yield. However, addition of a small amount of methanol to the carbon dioxide increased extraction yield significantly [3]. In the second phase, the oxidation of oleic acid (a common lipid component) was performed in supercritical carbon dioxide [4]. Ozonolysis of oleic acid is the most common oxidation reaction performed commercially to generate azelaic acid and pelargonic acid. Although solubility data for oleic acid are readily available, the solubility of azelaic acid and pelargonic acid had to be determined. Pelargonic acid had a much higher solubility in supercritical carbon dioxide than azelaic acid [5, 6]. In order to fully understand the solubility results, several models were utilized, including the Peng-Robinson equation of state and Chrastil's equation. The effectiveness of these models was evaluated in predicting the solubility behavior of a range of compounds [7].

The author's primary research interests are in the areas of renewable fuels and chemicals, supercritical fluids, and lipids. As part of completing the Ph.D. qualifying exam, the author wrote and successfully defended a NSF-style proposal. As a result, the author has several proposal ideas in the aforementioned areas. Potential funding sources for these areas are the Department of Energy, the Department of Agriculture, and the National Science Foundation.

Though confident in the ability to teach any chemical engineering course at both the undergraduate and graduate levels, the author's teaching interests include mass transfer operations, reaction kinetics, thermodynamics, and elective courses in renewable fuels and lipid chemistry. The author served as a tutor for Mass and Energy Balances for several years and has taught a junior-level undergraduate course entitled Mass Transfer Operations multiple times. Additionally, the author developed a new chemical engineering elective course at Mississippi State University on bio-analytical research methods. Hence, the author has experience interacting with students and is excitedly anticipating the opportunity to teach students on a permanent basis.

[1] Sparks, D., R. Hernandez, M. Zappi, D. Blackwell, and T. Fleming, 2006, Extraction of Rice Bran Oil Using Supercritical Carbon Dioxide and Propane, *JAOCS* 83: 885-891.

[2] Zappi, M., French, T., Hernandez R., Dufreche S., and Sparks, D., Production of Biodiesel and Other Valuable Chemicals from Wastewater Treatment Plant Sludges, United States Patent Application 20050112735.

[3] Dufreche, S., R. Hernandez, T. French, D. Sparks, M. Zappi, and E. Alley, 2007, Extraction of Lipids from Municipal Wastewater Plant Microorganisms for Production of Biodiesel, *JAOCS* 84: 181-187.

[4] Sparks, D.L., L.A. Estévez, and R. Hernandez, 2009, Supercritical-Fluid-Assisted Oxidation of Oleic Acid with Ozone and Potassium Permanganate, *Green Chemistry*, Accepted for Publication.

[5] Sparks, D.L., L.A. Estévez, N. Meyer, and R. Hernandez, 2007, Solubility of Azelaic Acid in Supercritical Carbon Dioxide, *J. Chem. Eng. Data* 52: 1246-1249.

[6] Sparks, D.L., L.A. Estévez, K. Barlow, and R. Hernandez, 2008, Solubility of Pelargonic Acid in Supercritical Carbon Dioxide, *J. Chem. Eng. Data* 53: 407-410.

[7] Sparks, D.L., L.A. Estévez, and R. Hernandez, 2008, Evaluation of Density-Based Models for the Solubility of Solids in Supercritical Carbon Dioxide and Formulation of a New Model, *Chem. Eng. Sci.* 63: 4292-4301.

See more of this Session: Poster Session: Meet the Faculty Candidate
See more of this Group/Topical: Education

Controlling the Bond Scission Sequence of Methanol Decomposition as An Example of Rational Catalyst Design

Alan Lee Stottlemyer, Department of Chemical Engineering, University of Delaware, Newark, DE

Jingguang G. Chen, Department of Chemical Engineering, University of Delaware, Newark, DE

The so called "Holy Grail" of heterogeneous catalysis is a fundamental understanding of catalyzed chemical transformations which span multidimensional scales of both length and time, enabling rational catalyst design. Such an undertaking is realizable only with an atomic level understanding of bond formation and destruction with respect to intrinsic properties of the metal catalyst. In this study, we investigate the bond scission sequence of methanol on bimetallic transition metal catalysts and transition metal carbide catalysts, with the simplicity of methanol allowing us to follow the different reaction pathways both experimentally and with density functional theory (DFT) modeling. Additionally, methanol is of interest both as a hydrogen carrier for reforming to hydrogen and CO and as a fuel in direct methanol fuel cells (DMFC). In order to bridge the so-called "materials gap" and "pressure gap" this work adopted three parallel research approaches: (1) ultra-high vacuum (UHV) studies including temperature programmed desorption (TPD) and high-resolution electron energy loss spectroscopy (HREELS); (2) electrochemical studies including cyclic voltammetry (CV) and chronoamperometry (CA); (3) DFT studies including thermodynamic and kinetic calculations.

The Ni/Pt bimetallic system was studied as an example for using methanol as a hydrogen source. There are two well characterized surface structures for the Ni/Pt system [1]: the surface configuration, in which the Ni atoms reside primarily on the surface of the Pt bulk, and the subsurface configuration, in which the second atomic layer is enriched in Ni atoms and the surface is enriched in Pt atoms. These configurations are denoted NiPtPt and PtNiPt, respectively. TPD studies revealed that the NiPtPt surface was more active to methanol reforming than the Pt or PtNiPt surfaces. HREELS confirmed the presence of strongly bound reaction intermediates [2], including aldehyde-like species [3], and suggested that the first decomposition step was likely O-H bond scission. Thus, the binding energies of the deprotonated reaction intermediates are likely important parameters in controlling the decomposition of methanol.

Recent studies have suggested that tungsten monocarbide (WC) may behave similarly to Pt for the electrooxidation of methanol [4, 5]. TPD was used to quantify the activity and selectivity of methanol decomposition for WC and Pt-modified WC (Pt/WC) as compared to Pt [5]. WC appeared to be more active than Pt, but C-O bond scission on WC resulted in gas phase methane, an undesired reaction for DMFC. When Pt was added to WC by physical vapor deposition, the methane reaction pathway was eliminated, suggesting that Pt synergistically modifies WC to improve the selectivity toward C-H bond scission to produce hydrogen and CO. Additionally, TPD confirmed WC and Pt/WC to be more CO tolerant than Pt [6]. DFT calculations suggested that the bond scission sequence of methanol could be controlled using monolayer coverage of Pt on WC and that the resulting mechanism was different for Pt/WC as compared to either parent surface. HREELS results verified that surface intermediates were different on Pt/WC as compared to Pt or WC.

Both studies demonstrated that the bond scission sequence of methanol can be controlled using either bimetallic or carbide catalysts. Such success was only possible using a methodology that combines both calculations to predict catalytic properties and experiments to fine-tune theoretical predictions.

[1] J. G. Chen, C. A. Menning, M. B. Zellner, Surface Science Reports 2008, 63, 201.

[2] A. L. Stottlemyer, H. Ren, J. G. Chen, Surface Science 2009, submitted.

[3] O. Skoplyak, C. A. Menning, M. A. Barteau, J. G. Chen, Journal of Chemical Physics 2007, 127, 114707.

[4] H. Hwu, J. G. Chen, Chemical Reviews 2005, 105, 185.

[5] E. C. Weigert, A. L. Stottlemyer, M. B. Zellner, J. G. Chen, Journal of Physical Chemistry C 2007, 111, 14617.

[6] Z. J. Mellinger, E. C. Weigert, A. L. Stottlemyer, J. G. Chen, Electrochemical and Solid-State Letters 2008, 11, B63.

See more of this Session: Poster Session: Meet the Faculty Candidate
See more of this Group/Topical: Education

Quantifying Cellular Fluxes to Drive Metabolic Engineering for the Production of Biofuels

Patrick F. Suthers, Department of Chemical Engineering, The Pennsylvania State University, University Park, PA

One of the key descriptors of cellular physiology is the set of flux values that flow through the cell's metabolic pathways. With the rapid generation of genomic data, creating such metabolic models in a systematic framework becomes crucial. Once constructed, these models can be used to propose experimentally testable hypotheses and guide modifications of the organism, e.g., to improve product yield. In this poster, I present 1) methodologies for quantifying fluxes in large and genome-scale models using isotope labeling data and 2) an optimization framework that uses these flux values to suggest engineering strategies for the overproduction of the product at the desired target.

Extending the scope of isotope mapping models becomes increasingly important in order to analyze strains and drive improved product yields as more complex pathways are engineered and as secondary metabolites are used as starting points for new products. Here I present how the elementary metabolite unit (EMU) framework and flux coupling significantly decrease the computational burden of metabolic flux analysis (MFA) when applied to large-scale isotope mapping metabolic models. The combined use of EMU and flux coupling analysis leads to a ten-fold decrease in the number of variables in comparison to the original isotope distribution vector (IDV) version of the *Escherichia coli* model. The observed computational savings reveal the rapid progress in performing MFA with increasingly larger isotope models and enable the ultimate goal of handling genome-scale models of metabolism.

Once obtaining flux values and ranges, I present an integrated computational base to support pathway identification and strain optimization, with an emphasis on C4+ biofuels and other biochemicals. I describe the application of the OptFlux computational framework to pinpoint engineering modifications (knock-outs/up/down) that are required for the targeted biochemical overproduction. This is accomplished by classifying reactions (and combinations thereof) in the metabolic model depending upon whether their flux values must increase, decrease or become equal to zero to meet the pre-specified overproduction target. A "force set" can then be extracted that contains a sufficient and non-redundant set of reactions that need to be directly changed to meet the production requirements. I apply the integrated framework for the production of succinate and 1-butanol in *E. coli* using the most recent *in silico E. coli* model, *i*AF1260. The proposed computational workflow not only recapitulates existing pathways and engineering strategies but also reveals novel and non-intuitive ones that boost production by using and performing coordinated changes on sometimes distant pathways.

See more of this Session: Poster Session: Meet the Faculty Candidate
See more of this Group/Topical: Education

Development of New Catalysts for Energy Conversion and in-Situ Studies of Catalysis

Feng Tao, Department of Chemistry, University of California at Berkeley, Berkeley, CA

Heterogeneous catalysis is crucial for a wide spectrum of technological fields, such as energy conversion, pollution control, and production of most chemicals. So far almost all fuels used for various transportations have been refined from crude oil in a number of heterogeneous catalytic processes. Due to the environmental impact and diminishment of fossil fuels, there is a significant drive to relinquish our dependence on fossil fuels by searching renewable or sustainable energy sources and efficiently converting them to clean energy. The conversion of new energy sources to applicable energy such as electrical energy will need entirely new catalysts which are more active and selective, and are able to be prepared from earth-abundant elements. Toward efficient conversion of new energy sources to applicable clean energy, here the proposed projects will be (1) the design of new nanocatalysts for catalytic reactions in energy conversion, (2) the development of new technologies of energy conversion, and (3) the in-situ studies of important catalytic processes in new energy conversion by in-situ surface analytical techniques.

Overall, the design of new nanocatalysts, the in-situ studies of their catalytic processes in new energy conversion, and the development of new technologies of energy conversion will significantly contribute to the on-going efforts to overcome quite challenging energy-related issues.

See more of this Session: Poster Session: Meet the Faculty Candidate
See more of this Group/Topical: Education

Engineering Biomimetic Polymer-Hybrid Delivery Systems

Treniece L. Terry, Polymer Science & Engineering, University of Massachusetts Amherst, Amherst, MA

Progress in the fields of molecular biology has elucidated many of the constituent molecules responsible for the progression of various genetic and infectious diseases. These small molecules have been developed into biopharmaceuticals, whose therapeutic efficacy has been substantial limited due to the need for multiple injections. The current approach integrates biopharmaceuticals into synthetic delivery systems whose surface has been modified by polyethylene glycol (PEG) to create a stealth delivery system. PEG primarily functions by preventing protein adsorption via steric stabilization, an inherently passive process, in which, clearance is dependent upon polymer molecular weight and surface heterogeniety. The mechanisms used by cells and pathogens are much more evasive and effective in suppressing phagocytosis. Specific surface receptor-ligand interactions of healthy, native cells protect them from immune clearance. Many of these inhibitory recognition pathway are now known. By mimicking these biological processes, closely, with short peptides utilizing existing signaling pathways to deliver inhibitory singals to immune cells will thereby suppress recognition, inhibiting phagocytosis. The objective is to produce phagocytic-resistant delivery systems through conjugations of "stealth" peptides onto surfaces of synthetic microparticles. Such systems have applications in immunotherapy, biosensors and vaccine delivery, in which, prolonged cellular exposure is needed to create cellular memory.

See more of this Session: Poster Session: Meet the Faculty Candidate
See more of this Group/Topical: Education

Meeting in the Middle: Top-Down and Bottom-up Approaches to Metabolic Engineering

Keith EJ Tyo, Chemical and Biological Engineering, Chalmers University of Technology, Gothenburg, Sweden

Gregory Stephanopoulos, Department of Chemical Engineering, Massachusetts Institute of Technology, Cambridge, MA

Jens Nielsen, Chemical and Biological Engineering, Chalmers University of Technology, Gothenburg, Sweden

Metabolic engineering has been tasked with harnessing the amazingly specific and complex chemistry of biological life for the improvement of human existence. Unlike other fields of engineering, where mathematical models can accurately described the system and are central to design, metabolic engineering has and continues to rely on engineering strategies that must be robust in the face of large biological uncertainty. The understanding of both small sets of interacting molecules (e.g. a short metabolic pathway), and global systems (e.g. regulatory stress responses), present a formidable challenge to further unlock the usefulness of cells. In my research, I have contributed by both top-down and bottom-up approaches for metabolic engineering

Bottom-up approaches have been used in **(1)** improving genetic stability of recombinant expression and **(2)** analyzing the biopolymer, polyhydroxybutyrate (PHB), production in *E. coli*. **(1)** Recombinant productivity loss, as a result of plasmid instability, was studied by a simple probabilistic model of plasmid propagation. The model identified a new genetic instability pathway, termed "allele segregation." A novel high-copy genomic integration strategy, Chemically Induced Chromosomal Evolution (CIChE), was developed that avoided allele segregation and can improve genetic stability 10 fold. CIChE was implemented experimentally and significantly improved genetic stability and specific productivity over plasmids. We expect CIChE to find application in many industrial recombinant processes because of CIChE's superior properties compared to plasmids. **(2)** Systematic evaluation of the PHB metabolic pathway enzymes identified the acetoacetyl-CoA reductase as the major limitation in the pathway, despite the long standing hypothesis that production was substrate limited. The key insight to measure *flux* to PHB, rather than batch accumulation, delineated growth rate dilution effects from changes in pathway flux. Through systematic overexpression and chemostat studies, these biological observations could be made, and PHB flux was increased by 3.7 fold.

Top-down approaches were successfully used in **(3)** PHB improvement in photosynthetic organisms and are currently being applied to **(4)** yeast protein secretion. **(3)** Regulatory pathways in the photosynthetic bacterium Synechocystis PCC6803 are largely unknown, but strictly control PHB production. The development of a mutational library and screening strategy resulted in the identification of distal genes that improve PHB production, by as much as 3 fold. These biological results may be useful in the engineering of higher plants for biopolymer production, where such screening would be impossible. **(4)** I am currently expanding my top-down competence to include high-throughput data analysis (X-omics) in the important eukaryotic organism, *S. cerevisiae*. Work is focused on improving throughput and trafficking of secreted proteins in the yeast secretory pathway. This will be valuable for the production of protein therapeutics and industrial enzymes, such as cellulases.

In future challenges, the flexibility to approach problems from a "top-down" or "bottom-up", or most importantly, the development of theories and methods for integration of the two approaches will be essential. Experience in both of these approaches should prove useful in metabolic engineering pursuits.

See more of this Session: Poster Session: Meet the Faculty Candidate
See more of this Group/Topical: Education

Integration of Materials and Biomedical Sciences: Assessing Structural Changes in Biomaterials

Efrosyni Themistou, Department of Chemical and Biological Engineering, University at Buffalo - The State University of New York (SUNY), Buffalo, NY

Developing a research program at the interface of materials science and nanotechnology, and biomedical sciences and medicine is expected to give new inside in basic science research and yield novel therapeutics. By having extensive training in synthesis, isolation and characterization of polymers, nanostructured materials, soft materials, biomaterials and functional molecules by using up-to-date methods, techniques and instrumentation, I intend to develop an innovative research and education program to examine new nanomaterials that can find various applications: (1) in the biomedical field as matrices for drug/protein/DNA release and delivery, and enzyme immobilization, and in the cardiovascular research (interactions of blood components and polymers) and (2) in the electronics industry as thermolyzable materials for removal and re-attachment of components.

The constituent chemical elements and the way in which a material has been processed into its final form determine its structure. The architecture, composition and the stability of a biomaterial (polymer, protein or blood component) affect its properties and its performance in a specific application. To this end, characterization of the material and investigation of its detailed structure and any possible conformational changes under different conditions (acidic, basic and enzymatic hydrolysis, thermolysis, shear, and change in solution properties) is important since it can relate the properties and performance (function) of the material in an application to its structure.

In this poster, I describe the (1) synthesis of functional molecules and well-defined polymeric nanostructures, (2) isolation, purification and characterization (size, structure and properties) of organic molecules and biomaterials (polymers, proteins and blood components), and (3) investigation of structural stability of biomaterials under different conditions. The development of new tools such as small angle neutron scattering (SANS) and fluorescence dye-binding assays that can be used to examine conformational changes and stability of biomaterials, a major regulator of their function, is also presented.

See more of this Session: Poster Session: Meet the Faculty Candidate
See more of this Group/Topical: Education

Multi-Scale Simulation of Self-Assembly in Al(110) Homoepitaxial Growth

Yogesh Tiwary, Dept. of Chemical Engineering, The Pennsylvania State University, University Park, PA

Homoepitaxial growth on Al(110) exhibits nanoscale atomic self-assembly [1] due to an interplay between the thermodynamic interactions between atoms and the kinetics of atomic diffusion and deposition. After 30 ML deposition with a flux of 1 ML/min at 450 K, 3D hut-shaped nanostructures called "nanohuts" were seen in AFM images, which self-organize over the micron scale. Nanohuts have smooth and well-defined (100) and (111) facets and reach heights of up to 50 nm. Morphology during homoepitaxial growth under other conditions and structural evolution of nanohuts during deposition are presently not clear. We conducted multi-scale simulation studies to understand, predict and control self-assembly in this system.

We obtained accurate physical details of atomic diffusion and interactions in this system using ab initio calculations. We studied the key adatom diffusion moves responsible for the self-assembly and obtained their energy barriers [2] using the climbing-image nudged elastic band method [3]. We found several new mechanisms for adatom diffusion and also refined the energy barriers for the known mechanisms [4]. The energy barriers for adatom diffusion are sensitive to the local environment due to atomic interactions at the initial and transition states. Under certain configurations of neighboring atoms, the energy barriers for some diffusion moves are significantly reduced compared to that of single atom moves. Thus, we found that atoms assist each other to achieve moves that are kinetically difficult for single atoms, and lead to interesting self-assemblies. To accurately describe high-order many-body atomic interactions at surfaces, we developed the Connector Model [5] which can predict interactions at the initial states with accuracy close to ab initio calculations. We found that the connector model is more accurate and efficient in describing many-body interactions on Al(110) and Al(100) compared to the traditional lattice-gas approach [5,6].

The insights gained at the nanometer and pico-seconds scale were incorporated in a 3D kinetic Monte Carlo (kMC) model to extend the simulations to close to micron and minutes scale. Using our 3D kMC model, we simulated self-assembly under different deposition conditions in Al(110) homoepitaxial growth by modifying deposition flux and temperatures [7]. Further, we proposed innovative approaches to achieve the desired shapes and spatial distributions of nanostructures.

[1] F. Buatier de Mongeot, W. Zhu, A. Molle, R. Buzio, C. Boragno, U. Valbusa, E. Wang, and Z. Zhang, Phys. Rev. Lett. 91, 016102 (2003).

[2] Y. Tiwary and K. A. Fichthorn, to be published.

[3] G. Henkelman, B.P. Uberuaga, and H. Jonsson, J. Chem. Phys. 113, 9901 (2000).

[4] W. Zhu, F. Buatier de Mongeot, U. Valbusa, E. G. Wang, and Z. Y. Zhang, Phys. Rev. Lett. 92, 106102 (2004).

[5] Y. Tiwary and K. A. Fichthorn, Phys. Rev. B 78, 205418 (2008).

[6] Y. Tiwary and K. A. Fichthorn, Phys. Rev. B 75, 235451 (2007).

[7] Y. Tiwary and K. A. Fichthorn, to be published.

See more of this Session: Poster Session: Meet the Faculty Candidate
See more of this Group/Topical: Education

From Fundamentals to Functionality

O. Berk Usta, Chemical and Petroleum Engineering, University of Pittsburgh, Pittsburgh, PA

The interactions of rigid and deformable solids and a flowing fluid medium determine behavior and functionality in most biological and industrial systems. These interactions between different phases give rise to a variety of very complex phenomena and hence vital functions such as hearing, smelling, reproduction and threat sensing. Understanding these complex systems is not only important in solving problems with the imminent system but also to create artificial devices that can mimic similar functionality. In my research, I combine simple principles of fluid mechanics and solid mechanics to gain insight into the diverse mechanisms as a result of the fluid-solid interactions in physical and biological systems ranging from hard spheres and polymers to microcapsules and hair cells. My ultimate aim is to solve problems that are relevant to both the scientist in the lab and the broader community using fundamentally sound, state of the art yet simple approaches.

In this session, I will primarily focus on mesoscopic computational methods that help solve the afore-mentioned complex problems in a highly competitive manner and will describe results on some recent exciting problems that have garnered significant attention. These include the migration in dilute polymer solutions; sorting of microcapsules; chemical signaling and self propelled motion; and deformation of viscous and viscoelastic drops.

See more of this Session: Poster Session: Meet the Faculty Candidate
See more of this Group/Topical: Education

Bone Tissue Growth Mechanism Fundamentals Via Fluid Shear and Nutrient Transport Simulation in 3D Porous Scaffolds Cultured in a Perfusion Bioreactor

Roman S. Voronov, School of Chemical, Biological, and Materials Engineering, University of Oklahoma, Norman, OK

Samuel VanGordon, Bioengineering, University of Oklahoma, Norman, OK

Bonnie Landy, Bioengineering, University of Oklahoma, Norman, OK

Vassilios I. Sikavitsas, Chemical, Biological, and Materials Engineering, University of Oklahoma, Norman, OK

Dimitrios V. Papavassiliou, Chemical, Biological, and Materials Engineering, University of Oklahoma, Norman, OK

A wide variety of scaffold geometries utilized for culturing osteoblastic cells in a flow perfusion bioreactor for bone regeneration, yet the question remains unanswered as to what scaffold design is optimal for bone tissue growth. The answer depends on several factors, such as the flow regime under which the tissue is cultured, the amount of fluid shear forces experienced by the cells within the scaffold (mechanostimulation, detachment, bursting) and the efficiency of nutrient/waste transport to/away from the cells (osteoblast differentiation, proliferation, upregulation of angiogenic and osteogenic factors, and mineralized matrix production). Fundamental understanding of factors that lead to favorable cell differentiation can lead to design procedures for scaffolds that would maximize tissue growth. However, due to the inherently random architecture of porous scaffolds internal structure theoretical prediction of tissue growth is impractical. Therefore, the primary goal of this work is to obtain local shear force and mass transport distributions within typical bone engineering scaffolds (salt leeched and non-woven fiber mesh PLLA) via computation. The juxtaposition of the local tissue growth and computer simulation results is used to obtain insight into the tissue growth process with the ultimate goal of being able to predict where the tissue will grow a priori - based on the plain scaffold geometry, only.

Flows of osteogenic media through the scaffolds are modeled via fluid dynamics simulations Lattice Boltzmann Method. Macroscopic mass transfer is modeled using the Lagrangian Scalar Tracking method. High performance computing in conjunction with a house hybrid parallelized scheme is employed in order to take advantage inherent code parallelizability. Various scaffold design parameters are explored for optimal tissue growth and the obtained results are parametrized based on geometric characteristics of the scaffolds (such as tortuosity, surface area to volume ratio, and porosity). Micro-Computed Tomography (μCT) with 10 μm resolution is used to obtain 3D structure of the scaffolds. Modeling results are validated against experiments via μCT and histological imaging of tissue growth as the tissue culturing experiment progresses with time.

See more of this Session: Poster Session: Meet the Faculty Candidate
See more of this Group/Topical: Education

Protein Engineering for Biomedical Applications and Biofuels Production

Fei Wen, Chemical and Biomolecular Engineering, University of Illinois at Urbana-Champaign, Urbana, IL

Huimin Zhao, Chemical & Biomolecular Engineering, University of Illinois at Urbana-Champaign, Urbana, IL

The overall theme of my PhD thesis research is to apply cell surface display and directed evolution strategies to address challenges in biomedical research and biofuels production. Throughout my thesis study, I have been working with various types of proteins (human membrane proteins including T-cell receptors (TCRs) and major histocompatibility complexes (MHC), and enzymes from bacteria, fungi, and yeast) and a wide variety of expression systems (E. coli, yeast, insect cells, and mammalian cells) as described below.

1. Development of a new tool for T-cell epitope mapping. Identification of T-cell epitopes (antigenic peptides recognized by T cells) is a critical step in studying and modulating the immune responses to tumors, infectious agents, and autoantigens. We developed a facile, accurate, and high-throughput method for T-cell epitope identification using yeast displaying pathogen-derived peptide library. For the first time, we demonstrated that yeast cells displaying the peptide-MHC (pMHC) complexes could be used as artificial antigen presenting cells. Using human MHC class II molecule HLA-DR1 and influenza A virus (X31/A/Aichi/68) as a model system, we showed that this method can be used to rapidly pinpoint a 17-amino-acid-long epitope from the entire influenza A virus genome. Although the method was demonstrated by identifying a viral epitope, it should be generally applicable to the identification of T-cell epitopes from other systems such as cancer and autoimmune diseases.

2. Directed evolution of human T-cell receptor (TCR) ligands with improved affinity. The development of pMHC (TCR ligand) tetramer staining has revolutionized the field of T-cell research. However, pMHC tetramers are very time-consuming and labor-intensive to prepare and often show low avidity. To address these limitations, we sought to engineer MHC monomers with high TCR-binding affinity. The human MHC class II molecule DR2, which is associated with multiple sclerosis, was chosen to be our engineering target. By fusing the leucine zipper dimerization motifs, the wild-type DR2 was successfully displayed on insect cell surface and was capable of binding a candidate epitope involved in multiple sclerosis (MBP85-99) and activating specific T-cell hybridomas. More importantly, the surface displayed DR2 bound specific TCR tetramers in an epitope-dependent manner. Coupled with directed evolution and fluorescence-activated cell sorting (FACS), DR2 variants with improved affinity could be readily identified from a mutagenesis library in a high throughput manner. Currently, the validation of the described screening method is in process.

3. Lignocellulosic ethanol production using yeast displaying an engineered mini-cellulosome. In an effort to develop a consolidated bioprocessing-enabling microorganism, we created a recombinant cellulolytic yeast strain by displaying an engineered mini-cellulosome on the surface. The mini-cellulosome consists of a mini-scaffoldin derived from C. thermocellum and three types of cellulases - an endoglucanase, a cellobiohydrolase, and a beta-glucosidase. The surface expression and functionality of the mini-cellulosome was confirmed by various detection methods. The engineered mini-cellulosome showed both enzyme-enzyme synergy and enzyme proximity synergy. In addition, the engineered mini-cellulosome was stable at 4 C for at least four months. More importantly, yeast cells displaying the engineered mini-cellulosome could directly convert amorphous cellulose to ethanol; and they also showed activity towards crystalline cellulose. The evaluation of ethanol production from crystalline cellulose is in progress.

See more of this Session: Poster Session: Meet the Faculty Candidate
See more of this Group/Topical: Education

Reaction-Diffusion: From Molecular to Macroscopic Length Scales

Paul J. Wesson, Chemical and Biological Engineering, Northwestern University, Evanston, IL

Reaction-diffusion (RD) processes can occur on many length scales, and create natural structures ranging from the size of raidiolarians (~10^{-5} m) to the patterns on zebras (~10^{-1} m). Using these natural structures as inspiration, I have studied RD systems on several length scales - from molecular to macroscopic - to not only better understand the limitations of these systems but also to develop more technologically relevant RD processes.

On the molecular scale, I have used a lattice-gas model to predict the smallest size feature that can be produced using RD, and also present a nanoparticle based electronic memory device which relies on the diffusion of counterions to store information. Similarly on the macromolecular length scale, I have developed a method to measure protein-ligand dissociation constants that does not rely on any specific "signature" from the protein or ligand, shown that a RD process limits the resolution of a nanoparticle based self-erasing "ink," and also used RD to fabricate three dimensional core and shell particles where the core and shell do not necessarily display the same symmetry. On the macroscopic length scale, I have used RD to guide the motion of chemotactic droplets so that a droplet of oil may be used to solve a maze.

See more of this Session: Poster Session: Meet the Faculty Candidate
See more of this Group/Topical: Education

Enzymatic Synthesis of Simvastatin

Xinkai Xie, Chemical and Biomolecular, University of California at Los Angeles, Los Angeles, CA

Natural products are small molecules found in diverse natural sources such as bacteria, fungi, plants and so on. They are produced by dedicated biosynthetic pathways using enzymes that possess exquisite catalytic power. Enzymes discovered from these pathways collectively constitute a versatile molecular toolbox for use in drug discovery and drug synthesis. Simvastatin (Zocor), one of the best selling drug in the united states with annual sales ~$5 billion before the expiration of the patent on 2007, is a cholesterol-lowering drug due to its potent inhibitory activity towards the rate-limiting step of cholesterol biosynthesis. Currently simvastatin was semisynthesized from lovastatin, a natural product derived from Aspergillus terreus.

We enzymatically synthesized simvastatin by using of an acyl-transferase, LovD (46 kDa), which is involved in the lovastatin biosynthetic pathway. We first demonstrated that LovD has broad substrate specificity. We then developed an Escherichia coli based whole-cell platform, which can readily synthesize simvastatin from easily accessible substrates. The new process avoided the currently used multiple-step organic synthesis of simvastatin and the final yield was much higher. Later, multiple techniques, including screening of the substrates, metabolic engineering of E. coli, protein engineering of LovD, etc., were carried out to optimize the process All of these significantly improved the whole cell biocatalytic process. High cell density fermentation opens the door for industrial production of simvastatin at lower cost. Besides that, to understand the fundamental details of naturally happened lovastatin biosynthesis, the megasynthase LovF from the same pathway was expressed and purified from Saccharomyces cerevisiae. We showed that protein-protein interactions between LovF and LovD facilitated highly efficient biosynthesis of lovastatin. To understand the basis of LovD catalyzed reaction, X-ray crystallography was employed, which provided atomic resolution of the mechanism of catalysis, substrate and product binding, and a likely basis of beneficial mutations.

See more of this Session: Poster Session: Meet the Faculty Candidate
See more of this Group/Topical: Education

How Experiment-Driven Quantitative Modeling and Control Help Optimize Oncolytic Adenovirus Cancer Treatment

Neda Bagheri, Department of Biological Engineering, MIT, Cambridge, MA

Marisa Shiina, Division of Gastroenterology and Hematology/Oncology, UCSF, San Francisco, CA

W. Michael Korn, Division of Gastroenterology and Hematology/Oncology, UCSF, San Francisco, CA

Douglas A. Lauffenburger, Department of Biological Engineering, MIT, Cambridge, MA

Replication-selective adenoviruses replicate in cells containing certain mutations, motivating their use in targeted gene therapy. An example is ONYX-015, which aims to target and preferentially replicate in p53-defective tumor cells, potentially offering improved treatment options for certain cancers. Although recent studies demonstrate that ONYX-015 affects a variety of signaling pathways not necessarily involving p53, the adenovirus continues to be studied in a number of clinical trials.

The efficacy of adenovirus mediated therapy depends on three general processes: (i) cell entry via the coxsackievirus and adenovirus receptor, CAR, (ii) viral replication, and (iii) virus-mediated cell death. Experimental data suggests that CAR is down-regulated in highly malignant cells, hindering ONYX-015's ability to infect. Pharmaceutical intervention into the Raf-MEK-ERK pathway via the CI1040 MEK inhibitor has shown to counter this effect by up-regulating CAR expression. MEK inhibition, however, causes G1 cell cycle arrest thereby stunting viral replication and consequently virus-induced cancer cell death. Through the data-driven modeling of cancer cells subjected to CI1040 and ONYX-015, we aim to characterize and predict system dynamics, providing a means to optimize the efficacy of oncolytic adenovirus cancer treatment by manipulating the timing of drug treatment and infection. Validation of simulated results should provide a means to refine the current cellular level ODE model to include relevant sub-cellular phenomena that better characterizes both the mechanistic and dynamic behavior of adenovirus cancer treatment. Pending successful test of model predictions, our goal is to elucidate optimization strategies that could offer practical and effective means for minimizing cancer growth.

Engineering principles offer a variety of tools for the systematic investigation and prediction of complex behavior. I aim to refine control theoretic methodologies to address specific challenges in medicine and biology. One such challenge generally involves optimizing the combinatorial effect of therapeutic protocols to minimize cancer growth. As our pursuits in biology become increasingly complex, we create a greater demand for interdisciplinary research and maximally informative experiments/predictions. The refinement of tools provided by control theory and signal processing will be critical in shaping research strategies in medicine, experimental biology, and environmental stability.

See more of this Session: Poster Session: Meet the Faculty Candidate
See more of this Group/Topical: Education

Metabolic Engineering Microorganisms for Applications in Drug Discovery, Biofuel Development and Commodity Chemical Production

Yajun Yan, Chemical & Biomolecular Engineering, University of California, Los Angeles, Los Angeles, CA

Mattheos A. G. Koffas, Chemical and Biological Engineering, University at Buffalo, the State University of New York, Buffalo, NY

James C. Liao, Chemical & Biomolecular Engineering, University of California, Los Angeles, Los Angeles, CA

My Ph.D and postdoctoral research focuses on the field of metabolic engineering and synthetic biology with the general goal of programming microbial cells to achieve the unique capability of synthesizing important molecules for biofuel development, drug discovery and commodity chemical production. In this poster, I will present an overview of my Ph.D work on constructing microbial production platforms for the synthesis of natural drug candidate-flavonoids and my postdoctoral research on engineering metabolic systems for the production of commodity chemicals and advanced fuels from renewable sources.

See more of this Session: Poster Session: Meet the Faculty Candidate
See more of this Group/Topical: Education

Engineering Optimal Cell Culture Platform Using Micro- and Nanotechnology

Yong Yang, Biomedical Engineering, Duke University, Durham, NC

The microenvironment in living tissues comprises numerous cells, extracellular matrix (ECM) proteins and a variety of soluble and ECM bound factors. The ECM with which cells interact often includes topography at the nanoscale and provides, in concert with the spatio-temporally arranged signaling molecules and external stimuli, cues for cell adhesion, migration, proliferation and differentiation. Through the ECM, small fluid flows called interstitial flows, driven by dynamic stress, help to transport nutrients and play an important role in tissue maintenance and pathobiology. Knowledge of these interactions is crucial to the understanding of many fundamental biological questions and to the design of medical devices. This fact brings up the need to investigate cell-substrate interactions at the nanoscale in a dynamic environment. It is thus important to engineer dynamic 3D cell culture platform with nanostructures for advanced studies of cell-substrate interactions.

Based on polymer thin film technology, I have developed a cost-effective technique to generate a large nanopatterned surface from small nanostructured patterns, so that enough cells can be seeded for subsequent biochemical and molecular biology analyses. Taking advantage of polymer thin film and microcontact printing technologies, a simple technique has been developed to embed nanostructures into polymeric microfluidic platform. Moreover, I have also invented a biologically benign technique to process and assemble polymeric structures at the micro-/nanoscale. By using supercritical fluids (e.g., CO_2, N_2), the degree of enhanced chain mobility can be controlled to result in entanglement or patterning of surface chains without altering the integrity of the substrate. With this technique biomolecules and even cells can be incorporated into polymeric micro-/nanodevices without damaging their bioactivity. Preliminary studies with human mesenchymal stem cells (hMSCs) on the multifunctional microfluidic platform show that the nanotopography, flow-induced mechanical stress and chemical cue play a role in attachment, migration, spreading, and gene expression of hMSCs.

Equipped with tools to precisely control topographical, chemical and mechanical cues on cell culture, I am able to systematically investigate how these cues affect proliferation and differentiation of stem cells. In particular, I am interested in the expansion of hematopoietic stems cells (HSCs). I am expecting to a clear picture of the expansion performance of HSCs in the microfluidic cell culture platform with respect to the phenotypic analysis and engraftment assay. I hope to develop a clinically relevant culture technology for fully realizing the potential of HSC transplantation. Beyond expansion of HSCs, the broad objective of this research is to create an optimal microenvironment that can regulate the fate of stem cells and eventually expand specific stem cell phenotypes for clinical applications.

See more of this Session: Poster Session: Meet the Faculty Candidate
See more of this Group/Topical: Education

Metal-Organic Frameworks for Energy Applications

A. Ozgur Yazaydin, Chemical and Biological Engineering, Northwestern University, Evanston, IL

Randall Q. Snurr, Chemical & Biologial Engineering, Northwestern University, Evanston, IL

Metal-organic frameworks (MOFs) are a new class of nanoporous materials that have potential applications in energy storage, separation processes, catalysis, and sensing. MOFs are inorganic-organic hybrids that are synthesized using organic linker molecules and metal joints that self-assemble to form materials with well-defined pores, high surface areas, and tunable chemical functionalities. By judiciously choosing the metal cluster and organic linker, MOFs can be designed such that the materials have desired magnetic, optical, catalytic, or selective adsorption properties.

My research has focused on the application of MOFs in the energy field. I particularly worked on CO2 capture from flue gas and hydrogen storage. Concerns about greenhouse gases in the atmosphere have led to significant interest in removing CO2 from the exhaust streams of fossil fuel combustion as the first step in carbon sequestration. Revolutionary advancements are needed to improve the efficiency and cost effectiveness of CO2 capture technologies. By using a combined experimental and modeling approach we screened a diverse collection of MOFs for CO2 capture from flue gas. The diversity of the chosen MOFs was critical for improving our understanding of CO2 capture in MOFs. We also used this data to validate a generalized strategy for molecular modeling of CO2 and other small molecules in MOFs.

MOFs have also shown great potential for the storage of hydrogen, which is a potential energy carrier for various applications. One major shortcoming of these materials is that hydrogen is bound only weakly at room temperature due to the fact that in classical MOFs with fully coordinated metal ions the metal-hydrogen interaction is largely screened. This necessitates cryogenic operation temperatures to obtain a significant amount of hydrogen adsorption. One way of increasing the binding energy is to create MOFs with unshielded metal-hydrogen interactions. Using molecular modeling we tested several MOFs having coordinatively unsaturated metal sites or extraframework metal ions.

My presentation will include key findings from our research on how to improve CO2 capture from flue gas, and increase the binding energy between MOFs and hydrogen for improved room temperature storage.

See more of this Session: Poster Session: Meet the Faculty Candidate
See more of this Group/Topical: Education

"Gated" Ion Transport through Dense Carbon Nnaotube Membrane

Miao Yu, Chamical and Biological Engineering, University of Colorado at Boulder, Boulder, CO

"Gated" ion transport through hydrophobic protein domains in various cellular membranes was found widely, while these hydrophobic protein channels regulate ions transport through these membranes with both high selectivity and high flux. Hydrophobic and narrow carbon nanotubes (CNT) may serve as a simplified model to mimic these biological pores and thus reveal main important properties of these hydrophobic nanochannels. In our previous study, structurally well defined dense CNT membranes with both CNT pores and space between CNTs available for transport were prepared by a simple capillary-driven shrinking process. The average pore size of these two types of pores in dense CNT membranes was 3 nm, as indicated by nano-gold filtration and N2 porosimetry. In this study, we show that ion diffusion through dense CNT membranes with 3 nm hydrophobic pores has a "gated" behavior, possibly due to the poor water wettability on CNTs, as suggested by water vapor adsorption isotherms at room temperature and zero-current potential measurements. At 298 K, the ion diffusion channel was gradually closed gradually due to the discontinuity of the water phase in dense CNT membranes and thus ions cannot continuously diffuse through dense CNT membranes. Small perturbations, such as ultrasonication or small temperature change (< 20 K), open the gate for ions to diffuse. These small perturbations may change the water filling state in and around CNTs and create a continuous or temporarily continuous ion diffusion channel. When the ion diffusion channel was open, ions diffused through dense CNT membranes at a rate similar to their diffusion rate in bulk water. This water wettablity induced "gated" ion diffusion behavior in dense CNT membranes may have significant implications on "gated" ion diffusion in hydrophobic biological pores.

See more of this Session: Poster Session: Meet the Faculty Candidate
See more of this Group/Topical: Education

Biomimetic Microdevices for Clinical Therapeutic and Diagnostic Applications

Chong Wing Yung, Surgery - Vascular Biology Program, Harvard Medical School, Children's Hospital Boston & Wyss Institute, Boston, MA

Donald E. Ingber, Surgery & Pathology, Harvard Medical School, Children's Hospital Boston & Wyss Institute, Boston, MA

Sepsis is a major killer of man because a high microbial load in blood overcomes the immune system resulting in multi-systems failure. This is a particularly significant problem in premature infants and immunocompromised patients with fungal infections since antibiotic therapies are dose-limiting. Development of an extracorporeal blood cleansing device that can rapidly clear the blood of pathogens without loss or alteration of normal blood components could help remedy this problem. We have previously developed a prototype microfluidic-micromagnetic blood cleansing device that rapidly clears living Candida albicans pathogens from flowing human whole blood with high efficiency (80%) and throughput (20 mL/hr/device) in a single pass. Magnetic opsonins were created by coating 1-μm microbeads with pathogen-specific antibodies. One drawback of this device is that some blood is loss in the saline stream (~13%) during the separation process. More recently, we have developed an advanced design inspired by the native architecture of the spleen that virtually eliminates blood loss while further enhancing separation efficiency. This Biomimetic Spleen has three microfluidic layers that replicate the microvasculature, sinusoids, and lymphoid follicles of the organ. The result is a simple and yet robust device that can clear 85-95% of C.albicans at similar flow rates without any blood loss or dilution. (Funded by CIMIT, DoD, Wyss Institute)

See more of this Session: Poster Session: Meet the Faculty Candidate
See more of this Group/Topical: Education

From Genetic and Metabolic Engineering to Cell Mechanics and Genetic Diseases

Zhixia Zhong, Department of Chemical Engineering, Carnegie Mellon University, Pittsburgh, PA

My research initiated from genetic and metabolic engineering. By rationally design of functional genes and selection of host stains, we can produce and enhance the productivity of many natural products such as functional proteins and secondary metabolites. We successfully expressed and obtained functional host-toxic peptides (human beta defensins) in *E.coli* and cell-free system. Their antimicrobial function were tested in disease-causing bacteria. We also construct an *S. cerevisiae* stain with high productivity of S-Adenosyl methionine by gene modification and industrial scale-up, which has been used in industrial SAM-e production. The techniques I learned from those projects are systematic and can be applied into many exciting areas such as producing energy products, medical products and other functional molecules.

Beyond bacteria and yeast, my research extended to mamallian cell later on. Those researches focus on fundamental analysis of a family of elastic proteins (spectrin repeat proteins) *in vitro* and *in vivo*. We are also interested in their relationship with several genetic diseases especailly laminopathies. The detailed projects include: 1) the study of thermodynamic stalibity of nesprin proteins by using molecular clonging techniques and structure analysis equipment. 2) the study of mechanical properties of alpha II spectrin network in cell by using fluorescent immunostain & bioimaging techniques, micromanipulation equipment and force-generating apparatus. We found that nesprins are stablized by its central adaptive domain and alpha II spectrin is essential to the mechanical integrities of nuclear envelope and cell membrane. The techniques I learned from those projects can also be applied into many interesting areas such as biophysics, health care, pathology, *etc*.

My research experience spreads gene, protein, bacteria, yeast and mamalian cells and related techniques. In my future research, I will focus on applying those knowledge and techniqes into the areas of energy, healthcare and biophysics and hopefully creating new techniqes and tools.

See more of this Session: Poster Session: Meet the Faculty Candidate
See more of this Group/Topical: Education

Hydrolytically Degradable Poly(ethylene glycol) Hydrogel as a Tunable Scaffold for Neural Tissue Engineering

Silviya Petrova Zustiak, Chemical and Biochemical Engineering, University of Maryland Baltimore County, Baltimore, MD

Jennie B Leach, Chemical and Biochemical Engineering, University of Maryland Baltimore County, Baltimore, MD

The generation of replacement organs and tissues relies on the design of components that replicate the three-dimensional (3D) structures and physiologies found in the body. While great strides have been made in developing new biomaterials, the tissue engineering efforts focused on nerve repair have been limited by a rudimentary understanding of how neurons interact with their 3D environment. Our overall goal is to develop new 3D matrices that are specifically designed to a) mimic the properties of soft nervous tissues and b) enable new fundamental investigations of how scaffold physical properties (e.g., porosity and mechanical properties) impact neuronal response. Our hydrogel of choice is poly(ethylene glycol) (PEG), because it fulfills most of the requirements imposed on the scaffold's design such as biocompatibility, biodegradability, as well as, resistance to non-specific protein adsorption.

In this study we use PEG vinyl sulfone gels, cross-linked with PEG-dithiol ester cross-linkers. One benefit of this hydrogel is that it can be formed in physiological conditions and thus can be seeded with cells prior to gelation. Further, we explore the change in diffusion, mechanical properties and porosity with degradation by utilizing PEG polymers of varying molecular weight, varying location of the hydrolysable group and by manipulating polymer density. Thus far, we have synthesized PEG hydrogels that span a range of mechanical properties (G', 0.4-5 kPa) that mimics human nervous tissue (e.g., brain and spinal cord).

We will present current work with a focus on: (a) characterization of gel porosity and stiffness with gel degradation (degradation times spanned from several hours to several days and 20-fold change in gel physical properties was observed over time), (b) characterization of diffusion of various model solutes in the gels (De/Do=0.58 for Lysozyme; 0.18 for BSA; 0.1 for Ig), (c) addition of common adhesive ligands to the scaffolds to promote neuronal outgrowth and the associated change in gel mechanical properties (addition of RGDS or YIGSR lead to an increase in G' and decrease in swelling ratio and mesh size), and (d) investigation of neuronal and PC12 cells response to these novel scaffolding biomaterials with varying tunable properties (an average of 90% viability was achieved with all gel types).

This work will ultimately lead to a fundamental quantitative understanding of how to more effectively design tissue engineering scaffolds for directed neuronal response and may result in advanced model systems for in vitro neurobiology studies as well as improved materials to treat neural injuries.

See more of this Session: Poster Session: Meet the Faculty Candidate
See more of this Group/Topical: Education

Ionic Liquid-Based Materials for CO2 Capture and Polymer Engineering

Jason Bara, Department of Chemical and Biological Engineering, University of Colorado, Boulder, CO

Ionic liquids (ILs) are far more than non-volatile molten salts. ILs offer nearly infinite possibilities to create both liquid and solid materials not only for gas separations, but much broader chemical engineering and materials science applications. The ability to chemically "tune" the properties of ILs and related building blocks through chemical synthesis offers unprecented control over material performance and properties.

My ongoing research with ILs focuses on validating their use as industrially viable solvents for CO2 capture from flue gas and natural gas via reaction with amines. Additionally, I have explored their use in gas separation membranes, both in supported liquid membrane configurations and as monomers for fabricating poly(ionic liquid) and poly(ionic liquid)-ionic liquid composite membranes. Results to date and future directions will be discussed.

See more of this Session: Poster Session: Meet the Faculty Candidate
See more of this Group/Topical: Education

A New Physically-Based Parameterization of Ice Cloud Formation for Large-Scale Simulations

Donifan Barahona[1] and Athanasios Nenes[1,2]

[1] School of Chemical and Biomolecular Engineering, Georgia Institute of Technology
[2] School of Earth and Atmospheric Sciences, Georgia Institute of Technology

Abstract. This study presents a comprehensive ice cloud formation parameterization that computes the ice crystal number, size distribution, and maximum supersaturation from precursor aerosol and ice nuclei with any size distribution and chemical composition. The parameterization provides an analytical solution of the cloud parcel model equations and accounts for the competition effects between homogeneous and heterogeneous freezing, and, between heterogeneous freezing in different modes. The diversity of heterogeneous nuclei is described through a nucleation spectrum function which depends on the surface area, chemical composition, and size distribution of the precursor, insoluble aerosol. The parameterization reproduced the predictions of a detailed numerical parcel model. The average error in ice crystals number concentration was -2.0 ± 8.5% for conditions of pure heterogeneous freezing, and, 4.7 ± 21% when both homogeneous and heterogeneous freezing were active. Apart from its rigor, excellent performance and versatility, the formulation is extremely fast and free from requirements of numerical integration.

Key Words: Heterogeneous nucleation; homogeneous nucleation; Cirrus Clouds.

INTRODUCTION

Representation of ice clouds in climate and weather prediction models remains a challenge due to the limited understanding of ice cloud formation processes and the difficulties associated with their remote sensing (*1*). Ice clouds form by homogeneous freezing of liquid droplets or heterogeneous freezing upon ice nuclei, (IN) (e.g., *2*). Observational data show that the two freezing mechanisms are likely to interact during cloud formation (*3*); their relative contribution is however a strong function of IN, aerosol concentration, and cloud formation conditions (*4, 5*).

Several approaches have been proposed to parameterize ice cloud formation in atmospheric models. Empirical correlations derived from field campaigns are most often employed to express IN concentrations (e.g., *6*). These expressions are simple but only provide the availability of IN over a limited spatial region. Calculation of ice crystal number concentration, N_c, requires the knowledge of cloud supersaturation, s_i, and therefore the usage of a dynamical framework. Recent physically-based parameterizations address this issue (*7, 8*). They however do not allow direct incorporation of experimental freezing data within their framework, and rely on oversimplified assumptions regarding the nature of heterogeneous freezing.

This work presents a new physically-based, analytical scheme to parameterize ice cloud formation in a computationally efficient manner. The new scheme allows the usage of both empirical and theoretical IN data in a simple dynamical framework, and can consider the variability in aerosol and IN composition. The new parameterization builds upon the frameworks of Barahona and Nenes *(4, 9)* that combine homogeneous and heterogeneous mechanisms of ice formation, and explicitly resolves the dependency of N_c on conditions of cloud formation (i.e., T, p), updraft velocity, aerosol number and size, and the freezing characteristics of the IN population.

THE PARAMETERIZATION SCHEME

The ice crystal number concentration, N_c, is given by

$$N_c = \begin{cases} N_o e^{-f_c}(1-e^{-f_c}) + N_{het}(s_{\hom}) & f_c > 0 \\ N_{het}(s_{\max}) & f_c \leq 0 \end{cases} \quad (1)$$

where N_o is the deliquescent aerosol (i.e., droplet) concentration, $s_{\max}$ is the maximum supersaturation during cloud formation, f_c is the droplet freezing fraction, and $s_{\hom}$ is the homogeneous freezing supersaturation threshold. The homogeneous contribution to N_c in equation (1) is given by the term $N_o e^{-f_c}(1-e^{-f_c})$. Heterogeneous nucleation is described using a nucleation spectrum function represented as $N_{het}(s_i)$, which can be either experimentally or theoretically derived *(10)*.

The parameterization accounts for competition effects between homogeneous and heterogeneous nucleation. For this, the concentration of IN that completely inhibits homogenous nucleation, $N_{\lim}$, is used to scale the fraction of droplets that freezes homogeneously in the presence of IN. $N_{\lim}$ it is given by

$$N_{\lim} = \frac{N^*}{\sqrt{\left.\Delta s^*_{char}\right|_{s_{\hom}}}} \frac{\left(1+s_{\hom}\right)}{s_{\hom}} e^{\frac{2}{\lambda s_{\hom}}} \quad (2)$$

where N^* and λ are functions of cloud formation conditions (i.e., updraft velocity T, and p) and of the water vapor to ice deposition coefficient. Δs^*_{char} is given by

$$\left.\Delta s^*_{char}\right|_{s_i} = \left\{ \frac{\Delta s_{char}\left[\frac{4}{3}\Delta s_{char} + 2(s_i - \Delta s_{char})\right]}{(1+s_i-\Delta s_{char})} \right\}^{1/2} \quad (3)$$

The parameter Δs_{char} is characteristic of the IN population, and is given by,

$$\Delta s_{char} = \min\left[\left(\frac{d\ln N_{het}(s_i)}{ds_i}\right)^{-1}, s_i\right] \tag{4}$$

when applied in Eq. (3), Δs_{char} is calculated at $s_i = s_{\text{hom}}$.

Combined Homogeneous and Heterogeneous Freezing

When both, homogeneous and heterogeneous nucleation take place, heterogeneously frozen crystals tend to deplete water vapor before the onset homogeneous freezing (*11*). Thus, the droplet freezing fraction, f_c, depends on the IN concentration and on the freezing characteristics of the IN population; f_c is given by,

$$f_c = f_{c,\text{hom}}\left\{1-\left(\frac{N_{het}(s_{\text{hom}})}{N_{\text{lim}}}\right)^{3/2}\right\}^{3/2} \tag{5}$$

where $f_{c,\text{hom}} = \frac{\rho_a}{\rho_i}\frac{k_{\text{hom}}^{1/2}}{\beta N_o}\left[\frac{2\alpha V(s_{\text{hom}}+1)}{\pi\Gamma s_{\text{hom}}}\right]^{3/2}$ is the droplet freezing fraction in absence of IN, i.e., pure homogeneous freezing (*9*) which depends only on *T*, *p*, and updraft velocity.

Pure Heterogeneous freezing

If the IN concentration is high, heterogeneously frozen crystals may deplete enough water vapor so that $s_{\max} < s_{\text{hom}}$, inhibiting homogeneous freezing (*4, 5*). In this regime only heterogeneous freezing takes place. N_c and the maximum supersaturation, $s_{\max}$, are given by solving,

$$\frac{N_{het}(s_{\max})}{N^*} = \frac{1}{\sqrt{\left.\Delta s^*_{char}\right|_{s_{\max}}}}\frac{(1+s_{\max})}{s_{\max}}e^{\frac{2}{\lambda s_{\max}}} \tag{6}$$

Δs^*_{char} and Δs_{char} are calculated setting $s_i = s_{\max}$ in Eqs. (3) and (4), respectively.

EVALUATION

The parameterization was compared against the results of a detailed cirrus parcel model (4, 9), for different formulations of $N_{het}(s_i)$ (both theoretical and experimental) and over a wide range of *T*, *p*, aerosol concentration, and updraft velocity values (about 1200 runs). For pure heterogeneous nucleation, the overall error, with respect to parcel model results, for the calculation of s_{max} and N_c, was $-1.68 \pm 3.42\,\%$ and -2.0 ± 8.5 %, respectively. For combined homogeneous and heterogeneous freezing the error in the calculation of N_c was $4.7 \pm 21\,\%$. The low error indicates no biases in the parameterization. The new parameterization was also compared against the schemes of Liu and Penner (7, LP05) and Kärcher et al. (8, K06) (Figure 1). Results showed that the assumed freezing efficiency of the different IN populations is the main factor determining the effect of heterogeneous freezing on the total ice crystal concentration.

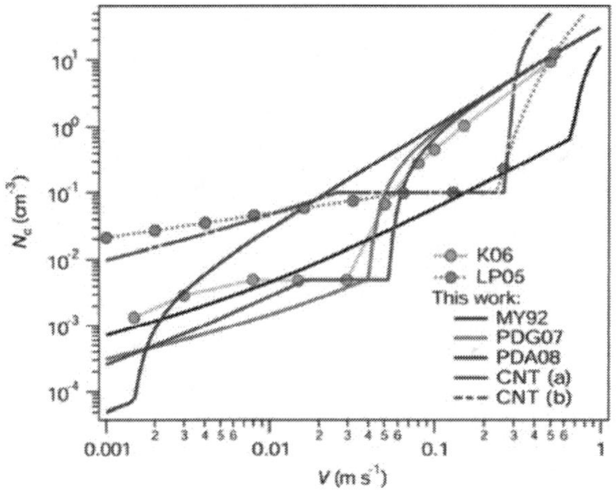

Figure 1. N_c vs. V calculated using the new parameterization using several published heterogeneous nucleation spectra (taken from Barahona and Nenes (*10*)). Also shown are results taken from Kärcher et al. (*8, K06*) and the parameterization of Liu and Penner (*7, LP05*) . Conditions considered were T = 206 K, p = 22000 Pa, α_d = 0.5. N_{soot} = 0.1 cm^{-3}, N_{dust} = 0.1 cm^{-3}, For CNT (b), $e_{f,soot} = 1.0$, and $\alpha_d = 0.1$.

Acknowledgments

This study was supported by NASA MAP, NASA ACMAP and a NASA New Investigator Award.

References

1. D. E. Waliser *et al.*, *Journal of Geophysical Research* **114**, doi:10.1029/2008JD10015 (2009).
2. H. R. Pruppacher, J. D. Klett, *Microphysics of clouds and precipitation* (Kluwer Academic Publishers, Boston, MA ed. 2nd, 1997), pp. 954.
3. P. J. DeMott *et al.*, *Proceedings of the National Academy of Sciences of the United States of America* **100**, 14655 (2003).
4. D. Barahona, A. Nenes, *Atmospheric Chemistry and Physics* **9**, 369 (2009).
5. K. Gierens, *Atmospheric Chemistry and Physics* **3**, 437 (2003).
6. M. P. Meyers, P. J. DeMott, R. Cotton, *Journal of Applied Meteorology* **31**, 708 (1992).
7. X. Liu, J. E. Penner, *Meteorologische Zeitschrift* **14**, 499 (2005).
8. B. Kärcher, J. Hendricks, U. Lohmann, *Journal of Geophysical Research* **111**, D01205 (2006).
9. D. Barahona, A. Nenes, *Journal of Geophysical Research* **113**, D11211 (2008).
10. D. Barahona, A. Nenes, *Atmospheric Chemistry and Physics* **9**, 5933 (2009).
11. P. J. DeMott, M. P. Meyers, R. W. Cotton, *Journal of the Atmospheric Sciences* **51**, 77 (1994).

Basic and Applied Studies in Biologically Relevant Polymer Thin Films

Brad J. Berron, Chemical and Biological Engineering, University of Colorado, Boulder, CO

Christopher N. Bowman, Chemical and Biological Engineering, University of Colorado, Boulder, CO

Here, we describe a novel approach towards cell and tissue staining based on polymer networks. Our approach exhibits exceptional detail of protein localization at antibody concentrations several orders of magnitude lower than typical immunofluorescent protocols. We will describe our methodology in detail as well as provide a few key results demonstrating a 3 order of magnitude reduction in antibody consumption. We will demonstrate the effectiveness of this approach in the staining of both large regions of cytoplasm and small localized protein structures on the nuclear envelope. We will also discuss characterization of the polymeric labeling system. While this approach would be effective in reducing antibody consumption in labs, it is particularly useful when paired with custom or rare antibodies.

In addition, I will describe the development of glucose oxidase initiated polymeric hydrogel films. An enzymatic polymerization initiator is a facile, noncytotoxic, room temperature approach for biomaterial generation. Glucose oxidase has been shown to possess exceptional stability and activity, making it particularly attractive as a potential polymerization initiator. Biologically, glucose oxidase is used to convert beta-D-glucose into D-glucono-1,5-lactone, and then regenerates FADH2 by the oxidation of molecular oxygen to hydrogen peroxide. We utilize the enzymatic generation of hydrogen peroxide to react with Fe2SO4 (as in Fenton's Reagent) to form hydroxy radicals, which initiate polymerization. We report our investigation into the both the bulk and thin film polymerization kinetics of hydrogel systems including polyethylene glycol diacrylate and hydroxyethylacrylate. We will discuss the concentration dependence of coinitiation and inhibition by the iron species. We will also highlight potential applications including biodetection and cell encapsulation for culture and immunoisolation applications.

See more of this Session: Poster Session: Meet the Faculty Candidate
See more of this Group/Topical: Education

A Microcapsule System That Mimics Heptotactic Cells

Amitabh Bhattacharya, Chemical Engineering, University of Pittsburgh, Pittsburgh, PA

Biological cells can signal to each other using biomolecules, and this signaling acts as a cue for motion. Often, this motion is driven by adhesion gradients on the substrate where the cells are located. In this work, it is shown that synthetic microcapsules, encapsulated with nanoparticles, can mimic cell behavior. The nanoparticles can act as 'signals', which modulate the adheshion gradient of the substrate on which the microcapsules are located, leading to autonomous motion of the microcapsules. Simulations and analysis of this system yield several interesting results. For a train of N microcapsules moving at a velocity U, we show that $U \sim N^{-1/2}$. Also, results on collective behavior of a system of microcapsules in 3D is presented.

See more of this Session: Poster Session: Meet the Faculty Candidate
See more of this Group/Topical: Education

Nitrogen-Containing Carbon Nanostructured (CNx) Catalysts for the Oxygen Reduction Reaction in PEM and Direct Methanol Fuel Cells

Elizabeth J. Biddinger, Chemical & Biomolecular Engineering, The Ohio State University, Columbus, OH

Dieter von Deak, Chemical & Biomolecular Engineering, The Ohio State University, Columbus, OH

Douglas S. Knapke, Chemical & Biomolecular Engineering, The Ohio State University, Columbus, OH

Umit S. Ozkan, Chemical & Biomolecular Engineering, The Ohio State University, Columbus, OH

PEM and direct methanol fuel cells (DMFCs) have the potential to be used as portable, fast start-up energy conversion devices, however, there are some technical and economic problems that still need to be addressed before full-scale commercialization can occur. One component that still needs improvement is the oxygen reduction reaction (ORR) cathode catalyst. The ORR at the cathode is kinetically slow, limiting the performance of the fuel cells. Due to the slow kinetics, large quantities of platinum are used as the commercial catalyst at the cathode in both PEM and DMFCs, significantly increasing the cost. Alternate catalysts to platinum are limited due to the acidic and oxidizing environment of the fuel cell cathode. Carbon-nitrogen catalysts have been studied as alternatives to platinum. Early on, these catalysts were transition metal containing macrocycles, which were selected because of their similarity to the hemoglobin-type centers. While the macrocycles were active for ORR, they were not stable in the acidic environment. It was then found that heat treating, or pyrolyzing, the macrocycles at temperatures of 400-1000°C in an inert atmosphere increased the stability of the catalysts while maintaining high ORR activity. Studies on the pyrolyzed macrocycle catalysts showed that the macrocycle center did not remain intact after heat treatment, opening up the debate on the nature of ORR activity for these catalysts. The debate was further enflamed when it was found that ORR catalysts could also be produced using simple sources of nitrogen, carbon and a transition metal, like ammonia, carbon black, and iron acetate. Currently, there are two main arguments for the nature of activity for the carbon-nitrogen ORR catalysts: (i) metal (usually iron) stabilized by nitrogen is the active site and (ii) the active site is transition metal free. If the nature of ORR activity of these carbon-nitrogen catalysts can be definitely determined, active site concentration can be increased, thereby increasing the ORR activity to levels competitive to commercial platinum.

At Ohio State, we have grown nitrogen-containing carbon nanostructured (CNx) catalysts over oxide supports with less than 1 ppm transition metals contamination to study the nature of activity. Metal-free CNx catalysts were found to be active for ORR, supporting the argument for a transition metal-free active site. For comparison, catalysts were also grown using iron, nickel and cobalt impregnated on the supports. The support and metals were washed away after growth of the carbon nanostructures using acid and base washes, leaving the CNx. CNx catalysts grown over Fe- and Co- containing supports were more active for ORR than the transition metal-free catalyst, while those grown over Ni-containing supports were less active for ORR. Characterization of the catalysts through TEM and XPS found that the catalysts with higher quantities of stacked cup nanofibers and pyridinic-type nitrogen were more active for ORR.

Present work involves the involves investigation of the nature of the activity through control of both the nanostructure and nitrogen content, effect of heteroatom incorporation into the carbon beyond nitrogen on the activity, and the application of these catalysts in a DMFC environment.

See more of this Session: Poster Session: Meet the Faculty Candidate
See more of this Group/Topical: Education

Tackling Energy and Health Issues with Systems Biology

Mark P. Brynildsen, Biomedical Engineering Department, Center for BioDynamics and Center for Applied Biotechnology, Howard Hughes Medical Institute, Boston University, Boston, MA

James J. Collins, Biomedical Engineering Department, Center for BioDynamics and Center for Applied Biotechnology, Howard Hughes Medical Institute, Boston University, Boston, MA

Systems biology is a versatile field that promises to revolutionize our understanding of and ability to engineer biological systems. The power of systems biology lies in its network-driven approaches that are capable of rendering complex phenotypes more tractable. My work has focused on both the development of these network-based techniques (1-6) and application of systems biology to energy and health issues (1).

With the ever-rising cost and unstable supply of petroleum, biology has become a prospective source of alternative energy (biofuels). However, major obstacles to the economic feasibility of biofuels exist, including the slow and expensive conversion of biomass into fermentable compounds and low product titer due to end-product or intermediate toxicity. Solutions to these problems most likely lie in the organisms we use to harvest solar energy and convert it into biofuel. Since the processes these organisms use are complex, involving multiple cellular systems (e.g., metabolism, signal transduction, etc.), systems biology would provide a greater understanding of what influences these processes and how we can engineer them to increase production. An example of this is from my work on biofuel stress (1), in which we discovered that membrane disruption from biofuels (isobutanol and n-butanol) leads to quinone malfunction (membrane electron carrier) and subsequent restructuring of metabolism through the action of the transcription factors ArcA, Fur, and PhoB. This information allowed us to predict and confirm that certain carbon sources were better growth substrates than others in the presence of isobutanol, which has important implications for feedstock selection for isobutanol production.

Systems biology also has many potential applications related to health. Cancer, bacterial persistence, and drug synergy or antagonism are all examples of where systems biology can be applied to improve treatment. Recently, my work has focused on the application of systems biology to bacterial infections in efforts to eradicate bacterial persistence, and predict how bacterial metabolism can synergize or antagonize the action of antibiotics.

In addition to these applications of systems biology, my research has addressed fundamental network biology questions important to systems biology. These include how transcription networks are structured, what capabilities these structures entail, how these networks respond to environmental stress, and how can we identify the direct targets of stimuli from gene expression data and other available sources. To this end, we characterized the ability of transcription networks to generate gene expression (2), developed a method to cope with uncertainty in transcription networks caused by network dynamics and noise (3), designed a method to identify transcription networks solely from gene expression data (4), developed techniques to infer functional roles for transcription regulators (5, 6), and identified the isobutanol response network of Escherichia coli (1).

The potential for systems biology to address energy and health concerns is great, and driven by the development of inventive network-based techniques, and the application of creative methods to approach these complex problems.

1. Brynildsen MP, Liao JC. (2009) An integrated network approach identifies the isobutanol response network of Escherichia coli. Mol Syst Biol. in press

2. Brynildsen MP, Tran LM, Liao JC. (2006) Versatility and connectivity efficiency of bipartite transcription networks. Biophys J., Oct 15;91(8):2749-59. (Epub 2006 Jun 30)

3. Brynildsen MP, Tran LM, Liao JC. (2006) A Gibbs sampler for the identification of gene expression and network connectivity consistency. Bioinformatics, Dec 15;22(24):3040-6 (Epub 2006 Oct 23)

4. Brynildsen MP, Wu TY, Jang SS, Liao JC. (2007) Biological network mapping and source signal deduction. Bioinformatics, (Epub May 15)

5. Yang YL, Suen J, Brynildsen MP, Galbraith S, Liao JC. (2005) Inferring yeast cell cycle regulators and interactions using transcription factor activities. BMC Genomics, Jun 10;6(1):90.

6. Tran LM, Brynildsen MP, Kao KC, Suen JK, Liao JC. (2005) gNCA: a framework for determining transcription factor activity based on transcriptome: identifiability and numerical implementation. Metab Eng., Mar;7(2):128-41.

See more of this Session: Poster Session: Meet the Faculty Candidate
See more of this Group/Topical: Education

Polymer Nanocomposites: From Hierarchical Ordering to Energy Application

Tirtha Chatterjee, Materials Research Laboratory, University of California, Santa Barbara, Santa Barbara, CA

Hierarchical structure of carbon nanotubes network and their linear and non-linear dynamics:

Multifunctional polymer nanocomposites represent the next generation of advanced materials. Two critical issues in their implementation of applications are: a) understanding and tailoring their structure and consequently properties and; b) critically evaluate the processability of such materials. We studied the structure of single walled carbon Nanotubes (SWNT) based polymer nanocomposites and related it to their mechanical properties and processability. Using neutron scattering studies (performed at NCNR, NIST) we identified the hierarchical fractal network of SWNTs dispersed in polymer matrices. With increasing SWNT concentration, the network grows self-similarly, the mesh size (nanometer length scale) decreases weakly, the floc size (micron length scale) remain constant and the number of flocs increases linearly.[1] The network growth is primarily through the formation of new flocs and inter-floc interactions mediated by the polymer dominate the elastic strength of the system.

As a direct consequence of the self-similar fractal network, the linear flow properties display 'time-temperature-composition' superposition. This superposability can be extended for non-linear deformations when the non-linear properties are scaled by the local strain experienced by the elements of the network.[2] More interestingly, under steady shear these nanocomposites show network independent behavior. The absolute stress value is a function of the nanotube loading but the characteristic time scales related to the process are independent of it. For fully grown network in viscous polymer, cluster dynamics under external shear controls the non-linear behavior of the system.[3]

Polymer Nanocomposites: Advanced polymer electrolyte materials for Li ion batteries:

Inspired by the ability of lithium to intercalate between carbon nanotubes, we prepared a well dispersed (geometrical percolation at 0.09 vol % SWNTs loading with an effective aspect ratio ~ 650) lithium dodecyl sulfate (LDS) compatibilized SWNTs-PEO nanocomposite. These nanocomposites demonstrate significant change in the melting and crystallization behavior of the PEO along with a decrease in percent crystallinity.[4] Incorporation of nanotubes acts as solid confinements to the PEO chains which not only make the chains stiffer but also provide huge steric hindrance to the chain transport. This effect coupled with high crystallization activation energy barrier leads to dramatic slowing down in PEO growth kinetics and makes the system more amorphous.[5] The observed changes in the SWNTs based nanocomposites far exceed those observed for equivalent Li+ ion concentration mixture and indicate that the SWNTs and LDS synergistically reduce PEO crystallinity. This nanocomposite with aligned SWNTs (described next) has been proposed as an alternative polymer electrolyte for Li ion batteries.

Hierarchical Polymer Nanocomposites:

The influence of well dispersed and aligned single walled carbon nanotubes to template the growth of polymer crystals from the unit-cell to lamellae is examined towards developing strategies for producing hierarchical materials. For the case where nanotubes nucleate the polymer crystals, a "shish-kebab" structure is realized with the nanotubes and polymer crystals acting as the shish and kebab respectively and results in nanocomposites that are highly reinforced.[6] On the other hand, for the case where the nanotubes disturb the formation of polymer crystals, the oriented nanotubes, because of the short inter-tube distances even at low nanotube concentrations, cause a templating of the polymer crystals with the lamellar-normals oriented orthogonal to the nanotube axes.

The research program that I plan to develop will largely build on my postdoctoral and doctoral research and specifically, I would like to develop a research program in the following areas:

1. Dynamics of polymer chains in presence of soft and hard confinement environment, effect of size and shape

2. Conjugated polymer based advanced materials morphology for organic photovoltaics application

3. Influence of colloid addition on the structure and properties of micellar solutions

References:

1.Chatterjee, T.; Jackson, A.; Krishnamoorti, R., Hierarchical structure of carbon nanotube networks. Journal of the American Chemical Society 2008, 130, (22), 6934-+.

2.Chatterjee, T.; Krishnamoorti, R., Dynamic consequences of the fractal network of nanotube-poly(ethylene oxide) nanocomposites. Physical Review E 2007, 75, (5).

3.Chatterjee, T.; Krishnamoorti, R., Steady shear response of carbon nanotube networks dispersed in poly(ethylene oxide). Macromolecules 2008, 41, (14), 5333-5338.

4.Chatterjee, T.; Yurekli, K.; Hadjiev, V. G.; Krishnamoorti, R., Single-walled carbon nanotube dispersions in poly(ethylene oxide). Advanced Functional Materials 2005, 15, (11), 1832-1838.

5.Chatterjee, T.; Krishnamoorti, R., Isothermal crystallization behavior of poly(ethylene oxide) in presence of carbon nanotubes. (manuscript under preparaion).

6.Chatterjee, T.; Mitchell, C. A.; Hadjiev, V. G.; Krishnamoorti, R., Hierarchical polymer-nanotube composites. Advanced Materials 2007, 19, (22), 3850-3853.

See more of this Session: Poster Session: Meet the Faculty Candidate
See more of this Group/Topical: Education

A Systematic Property Based Approach for Process and Molecular Design

Nishanth Chemmangattuvalappil, Department of Chemical Engineering, Auburn University, Auburn University, AL

Mario R. Eden, Department of Chemical Engineering, Auburn University, Auburn University, AL

The selection of product/product mixtures that give the optimum performance of a process is a critical issue for a design engineer. The process performance is usually understood in terms of physical properties and on many occasions, the physical properties of the product rather than their chemical structure determine the suitability of a specific product as the input to the process. For example, in the design of a blanket wash solvent, the primary focus of designer are the solubility parameter, flammability, vapor pressure etc. of the solvent. Molecular design algorithms generally require target properties to design the molecules. Therefore, for obtaining the optimal solution for this type of problem, it is necessary to have a methodology to represent the product performance in terms of measurable physical properties and identify the molecule/mixture that gives the property targets corresponding to the optimum process performance.

In spite of the relationship between the process design and product design problems, they have been traditionally considered as two separate problems because the product design part is generally considered as outside the scope of chemical engineering design. Therefore, the product identified by chemists may not provide the best one corresponding to the optimum process performance and this makes the process of identifying the suitable molecule/mixture an iterative process. Therefore, on most occasions, the chemists try to design the product based more on expert knowledge, trial and error, heuristics and experimentation based on intuition rather than any specific scientific reasoning. The attempts to develop a procedure to pass the flow of information between the process and product designer brought about the development of reverse problem formulations

In my PhD work, systematic algorithms have been developed for the design of molecules corresponding to the optimum performance of a process. The concept of property clustering has been extended to the formation of molecular clusters based on second and third order group contribution methods. An algebraic approach has been developed utilizing higher order groups built from first order groups subject to the constraints of overlapping. The advantage of using an algebraic approach is that it can handle any number of molecular groups or properties and can generate all possible compounds within the required range of properties. The most significant aspect of the aforementioned method is that both the application range and reliability of the molecular property clustering technique are considerably increased by incorporating second and third order estimation. A methodology has been developed for incorporating the property contribution predicted using combined group contribution and connectivity indices into the design framework in case the property contributions of any of the molecular groups of interest are not available in literature. For the design of simple monofunctional molecules, a modified visual approach has been used where as for the design of more complicated structures and/or for treating more than three properties at a time, an algebraic method has been used.

The developed algorithm can be used as a tool for the solution of integrated process and product design problems. Since an algebraic approach for the solution of a process design problem is already available, it is now possible to solve the integrated problem irrespective of the number of properties involved. The process design problem can be solved in terms of constitutive variables without committing to any component beforehand. The property targets estimated in the process design step along with the molecular groups will form the input to the molecular design algorithm. The algebraic equations formed using the molecular operators are solved simultaneously to generate the possible molecular structures that match the optimum process performance defined by the property targets.

However, apart from the targets set by the process design, the molecule has to satisfy a number of other environmental and safety constraints in order to be used in an actual process. In addition, there are many properties which can not be estimated through group contribution methods because it is not always possible to find a correlation between the molecular groups and properties. Similarly, not all possible atomic arrangements are represented in group contribution methods. So, there is a need of an efficient methodology for the design of molecules with a more diverse property targets. In my thesis, I have developed an algorithm that uses the concept of molecular signature descriptors for molecular design. The signature is a systematic coding system of atom types and the signature of a molecule can be obtained as a linear combination of its atomic signatures. It has been proved that any of the topological indices of molecules can be represented in terms of molecular signatures and it is possible to correlate the topological indices to the actual properties and biological activities. Here, the new algorithm utilizes molecular property operators based on signatures for solving the reverse problem of obtaining the molecular structures that satisfy the property targets estimated in the process design step. A new set of equations will be employed to ensure that the molecule meets the safety and environmental constraints as well. The principles in graph theory are incorporated to avoid the generation of infeasible structures. Since the molecular operators are formed based on molecular signatures, the property models based on different topological indices can be represented on the same property platform. Since different properties are described using different topological indices, the height of molecular signature required to describe those topological indices will be different. However, techniques have been developed to describe all topological indices with a single signature height. The final solution will be obtained as a set of the biggest signature height used to describe the topological indices. Therefore, the accuracy of prediction will not be compromised because off this transformation. The property models required to describe the target properties may be based on QSPR/QSAR (quantitative structure-property/activity relationship) models or in the form of group contribution methods. In the developed algorithm, the group contribution models can also be represented using the signatures which allow for the simultaneous consideration of all property targets irrespective of whether the targets are based on process design or other constraints. The accuracy of this method depends only on two factors- how well the actual property-topological index relationships are estimated and the height of atomic signatures used to describe the topological indices. Since, many topological indices can be used to describe each property, this algorithm generally provide reliable results.

The aforementioned algorithms provide excellent tools for the design of molecules corresponding to the process requirements. The current research in this area is to extend these algorithms into the design of process flow sheets. The recently developed computer aided flowsheet design techniques will be utilized along with the molecular design techniques to obtain the required process performance.

See more of this Session: Poster Session: Meet the Faculty Candidate
See more of this Group/Topical: Education

Global Optimization in Systems Biology

YoungJung Chang, Chemical Engineering, Carnegie Mellon University, Pittsburgh, PA

Nikolaos V. Sahinidis, Chemical Engineering, Carnegie Mellon University, Pittsburgh, PA

Patrick F. Suthers, Chemical Engineering, The Pennsylvania State University, University Park, PA

Costas D. Maranas, Chemical Engineering, The Pennsylvania State University, University Park, PA

During the last decade, there have been major advances in technologies for acquiring biological data and engineering desired features into organisms. Concurrently, the gap between data collection and implementation motivated many mathematical and computational approaches in analyzing biological data, inferring useful hypotheses, and systematically (re-)designing biochemical networks. Among numerous such mathematical and computational frameworks, optimization approaches based on mathematical programming can provide the most comprehensive coverage in solving biology motivated problems. This optimization approach became feasible initially due to the development of efficient local optimization algorithms, and can become an even more important approach now that rigorous global optimization techniques have been developed. This presentation reviews selected problems related to systems biology that are addressed by the authors using global optimization.

Metabolic networks are often redesigned for specific purposes such as productivity of microorganisms through steady-state optimization formulations. One of the major considerations in designing a dynamic system is stability. The stability is determined through the computation of the eigenvalues of Jacobian matrix. While it is straightforward to analyze the stability of a given system, the challenge is to redesign a metabolic network in a way that guarantees that the system will be stable around the new steady state. For this purpose, we propose to model metabolic networks through classical optimization formulations with an additional constraint to enforce stability within a prespecified neighborhood of the solution point. The proposed formulation is a bilevel optimization problem that is very difficult to solve. We develop suitable global optimization algorithms to solve this problem after transforming it to a semi-infinite optimization problem. This approach is demonstrated with a tryptophan biosynthesis problem in bacteria.

Metabolic fluxes are the key descriptor of cell's physiology, and metabolic flux analysis infers intracellular fluxes from isotopic experiments. A key question in this analysis is if the flux distribution can be uniquely determined from the measurements. While this identifiability question can be analyzed by some established methods, the challenge is to design the isotopic experiments that is capable of yielding a unique flux distribution. We propose to use incidence structure analysis to enforce the flux identifiability while designing isotopic experiments. This results in a large scale integer linear programming problem, and we exlpoit its structure to efficiently solve it. The proposed methodology was applied to both stationary and isotopically dynamic MFA experiments for E. coli strain capable of producing 1,3 propanediol to suggest critical measurements that are essential for the flux identifiability.

See more of this Session: Poster Session: Meet the Faculty Candidate
See more of this Group/Topical: Education

Elucidating Protein Binding and Aggregation

Naresh Chennamsetty, Chemical Engineering, Massachusetts Institute of Technology, Cambridge, MA

My overall research goal is to understand the mechanisms behind protein binding and aggregation using multi-scale modeling and simulation techniques in close collaboration with experiments. Protein binding is vital to its function in a variety of biological processes such as cellular communication, metabolism, gene expression, and immune responses. On the other hand, aggregation disrupts protein function and is associated with a number of neurodegenerative disorders such as Alzheimer's and Creutzfeldt-Jakob diseases, and type II diabetes. Furthermore, aggregation of proteins in therapeutic formulations leads to loss of their activity and raises concerns of potential immunogenicity. My research is aimed at understanding the mechanisms behind these protein binding and aggregation processes, and to design proteins with improved binding and stability.

I. Mechanisms for Binding and Aggregation of Therapeutic Proteins

In my current post-doctoral position with Prof. Bernhardt L. Trout at the Massachusetts Institute of Technology, we study the mechanism behind binding and aggregation of therapeutic antibodies and we design antibodies with enhanced stability.[1,2] Antibodies constitute the most rapidly growing class of human therapeutics for use in the treatment of numerous cancers, chronic inflammatory diseases, and infectious diseases. These antibodies are, however, thermodynamically unstable under conditions required for storage and administration and degrade due to aggregation. We developed a molecular simulation tool called spatial-aggregation-propensity (SAP) in collaboration with experimental techniques to identify the regions of the antibody most prone to aggregation. Subsequently, we perform mutations in these specific aggregation prone regions to engineer antibodies with enhanced stability. We also developed methods for predicting protein-binding regions with good accuracy.

II. Development of Multi-scale Modeling and Simulation Techniques for Soft Condensed Matter Systems

During my PhD work with Prof. Keith E Gubbins at North Carolina State University, we developed a multi-scale methodology connecting the atomistic and meso-scale models useful in understanding surfactant self-assembly, polymers, colloids and protein solutions.[3] We also developed a novel coarse-graining technique to determine rigorous effective potentials.[4] In addition, we used meso-scale models to explore the effect of cosolvents and cosurfactants such as alcohols on surfactant self-assembly in supercritical carbon dioxide with applications in separation processes in the food industry such as extraction of water-soluble vitamins and proteins, dry cleaning and polymerization.[5]

References:

1. N. Chennamsetty, V. Voynov, V. Kayser, B. Helk and B. L. Trout, "Design of therapeutic proteins with enhanced stability", **Proc. Natl. Acad. Sci. U.S.A.,** 106, 11937 (2009). This also appears as research highlight, "Unstuck by design", in **Nature**, 460, 155 (2009)

2. N. Chennamsetty, B. Helk, V. Voynov, V. Kayser, and B. L. Trout, "Aggregation Prone Motifs in Human Immunoglobulin G", **J. Mol. Biol**., 391, 404 (2009)

3. J. R. Silbermann, S. H. L. Klapp, M. Schoen, N. Chennamsetty, H. Bock and K. E. Gubbins, "Mesoscale modeling of complex binary fluid mixtures: Towards an atomistic foundation of effective potentials", **J. Chem. Phys**., 124, 074105 (2006)

4. N. Chennamsetty, H. Bock, K. E. Gubbins, "Coarse-grained potentials from Widom's particle insertion method", Special issue in honor of Ben Widom, **Mol. Phys.**, 103, 3185 (2005)

5. N. Chennamsetty, H. Bock, L. F. Scanu, F. R. Siperstein, and K. E. Gubbins, "Cosurfactant and cosolvent effects on surfactant self-assembly in supercritical carbon dioxide" **J. Chem. Phys.**, 122, 94710 (2005)

See more of this Session: Poster Session: Meet the Faculty Candidate
See more of this Group/Topical: Education

Structural Requirements On Catalyst Surfaces During Chemical Turnovers: Hydrogen Production and C1 Conversion Chemistry

Ya-Huei (Cathy) Chin, Department of Chemical Engineering, University of California, Berkeley, Berkeley, CA

Enrique Iglesia, Department of Chemical Engineering, University of California, Berkeley, Berkeley, CA

Chemical conversions of natural gas (primarily methane) to transportation fuel and high energy density chemicals are promising technologies important for harvesting stranded gas at remote reservoirs. Strong carbon-hydrogen bond in methane relative to other hydrocarbons renders the direct conversion pathway unselective. The indirect methane conversion pathway, for this reason, remains to date as the commercially viable gas-to-liquid route. The economic feasibility of this route is determined by the efficiency of the initial methane conversion step to synthesis gas as the molecular building blocks for downstream chemical synthesis. Additionally, methane conversion is important for applications in natural gas catalytic combustor and for on-board hydrogen production at distributed sites to deliver hydrogen as a clean energy carrier for next generation power systems. These methane reactions (reformings, catalytic partial oxidation, and combustion) required high temperatures (> 873 K); they are energetically intensive and in the case of reformings and apparent partial oxidation, constrained by thermodynamics. These technology challenges continue to spark interests in understanding the fundamental reaction dynamics and the structural requirement of catalytic surface during CH_4 chemical turnovers.

Kinetics for methane steam and dry reformings have been previously established. The mechanistic interpretation of the reaction pathways and selectivity during CH_4-O_2 reaction, however, remain contradictory despite significant research efforts. The disagreement is in part related to the highly exothermic and endothermic nature of combustion and reforming reactions, which occurred at reaction zones within the length scales of heat conduction. Rate and selectivity data, as a consequence, are corrupted by the inherent temperature and concentration gradients across the catalyst pellets and reactor bed, which lead in turn to disagreements on the microscopic reaction pathways and on the kinetic functions of active sites during CH_4 reactions. The mechanistic interpretation of rate and selectivity data are made even more complex as the identity of the most abundant surface intermediates and the thermodynamic phase stability of metal or oxide vary along the reactors in response to changing local reactant pressures and temperatures. Understanding the surface kinetics is nevertheless one of the requirements to predict the overall reactor performance. Our approach is to carry out kinetic measurements under strict kinetic control at relevant operating temperatures on a series of supported Group VIII metal cluster catalysts. This, combined with quantum chemical calculations of transition states, reaction energies and barriers, allows us to derive a comprehensive description of reaction paths at the molecular scale level, to elucidate the site-requirements, periodic trends, and cluster size effects of CH_4 reactions on metal clusters.

We propose here, based on rigorous kinetics and isotopic studies, a simple yet complete sequence of elementary reaction steps to describe the CH_4 combustion and the apparent partial oxidation on metal clusters. CO is not detected before the complete O_2 consumption, which indicates that the direct partial oxidation does not occurred at the molecular scale. CO and H_2, by inference, formed via sequential reforming of CH_4 with combustion products (H_2O and CO_2). Four distinct kinetic regimes, each with unique reactant pressures dependences and kinetic isotope effects, were detected depending on the coverages and reactivity of the chemisorbed oxygen. Kinetic analyses show that the seemingly complicated response in CH_4 conversion rates with reactant pressures is a single-valued function of the virtual oxygen pressure $(O_2)_v$, a surface oxygen chemical potential term that is a function of gas phase composition and the elementary rate constants. The (O) determines the relative concentrations of oxygen-oxygen, oxygen-

2 v

oxygen vacancy, and vacancy-vacancy site-pairs available on metal surfaces for the kinetically-relevant C-H dissociation steps; it is therefore an appropriate descriptor which allows us to predict both the catalytic activity and also the thermodynamically stable state of the metal clusters during reactions. Density functional theoretical calculations are used to provide molecular level details on the catalytic functions of chemisorbed oxygen and oxygen-vacancy during the kinetically-relevant C-H bond activation steps. C-H bond dissociation proceeds on oxygen-vacancy sites via a metal-atom insertion transition state similar to that on metal surfaces. In contrast, the methyl group on surfaces saturated with oxygens acquired free-radical like character in the transition state which indicates a homolytic C-H bond cleavage route. Detailed analyses show that the elementary C-H and O=O bond activation barriers for methane and oxygen are functions of the oxygen binding energy on metal following the prevalent BrØnsted-Evans-Polanyi relation. Combining experimental and theoretical investigations allow us to correlate the CH_4 oxidation activity to oxygen binding strength on metal clusters and derive the periodic trend to describe the reactivity Group VIII metal and their cluster size effects during CH_4 conversion reactions.

CH_4-O_2 reactions on Group VIII metal clusters are described by an identical catalytic cycle consisted of a closed sequence of elementary steps. The chemical turnover rates, selectivity, and structure sensitivity vary with the surface oxygen chemical potential because it dictates the identities of the kinetically-relevant step and of the most abundant surface intermediates. The chemistry is of practical significance and can be extend to describe the C-H bond activation on metal clusters involving chemisorbed oxygen species.

See more of this Session: Poster Session: Meet the Faculty Candidate
See more of this Group/Topical: Education

Fundamental Understanding of Zeolites towards Engineering of Zeolite Membranes

Jungkyu Choi, Chemical Engineering, University of California, Berkeley, Berkeley, CA

Separation processes currently consume 15 % of global energy (60 vs. 400 1015 Btu/Year). With the global commodity production expected to increase six-fold by 2040, a business as usual scenario is not sustainable and an order of magnitude increase in separation efficiency is a necessary step toward sustainable global prosperity. Active nanoporous (10-9 ~10-7 m pore size) adsorbents (e.g., zeolites) and membranes enabling separations with molecular resolution to replace thermally driven processes like distillation and crystallization meet this goal and are re-emerging as an area of nanotechnology research. These highly selective zeolite membranes could not only efficiently purify products but also be integrated with chemical reactors to increase the yield of the desired product eliminating or reducing the need for energy intensive separation steps. Zeolites are crystalline inorganic oxides built up by connected TO4 (T= Silicon or Aluminum) tetrahedra resulting in unique pore sizes and structures. Zeolites are used in a number of applications such as detergents, ionic exchangers for water softening and purification, and catalysts in the specialty chemical and petrochemical industry. A breakthrough technology for energy efficiency in the field of separations is that of molecular sieve or zeolite membrane technology.

Through my doctoral (Dept. of Chemical Engineering & Materials Science, University of Minnesota with Professor Michael Tsapatsis) and postdoctoral (Dept. of Chemical Engineering, University of California, Berkeley with Professor Enrique Iglesia) work, we have focused on resolving several issues in the field of zeolite technology including: (1) controlling morphology of zeolite particles via an appropriate choice of structure directing agents, (2) developing a methodology for oriented molecular-sieve zeolite and composite films, (3) engineering defects present on most polycrystalline films, and (4) characterizing transport inside zeolite crystals using frequency modulation flow protocols.

1. Methodology development for oriented zeolite film formation (Ph.D. work with Prof. Tsapatsis).

Zeolite membranes are physical barriers that can separate molecules based on even minute differences in molecular size and/or shape. As a result, much research has been focused on the field of zeolite membranes with respect to control of pore orientation and film thickness on diverse supports. However, the use of zeolite membranes in industry has been limited largely because the reproduction of defect-free zeolite membranes on a large scale is difficult. In the first efforts to make zeolite films on porous supports, many researchers followed the so called in-situ method in which the supports were placed in direct contact with the growth solution. Although researchers contrived to find optimal synthesis conditions by exhaustive investigation of all possible parameter combinations, the success was limited because it was not easy to control both nucleation and crystal growth at the same time.

Therefore, nucleation and crystal growth need to be decoupled for practical controllability. One possible way is the so called secondary growth method.[1, 2] That is, a seed layer is formed on supports and layered crystals are inter-grown by closing the gaps between crystals. The key for successful secondary growth for continuous oriented films is to promote in-plane (parallel to the support) growth of the deposits, while minimizing out-of-plane (perpendicular to the support) growth and avoiding new nucleation of grains. We found that the secondary growth conditions need to be optimized to induce comparable in-plane and out-of-plane growth rates.[1] This requirement is satisfied mainly by an appropriate choice of SDAs. We have shown conclusively that the SDAs can play a dual role in the synthesis of membranes: they template the desirable crystal structure and they allow for systematic control of growth rates along different crystallographic directions.[1]

Along with secondary growth, we also demonstrated that composite films consisting of zeolite particles and a matrix between them (MCM-22/silica films [3] and MFI/PTMSP films [4]) showed promise in gas and vapor separation applications. In particular, no requirement of hydrothermal growth, known as a time- and energy-consuming step, during film fabrication in such methods made these membranes attracted for practical use at the large scale.

2. Engineering grain boundary defects on c-oriented MFI Membranes (Ph.D. work with Prof. Tsapatsis).

The overarching goal of my research was to develop a hierarchical manufacturing process for the production of high performance zeolite membranes. At the more applied level, the goal was to develop scalable processing strategies that allow the fabrication of high performance zeolite membranes on practical industrial supports like cylindrical porous stainless steel tubes.

Defects frequently present between and/or on polycrystalline zeolite films could not be well controlled by any means up to now, though poor separation performance was already attributed to non-selective pathways provided by these defects. Rather, they were concomitantly generated during membrane synthesis and/or calcinations steps. Rapid thermal processing (RTP) was introduced for the first time to calcine c-oriented MFI membranes, because any efforts to reduce defect formation based on a slow heating process did not work. This RTP resulted in the dramatic improvement of separation performance for both aromatic compounds and linear/branched alkanes.[5] The remarkable improvement was attributed to the lowered density of grain boundary defects on RTP-treated c-oriented MFI membranes, as envisaged by Fluorescence Confocal Optical Microscopy at the molecular level. Despite the virtually identical microstructures (e.g., out-of-plane orientation, film thickness, grain size, Si/Al ratio, etc.) to conventionally calcined c-oriented MFI membranes, reduced defects could lead to approaching towards the expected high separation performance. The easy, but robust fabrication protocol for c-oriented MFI films makes RTP-treated ones feasibly adopted as alternative or supplemental to industrially important, but energy-intensive separation system.

3. Characterization of transport inside zeolites by frequency modulation flow protocols (Postdoctoral work with Prof. Iglesia).

Through post-doctoral research, my knowledge of separations during Ph.D. course is being complemented with experiences in catalysis. The effective integration of separation and reaction systems has the potential to be of great industrial value. For instance, the combination of reaction and separation can lead to the realization of the membrane reactor concept which could save even more energy than a separation membrane alone.

The current project is aimed to perform the measurement of diffusion rates and isotopic exchange rates for probing the role of intra-zeolite channels on the traffic and reactive encounters of hydrogen-containing molecules. Prior to an intensive work on this ultimate goal, initially we are attempting to prove the concept of the frequency response technique by employing a diffusion system without a reaction being involved. As an interesting example, the separation of $CO2$ and $CH4$ during natural gas extraction attracts researchers because of its technical and economical effectiveness. By doing so, one can eliminate or at least reduce the need of additional $CO2$ gas separation resulting in energy saving. Concomitantly, this approach will help to lower $CO2$ concentration in the atmosphere as global climate change becomes intensively concerning to many. We believe that the frequency response technique has a strong potential to contribute to understanding the separation of $CO2$ and $CH4$ in powder-based approaches. More specifically, we could characterize the diffusion coefficients of these and find optimal operating conditions for maximizing separation performance. Now, we are measuring the diffusion coefficients of $CH4$ and $CO2$ single gas components and the corresponding mixture on MFI type zeolites using the frequency response technique. This approach will be easily extended to the examination of other types of zeolites and direct a guideline to seek out appropriate materials for this specific $CO2/CH4$ separation. Given that a model is well established that reflects this system, one could take advantage of this technique to characterize the properties of diverse materials systematically and further select proper ones for a specific purpose.

References

1. J. Choi, S. Ghosh, Z. Lai, and M. Tsapatsis, Angew. Chem., Int. Ed., 2006. 45(7): p. 1154-1158.

2. J. Choi, S. Ghosh, L. King, and M. Tsapatsis, Adsorption, 2006. 12(5-6): p. 339-360.

3. J. Choi, Z. Lai, S. Ghosh, D.E. Beving, Y. Yan, and M. Tsapatsis, Ind. Eng. Chem. Res., 2007. 46(22): p. 7096-7106.

4. M. Woo, J. Choi, and M. Tsapatsis, Micropor. Mesopor. Mater., 2008. 110(2-3): p. 330-338.

5. J. Choi, H.-K. Jeong, M.A. Snyder, J.A. Stoeger, R.I. Masel, and M. Tsapatsis, submitted.

See more of this Session: Poster Session: Meet the Faculty Candidate
See more of this Group/Topical: Education

Engineering Environmental Application of Adsorption and Membrane-Based Separations Toward a Green Energy Economy

Sunho Choi, School of Chemical & Biomolecular Engineering, Georgia Institute of Technology, Atlanta, GA

Despite extensive research on alternative or renewable energy sources such as wind, solar geothermal and biomass-derived energy, it is expected that fossil fuel use, typically the most economic energy option, will continue to supply the majority of energy for the next several decades, as these emerging alternative technologies mature and are commercialized. With a rising population and energy demand and a continued reliance on fossil energy, atmospheric CO2 levels will continue to rise, potentially contributing to continual climate change. Thus, these two seemingly contradicting societal demands of cheap energy and minimal environmental impact present a striking scientific and engineering challenge for our time. Clearly, it is critical to develop efficient means of using fossil energy.

A promising practical approach for this is the introduction of separation and purification techniques that minimize detrimental environmental effects by reducing emission of anthropogenic contaminants, ideally via in both new facilities and via retrofitting into current industries. In the light of this, this poster gives an overview of my recent studies on this theme, including promising routes to manage anthropogenic CO2 associated with the combustion of fossil fuels. It includes investigations of (i) membrane-based separations targeting hydrogen separation from pre-combustion gas streams at high temperature operating conditions, (ii) adsorptive removal of anthropogenic CO2 from flue gases using rationally designed solid sorbents, and (iii) selective capture of atmospheric CO2 from ambient air. Inorganic-organic hybrid materials, such as nanoporous layered materials swollen by primary amines and hyperbranched aminosilicas, are proposed as novel materials that can facilitate these objectives. Experimental results assessed in this study suggest that these strategies will be among the effective pathways facilitating environmentally-benign application of fossil fuels as a green energy source.

See more of this Session: Poster Session: Meet the Faculty Candidate
See more of this Group/Topical: Education

Biophysical Engineering Strategies for the Development of Protein-Based Solar Convertors

Juan C. Cruz, Chemical Engineering, Kansas State University, Manhattan, KS

Peter H. Pfromm, Chemical Engineering, Kansas State University, Manhattan, KS

Mary Rezac, Chemical Engineering, Kansas State University, Manhattan, KS

We have engineered a novel nano-platform for the cost-effective production of enzyme based biocatalysts for applications in organic solvents. We incorporated an inexpensive nanostructured material as support in a two-step procedure only making use of simple electrostatic interactions. This will potentially help to expand the application of enzymes to a variety of processes of commercial interest including the production of commodities and pharmaceuticals. To fully understand the intricate surface-protein and protein-protein interactions as well as optimize our immobilization protocol, a number of state-of-the-art spectroscopic and nano-imaging tools were applied. The secondary and tertiary structural changes of the immobilized enzymes were monitored with Circular Dichroism, FTIR, and Fluorescence spectroscopy. Direct observation of the physical state of the enzyme molecules was achieved with AFM. Moreover, the mobility and adsorptive properties at the nanoscale were studied with Fluorescence Recovery After Photobleaching (FRAP) and Confocal Microscopy. A robust multiparametric model will be generated to adjust the immobilization conditions to prepare highly efficient biocatalysts for customized applications.

As an independent researcher, I expect to apply this knowledge for the development of protein-based nanobiophotonic devices capable of efficiently converting solar energy. These nanobiophotonic devices capture solar energy through immobilized protein photosystems isolated from natural sources such as bacteria or plants. The energy is transferred by specialized reaction centers involving a concerted cascade of chromophores. Further electrochemical reactions can be coupled to the transfer cascade to generate electricity. The major limitation for the commercialization of this technology is the instability of the biological components. This instability results in low conversion efficiency when compared with photovoltaic devices. The natural photosystems are highly sensitive multi-protein systems. A rational design of prototypes must be focused in properly tuning the protein-surface interactions responsible for the final protein conformation and consequently energy transfer properties. The success of this technology will assure a greener energy supply for a world with an increasing energetic demand.

See more of this Session: Poster Session: Meet the Faculty Candidate
See more of this Group/Topical: Education

Electrochemical Principle-Enabled Nanostructure Synthesis for Catalysis, Hydrogen Generation, and Functional Nanodevices

Qingzhou Cui, Chemical Engineering, University of Massachusetts Lowell, Lowell, MA

Julie Chen, Mechanical Engineering, University of Massachusetts Lowell, Lowell, MA

Zhiyong Gu, Chemical Engineering, University of Massachusetts Lowell, Lowell, MA

Traditional electrochemistry-based synthesis, as a powerful tool, has shown great potential to guide many novel nanomaterial fabrications. In the present study, we demonstrate two such examples and their related applications. First, a new template material of Aluminum nanoparticles is applied to fabricate nanostructures through galvanic replacement reaction. Nanoshells and nanoporous structures from nickel, cobalt, copper, and precious metals are generated from such process. The underlining mechanism for the replacement reaction is believed to be a galvanic replacement reaction process. This means that a galvanic cell is formed within each individual nanoparticle, and electron flow within the metallic nanoparticles, oxidizing the inner Al material and reducing metal ions on the nanoparticle outer shell. These novel nanostructures have great potential in catalysis and fuel cell applications due to their large surface area. Second, porous template materials were used for electrodeposition in fabricating functional nanorods and nanowires. These nanorods/nanowires are promising building blocks to be used for nanoelectronic device fabrication through low melting point soldering materials. These nanosolder materials could form robust interconnects between nanocomponents and nanocomponents and also between nanocomponents and patterned substrates.

See more of this Session: Poster Session: Meet the Faculty Candidate
See more of this Group/Topical: Education

Modeling and Simulation of Soft Matter: Biomimetic and Energy Applications

Pratyush Dayal, Department of Chemical and Petroleum Engineering, University of Pittsburgh, Pittsburgh, PA

The need for miniaturization has compelled the current technology to have a paradigm shift from a macro-scale to a nano-scale. It has been known that the fibers of nanometer size diameters can be produced by electrospinning, a technique based on deformation of the surface induced by the amount of electrical charge on the fiber surface. Although fibers of diverse morphologies have been produced using electrospinning, the in-situ experimental monitoring of the micro-/nano-structure development in fibers is still not feasible due to the extremely fast nature of the process. Therefore, the theoretical modeling and simulation is of paramount importance in probing the spatio-temporal evolution of microstructure and form a basis to determine structure-property relationships, which in turn provides guidance to the electrospinning process. These relationships can then be utilized in tailoring micro-/nano fibers according to the desired application. My future research is aimed at modeling and simulation of electrospinning process by incorporating morphology development in polymer fibers and subsequently, extending the electrospinning process to electro-extrusion for producing tapes and films. The knowledge acquired by modeling and simulation of these processes will be applied to facilitate the manufacture of charge storage devices for energy applications and diverse micro-structured materials, that would have significant impact on industries of specialty polymers.

See more of this Session: Poster Session: Meet the Faculty Candidate
See more of this Group/Topical: Education

Novel Experimental Models for Cancer Research and Their Applications

Zoe N. Demou, Northwestern University, Chicago, IL

The tremendous challenges in understanding and treating cancer are addressed more effectively through multidisciplinary research, in which bioengineering can contribute significantly. This presentation focuses on hardware and software I have designed to model specific aspects of tumor biology. These systems emphasize three-dimensional substrates, long term cell culture under physiologic conditions, automation, and compatibility with time-lapse imaging and cellular and molecular analyses. Specifically, a fully automated system that tracks cellular motion in 3D with parallel statistical analysis of the cell trajectories enabled measurements of the migratory and invasive potential of tumor cells, and can be used as a screening tool for cancer motility and potentially therapeutic drugs. Another device, the cell-pressor, simulates the stress built-up in a growing tumor and allows quantification of the effects of mechanical compression on cancer cell morphology, metastatic potential and in vivo aggressiveness. Another chamber for time-lapse topography and laser capture microdissection in differentiating living 3D cultures, led to identifying via genomics and proteomics analyses, a set of molecules that could enable autopoiesis of vascular-like networks of tumor cells with potentially significant clinical implications, such as response to anti-angiogenic treatments. The systems used independently or in tandem, enable novel insights into the individual roles and interplay between tumor cell phenotypic plasticity and structural and mechanochemical components of the microenvironment in the process of cancer metastasis.

See more of this Session: Poster Session: Meet the Faculty Candidate
See more of this Group/Topical: Education

Electrochemistry and Self-Assembly of Complex Single-Walled Carbon Nanotube (SWNT) Nanostructures

Juan G. Duque, Department of Chemical and Biomolecular Engineering, Rice University, Houston, TX

Matteo Pasquali, Department of Chemical and Biomolecular Engineering, Rice University, Houston, TX

This work explores the optical and electrochemical properties of surfactant stabilized and individually-suspended SWNTs in aqueous media via application of various external stimuli. As a result, novel nanostructures are obtained, and this allows for the study of the inherent nanotube electronic and optical properties. The goal is to engineer SWNT systems which can be tuned by understanding the mechanism of the electrochemical and environmental reactions so that applications in nanophotonics, photovoltaics, and electronics can be effectively exploited.

A strategy to obtain a surfactant/polymer protective "shell" that improves the stability and luminescence signal of individual SWNTs is presented. We used literature evidence of fluorescence emission shifts to understand the interactions between polymers and surfactants and show how morphological changes induced by extrinsic factors distort the SWNT luminescence. We developed an in-situ polymerization which creates an outer shell around the SWNT micelle that resulted in suspensions with stable luminescence at all pH, in saline buffers, and on the surface of living cells.1

Nanoparticle-nanotube structures were also fabricated by exploiting the electrochemical properties of SWNTs upon activation with alternating electromagnetic fields (EM). Under EM, metallic SWNTs will preferentially polarize creating a localized apparent field enhancement at the tips due to their high aspect ratio and electronic properties -antenna behavior. This field enhancement can generate sufficient current densities which drive electron transfer reactions preferentially at their tips, reducing transition metal salts into nanoparticles. The selectivity is assessed using microscopic and spectroscopic techniques such as atomic force microscopy, UV-vis absorbance, and Raman spectroscopy. At the end we obtain a system with tunable reaction conditions where spontaneous particle formations are controlled and selective deposition is obtained to form novel nanostructures.2

References

1. Duque, J.; et, al; JACS, 2008, 130, (8), 2626

2. Duque, J.; et, al; JACS, 2008, 130, (46), 15340

See more of this Session: Poster Session: Meet the Faculty Candidate
See more of this Group/Topical: Education

Chemical Engineering and Engineering Education

Laura Dykes, University of Maryland, Baltimore County, Baltimore, MD

My career objective is to work in a chemical engineering department at college or university. I would like to teach traditional chemical engineering courses and perform research in engineering education with a focus on mentorship, engineering in the media, and engineering identity.

In my PhD research performed at Northwestern University, I used in situ x-ray techniques in conjunction with complementary rheological testing to probe links between nanocomposite microstructure and rheology in simple shear and complex flows.

I am currently working on post-doctoral research in Engineering Education at the University of Maryland-Baltimore County. My research investigates the student/advisor relationship and its impact on the career decisions of doctoral students.

See more of this Session: Poster Session: Meet the Faculty Candidate
See more of this Group/Topical: Education

Nanobiotechnology: Fundamental Physical Studies of Biological and Engineered Soft Matter for Applications in Nanomedicine

Prajnaparamita Dhar, Chemical Engineering, University of California, Santa Barbara, Goleta, CA

Joesph A. Zasadzinski, Department of Chemical Engineering, University of California, Santa Barbara, CA

Thomas Martin Fischer, Non-linear dynamics, Universitat Bayreuth, Bayreuth, Germany

Nanomedicine, an offshoot of nanotechnology, seeks to develop highly specific medical intervention at the molecular scale by developing smart nanomaterials and nanomachines that can autonomously detect, cure or replace damaged and diseased cells and tissues. A major challenge in nanomedicine is to develop a complete and thorough understanding of the physical and chemical properties of biological structure and design as well as developing smarter strategies for designing artificial autonomous nanomachines that can work synergistically within the human body to provide more efficient medical therapies. My ultimate research goal is to address this challenge by developing novel biophysical and biochemical tools to probe and manipulate Nature's intricate biological structures to understand and treat diseases. I am also interested in designing efficient engineered nano-carriers of drugs that will work effectively within the body for specific treatments of diseased cells and tissues.

See more of this Session: Poster Session: Meet the Faculty Candidate
See more of this Group/Topical: Education

Whey Protein Isolate Sol-Gel as a Novel Scaffold for Bone Tissue Regeneration

Mia Dvora, Cain Department of Chemical Engineering, Louisiana State University, Baton Rouge, LA

James E. Henry, Gordon A. & Mary Cain Department of Chemical Engineering, Louisiana State University, Baton Rouge, LA

Whey protein isolate (WPI) sol-gel is a promising base material for use as a bone tissue regeneration scaffold. The characterization of the material is presented here and serves to highlight the potential of such a scaffold. The successful scaffold measures within a window of suitable mechanical properties, has a high porosity and interconnectivity, and its surface promotes the growth and differentiation of osteoprogenitors into bone-forming osteoblasts. This set of characteristics can be achieved by using WPI as a dispersing phase in a two-phase composite. The dispersed phase enhances mechanical properties while increasing the growth rate of proliferating cells. Material composition and processing serve to tune the pore network structure of the scaffold, as determined by scanning electron microscopy and microcomputed tomography. In vitro testing has shown the ability of WPI to support the adhesion and proliferation of mouse preosteoblasts, and their differentiation on the scaffold surface. The merits of the material and the optimization process involved in its design are presented.

See more of this Session: Poster Session: Meet the Faculty Candidate
See more of this Group/Topical: Education

Modeling Nonlinear Dynamics in Biological Systems: Effects of Uncertainty

Joshua A. Enszer, Chemical and Biomolecular Engineering, University of Notre Dame, Notre Dame, IN

Engineering analysis and design problems frequently involve uncertain parameters and inputs, and propagating these uncertainties through a complex dynamic model can be a challenging problem. In this work, we demonstrate the use of Taylor model methods for propagating interval uncertainties through nonlinear ODE models. Intervals provide upper and lower bounds on uncertainties, but often more information is available, and this information may include measures of central tendency, measures of variation, or even a specific probability distribution. In such a case, we may use Taylor model methods combined with probability boxes (p-boxes) to propagate all provided information through the dynamic model. This allows us to obtain p-box representations of the uncertainties in the state variables and outputs of a nonlinear ODE model based on interval and p-box representations of initial states and parameters. Examples are used to demonstrate the potential of this approach for studying the effect of uncertainties with imprecise probability distributions.

See more of this Session: Poster Session: Meet the Faculty Candidate
See more of this Group/Topical: Education

Effective Low-Cost Numerical Problem Solving Software for ChE Education in Developing Countries

Michael B. Cutlip, Dept. of Chemical, Materials & Biomolecular Engineering, University of Connecticut, Storrs, CT 06269-3222

Engineering education increasingly involves problem solving that requires numerical solutions of complicated systems of equations and/or regressions/curve fitting of extensive data. Software required for these types of problems can be expensive for academic departments, and typically it can also be very expensive for individual graduates when they start their professional careers.

Many engineering programs in developing have limited resources to provide this type of software for their faculty and students, and the latest personal computers that enable this software problem solving are in limited supply.

A possible solution to this challenge in the advancement of global engineering education is to be enabled by philanthropic cooperation of a non-profit educational corporation, the CACHE Corporation[1] (Computer Aids for Chemical Engineering Education) and Polymath Software[2]. These two organizations have been cooperating to provide low-cost site licenses of POLYMATH to mainly chemical engineering departments since 1990, with currently some 120 site licenses in use by universities world wide with most institutions being located in the USA.

A new initiative that provides these site licenses, at no cost to interested engineering departments or programs in developing countries is presented in this paper. This initiative will enable interaction numerical problem solving on a wide range of personal computers with Microsoft[3] operating systems of Windows 2000[3] through Windows 7[3]. These site licenses will allow distribution of individual copies of the software to all students, faculty and staff. In addition, the software may be installed in engineering computer labs and laboratories. Initial distribution of this software is being focused in Africa a this time.

POLYMATH Software – A Brief Overview

POLYMATH enables the user to apply effective numerical analysis techniques during interactive problem solving on personal computers with Microsoft operating systems. This software is extremely easy to use as there is no command language to remember. Results are presented in tabular or graphical formats for easy understanding and for incorporation into papers and reports. Capabilities now include a color-coded full-screen editor and allow for solution of problems involving Linear Equations, Nonlinear Equations, Differential Equations, Data Analysis and various types of Regression. Capabilities of POLYMATH are summarized in Table 1. The installation file is relative small, about 17 Mb, and the installed code requires only 9 Mb.

POLYMATH Software – A Very Simple Example

Let's consider a simple example of a model of a well-mixed batch reactor that is described by a system of ordinary differential equations. The batch reactor initially has only reactant A. The reactions are first order and irreversible. This example will be solved with all three software packages.

$$A \xrightarrow{k_1} B \xrightarrow{k_2} C$$

Table 1 – POLYMATH Educational Version

- Linear Equations - up to 100 simultaneous equations.
- Nonlinear Equations - up to 30 simultaneous nonlinear and 40 explicit algebraic equations
- Differential Equations - up to 30 simultaneous ordinary differential and 40 explicit algebraic equations
- Data analysis and Regression - up to 300 data points with capabilities for linear, multiple linear, and nonlinear regressions
- Automatic Creation of Excel Spreadsheet Solution for Problem (include unique ODE_Solver Add-In for solving simultaneous first-order differential equations)
- Generation of Ordered MATLAB[4] Code for Problem

The differential mass balances on a batch reactor yield

$$\frac{dC_A}{dt} = -k_1 C_A \qquad \text{I. C. } C_A\big|_{t=0} = 1$$

$$\frac{dC_B}{dt} = k_1 C_A - k_2 C_B \qquad \text{I. C. } C_B\big|_{t=0} = 0$$

$$\frac{dC_C}{dt} = k_2 C_B \qquad \text{I. C. } C_C\big|_{t=0} = 0$$

$$k_1 = 2 \qquad k_2 = 3$$

POLYMATH Solution – A Brief Look

Entry of this sample problem is accomplished within the POLYMATH full screen editor as shown in Figure 1. Note that entry is facilitated for new users by use of templates for the input of differential equations or explicit algebraic equations.

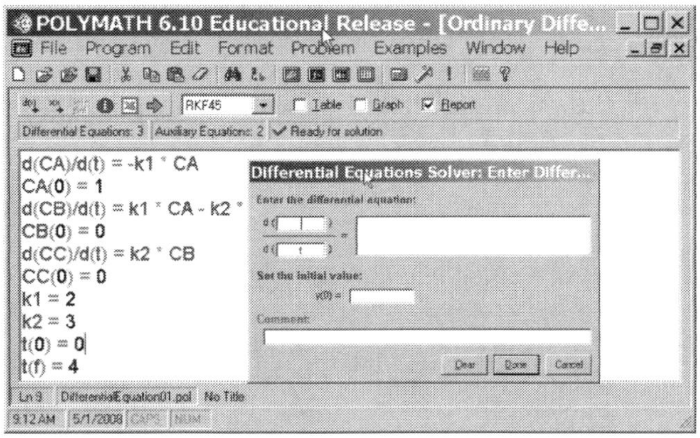

Figure 1 – Entry of Differential Equations into POLYMATH

Selection of the desired integration algorithm and the desired output followed by a mouse click on the highlighted arrow yields the numerical solution. The graph output option gives the expected result shown in Figure 2. Additional options for a Table of results and for a Report of the problem solution are not shown. The POLYMATH program can handle system of nonlinear ordinary differential equations with a variety of standard integration algorithms including those for stiff systems.

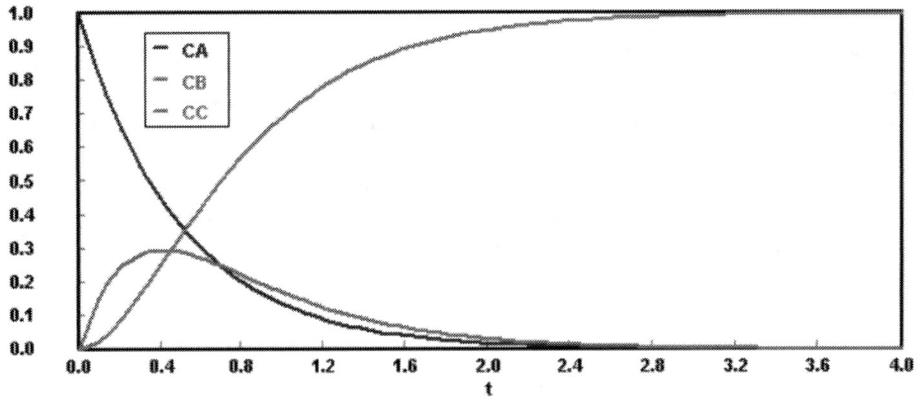

Figure 2 – Graph Output of the Differential Variables from POLYMATH

Exporting Problems from POLYMATH for Solution in Excel[2] – A Brief Look

Problems that can be solved completely in POLYMATH can also be automatically exported to Excel for solution. (A working spreadsheet is generated.) A POLYMATH ODE_Solver Add-In is included with POLYMATH for solving ordinary differential equations in Excel. Available Excel capabilities are used wherever they are available. For example, the provided Add-In ***solver*** is used to solve systems of simultaneous nonlinear equations in Excel.

The Excel solution for this sample problem is achieved by clicking the Excel icon from the POLYMATH editor. This then automatically exports the problem to an open Excel spreadsheet as shown in Figure 3. Note that all intrinsic functions and logical variables are automatically converted and placed in the Excel spreadsheet. The POLYMATH variable names and equations are shown for the convenience of the user. Since Excel does not have a differential equation solver, an Add-In, ODE_Solver.xla, is provided along with POLYMATH that will solve simultaneous ordinary differential equations. This additional software runs completely independently of POLYMATH.

The Excel solution is automatically generated in a new worksheet as shown in Figure 4. Note that a Table is given that summarizes the solution over the integration interval. Also output of the desired variables by tabulated column give the user options for use or plotting within Excel.

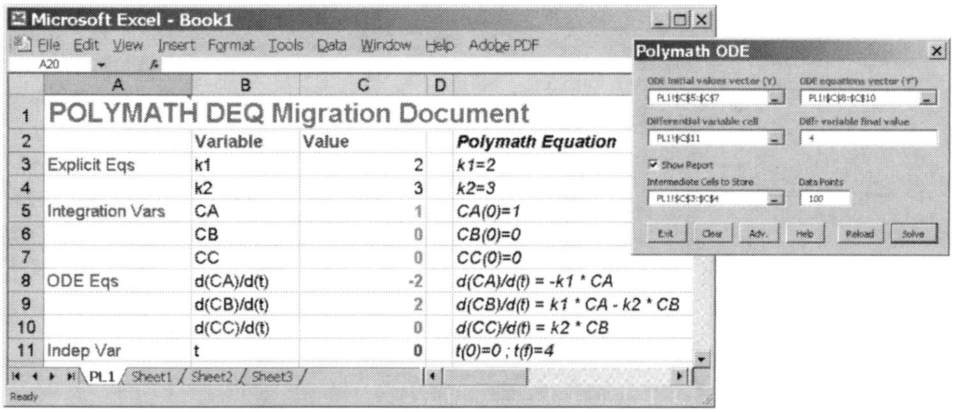

Figure 3 – Excel Spreadsheet with POLYMATH ODE_Solver Add-In

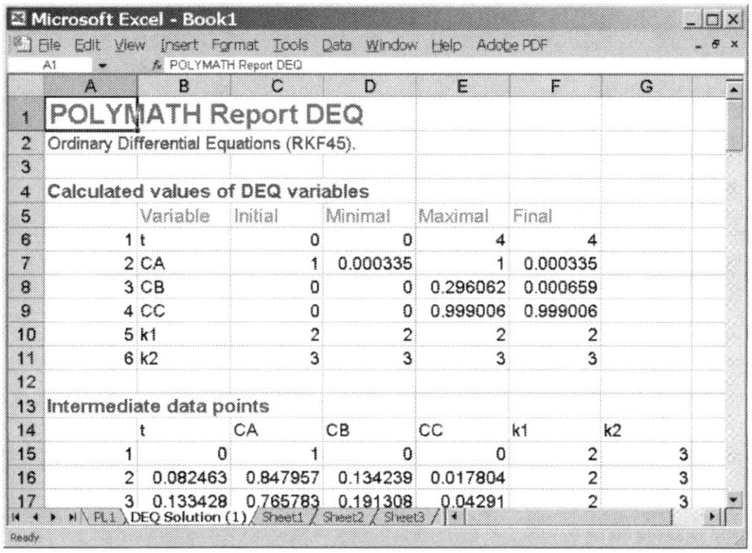

Figure 4 – Worksheet Automatically Generated During Sample Problem Solution

Exporting Problems from POLYMATH to MATLAB for Solution – A Brief Look

The MATLAB solution for this sample problem is obtained by setting an option within POLYMATH that will output the problem code for MATLAB to utilize. This option leads to code which contains the translation of the intrinsic functions and logical expressions from POLYMATH into MATLAB code automatically. The generated code for this sample problem is shown in Figure 5 as is provided at the end of the POLYMATH program report.

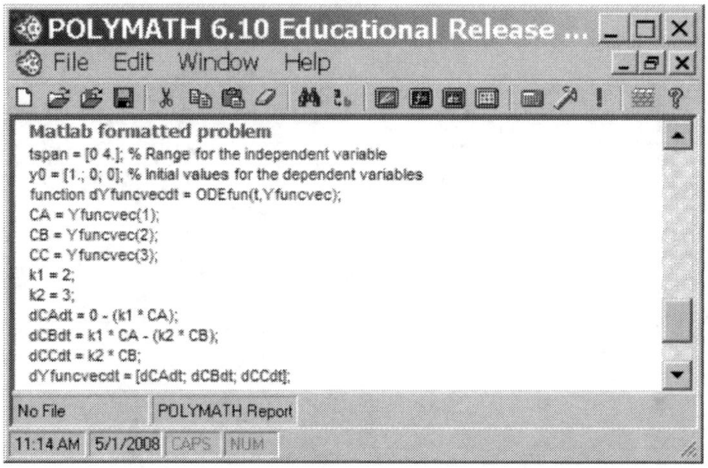

Figure 5 – POLYMATH Generated MATLAB Code

Incorporation of the generated code into a supplied MATLAB Template for simultaneous ordinary differential equation gives the m-file given in Figure 6 where the yellow bars indicate the placement of the code generated by POLYMATH.

```
function MATLAB01
clear, clc, format short g, format compact
tspan = [0 4.]; % Range for the independent variable
y0 = [1.; 0; 0]; % Initial values for the dependent variables
disp(' Variable values at the initial point ');
disp([' t      = ' num2str(tspan(1))]);
disp('            y                  dy/dt          ');
disp([y0 ODEfun(tspan(1),y0)]);
[t,y]=ode45(@ODEfun,tspan,y0);
for i=1:size(y,2)
    disp([' Solution for dependent variable y' int2str(i)]);
    disp(['             t                    y' int2str(i)]);
    disp([t y(:,i)]);
    plot(t,y(:,i));
    title([' Plot of dependent variable y' int2str(i)]);
    xlabel(' Independent variable (t)');
    ylabel([' Dependent variable y' int2str(i)]);
    pause
end
%- - - - - - - - - - - - - - - - - - - - - - -
function dYfuncvecdt = ODEfun(t,Yfuncvec);
CA = Yfuncvec(1);
CB = Yfuncvec(2);
CC = Yfuncvec(3);
k1 = 2;
k2 = 3;
dCAdt = 0 - (k1 * CA);
dCBdt = k1 * CA - (k2 * CB);
dCCdt = k2 * CB;
dYfuncvecdt = [dCAdt; dCBdt; dCCdt];
```

Figure 6 – Code for MATLAB m-file

The working m-file gives both tabular results (not shown) and graphical results (partly shown in Figure 7).

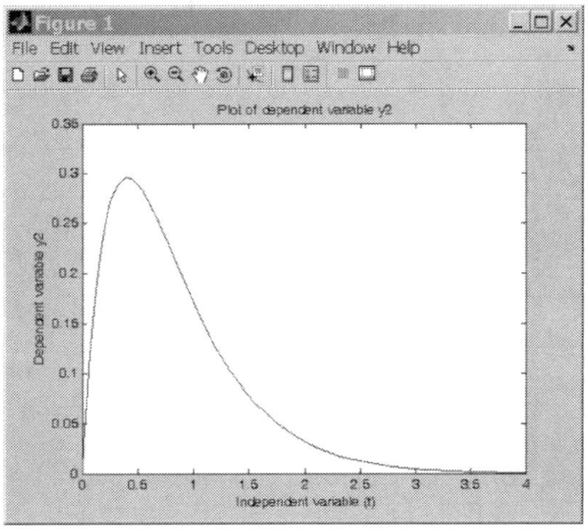

Figure 7 – Partial Graphical Results as Generated by MATLAB for Concentration of B

Integrating the Use of POLYMATH, Excel, and MATLAB

The POLYMATH software enables a programming and problem solving environment that also can utilize Excel and MATLAB as is shown in Figure 8. In this environment, the problem is first entered into POLYMATH and then can be automatically exported to Excel. From the initial POLYMATH program, the capability to generate MATLAB code enables the creation of a problem m-file.

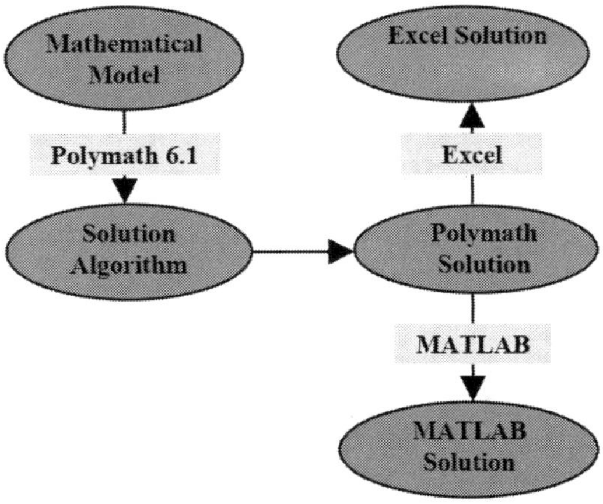

Figure 8 – Desktop Programming/Problem Solving Environment

This environment can be provided in a computer laboratory or to students and faculty for use on their own personal computers. The POLYMATH site license allows distribution of POLYMATH to all students, faculty, and staff plus use in all computer labs. Excel is widely available as a component of the Microsoft Office software, and this is usually available in computer labs and on individual computers. MATLAB provides inexpensive site licenses to most colleges and universities and also markets student versions.

Recent African Initiatives

A major connection with all of the engineering programs in Nigeria was made as a result of the presentation of a POLYMATH workshop at the 7th Annual ASEE Global Colloquium on Engineering Education[5] in Cape Town, South Africa in October of 2008. One of the attendees was Felix Atume who is the Registrar of Council for the Regulation of Engineering in Nigeria, COREN. This organization accredits all the academic engineering programs in Nigeria and also maintains required registration of all practicing engineers in the country. This connection to COREN in South Africa resulted in this author being invited to Nigeria in February of 2009 to present a six-hour workshop to most of the deans and the various department heads, not just chemical engineering, of the 28 engineering programs in the country (~ 110 individuals).

The workshops in Nigeria were supported by the preparation of a web site that provided all of the participants with access to the all of the workshop materials. This also included the installation files for POLYMATH software so that the software could be freely distributed to students, faculty and staff. Also the software can be installed on any computers in the college or university. The MathWorks, Inc. also generously provided a two-month license for MATLAB to the participants. In addition, a CD-ROM with all the workshop materials was given out to each of the faculty participants at the workshop to take back to their home departments.

In September of 2009, the website was updated and the new POLYMATH software installation file that will execute for another year was made available. The workshop materials were given in Acrobat pdf files that contain the actual program files that are used during the workshop. Thus faculty members are able to present workshops using these materials. There have been some preliminary discussions regarding some additional workshops for COREN to be held in Nigeria in February of 2010.

Future African Initiatives

Among other connection made at the ASEE Global Symposium was one with Prof. Funso Falade who is the current President of the African Engineering Education Association[6]. There may be a possible opportunity for a workshop and/or presentation at the next biannual meeting in September of 2010. It is hoped that this conference would be a good venue to promote the use of POLYMATH in developing countries in the African continent.

Assessment of Current Activities

There is strong encouragement from presentations, workshops and contacts that have been made at international meetings regarding the use of POLYMATH in many engineering disciplines within academic institutions. The necessary connections and interactions do require some international travel and the delivery of workshops and presentations within countries where POLYMATH and other related software can be used to enhance the problems solving capabilities of engineering students who have some even very limited access to personal computers. The use of Internet web sites as well as CD-ROMs and DVDs helps to disseminate the software and related educational materials. Hopefully this will contribute to the improvement of engineering education in developing countries. Much future work continues to be necessary in this area.

Conclusions

- POLYMATH Software provides personal computers with extensive interactive software that can be widely used for engineering and scientific problem solving.
- The automatic export of problems from POLYMATH to Excel enables the efficient use of this spreadsheet software in the solution of a variety of problems.
- The generation of ordered MATLAB code by POLYMATH enables this high-level and programmable software to solve many problems and enable extensions to higher level capabilities where programming is involved.
- POLYMATH software is well received by faculty in developing countries in various engineering disciplines in addition to chemical engineering.
- Future use of POLYMATH site licenses for developing countries will continue to be promoted by POLYMATH Software and the CACHE Corporation, particularly to departments in developing countries, including those with limited financial resources.
- Efforts to widen the use of POLYMATH and related software in engineering problem solving will continue to be made by workshops and presentations at appropriate meetings and by the preparation and dissemination of materials via the Internet and by CD-ROMs and DVDs.

References

1. CACHE Corporation, http://www.cache.org/. CACHE, which stands for Computer Aids for Chemical Engineering, is a not-for-profit organization whose purpose is to promote cooperation among universities, industry and government in the development and distribution of computer-related and/or technology-based educational aids for the chemical engineering profession. CACHE was initially sponsored by the Commission on Education of the National Academy of Engineering and funded by the National Science Foundation. The current Executive Officer of CACHE is Professor Thomas F. Edgar of the University of Texas at Austin.
2. Polymath Software, http://www.polymath-software.com/. The various POLYMATH programs allow the user to apply effective numerical analysis techniques during interactive problem solving on personal computers. The CACHE Corporation provides site licenses for POLYMATH to colleges and universities.
3. Microsoft Corporation, http://www.microsoft.com. Windows 95, Vista, Excel, and Microsoft Office are products and trademarks of the Microsoft Corporation.
4. MATLAB is a product and trademark of The MathWorks, Inc., http://www.mathworks.com.
5. 7th Annual ASEE Global Colloquium on Engineering Education, http://asee.org/conferences/international/2008/index.cfm.
6. African Engineering Education Association (AEEA), http://www.aeea.co.za/.

Sustainability of Global Humanitarian-Based Engineering Projects

Randy S. Lewis, Chemical Engineering, Brigham Young University, Provo, UT

In 2007, xxx University organized an Engineers Without Border (EWB) Student Chapter and a Global Projects in Engineering course. The course was designed to complement the EWB Chapter by allowing students to work on humanitarian-based engineering projects that offered the opportunity for engineers to work on global problems, researching not only technical but economical and socio-cultural issues. In the course, students have the opportunity to work on teams with engineers from different disciplines. Each student learns that out in the real world, boundaries between engineering disciplines fade and all engineers at the simplest levels are problem solvers, each contributing a unique viewpoint or technical background to solve the problem at hand.

The course has recently finished its third year. During the first year, the project involved the demonstration of converting coconuts to biodiesel in the Kingdom of Tonga. The second year involved water distribution, water storage, water heating, cooking, and biogas production in the village of Salkantay, Peru. This past year, the course involved projects in both Ghana and Peru. In Ghana, the projects involved electricity from windmills to charge lanterns for schoolwork, solar heating to purify water, and improved development of a biofil toilet. In addition to the windmill and solar heating, the Peru projects also involved water heating.

During all three years, various lessons were learned regarding design, implementation, and sustainability. Although the first two issues will be briefly discussed, this work will focus on the sustainability of the implemented projects. The discussion will include aspects associated with in-country contacts, project selection, local feedback mechanisms, testing of projects during implementation, and long-term use and maintenance. A summary of important sustainability principles will also be presented.

See more of this Session: Global Awareness and Ethics in Chemical Engineering
See more of this Group/Topical: Education

Global Cooperation in Laboratories: Armenia, Mexico, US

Jim Henry, Chemical Engineering, University of Tennessee at Chattanooga, Chattanooga, TN

Oleg Gasparyan, State Engineering University of Armenia, Yerevan, Armenia

Benito Serrano-Rosales, Zacatecas University, Zacatecas, Mexico

International collaboration on developing best practices in remote laboratory experimentation is described. UT-Chattanooga and U-Zacatecas have collaborated on the experimental work for a senior thesis on multi-tank water-level control. UT-Chattanooga and SEU-Armenia have collaborated on remote education in controls laboratory. This presentation will discuss the opportunities that remote, Internet experimentation offers and will report on effective techniques in regular class/laboratory courses and in special student research projects.

See more of this Session: Global Awareness and Ethics in Chemical Engineering
See more of this Group/Topical: Education

Project-Based Learning and Education Abroad: A Poorly Exploited Synergy?

David DiBiasio, Chemical Engineering Department, Worcester Polytechnic Institute, Worcester, MA

Terri Camesano, ChE, Worcester Polytechnic Institute, Worcester, MA

Susan Zhou, ChE, Worcester Polytechnic Institute, Worcester, MA

For several years, undergraduate chemical engineering students from WPI have conducted biochemical engineering research in the labs at l'Ecole Nationale Supérieure des Industries Chimiques in Nancy, France. They worked for some of Europe's foremost researchers in life science and bioengineering. Projects included kinetics of drug release from alginate-protein matrices, polymerization in micro emulsion reactors, and pollutant tracking in the Meurthe and Moselle rivers. Students returning from this program frequently win local academic competitions and become co-authors on papers published in research journals.

This past year, three chemical and environmental engineering students traveled to Shanghai and conducted research at Shanghai Jiao Tong University. They helped contribute to SJTU's long standing commitment to solving the daunting pollution issues in China's most prosperous and eclectic city. Their work examined reactor engineering problems in up-flow anaerobic sludge digesters, drinking water purification with ultrafiltration membranes, and ammonia removal from landfill leachate. While working fulltime with SJTU faculty and graduate students these students gained the well-known benefits of undergraduate research while at the same time learned about Chinese culture, struggled with the language, and negotiated life in suburban Shanghai.

In another part of Asia, a chemical engineering student working alongside WPI students from other majors investigated risk communication issues in the Rayong province of Thailand. They examined three public health issues-HIV/AIDS, motorcycle accidents, and petrochemical pollution. Their purpose was to understand how communication problems shape local understanding and possible solutions to these important public health issues. Rather than work in labs, they were in the local community interviewing government officials, public health organizations, and local residents. By synthesizing and analyzing a large amount of qualitative data from disparate sources, they could make informed recommendations to their Thai sponsors.

A continent away in South Africa two chemical engineering students and two biomedical engineering students studied flooding problems in an informal settlement outside of Cape Town. The settlements, remnants of apartheid, are vulnerable to increasing flooding incidents that result in serious health problems, affect livelihoods, and cause property damage. Using case studies, interviews, and field work the team helped the local community begin to understand the complex social, geographical, political, and economic forces contributing to the problem. Recommendations included specific means for building community capacity, more efficient flood risk data collection, and improved avenues of communication to minimize flood damage.

These examples represent a variety of experiences, contexts, and outcomes. They are not isolated anecdotes from a select few chemical engineers, fortunate enough to travel abroad, but they are the norm for several hundred WPI students who participate each year in our Global Perspectives Program (GPP). And, despite their variety they have elements in common---they are all project-based learning (PBL) experiences conducted in international locations. Although PBL has been part of WPI's curriculum for over 35 years, it is only in the last 10-12 years that the global program expanded to become the largest in engineering education. During that time we've learned a great deal about necessary elements of program design and implementation, and the synergy that comes from doing PBL abroad.

As described elsewhere: "The GPP instructional design is based upon situated learning theory that includes authentic activities, contexts, and assessments. It provides collaborative knowledge construction and opportunities for explicit articulation of knowledge during the learning process (Herrington & Oliver, 2000; Brown & Palinscar, 1989; Brown, et al, 1989). Authentic learning environments seek to place students in situations that mimic the way knowledge will be used in professional practice. Learners have access to both WPI and host country experts, and in some sense are engaged through a process of initiation much like the apprentice-learning model (Dewey, 1974). Collaborative activities provide multiple roles, and multiple opportunities to engage material (Lave & Wenger, 1991; Rogoff & Lave, 1984; Bruer, 1993). For students to become full members of a community of practice, it is essential that they have opportunities for legitimate participation in the practices of that community. Our off-campus experience is designed to start that process.

Assessments that are consistent with this instructional design are usually performance-based (Loacker, 2000; Mentkowski, 2000). Most other colleges provide these elements in senior level courses. Providing them at lower levels of the curriculum can be problematic since the traditional assumption is that students must learn fundamentals before they can successfully attack significant open-ended problems. How can students solve difficult open-ended interdisciplinary problems before they've actually learned some of what they need to know in order to solve them? How can they do this in foreign culture when a significant language barrier exists? The answer lies in proper preparation, project and team management, and in providing multidimensional assessments that support the academic enterprise. The assessment network functions at multiple levels, and absence of any one level seriously degrades the student learning process. Assessment is used for continuous improvement of all program aspects." (DiBiasio, D. and N. Mello, Assessing a Nontraditional Study Abroad Program in the Engineering Discipline, Frontiers: The Interdisciplinary Journal of Study Abroad 10, 237-252, 2004).

We will briefly describe the GPP and its relation to WPI's learning outcomes and those explicated in ABET's Criterion 3. Although transformational learning during study abroad is well-known, we will present additional assessment results that show how international PBL satisfies student learning outcomes at levels significantly higher than on-campus PBL. These results are consistent with those recently presented by others (VandeBerg, M. Assessment Research for Optimizing Student Learning Abroad: the Georgetown Consortium Study. in NSF Grantee Conference. Purdue University, October 31 2007 and the Study Abroad for Global Engagement work at U. Minnesota: http://cehd.umn.edu/projects/sage/news.html) regarding the value of experiential education abroad programs. These studies describe broader and larger assessment projects that validate the need for proper design and facilitation, the value of non-classroom learning, and the positive impact of short-term programs with those attributes.

Our presentation's goal is not to simply review the WPI program, but to communicate elements of structure, instructional design, student preparation, and student mentoring that are transferable to others wishing to include effective global experiences in their curriculum. The abroad experience itself can be transformed from the "ordinary" to one that advances student learning in many dimensions, technical, professional, and personal.

ChE students at WPI have the highest participation rate on campus with 80-85% going off-campus each year. This comes from the discipline with the most restrictive degree requirements, but we believe is further illustration of how multiple outcomes can be met with global experiences. Additionally we will present other demographic data with interesting consequences. For example, women at WPI participate in the GPP at rates nearly double the on-campus demographic. This has consequences for recruiting but also for addressing what is currently a target underrepresented minority group in study abroad: male, engineering students.

The beauty of well-designed, well-managed, project-based international experiences is that they incorporate many of the outcomes we want to add to the curriculum: open-ended problem solving, teaming, project management, communication skills, cultural awareness, critical thinking, developing research skills, and dealing with ambiguity, ethics, and economics. The challenge is making it all happen in an already crowded curriculum. We hope the presentation will generate discussion and thinking about how that can be accomplished.

See more of this Session: Global Awareness and Ethics in Chemical Engineering

See more of this Group/Topical: Education

Chemical Engineering Demographics of Hispanic American Students in Higher-Education Institutions

Tom R. Marrero, Chemical Engineering, University of Missouri, Columbia, MO

The purpose of this review is to assess statistics on enrollment and awarded degrees --- BS, MS, and PhD --- in chemical engineering from colleges in the United States for Hispanic American students. The source of the statistics is primarily published reports by the National Science Foundation. From these data annual trends are determined for the last ten years, or more. The trends for BS degrees and graduate degrees indicate a small but steady increase for all Hispanic Americans; but for females the number of degrees is accelerating. This review indicates that changes are warrented to encourage more K-12 students to consider careers in chemical engineering.

See more of this Session: Global Awareness and Ethics in Chemical Engineering
See more of this Group/Topical: Education

Incorporation of Engineering Ethics in a Senior Unit Operations Laboratory

James P. Abulencia, Department of Chemical Engineering, Manhattan College, Riverdale, NY

An unfortunate realty of today's undergraduate engineering environment is that engineering students often scoff at the idea of taking an ethics course. A quick survey of students reveals that there are several reasons behind this. First, many believe that engineering ethics cannot be taught. They believe that their peers are sufficiently savvy that a person's behavior cannot change by teaching right and wrong behavior (i.e. the "you can't teach an old dog new tricks" argument). Second, many believe that a course in ethics is not the reason they went to engineering school. These students would rather be calculating the number of theoretical trays in a distillation column, or the best way to design a heat exchanger (i.e. the "I don't need these silly humanities courses as an engineering major" argument). Despite these reasons, it is difficult to imagine an area of study that has a greater importance to society than understanding the "rights and wrongs" behind difficult decisions.

In this project, the use of a real-world engineering ethics case study was integrated into the Senior Unit Operations Laboratory course at Manhattan College. The motivation behind this was twofold. First, the assignment provides the opportunity to develop a student's oral communication skills. Although oral presentations were an existing component of the course, this assignment provided a separate and distinct venue for a student to use this skill. More specifically, students were required to use reason and analysis in presenting their final argument. Second, an ethics component in this course would help bridge the transition between college and the workplace. An ethical dilemma involving the choice between doing what is right and maintaining a positive bottom line is often seen in industry. Exposure to previous examples and the decision-making process involved would be beneficial for students.

The data for the case study was compiled by the Laboratory for Innovative Technology and Engineering Education (LITEE) at Auburn University, and incorporates several ethical issues and moral challenges that emerge from the interplay of manufacturers, employers, and employees. Before presenting this scenario, a short lecture on basic ethics was delivered by the instructor, thus creating a framework of knowledge for students to process the information provided in the case study. Subsequently, students were divided into four constituencies, and asked to review the relevant data from the LITEE website. Each group offered their positions through an oral presentation, and followed up with other questions after all groups have finished.

The ultimate questions in this study are: 1) Do students find it important to study ethics in the context of an engineering program, and 2) Is senior laboratory the best course in the chemical engineering curriculum to insert an ethics project. Results obtained from surveys distributed after the assignment show that students agreed that it is important to study ethics, but senior lab is not the best venue. Moreover, they felt that the assignment improved their presentation skills. Overall, students felt that the assignment was an additional load in a course that is already work intensive.

See more of this Session: Global Awareness and Ethics in Chemical Engineering
See more of this Group/Topical: Education

STEPS OF INTERNATIONAL ENGINEERING CHALLENGE

Fernanda E. S. Duarte[1], Fabio Sousa do Fundo[1], Flavia N. David[1], Flavio Waltz[1].
[1]*Special Projects Division – CHEMTECH – A Siemens Company*
Rua da Quitanda, n. 50, CEP 20011-030, Rio de Janeiro - RJ, Brasil.
E-mail address: fernanda.duarte@chemtech.com.br

The Chemtech International Engineering Challenge 2009 aims to disseminate knowledge of advanced technology to students at South America universities. The number of engineering projects is growing as never before that motivates CHEMTECH to train engineering students in crucial areas from process industry, contributing to their development, improving their acceptance in the labour market, and guarantying the availability of better professionals to the industry. In 2008, the National Engineering Marathon included Computational Fluid Dynamics and Process Control Theory, two important engineering areas. In this year, the scope of the competition involves Process control analysis and thermodynamics. This challenge was organized in three steps: call for the contest, training of the attendees and the contest itself, which take place during the 10th International Symposium on Process Systems Engineering (PSE'09). During the first step, CHEMTECH invited the best Brazilian universities and five South American universities, each one represented by their best students and an advisor professor. In the next step comprises two phases: Conceptual Design and a Web Course. The students should submit conceptual designs proposing economically feasible solutions to some of the typical problems encountered in Brazil's natural gas production chain. The challenge is to draw maximum profit from the Gas Processing Units. During this time, they will be trained through the internet using the methodology of Distance Learning. This training also gives the students an opportunity to be in contact with real industrial applications. In the end of this step, a time with two students from each university will be qualified for the final contest. During the PSE'09, they will have to solve a real industrial problem using the learned tools. Their solutions will be evaluated by professionals from CHEMTECH and from some of its remarkable clients. Beyond the knowledge and recognition, the winners will receive prices and their University computers donation. Moreover, CHEMTECH is contributing to improve South America engineering and quality of projects in world process industry.

1. Introduction

The number of engineering projects is growing as never before that motivates Chemtech to train engineering students in crucial areas from process industry, contributing to their development, improving their acceptance in the labor market, and guarantying the availability of better professionals to the industry.

CHEMTECH has been promoting and sponsoring events involving Brazilian university students since 2004. This kind of competition develops the technical knowledge of the students and helps them to apply this knowledge in real problems and common industrial situations.

The first edition was promoted in 2004. A regional competition was conduced as a first step, in three Brazilian cities: Belo Horizonte (Minas Gerais state group), Rio de Janeiro (Rio de Janeiro state group), and São Paulo (São Paulo state group). The competition was a really success, fact that encouraged the organization of a national competition.

In 2006, Chemtech organized the first National CFD challenge. The event counted with the participation of thirteen Brazilian universities and more than thirty students. The competition took place in the Rio Oil & Gas Expo of that year. This well-succeeded experience was expanded for other areas of the knowledge, as process control analysis. The first process control challenge happened in 2007.

In 2008, an innovation was proposed: Organize a challenge mixing both areas, the National Engineering Marathon, involving eighteen universities and more than two hundred students.

This year, Chemtech is promoting an International Engineering Challenge that involves students at South America universities.

2. Competition

The International Engineering Challenge 2009 included Thermodynamics and Process Control Theory, two important engineering areas. Process control analysis is as important as Thermodynamics because an efficient control system guarantees stable and confident processes, improving the quality of the final product.

This multidisciplinary competition contributes to student's development, guarantying the availability of better professionals to the industry. This challenge was organized in three steps: call for the contest, training of the attendees and the contest itself, which took place during the 10th International Symposium on Process Systems Engineering (PSE'09).

2.1. Call for the contest

During the first step, CHEMTECH invited five Brazilian universities: ITA - Technological Institute of Aeronautical, UFF - Fluminense Federal University, UFSCar- Federal University of São Carlos, USP- São Paulo University and UFBA - Federal University of Bahia; and five South American universities, each one represented by an advisor professor and their best students, which means 100 graduation students.

The Brazilian universities were selected based on the last competition, where ITA, UFF and UFSCar were on the first, second and third place on the podium, respectively. The Federal University of Bahia and São Paulo University were hosts of the congress. Likewise the South American universities were selected based on their highest reputation in those areas that the competition was promoted. The students are in different period in their graduations. This distribution is showed in **Figure 1**.

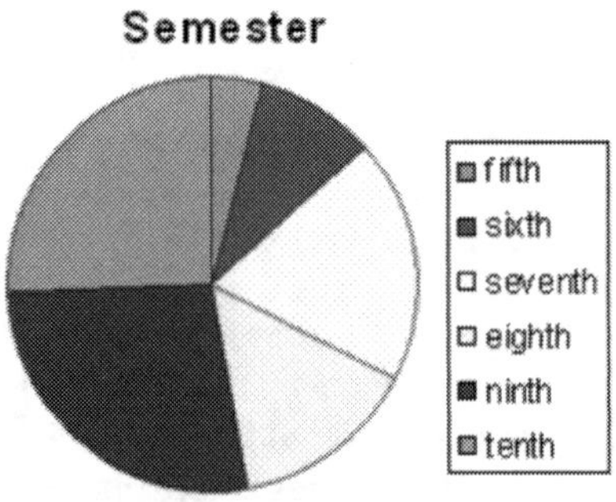

Figure 1: Distribution of the semester that students are enrolled.

2.2. Training

The next step comprises two phases: Conceptual Design and a Web Course.

First, the participants should submit conceptual design proposing economically feasible solutions to some of the typical problems encountered in Brazil's natural gas production chain. The concepts proposed need not, however, be limited to solving problems, but may also contain propositions to maximize profit by optimizing the quantities produced of each product, within the projected demands and quality requirements, based on the economic value of each product. The challenge was to draw maximum profit from the Gas Processing Units and the choice of process (Joule-Thompson, simple refrigeration, refrigerated absorption, turbo expansion, etc.) should be justified by a study comparing the profits and variable production costs of each process. Many students never been in contact with a real conceptual design and in this competition they had this opportunity. The groups were formed by four students and they had to complete all these following topics from the Gas Processing Unit: Purpose of process, Description of proposed alternative, Preliminary flow diagrams, Control loop descriptions, Interlock Systems, Equipment and Instrument List, Technology Challenge, Risk Analysis, Safety and Environment, Effluents generated, Demonstration of project feasibility and Reference documents.

After that, the students were all trained through the internet using the methodology of Distance Learning. The course material was composed by electronic books, divided in modules, available for download to all students. Each module corresponded to a week of course. During this time, chats among the students and the monitors (CHEMTECH professionals, with experience in thermodynamics and Control fields) were scheduled. Moreover, the student's doubts and questions could be discussed through forums on the internet or trough email, directed to the monitors.

The most difficulty task in this step, and in this teaching methodology itself, is to keep the attendance and participation levels. In order to try to keep the levels, the conceptual design and the proposed exercises were all based in real industrial problems.

2.3. Final Contest

The final contest took place in the 10th International Symposium on Process Systems Engineering (PSE'09). The selected students and their professors were brought to Salvador from many places in Brazil and South America during the exposition week.

This final step was divided in two parts. The first one was constituted by eliminatory tests, which were done by all the ten teams selected. The five better teams, ranked by the grade in the tests, were classified to the second step. With the best ten from the initial a hundred students the second step was conducted in the last day of the PSE'09. The final test lasted three hours and a problem about a compression system was proposed to the students.

Their solutions were evaluated by professionals from CHEMTECH and from some of its remarkable clients and the best group was the champion. Beyond the knowledge and recognition, the students and professors of the winning group also received prizes and their university received a donation of computers.

3. Conclusion

The International Engineering Marathon promoted an exchange among universities, their students, and the engineer companies, identifying new talents. Moreover, CHEMTECH is contributing to improve the international engineering and quality of projects in world process industry.

Some recognition has already resulted from this educational initiative and the company was awarded the 2004 Social Responsibility Prize, granted by CREA (National Engineering and Architecture Council).

The competition has also awaked the interest of a lot of undergraduate students in the most different engineering areas and has revealed some great potential future engineers. This successful experience is now being expanded to other countries, in the International Engineering Marathon that will happen in this year.

whynotchemeng – securing the talent pipeline of chemical and process engineers in the United Kingdom

Andrew Furlong
Director of Policy & Communication
Institution of Chemical Engineers
Davis Building, Rugby, UK
Tel: +44 1788 534484
afurlong@icheme.org

Extended Abstract for presentation on **Monday, November 9, 2009: 10:36**
Canal D (Gaylord Opryland Hotel)

Background

In 1997, IChemE conducted a detailed assessment of the state of chemical and process engineering in British universities. As a result, the Institution's governing Council established a project team under the Chairmanship of IChemE Past President, Bill Wilkinson to review the situation and report on factors affecting the viability of the discipline within the university system and propose actions that might alleviate problems as well as promoting and safeguarding the longer term health of the chemical engineering profession.

The findings were published in May 1998 in a report entitled *No immediate crisis – But more than a little local difficulty* [1](Described hereafter as simply The Wilkinson Report) highlighting a series of issues that required the Institution's attention:

1. The undergraduate first degree intake
2. A-Level entry standards
3. The relevance of undergraduate course content
4. University staff remuneration and morale
5. Chemical engineering research
6. Department size and regionalisation

This paper focuses on the initiatives and interventions developed in response to issues 1 and 2 and will examine the resultant trends and developments over the last decade.

The wider social context is also important. The post WWII period, and the 1950s, 1960s and 1970s in particular, have been described by some as the 'heyday' of chemical engineering in the developed world. Applications remained buoyant, competition for places remained healthy and the academic performance of applicants was not generally a cause for concern by admission tutors.
Social, technological and demographic changes in the 1980s brought an end to this comfortable state of affairs. In common with many Science, Technology, Engineering and Mathematical

---

[1] The Report of the Council Review Group on the State of Chemical Engineering in British Universities, 'The Wilkinson Report', IChemE, May 1998

disciplines, chemical engineering began to lose its lustre and this perception was heavily reinforced by marked deterioration in public support for the chemical and process industries in the UK. The reputation survey, conducted annually[2] by the UK trade body, The Chemical Industries Association, reveals that 'unfavourable' public attitudes began to dominate after 1988 and the chemical industry struggled to maintain any kind of positive public perception throughout the 1990s. The impact of this trend on the careers choices made by young people in the UK cannot be overlooked, but at the same time the UK government was reporting that demand for qualified chemical engineers would remain strong[3] and salary progression for new graduates entering the process industries suggested that demand was now outstripping supply[4].

It was clear that the decline in support for, and interest in, chemical engineering was contributing to a decline in course applications thereby presenting a real threat to the continued offer of many chemical engineering first degree programmes in the UK. A detailed analysis of the entry grade point standards, required for progression to the professional engineer status by the UK Engineering Council under its SARTOR 97[5] scheme, indicated that at least three, and potentially as many as seven of the 23 university departments offering IChemE accredited first degree programmes in chemical engineering, would be pressed to maintain their accredited status. This did not augur well for the future of the discipline in the UK nor, by extension, to the competitiveness of the UK in the chemical and process sector.

The Wilkinson Report Recommendations

Wilkinson and his colleagues recommended that IChemE, in partnership with employers, university departments and other stakeholders, relaunch its schools liaison activities to establish a single stop resource for advice and information on all aspects of the career opportunities available across the broad spectrum of chemical and process engineering. This was intended as a first step in stemming the fall off in applications to study on first degree programmes. It was accepted that the Institution would be required to devote substantial additional financial and staff resources to such activity and measures would be implemented to secure matched funding from other interested parties.

Specific conclusions included a recommendation that the principal focus of activity should be school teachers, including both subject teachers (chemistry, physics, biology and mathematics) as well as careers teachers. Parents were also identified as key influencers and a central role for appropriately trained volunteers drawn from the ranks of the Institution's membership, working closely with the secondary school community, was proposed. The increasing significance of the Internet as a communication tool was also noted.

The Wilkinson Report also recognised for the first time that IChemE's focus on the traditional A-Level (and equivalent) entry route to the chemical engineering profession should be expanded to embrace new education offerings from the growing further education sector in the UK.

---

[2] UK Chemical Industries Association Reputation Surveys 1978 – 2000,

[3] The Supply, Education & Training of Process Engineers in the UK, EPSRC/DTI Report 1998

[4] IChemE Salary Surveys 1996, 1998, 2000

[5] Standards and Routes to Registration (3rd Edition, 1997) http://www.engc.org.uk/ukspec/sartor/default.aspx

Significantly, the report also urged IChemE's professional community to proactively 'sell' chemical engineering as being relevant to an ever broadening range of rewarding, well-paid careers with a strong ethical grounding. It proposed that chemical engineering should be positioned as being of particular relevance to the emerging challenge of achieving sustainable development. Chemical engineering departments and those with responsibility for course design were urged to enthusiastically embrace sustainability as a part of a positive positioning to achieve a distinctive and attractive external image. It was proposed that IChemE capitalise on the success of the *Future Life*[6] report published in 1997 to celebrate the 75th Anniversary of the Institution's foundation. *Future Life* had already attracted widespread interest in the press, and in the UK parliament, as a result of its stimulating portrayal of the many opportunities for chemical engineering to shape a more sustainable future for humanity.

In essence the challenge facing the Institution was a simple one. An entirely new approach to its schools and careers campaign work was needed. Fostering a better understanding of chemical engineering and the opportunities it presented would surely drive up application and intake numbers. Quantity would drive quality and a healthy competition for places would secure the future of chemical engineering departments viewed as 'at risk', thereby securing a healthy supply of talented chemical engineers.

The *whynotchemeng* campaign

Developed with the support of a specialist communications consultancy in the UK and following extensive market research conducted via the leading opinion research organisation MORI, *whynotchemeng* brought about a dramatic change in IChemE's schools and careers work. The traditional emphasis of classic process engineering imagery was abandoned. Pumps, pipes and fractionation towers, accompanied by the inevitable earnest looking male figures in hard hats were discarded in favour of colourful new outcome focussed 'product icons'. Mobile phones, training shoes, snowboards and beauty products were chosen as the defining images for the first *whynotchemeng* campaign which ran from 2000 to 2004. These images featured prominently in a range of integrated campaign materials, which included a website, posters and a range of direct mail pieces that were designed to be used in a targeted manner with specially selected schools.

In 1997 there were approximately 3,500 secondary schools catering for 14-18 year old students and offering the UK's 'General Certificate of Education' (GCSE) and the 'Advanced Level General Certificate of Education' (A Level), which has traditionally been viewed as the *de facto* university entrance examination in the UK. IChemE recognised that a blanket approach to all secondary schools would not constitute the most effective use of resources, and thus an early decision was taken to specifically target just 500 secondary schools with an established track record of achievement at A-Level. The existence of school league tables[7] throughout the UK made the process of selecting schools for inclusion in the target list a relatively simple task. The target school list has been regularly reviewed and expanded since the beginning of the campaign, and it should be noted

[6] Future Life, IChemE 1997

[7] See http://www.dcsf.gov.uk/performancetables/ for a detailed overview of school league tables in the UK

that IChemE will happily support any school that is not included on the target list. However this relationship will be reactive rather than proactive.

Measures were implemented to establish and build personalised relationships with the target schools through identifying named members of teaching staff and the allocation of a *whynotchemeng* volunteer to as many target schools as possible. The volunteers were provided with basic guidance on communicating effectively with 14-16 year old students and presentation materials including slide packs and handouts. Most volunteers are linked to one or two local schools and they visit the schools on a regular basis to participate in careers event, class presentations and industry days.

The target schools receive direct mail communication on a regular basis timed to coincide with key dates in the academic calendar and all literature is designed to drive pupils as well as teachers, parents and other influencers to the campaign website[8] where a varied range of resources can be accessed and downloaded including:

- Case studies on real life chemical engineers, including student and graduate blogs
- Industry profiles, including links to company websites
- Links to universities offering IChemE accredited courses
- Classroom resources for teachers tailored to the UK curriculum
- Advice and information for parents
- Guidance and support materials for volunteers working in schools

The materials are reviewed and updated on an ongoing basis and in 2005 the launch of the second phase of the *whynotchemeng* campaign saw a full overhaul of the website and the campaign literature with a new focus on four Future Life themes; Healthcare, Energy, The Environment and Food & Water.

The most recent addition to the range of resources available, via the website, has been the *Top Ten Flash Bang Demonstrations* [9]– a series of practical classroom demonstrations illustrating some core chemical science concepts in an exciting, yet safe, manageable and fully risk-assessed fashion. The demonstrations contributed to a substantial surge in traffic to the website, upon their launch in January 2009, and are now widely accessible via YouTube and other online forums. The *whynotchemeng* website is one on the most heavily trafficked chemical engineering websites on the Internet and is consistently returned in the top three results for web searches conducted using the most popular search engines.

[8] www.whynotchemeng.com
[9] http://www.whynotchemeng.com/teachers/top-ten-flash-bang-demos

Results

1. Since the launch of *whynotchemeng* in 2001 the UK chemical engineering intake has risen by almost 75% and the 2008 intake of 1640 students substantially exceeds the mid-1990's peak of 1364[10]

THE GRAPHIC ON THIS PAGE WAS NOT AVAILABLE ON THE E-MEDIA VERSION OF THIS PAPER.

Figure 1. UK Undergraduate intake to chemical, process and energy engineering first degree programmes 1996 – 2008

2. Since the launch of *whynotchemeng* in 2001 individual applications to UK chemical engineering first degree programmes has almost doubled to 10,068[11], notwithstanding a reduction in the number of course applications per individual applicant from six to five in 2008.

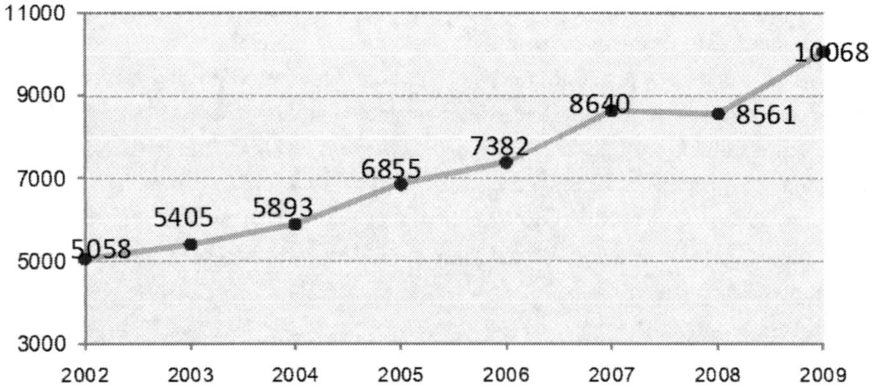

Figure 2. Applications to study chemical, process and energy engineering first degree programmes 1996 – 2008

[10] Source: UCAS www.ucas.com (Universities and Colleges Admissions Service)
[11] ibid

Conclusions

Whilst higher education in the UK has witnessed considerable expansion over the last decade, growth in application rates to study chemical engineering first degree programmes has significantly outperformed both the general trend and the performance of benchmark subjects including chemistry and mechanical engineering. Since 2003 the overall number of applications received by UCAS, the UK central organisation through which applications are processed for entry to higher education has increased by 24.5% from 374,307 to 465,862[12]. Over the same period, chemical engineering has witnessed a growth in applications exceeding 86% and UK provision is now largely at capacity.

IChemE believes that *whynotchemeng* has contributed substantially to this growth trend and survey work is ongoing to assess the profile and effectiveness of the campaign. In 2008, 49% of the UK domiciled undergraduate intake cohort indicated that they were familiar with the *whynotchemeng* brand and this figure increased to 56% for undergraduates who had attended target schools[13].

The worst case scenarios predicted by the Wilkinson project group in 1997 have not been realised and the intervening period has only seen the closure of one chemical engineering department offering an IChemE accredited course. Bradford University stopped offering chemical engineering in 2003; however this withdrawal has been attributed to a general downturn in interest in study programmes at Bradford following social unrest in[14] the city during 2001. Conversely, expanded provision has been reported at a majority of accredited departments and a new course has been offered at Paisley in Scotland and at Teesside University in 2009[15].

---

[12] Source: UCAS statistical service http://www.ucas.ac.uk/about_us/stat_services/
[13] IChemE student member questionnaire returns 2007
[14] See http://news.bbc.co.uk/1/hi/education/1654845.stm
[15] http://www.abdn.ac.uk/engineering/profgrps/chemgrp.php

Girl Scout STEM Workshop: Experience with 6th-10th Grade Girls in Rural Oklahoma

Ashlee N. Ford, Chemical & Biomolecular Engineering, University of Illinois, Urbana, IL

Recently, Girl Scouts of the U.S.A. has launched a science, technology, engineering, and math (STEM) initiative to engage girls in hands-on activities that make STEM fun and accessible and to promote gender equity in these fields by providing information and experiences highlighting the positive aspects which make careers in STEM fields attractive rather than just focusing on the need for science and math skills. On November 26, 2008, I led a STEM workshop in Altus, OK through Girl Scouts of Western Oklahoma Council. Seventeen girls aged 11-16 from five southwestern Oklahoma schools participated in the seven hour event focused on promoting interest in chemistry and engineering. The girls were students in towns and rural communities with populations ranging from 16 to 21,500. Only 3 of the girls said that they had any prior knowledge of engineering applications and careers. This is not surprising as western Oklahoma does not have a lot of engineering jobs. The students consisted of three African American girls, 1 Native American girl, and 13 Caucasian girls. Two of the girls had learning impairments (ADHD and autism). The majority of the students were in the 7th and 8th grades (the advertised target group for the event) with just two students each in 6th and 10th grades.

I introduced the engineering design process and explained different engineering disciplines to the girls. The girls formed teams and did some team building activities before moving to the engineering activities. The first engineering challenge was to work in teams to build 8-inch tall tables out of newspaper and masking tape. Information about this activity can be found on the PBS Design Squad Web site (http://pbskids.org/designsquad/projects/paper_table.html). The girls were very creative with this and were able to build very solid structures incorporating ideas about strength of materials. Next, the girls made water filtration systems using two liter pop bottles to explore concepts of wastewater treatment. Guidelines for this activity can be found on the PBS kids Web site (http://pbskids.org/zoom/activities/sci/waterfilter.html). On their evaluations of the event, several girls made comments about the usefulness of learning to filter water. They could see how this type of engineering application makes a difference in their lives everyday.

After a social lunch with team members, the girls had two more activities. They first designed rubber band cars with CDs and cardboard according to the activity guide directions supplied on the PBS Design Squad Web site (http://pbskids.org/designsquad/projects/rubber_band_car.html). No group was able to successful construct a car that would actually operate according to the design. Students responded to this activity on their surveys saying they learned to Take Action "by never giving up" and "by trying to do things I think are impossible." The day's final event was an egg drop competition. The girls were given marshmallows, spoons, straws, masking tape, popsicle sticks, and a zip top bag. They had to protect two eggs at one time, and the contraptions were dropped off the top of the football bleachers (about 30 feet). The girls loved this activity as they got to be creative, and they were eager to test the performance of their egg safety designs.

Fourteen girls and six adults were surveyed at the conclusion of the event. Rating scale questions included "How much did you know about science and engineering before today?", "How interested were you in pursuing careers in science and engineering fields before today?", "How much have your opinions about science and engineering changed as a result of this workshop?", "How interested are you in exploring opportunities for careers in science and engineering fields now?", and "How much did the volunteers positively contribute to your experience today?". The rating scale answer choices were "a lot," "some," "a little," and "none." Participants were also asked to evaluate the specific activities of the day with a five-point rating scale. Open-ended questions were also given on the survey in order to obtain any other feedback the students or adults had about the event. The open-ended prompts included favorite and least favorite things about the event, ideas for what to add to or remove from the event in the future, things learned the event that could be used in the participant's future, and additional comment free-response. GSUSA has recently a new leadership development strategy for Girl Scouts focused around three themes:

Discover, Connect, and Take Action. These themes were incorporated into the survey through the following questions: "What did you Discover about yourself or the world around you today?", "In what ways did you Connect with others today?", and "How can you Take Action with ideas that were discussed today?". The main results of the ratings portion of the survey were that most of the girls expressed increased interest in engineering and science careers as a result of the workshop. Three girls indicated that they had no interest in engineering careers and had no change in their career interests. The sample size was very small for this survey so it should be repeated with a larger group of students and with an open-ended question asking students why they chose their level of interest in science and engineering careers (e.g. not interested because already focusing on some other career field, particularly for older girls). Responses from the open-ended questions included the following statements: "some things you think are boring are fun," "new things are fun to try," "[I can Take Action] by becoming an engineer," "[I Discovered about myself that] I can be an engineer," and "[The event] was awesome!". These responses affirm the observation that the girls seemed to enjoy themselves and to be more motivated and interested in exploring science and engineering as a result of the event. Other student responses like "I learned to use teamwork" and "I learned to communicate with others" and parent/leader responses like "girls actually got involved with each other" and "girls learned the value of teamwork" emphasize the importance of teamwork and positive teamwork experiences.

See more of this Session: Interactions with K-12
See more of this Group/Topical: Education

Working with a Middle School to Build Curriculum Dedicated to Engineering Principles

J. Alex Lee, Department of Chemical Engineering and Materials Science, University of Minnesota, Minneapolis, MN

Gillian Rhoerig, College of Education and Human Development: Curriculum and Instruction, University of Minnesota, Minneapolis, MN

Tamara Moore, College of Education and Human Development: Curriculum and Instruction, University of Minnesota, Minneapolis, MN

Angel Brown, Columbia Heights School District, Columbia Heights, MN

Emily Christianson, Columbia Heights School District, Columbia Heights, MN

Eray S. Aydil, Department of Chemical Engineering and Materials Science, University of Minnesota, Minneapolis, MN

This talk will describe the origins, development, and current progress of a class dedicated to incorporating engineering principles at a middle school. Two middle school teachers---one a science teacher and the other a math teacher---have combined efforts this past year to implement STEM integration at the school through the initiation of an elective ``Engineering Course''. Through a small-scale university-funded program emulating the NSF's GK 12 programs, the author has had opportunity to work alongside these teachers both in and out of the classroom to develop and execute new design-oriented curricula based on sound principles and an emphasis on systematicness. The successes, inadequacies, and challenges will be discussed in terms of students' and teachers' enthusiasm, aptitudes, and expectations, as well as the current and future relationship between K12 school teachers and universities.

See more of this Session: Interactions with K-12
See more of this Group/Topical: Education

Chemical Engineering in the 8th Grade Classroom

Benjamin J. Davis, Chemical Engineering, The Cooper Union, New York, NY

Background / Motivation

As part of a National Science Foundation GK-12 Program, I spent the past academic year as a teaching fellow in Culver City Middle in Los Angeles, California. I split my time between two teachers, both of whom had 5 classes of 30-35 8th grade (physical science) students. I came during and after school two days a week, accounting for 10 hours of classroom time per week. I was also involved with other programs at the school, including an "in-school field trip" and the middle school science club. On most days, I helped to engage students in activities, added to the lecture or discussion of the day, helped set up lab activities, made observations about how students seemed to be responding to their activities, and answered student questions. On some days, I took charge of the whole class and gave a lesson that I had prepared that was related to the 8th grade science standards and incorporated a theme from my own Ph. D. research or the general theme of "Engineering of the Environment of Los Angeles". My goal was to create a novel, inquiry-based lesson which added some depth to the principles being taught in class.

Program goals

The goals of the GK-12 program at UCLA are many: primarily, the goal was to develop the teaching and communication skills of graduate student fellows (me) by having them give lectures, develop activities, and take a class specifically aimed towards preparing them for future employment in academia. The experience of giving three 1.5 hour lessons per day (or 6 45 minute lessons) is invaluable. Getting contact time with students and learning to present things in a way they will understand them is something that there is no substitute for. Developing the activities and lessons for the students also required great patience, flexibility, and skill. Learning how to write a good lesson which engages students, covers the material you would like to cover, and starts and ends on time is a required skill for any student who wishes to be an educator after graduate school. One of my favorite parts of the experience was our requirement to take a "Preparing Future Faculty" course for one hour per week. There we learned about different educational philosophies, institutional missions, and presentation methods. We also got to hear guest speakers from local colleges and universities who spoke about their experiences in academia. This kind of training is not typically required of graduate students (at UCLA at least) but I would consider it to be essential to anyone who wants to make the leap to academia.

As a secondary goal of the GK-12 program, the fellows tried to enhance the learning experience of the K-12 students by their presence as a practicing scientist in the classroom. As a side product of this goal, the fellows developed (publicly-available) lessons for their teachers to use in their classrooms which incorporate themes of the graduate student fellows' research. I feel that these lessons are great, but would be hard to use as a resource by teachers who don't have a hard science background. Most of the benefit of having graduate students write the lessons is that they are experts on their particular subject matter. Translating that knowledge to someone without a background in the field is incredibly difficult to do in the absence of a good teacher and the lessons that would perhaps be the most useful to another teacher might also be the most difficult for them to understand and get right. I am not yet convinced that these lessons, as currently constituted, provide a great benefit to K-12 teachers at large, but they are hugely successful in forcing the graduate student fellows to crystallize their thoughts on their topic and learning to communicate their work to a broad audience.

Lessons

Some lessons I had the 8th grade students participate in this past year were:

1. Hydrogen as an energy carrier / electrolysis of water (Electrochemistry)

The goal of this lesson was to teach students about hydrogen and its role as a fuel and energy carrier. Students should learn what hydrogen is, who discovered and named it, what it is used for, and how you make it. In learning about hydrogen, students should locate its place on the periodic table, identify how many protons, electrons, and neutrons it can have, and be able to tell you its physical and chemical properties. They will learn that you can use hydrogen to run an internal combustion engine like the one in their car, use it to run a "fuel cell" to make electricity, use it to propel the space shuttle up into space, and use it to make energy like in the sun. This lesson gives background to students so that they can do an activity where they produce hydrogen from water using electricity (electrolysis).

2. Make a "Cartesian Diver" (Fluid mechanics)

The goal of this lesson was to teach students some basic concepts from fluid mechanics. Students get hands on experience with buoyancy, density, and Pascal's principle. The activity has the student put an object in water which normally would float but has a small air bubble in it. By closing the bottle of water and squeezing it, the force of the student's hands is transferred to the bottle, the force of the bottle turns into pressure on the (incompressible) water, which in turn exerts a pressure (Pascal's principle) on the air bubble in the object. The (compressible) air decreases its volume under increased pressure (Boyle's Law) and the total density of the object therefore increases, causing it to sink in the water (buoyancy). This usually gets a good "Oh, wow!" out of middle school students and is a good way to teach all of these (generally unintuitive) concepts.

3. How a PEM fuel cell car works (Chemical Engineering)

This lesson (which builds off of the lesson on Hydrogen as an energy carrier) is on the nature of hydrogen and how it can be used as a clean energy carrier. It was intended to focus on how hydrogen can be used as a fuel and how one particular energy conversion device (a polymer electrolyte membrane fuel cell) works. As part of this activity, some classes were able to see a real fuel cell vehicle. Hopefully students next year will be able to do a hands-on activity with real (demonstration-scale) PEM fuel cells.

4. Where plastics come from (Petroleum Engineering)

This lesson aimed to teach students about plastics and how they are made. Students should learn what typical beverage containers they use are made of, what chemicals are required to make them, and what those chemicals are made from. In learning about plastics, the students should also learn about density, the difference between a physical and chemical change, data collection, and how a chemical reaction changes one substance into a completely different one. This lesson gives background to students so that they can do an activity where they collect data on the recyclable containers in their school and also make the supply chain for PET and PP. See Figure 1 below.

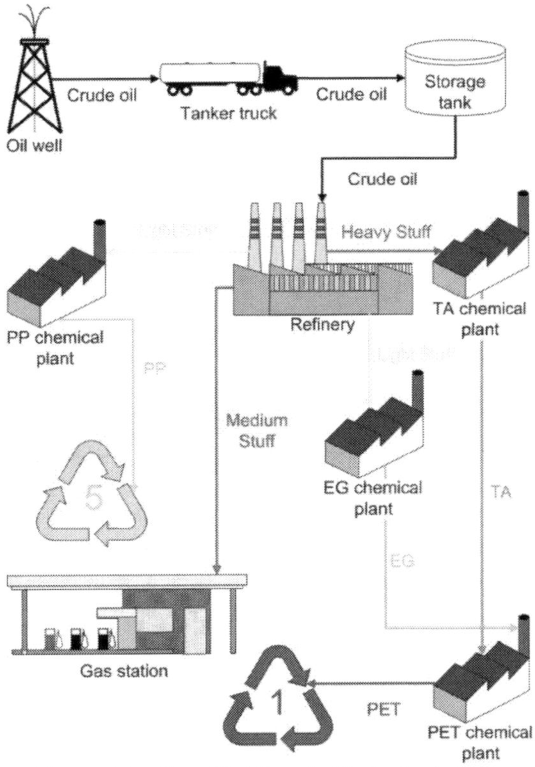

Figure 1: PET and PP Supply Chain for Activity

5. Units and orders of magnitude in the metric system (General Engineering)

The goal of this lesson was to try to give students some insight on orders of magnitude and what they mean for common units of measure like weight, distance, volume, etc. The students got to see different sizes of physical objects (metal weights, cups of water, marks on a counter) and try to relate that to factors of ten (kilograms, centimeters, etc.)

6. Cryogenics and liquid nitrogen

This lesson aimed to teach students about the properties of matter and how molecules behave at low temperatures. I brought in liquid nitrogen and did a few demonstrations for the students and explained how each one worked using chemistry and physics concepts. I also explained how liquid nitrogen is made and took student questions.

The lessons took anywhere from 15 minutes to multiple 1.5 hour class periods and ideally incorporated some sort of hands-on activity so students would have context to engage in inquiry based learning. I was also heavily involved in the middle school science club, designing engineering-themed activities for students such as an egg drop, a catapult-building competition, and designing basic circuits. These were typically the better science students, so I formed their "lessons" as competitions where they had to engineer / design a device which competed with other students'. For example, I had the students design a paper airplane that could fly as accurately as possible towards a target multiple times. Science club was a great opportunity to talk to interested students about topics a little beyond the scope of their standards in a small group setting (where it is easier to accomplish a lot of activities which require supervision).

Lessons Learned

1. GK-12 is a very valuable program for giving graduate students the opportunity and the focus to learn to teach. I think it is one of the primary reasons I found the job I did.

2. Inquiry-based learning is hard with 35 students. The options for activities with so many students are very limited due to safety, supervision, and engagement of students. Things which work well with 15 students might not work at all with 30.

3. Teaching engineering is hard when the students don't know math. It really makes you think about the physical underpinnings of what you know and use analogies that the person you're communicating with can understand.

4. Urban schoolchildren are lacking in hands-on learning experience activities. This is related to the large class size issue. A lot of students in my middle school don't build things, make things, cook, clean, walk through the park, collect bugs, bird watch, etc. It not only makes it tough to get them focused on and interested in science, it hampers their growth as potential scientists.

5. Science club was a good way to add to the standards-based curriculum. Some things I feel are terribly important that an 8th grader know about before they get to high school are largely ignored if their teachers "stick to the standards". Science clubs or after school activities are a great way to incorporate these things, give students hands-on activities, and interact with students in smaller groups.

6. Teaching middle school is really, really hard. Sounds obvious, but it has to be said.

Recommendations

To a college professor, I would say that finding a local school and just snooping around would be a good way to see how you can help out. Likely the most beneficial activity for someone who already teaches full time is to try to give students at a local school the opportunity to meet you and know what your field is all about.

To a university professor, I would say that offering up equipment, materials, and other resources that you uniquely have access to is a priceless thing to a local student or teacher. One of those resources is your graduate students; consider your obligation to them to help them succeed in their / your field and the quality of the training they receive outside of the laboratory.

To a graduate student, I would say that being a teaching assistant, tutor, volunteer, teaching fellow, guest lecturer, etc. is a very rewarding thing. Regardless of your ambitions after graduate school, improving your communication skills and learning about what is out there in the world is always going to make you more marketable.

Conclusions

The GK-12 program inspired me to participate in science and engineering education in this country. I am very grateful for my experience there and am glad I could have this opportunity to tell others about it.

How to Effectively Engage the Next Generation in STEM Careers

Colleen Gorey, Department of Chemical and Environmental Engineering, University of Toledo, Toledo, OH

It is becoming ever important on the world stage to produce successful scientists and engineers, but how do we make these fields more appealing to young Americans? The lessons learned from successful and unsuccessful classroom practices will help shed light on steps that need to be taken to reach a classroom of high school students. A National Science Foundation GK-12 grant at the University of Toledo titled "Graduate Teaching Fellows in STEM High School Education: An Environmental Science Learning Community at the Land-Lake Ecosystem Interface" pairs eight graduate students in STEM disciplines with eight teachers from local high schools. Graduate Fellows exchange their STEM knowledge and research experience through high school classroom and field exercises to improve their teaching competence and develop communication, mentoring, and team-building skills. High school teachers gain professional development, enduring relationships, and experience in inquiry-based state-of-the-art science content. High School Students develop their STEM skills, gaining greater interest in and understanding of the scientific method and potential for environmental science and engineering careers. The objectives for the program are to: 1) Generate student enthusiasm for STEM careers by engaging them in hands-on research into environmental problems, 2) Exchange STEM knowledge and pedagogies between graduate students and high school teachers resulting in cutting-edge environmental science content and increased teaching and communication skills, and 3) Develop hands-on solutions to environmental problems along schoolyard stream ecosystems feeding the Great Lakes. The UT GK-12 program thus embraces the public scholarship philosophic concept of merging scientific research with civic responsibility to benefit the public and the community and part of that concept is sharing the knowledge and experiences gained with the rest of the scientific community to prepare the next generation of Americans.

See more of this Session: Interactions with K-12
See more of this Group/Topical: Education

Bringing Outreach Into the Engineering Classroom - A Heat and Mass Transfer Student Class Project

Omolola Eniola-Adefeso, Chemical Engineering, University of Michigan, Ann Arbor, MI

One major contributing factor to the low number of students receiving degrees in engineering is the two decades of steady decline in the number of students enrolling in engineering disciplines. While engineering undergraduate enrolment across the US has seen recent increase due to the current global recession highlighting the apparent stability of engineering jobs, the US is still projected to face a shortage of up to 70,000 engineers by 2010. Recent surveys by the American Society for Quality (ASQ) and the National Academy of Engineering (NAE) suggest that this shortage and decline in engineering enrollment is linked to K-12 students' lack of knowledge of engineering careers and their perception of engineering as "boring" or just for geeks. The NAE study further concluded that though many engineering organizations have put money and effort to this image problem, their efforts have had little impact in changing the public image of engineering (NAE book). Overall, what is clear from the literature is that boosting enrollment and increasing the number of graduates in engineering is one key parameter in keeping the US competitive in the global economy of the 21st century and beyond. Thus, additional effort needs to be put into making the new generation of K-12 students aware of the engineering profession and its versatile contribution to solving the many critical global issues. This paper present a different approach to introducing chemical engineering to K-12 student involving the chemical engineering classroom and a science fair style presentation to local high school students. Specifically, students enrolled in the "Mass and Heat Transfer" chemical engineering (ChE) course at the University of Michigan in groups of 5 were asked to design and present one original experiment suitable for a high school teacher to use in introducing a basic heat or mass transfer concept of their choice to his or her high school science class. ChE students were given a $25 budget to build their experiments and local high school students from the Ypsilanti school districts were invited into the classroom to experience and judge the design outcome on its ability to engage their interest in engineering as a career choice. My talk will discuss the benefit and success of this "each-one-teach-one" approach to K-12 outreach and suggest how this can be replicated in other ChE courses.

See more of this Session: Interactions with K-12
See more of this Group/Topical: Education

Principles of Nanotechnology for Middle and High School Students

Tamara Floyd-Smith, Chemical Engineering Department, Tuskegee University, Tuskegee, AL

Kala Bean, Chemical Engineering Department, Tuskegee University, Tuskegee, AL

Dwayne Vickers, Chemical Engineering Department, Tuskegee University, Tuskegee, AL

David Baah, Materials Science and Engineering Program, Tuskegee University, Tuskegee, AL

An outreach activity to teach the basic principles of nanotechnology to middle and high school students was developed and assessed. Over two hundred middle and high school students participated in the activity that consisted of a lecture, hands on activities and - in the later stages - video presentations. Of the six groups that participated, all of the groups showed a statistically significant increase in knowledge of nanotechnology when compared to baseline levels established using a pre-workshop assessment. However, only the last group of middle school students showed a statistically significant increase in interest in science and engineering when compared to baseline levels also established using the pre-workshop assessment. The increase in interest was attributed to several modifications made to the activity including adding video presentations, adding a hands-on activity to test commercially available fabrics and increasing the number of trained volunteers available to assist students.

See more of this Session: Interactions with K-12
See more of this Group/Topical: Education

A Lecture, Demonstration, and Simulation to Teach Chemical Process Control Basics to High School STEM Instructors

Dale E. Weber, Isermann Department of Chemical & Biological Engineering, Rensselaer Polytechnic Institute, Troy, NY

B. Wayne Bequette, Department of Chemical and Biological Engineering, Rensselaer Polytechnic Institute, Troy, NY

Deborah Kaminiski, Mechanical, Aerospace, and Nuclear Engineering, Rensselaer Polytechnic Institute, Troy, NY

Recent data from the NSF and other agencies indicate declining interest in science and engineering at the high school level. The growth of undergraduate engineering degrees is almost a third of the level of growth of employment in science and engineering. Additionally, one quarter of the engineering degree holders in the labor force are over 50 years old. There is clearly a need to increase the number of quality engineers trained by colleges and universities in the United States swiftly, but this goal is intractable without a significant resurgence of enthusiasm in the fields of science, engineering, technology, and mathematics (STEM).

The NSF GK-12 program creates relationships between higher learning institutions and surrounding area high schools. Participating high school STEM instructors are paired with a graduate student to work in the classroom, create curricula, and spark interest in STEM undergraduate majors and careers. The graduate fellows are also tasked with exposing the students to their research and other campus facilities. In addition to direct interaction of graduate fellows and high school students, the program goals include fostering the teaching skills of the graduate with fellows with meetings and workshops. Another tenet of the program is acquaint the high school teachers with graduate research and provide them the opportunity to work alongside a graduate fellow in the laboratory.

Rensselaer Polytechnic Institute (RPI) has ten NSF GK-12 fellows from various departments working with teachers at six Albany, NY area high schools. To introduce all the teachers to all the research of the fellows, RPI is hosting a four-day immersion with two three-hour sessions daily for the fellows to provide a tour, give a demo, lead an activity, etc. This paper will detail the session led by fellow Dale Weber introducing chemical process control theory with a demonstration in the Chemical Engineering Senior Laboratory and a simulated experiment that can be taken back into the STEM classroom by the teachers.

See more of this Session: Interactions with K-12
See more of this Group/Topical: Education

Catalytic Activation of Carbon Dioxide On Cerium Oxide for Conversion to Liquid Fuels

Brent J. Sherman, Energy, Environmental and Chemical Engineering, Washington University in St. Louis, St. Louis, MO

Cynthia S. Lo, Energy, Environmental and Chemical Engineering, Washington University in St. Louis, St. Louis, MO

Rising atmospheric levels of carbon dioxide and methane contribute to global warming. While sequestration would reduce these levels, turning the unwanted gases into valuable chemicals would close the carbon cycle. The direct conversion of carbon dioxide and methane to liquid fuels using an integrated nanocatalyst of platinum nanoparticles on a cerium oxide support is the focus of this research. Computer modeling will be used to predict the catalyst's behavior and design improved catalysts.

Preliminary results obtained using density functional theory calculations indicate strong chemisorption of methane onto platinum and weak physisorption of carbon dioxide onto stoichiometric ceria. Since other researchers have suggested that carbon dioxide will be strongly chemisorbed onto a reduced ceria surface, we anticipate that ceria would be a good candidate for activating carbon dioxide in the desired reaction.

See more of this Session: Student Poster Session: Catalysis and Reaction Engineering
See more of this Group/Topical: Education

Reduction of Iron(III) Oxide Deposited On Silica Aerogel by ALD

Alan Izar, Departamento de Ingeniería Química, Instituto Tecnológico y de Estudios Superiores de Monterrey, Monterrey, Mexico

Victoria J. Aston, Chemical and Biological Engineering, University of Colorado at Boulder, Boulder, CO

Alan Weimer, Chemical and Biological Engineering, University of Colorado, Boulder, CO

One form to store and/or purify hydrogen is through the steam-iron process, based on the oxidation and reduction of iron to split water obtaining hydrogen as a product. This process can be limited by the fact that iron sinters when reduced from iron(III) oxide, reducing its surface area and with so, limiting its reactivity. Instead, it is proposed to use silica aerogel coated with iron(III) oxide by atomic layer deposition as a reactant. In this project the isothermal reduction of iron(III) oxide coating silica aerogel was studied by using a Thermogravimetrical Analyzer to study mass changes and compare them to the isothermal reduction of bulk iron(III) oxide. A model of three steps proposed for the reduction of bulk iron(III) oxide was followed to study both cases at two different temperatures. It appears that bulk iron(III) oxide as well as the coating iron(III) oxide follow such model, with a nuclei-growth mechanism for the first two steps followed by a diffusion mechanism. Differences between the models were the index of reaction, which explains geometry of nuclei growth, being a mixture of a one-dimensional and a two-dimensional model for the bulk iron(III) oxide, and one-dimensional for the coating iron(III) oxide; and the diffusion model exponent, being larger for the coating iron(III) oxide. Physical differences between the products of the reductions were the sintering of bulk iron, while the coated particles remained without sintering.

See more of this Session: Student Poster Session: Catalysis and Reaction Engineering
See more of this Group/Topical: Education

Characterization of Platinum Nanoparticle Electrocatalysts Supported On Graphene Sheets

Christine M. Jackson, Chemical Engineering, Case Western Reserve University, Cleveland, OH

Kaitlyn Zolton, Chemical Engineering, Case Western Reserve University, Cleveland, OH

Robert Savinell, Chemical Engineering, Case Western Reserve University, Cleveland, OH

This research addresses the development of a model electrocatalyst of platinum nanoparticles on a graphene support structure in order to study substrate-electrocatalyst interactions in important reactions like the limiting fuel cell oxygen reduction reaction (ORR) and to better understand support stability. Graphene is a two-dimensional crystalline structure with high electron mobility and mechanical strength made up of a single layer of carbon atoms. In this research, nanoparticles of platinum are deposited on synthesized graphene sheets by exploiting modified published methods for both graphene and platinum nanoparticle synthesis. The non-platinized synthesized graphene is examined and characterized using Raman spectroscopy, and platinum nanoparticle size and distribution are characterized by microscopy techniques. The oxygen reduction reaction (ORR) on these electrodes is studied using voltammetry and thin-film rotating disc electrode techniques. The results of these tests are compared to data obtained using commercial carbon-supported platinum nanoparticle electrocatalyst. This poster describes our research goals and approach, and presents our results and findings to date.

See more of this Session: Student Poster Session: Catalysis and Reaction Engineering
See more of this Group/Topical: Education

Motion of Self-Propelled Janus Particles Inside Double Emulsion Drops

Nitza B. Rodriguez, Chemical Engineering Department, University of Puerto Rico Mayagüez Campus, Mayagüez, PR

Ubaldo M. Cordova-Figueroa, University of Puerto Rico Mayagüez Campus, Mayagüez, PR

Carlos J. Martinez, School of Materials Engineering, Purdue University, West Lafayette, IN

There is growing interest in the self-propelled motion of particles with two distinct hemispheres, also known as Janus particles. Previous work has shown that motion in Janus particles can be accomplished by having a catalytic material on one side of the particle and leaving the other half-intact. If a chemical reaction occurs at one hemisphere of the object, it will generate a gradient in concentration causing the particle to move by a process called self-diffusiophoresis. Recent experiments performed on an aqueous system composed of polystyrene microspheres half-covered with platinum, and using hydrogen peroxide as a reactant, showed that the particle velocity depends strongly on reactant concentration. It has been theorized that directed particle motion could be achieved in the presence of a concentration gradient and that confinement within a small volume could lead to the assembly of particles at an interface. The present study aims to corroborate these theories by characterizing the motion of polystyrene particles half-coated with platinum inside double emulsion drops in the presence of a hydrogen peroxide concentration gradient. A microcapillary microfluidic device was used to generate water/oil/water double emulsion drops composed of a polydimethilsiloxane (PDMS) oil drops containing a diluted aqueous suspension of Janus particles in the internal droplet. The PDMS oil drop acts as a semipermeable membrane that regulates the concentration of hydrogen peroxide inside the inner droplet. A baseline test was performed to track the motion of free-flowing Janus particles in the presence of different concentrations of hydrogen peroxide. In a second experiment, double emulsion drops containing the Janus particles in the inner droplet were placed in different solutions of hydrogen peroxide and the particle motion was tracked as a function of time. Particle velocities showed a strong dependence with hydrogen peroxide concentration both outside and inside the double emulsion drops. The results in the form of velocities as function of concentration showed indeed that this is correct and it is expected that having a gradient of concentration within the double emulsion will generate autonomous motion of the Janus Particle.

See more of this Session: Student Poster Session: Catalysis and Reaction Engineering
See more of this Group/Topical: Education

Predicting the Catalyst-Assemblage Dynamics in a CSTR by Linear Operator Method Approach

Jennifer Anne Pascal, Chemical Engineering, Tennessee Tech University, Cookeville, TN

Pedro Arce, Chemical Engineering, Tennessee Technological University, Cookeville, TN

Continuous stirred tank reactors (CSTRs) are very versatile, since they are used in a wide variety of applications, including product processing, product development and laboratory studies of kinetic parameters. Some of the advantages of CSTRs include, saving energy, higher/est conversion of reactants and relatively easy control. When reactors are modeled by recognizing the catalyst assemblage nature of the system, interesting and more realistic behaviors can be determined. For example, catalytic chemical reactors, modeled by an assemblage behavior, have shown new steady states that are highly beneficial for the catalyst life, and the selectivity and durability of the reactor (Arce and Ramkrishna, 1986) and when interaction of mass among particles has been included, predictions based on ordinary reactor models become invalid (Ramkrishna and Arce, 1986).

Now, if CSTRs are used in the determination of kinetics and transport parameters under dynamic conditions, the possibility of obtaining different limiting cases is within the range of practical performance of such reactors. These different limiting states may lead to different parameter values. In this study and by using a linear operator method approach (LOMA), the authors will outline the different possibilities of a heterogeneous CSTR and will relate them to the experimental conditions needed to determine correct kinetic parameters. Through LOMA, the eigenvalue problem for the CSTR can be determined, and the spectrum of this problem can be used to find, or identify such potentially limiting behaviors as indicated, for example in Arce and Ramkrishna (1986) for the case of other chemical reactors.

See more of this Session: Student Poster Session: Catalysis and Reaction Engineering
See more of this Group/Topical: Education

Methane Autothermal Reforming in a Reverse Flow Reactor with Rh/Al2O3 Catalyst

Lucia Salemme, Chemical Engineering, University of Naples Federico II, Naples, Italy

Marino Simeone, Chemical Engineering, University of Naples Federico II, Naples, Italy

Laura Menna, Chemical Engineering, University of Naples Federico II, Naples, Italy

Gennaro Volpicelli, Chemical Engineering, University of Naples Federico II, Naples, Italy

Catalytic Partial Oxidation of Methane (CPOM) in a reverse flow reactor is a promising process for decentralized syngas and hydrogen production. This reactor configuration couples feed preheating and syngas production in the same unit through a periodic inversion of the feed. Thanks to reactor compactness, this technology is particularly indicated for small and medium scale plants.

However, the higher catalyst entrance temperature obtained thanks to the internal heath exchange between the hot effluent gas and the cold feed, although positive in terms of product composition, can thermally damage the catalyst and reduce its performance and lifetime. Accurate measurements of catalyst temperature profile provide an important insight to assess the thermal stress developed during reverse flow operation.

In this work, we present a systematic investigation on reactor performance and catalyst temperature profile during CPOM in a reverse flow reactor with commercial Rh/Al2O3 catalyst in pellets. Temperature profile of the catalyst bed was measured by IR thermography and product composition was measured with a continuous gas analyzer.

The effect of internal heat recovery on reactor performance and catalyst thermal stress is presented and compared to steady state operation.

The effect of water addition (Autothermal Reforming) on reactor performance and catalyst thermal stress is presented both in reverse flow and steady state operation. Feed direction switching time and water to methane were investigated as process operating parameters.

Data of catalyst bed temperature evolution during the flow cycle are presented and discussed.

See more of this Session: Student Poster Session: Catalysis and Reaction Engineering
See more of this Group/Topical: Education

Experimental and DFT Study of the Mechanism of Acetone - Peroxide Reaction Using Raman and Nuclear Magnetic Spectrometry

Migdalia Hidalgo, Chemical Engineering, University of Puerto Rico, Mayaguez Campus, Mayaguez, PR

Eduardo A. Espinosa, Chemistry, University of Puerto Rico, Mayaguez Campus, Mayaguez, PR

Leonardo Pacheco Lodoño, Chemistry, University of Puerto Rico, Mayaguez Campus, Mayaguez, PR

Samuel P. Hernandez Rivera, Chemistry, University of Puerto Rico, Mayaguez Campus, Mayaguez, PR

In this research has been proposed a mechanism for the Acetone - Peroxide formation reaction based on Raman and NMR spectrometry measurements supported by computational ab-initio Density Functional Theory (DFT) methods at 6-311g**d++ basis set. It was found that the mechanism comprises three steps: monomer formation, polymerization of the 2-hydroperoxipropan-2-ol monomer and cyclation. Rate constants were measured and the thermodynamic states were determined and proved to be in accordance with theoretical calculations. The activation energy for the polymerization reaction was proven to be more energy favored in comparison to other possible reactions. In fact it is threefold lower than the Acetone-monomer reaction and Peroxide-monomer reaction according to experimental and theoretical measurements, which was also proposed by Torent'ev and collaborators.

See more of this Session: Student Poster Session: Catalysis and Reaction Engineering
See more of this Group/Topical: Education

Hemicellulose Pre-Extraction From Red Maple Wood

Diego J. Rosso, Chemical Engineering, University of Puerto Rico - Mayaguez Campus, Mayaguez, PR

Rory Jara, Chemical Engineering, University of Maine, Orono, ME

Adriaan van Heiningen, Chemical Engineering, University of Maine, Orono, ME

Aqueous hemicellulose extracts from wood are considered a viable resource for the production of higher value products such as ethanol. The proposed process extracts hemicelluloses mainly as xylan-oligomers from wood material. Red maple (Acer rubrum L.) wood will be used for the pre-extraction of hemicellulose using hot water extractions. The extractions will be made in both batch reactor and continuous flow reactor processes. The hemicelluloses extraction yields and extraction rates of both processes will be compared. Temperature and time of exposure for the batch reactor process will vary in the ranges of 140°C-170°C and 45-90 minutes, respectively. The severity factor R_0, a function of time of exposure and temperature, will be used to measure the intensity of the reaction conditions in the batch reactor process. High Performance Liquid Chromatography and High Performance Anion Exchange Chromatography will be used to quantify the amount of oligomeric sugars, acetic acid, formic acid, and furfural in the extracted liquor. A UV-VIS detector will be used to determine the amount of lignin in the continuous flow reactor process. The sugar analysis will focus on the quantification of xylan, arabinan, galactan, glucan, and mannan. The extraction rate of red maple wood strands, wood chips, and wood meal will be analyzed and compared.

See more of this Session: Student Poster Session: Catalysis and Reaction Engineering
See more of this Group/Topical: Education

Electrostatic Segregation in Particulate Mixing

Ted Z. Wang, Chemical and Biochemical Engineering, Rutgers University, Piscataway, NJ

Frank Romanski, Chemical and Biochemical Engineering, Rutgers University, Piscataway, NJ

M. Silvina Tomassone, Chemical and Biochemical Engineering, Rutgers University, Piscataway, NJ

Troy Shinbrot, Biomedical Engineering, Rutgers University, Piscataway, NJ

The segregation of particulate media during mixing has been studied for years. Particle segregation is typically caused by variations in fill level, rotational speed, and particle size/density disparity between species. However, new insights into segregation suggest that electrostatic interactions between particles may have a more profound effect than previously understood. This study was designed to explore the effects of electrostatics on particulate blends. The aim was to establish and explore various prevalent effects that are potentially detrimental to several industries that frequently blend particulate matter such as the catalyst and pharmaceutical industries. Particles used in this study were made of various polymers and selected from the triboelectric series with a known charge affinity. Particles were typically between 500 μm and 1 mm in diameter; any larger particles tend to allow inertial forces to dominate over electrostatic forces. Particle charging was typically caused by contact with the walls of the blender as well as particle-particle contact. Binary mixtures were created using various polymers and tested for fill level at 1:1, 1:2, and 2:1 mass ratios. Additionally, binary mixtures were tumbled at three different speeds with a maximum of 56 rpm. Particle blends were mixed in an acrylic double cone blender with a 10 cm diameter. The use of acrylic allowed not only for known electrostatic properties of the rotating wall, but also allowed for visual inspection of pattern propagation. Several of the segregation and electrostatic effects observed included several unique banding patterns, particle-wall clinging, particle coating phenomena, and even the "snaking" of particles. Further experiments were conducted utilizing a Van der Graaff generator as a way to engraft additional electrostatic effects and interactions. Several unique banding patterns and aggregate patterns emerged as a result of the additional electric field. Additional experiments were conducted to isolate the electrostatic effects of the blender wall by flowing binary polymer blends down a 30 degree acrylic chute. The results of this work aim to elucidate the problems that plague several particulate industries and explore ways to predict and limit electrostatic effects during particulate blending.

See more of this Session: Student Poster Session: Catalysis and Reaction Engineering
See more of this Group/Topical: Education

Design Studies of Heterogeneous Catalysis in Microchannels

Jennifer Lewis, Chemical Engineering, University of Tennessee at Chattanooga, Chattanooga, TN

James Hiestand, Mechanical Engineering, University of Tennessee at Chattanooga, Chattanooga, TN

Frank Jones, Chemical Engineering, University of Tennessee at Chattanooga, Chattanooga, TN

This is a numerical study of the fluid flow and catalytic behavior in microreactors to produce biodiesel fuel. Vegetable oil and alcohol feed flows into a very small channel about the size of a human hair. The microchannel cross dimensions are 200 microns or smaller. The channel walls contain obstructions to add reactive surface area and to redistribute the fluid. These microreactors have a metal oxide catalyst coated on all surfaces. Due to the micro scale and heterogeneous catalysis, processing time is reduced from hours to seconds, and product purity is greatly increased relative to conventional approaches. A computer simulation study was conducted to understand how changes in the cross sectional area, flow rate, and reaction rate affect conversion. The microreactors were simulated using CFD-ACE+ software. The arrangement of obstructions was the result of a previous study. The focus of this study is to optimize the cross sectional area and flow rate to more efficiently meet conversion goals, taking into account micro fabrication and operational constraints, such as pressure drop. Simulations show that high levels of conversion are obtained in seconds.

See more of this Session: Student Poster Session: Catalysis and Reaction Engineering
See more of this Group/Topical: Education

New Enzyme Catalyst for the Production of Functionalized Molecules Using Escherichia Coli

Yadi Lopez Ferrer, Chemical Engineering, University of Puerto Rico Mayaguez Campus, Mayaguez, PR

Tracie Hennen-Bierwagen, Biochemistry, Biophysics, and Molecular Biology, Iowa State University, Ames, IA

Tom Bobik, Biochemistry, Biophysics, and Molecular Biology, Iowa State University, Ames, IA

One of the greatest concerns in the world right now is the use of non-renewable chemicals. In an attempt to diminish their use in the chemical industry, a group of researchers got together to produce bio-renewable materials through biological and chemical catalysis systems integration. Progress has been made by conducting fundamental research to address the underlying technical challenges in the development of new integrated catalytic systems. These will aid in the conversion of bio-based feedstock to industrial chemicals. The program behind this initiative has been divided in three Thrusts, and our part of Thrust 1 involves the use of genetically engineered Escherichia Coli for the creation of novel enzyme catalysts capable of (efficiently) converting glucose to the suite of molecules shown below. We have cloned all the genes known to catalyze reactions A-H shown below. In addition, the first four protein expressions have been characterized.

See more of this Session: Student Poster Session: Catalysis and Reaction Engineering
See more of this Group/Topical: Education

Role of Catalyst Nanoparticle Size On the Growth of Single-Walled Carbon Nanotubes

Humberto Reyna, Artie McFerrin Department of Chemical Engineering, Texas A&M University, College Station, TX

Juan C. Burgos, Artie McFerrin Department of Chemical Engineering, Texas A&M University, College Station, TX

Boris Yakobson, Department of Mechanical Engineering and Materials Science, Rice University, Houston, TX

Perla B. Balbuena, Artie McFerrin Department of Chemical Engineering, Texas A&M University, College Station, TX

This study investigates the effect of metal catalyst size on the growth of single-walled carbon nanotubes (SWNT). The synthesis process of SWNTs is a target of many studies. An understanding of how they grow will open more possibilities for controlling the structure and properties of SWCTs. During the synthesis of SWNTs, numerous variables such as temperature, flow rate of the precursor gas, and diffusion of carbon inside and outside the surface play a vital role in the growth of SWNTs. Reactive Molecular Dynamics simulations are performed in order to determine which catalyst sizes are favorable for growth and which are not. The use of MD simulations allows for observation at initial stages of the synthesis process which provides the work of adhesion between the metal nanoparticle catalyst and graphite structure. Our results illustrate the diameter of the fullerene corresponds to the size of the nanoparticle, hence, its curvature energy. When a catalyst size is favored for growth, the SWNT is characterized by a cap lift-off; otherwise, a graphitic encapsulation occurs. Therefore, the work of adhesion must be weak enough to obtain curvature energy similar to a cap lift-off.

See more of this Session: Student Poster Session: Catalysis and Reaction Engineering
See more of this Group/Topical: Education

The Production and Characterization of Novel Bimetal Catalysts

Ian Moltrup, Chemical Engineering, University of Rochester, Chaffee, NY

Benjamin Kremkow, Chemical Engineering, Michigan State University, Hudsonville, MI

Most industrial catalysts involve distributing metal particles onto a support structure. However, this method suffers from the phase separation of metals after deposition on the support. The present study created an aerosol reactor which aims to rid the catalysts of this phase separation by creating metal nanoparticles from a molten pool of metals. Previous experiments with the reactor focused on making single metal nanoparticles and did not have an accurate way to measure temperature. A thermocouple was integrated into the reactor and was continuously tested and adapted to increase the accuracy of temperature measurements. Since catalysts most often involve two metals, tests were performed with multiple metals including titanium, gold, iron, and palladium particles. These bimetal catalysts are used in many petrochemical and chemical industries, including water-gas shift reactions. Various ratios of these metals were tested to find an appropriate mixture for the most efficient titanium and gold catalysts and iron and palladium catalysts. Several preparation techniques for Inductively Coupled Plasma mass spectroscopy analysis, ICP, were analyzed to determine the most accurate preparation procedure. A dissolution solution of aqua regia solution and a dilution solution of 4% HCl produced the most accurate ICP results. Transmission Electron Microscope, TEM grids were analyzed at different magnifications to determine the particle size distribution and lattice spacing of the molecules. Catalyst experiments are performed for the catalysts in the water-gas shift catalysis reactor. The ICP and TEM analysis methods determined the optimal quench gas and the optimized flow rates that minimize the size distribution to be argon gas, with a stream flow rate of 3.94 L/min and a quench flow rate of 4.89 L/min. The results from this study lay down a solid foundation for future experiments to refine the process and produce more efficiently active catalysts.

See more of this Session: Student Poster Session: Catalysis and Reaction Engineering
See more of this Group/Topical: Education

Synthesis and Characterization of Gold-Platinum Bimetallic Catalyst Prepared by Electroless Deposition Method

Abraham A. Rodriguez, Chemical Engineering, University of Puerto Rico, Mayagüez, Mayaguez, PR

Jayakiran Rebelli, Chemical Engineering, University of South Carolina, Columbia, SC

Christopher T. Williams, Chemical Engineering, University of South Carolina, Columbia, SC

John R. Monnier, Chemical Engineering, University of South Carolina, Columbia, SC

Electroless deposition (ED) has been used for the preparation of continuous, thin film metal coatings, but also has the potential for the preparation of novel, bimetallic catalysts by adjustment of the deposition parameters. Unlike traditional methods for the preparation of bimetallic catalysts, ED permits the selective deposition of a secondary metal (in this case Au) on the surface of a catalytically active, primary metal surface. In order to prepare catalysts using this technique it is necessary to design a reactive, yet stable ED solution by the careful selection of the secondary metal salt and reducing agent(s), as well as the pH and temperature of the ED bath. In the past, stable electroless solutions have been developed for temperatures close to room temperature (25°C). Now, electroless solutions have also been studied for higher temperatures (60°C) for deposition of Au on Pt/SiO_2. Stable electroless solutions have been developed using $AuCl_4^-$ and $Au(CN)_2^-$ as metal sources and HCHO and N_2H_4 as the reducing agents, respectively. In order to determine the stability and specificity of the bath, the ED solutions were exposed to the support (in this case SiO_2) to ensure that neither metal salt adsorption nor spontaneous reduction and precipitation of the secondary metal from solution occurred. Thermodynamic stability of the bath over time has been assessed for the different bath compositions using UV-Vis Spectroscopy. Bath stability is observed to decrease over time and with increasing temperature and increasing reducing agent concentration and initial Au concentration. A series of $Au-Pt/SiO_2$ catalysts have been prepared and characterized using chemisorption (O_2-H_2 Titration). A decrease in H_2 uptake on Pt with increasing Au content corroborate that the Au is being deposited on the primary metal. Effects of temperature and composition in deposition and deposition rate have also been discussed.

See more of this Session: Student Poster Session: Catalysis and Reaction Engineering
See more of this Group/Topical: Education

Impregnation of MCM-41 and Carbon Framework with Fischer Tropsch Catalyst

Lauren A. Pulley, Chemical and Biomolecular Engineering, Vanderbilt University, Nashville, TN

Fischer-Tropsch Synthesis is used to produce liquid hydrocarbons from a synthesis gas mixture of carbon monoxide and hydrogen. Fischer-Tropsch employs the use of transition metal catalysts, such as Ni, Fe, and Co.

This project aimed to produce a Fischer Tropsch catalyst. A carbon framework was created in MCM-41 using a sucrose solution. Next, the MCM-41 and carbon framework were impregnated with an iron oxide using supercritical carbon dioxide. The iron was then reduced to a metallic state and Fischer Tropsch Synthesis was performed to produce long chain hydrocarbons, such as diesel fuel and gasoline.

See more of this Session: Student Poster Session: Catalysis and Reaction Engineering
See more of this Group/Topical: Education

Modification of the BINOL Skeleton: Application to Enantioselective Catalysis

Jorge L. Santiago-Ortiz, Chemical Engineering, University of Puerto Rico, Mayagüez Campus, Corozal, PR

Bruno Andrioletti, Chemistry, Institute de Chimie et Biochimie Moléculaires et Supramoléculaires, Université Claude Bernard Lyon 1, Villeurbanne, France

A new 1,1'-binaphthyl-2,2'-diol (BINOL) derivative that is soluble in water was successfully synthesized and characterized by performing several modifications on the BINOL skeleton, including methylation of the OH groups, and bromination, Sonogashira coupling, and quaternization in the 6 and 6' sites of the molecule. The product was preliminarily studied in a catalytic test and proved to act as an effective phase transfer catalyst in an alkylation reaction, yielding conversions of approximately 85%. Studies are currently underway to further understand the effectiveness of the compound as a catalyst and to investigate the presence of enantioselectivity.

See more of this Session: Student Poster Session: Catalysis and Reaction Engineering
See more of this Group/Topical: Education

Study of Isopropanol Oxidation On Au-Pd/SiO2 Catalysts Using Temperature Programmed Techniques

Jose Javier Navarro, Chemical Engineering, University Of Puerto Rico, Mayaguez, Mayaguez, PR

Jayakiran Rebelli, Chemical Engineering, University of South Carolina, Columbia, SC

Christopher T. Williams, Chemical Engineering, University of South Carolina, Columbia, SC

J.R. Monnier, Chemical Engineering, University of South Carolina, Columbia, SC

Au-Pd/SiO2 catalyst are bimetallic catalyst that are prepared using electroless deposition, in which the second metal, in our case Au, will be deposited on the surface of a catalytically primary, primary metal surface (in our case Pd). Catalyst with different Au coverage will be synthesized in order to study the interaction of isopropanol and acetone with the catalyst. Temperature program desorption (TPD) will be performed in order to study the adsorption properties of the catalyst towards isopropanol and acetone. TPD studies will also be performed for the dehydrogenation of isopropanol into acetone for the same bimetallic catalyst. A TPD study for isopropanol and acetone are carried out since is desired to understand the interaction with the catalyst before studying the reaction. The TPD studies for isopropanol and acetone shows maxima adsorption at certain Au coverage, then the adsorption decreases as the Au coverage continue to increase. The same behavior is noticed for the temperature of maximum desorption, for isopropanol and acetone. It was determined form the TPD that, the amount of acetone adsorbed on the catalyst is greater than for isopropanol in which the fractional Au coverage is less than 0.5, for greater coverage, isopropanol adsorption is greater. The dehydration of isopropanol was studied on Pd/SiO2 at different reaction temperature. After the reaction being carried out, TPD studies where performed in order to determine if materials strongly adsorbed on the catalyst. For low reaction temperatures (lower than TMD of acetone for the catalysts studied), the activity of the catalyst is low and also deactivation is present, since acetone is adsorbed on the catalyst. For reaction temperatures higher than TMD higher catalyst activity is achieved and few or no appreciable deactivation is shown, which means that little or none acetone was present on the catalyst when the TPD was performed.

See more of this Session: Student Poster Session: Catalysis and Reaction Engineering
See more of this Group/Topical: Education

Optimizing the Catalytic Cycle for the Dehydration of Biobased Glycerol to Economically Viable C3 Compounds

Luke Richardson, Chemical and Materials Engineering, University of Kentucky, Paducah, KY

Jeffrey R. Seay, Department of Chemical and Materials Engineering, University of Kentucky, Paducah, KY

Glycerol is a byproduct of biodiesel and has a limited amount of uses in industry. However, numerous economically viable C3 compounds such as acrolein, acrylic acid, 1,2-propanediol, 1,3-propanediol, propionaldehyde and hydroxyacetone can be produced from glycerol. If the biodiesel industry continues to grow at current rates there will be a surplus of glycerol. A catalytic process has been developed for dehydrating glycerol, but recently published research shows that the catalyst life is unquestionably too short for a typical industrial process. The catalyst used for this process has such a short life because solid carbon builds up and coats the catalyst carrier particles. The catalyst is effective over short time spans but since the life is so short it is not practical for an industrial process. There are a number of things that can be tried to extend this life span. The reaction with the current catalyst is very fast and produces a large quantity of solid carbon. This is the major problem that needs to be addressed. One possible approach is to utilize a less active catalyst that meets the conversion and yield objectives without completely reducing the products to solid carbon. It is hypothesized that this will increase the useable life of the catalyst and increase efficiency and productivity of the process due to minimizing costly catalyst changes. In order to increase the total carbon utilization, steam and hot air will be supplied to the carbon build up on the catalyst in order to produce syngas. If this is successful the syngas can be used as an energy source for the process. The experiment uses a packed bed reactor to house the catalyst and measure kinetics of the catalytic reaction. Integral reactor analysis will be used. It is subject to temperature variation, in order to get valid data the process has to run until it reaches steady state. At steady state the measurements of concentrations are read at the inlet and exit and used to calculate the conversion rate, yield rate and selectivity. Currently the selectivity and yield rates are insufficient for a long time period. By changing and comparing one variable at a time the limiting stage will be determined. The limiting stage is hypothesized to be either diffusion of glycerol to the catalytic site or the diffusion of product out of the pore and away from the site. In order to determine if this is the problem the catalytic carrier will be crushed into a powder, this should have a dramatic impact on the selectivity of the process. This research will be profitable to the biodiesel manufacturing industry in order to improve over all carbon utilization of the process. Also, due to a recent tax-credit for biodiesel manufacturing if the cost of manufacturing this renewable energy source can be decreased, it will be possible to decrease the impact of energy the global climate by the use of fossil fuels. The results of this research will result of farther development of domestic resources. This research is an important and viable step in enabling the conversion of the abundant by-product glycerol to marketable C3 compounds.

See more of this Session: Student Poster Session: Catalysis and Reaction Engineering
See more of this Group/Topical: Education

Use of Electrochemical Processes to Regenerate Sodium Hydroxide From Brine

Nicholas Salvetti, Chemical Engineering, Rowan University, Freehold, NJ

Craig Wiltsey, Chemical Engineering, Rowan University, Freehold, NJ

Matthew Angelow, Chemical Engineering, Rowan University, Freehold, NJ

A lab scale chlor-alkali system was designed in order to demonstrate the feasibility of the process for recycling a scrubber product brine solution to regenerate caustic for the Johnson Matthey company. An additional goal was to optimize the technology for use with this solution. An electrolytic cell containing a Nafion 324 membrane, a RuO2 based coating on Ti substrate cathode, a Steel or Ni based catalytic coating on Ni anode, tubing, and components resistant to corrosion were purchased. From the laboratory studies using this cell it was determined that the spent scrubber fluid can be economically converted to NaOH. Additionally, the optimum temperature for this process was found to be 90°C. The efficiency, based on the amount of electrical energy input into the cell electrodes, is 85% at room temperature (20°C) and is a maximum of 95% at 90°C. From the experiments performed it was determined that an electrode area of 100cm2 was able to produce NaOH at a rate of 0.1422 mol/h. Scaling this to the JMI processing requirements to produce the 800 gallons of NaOH at a 15%w with the spent scrubber solution treatment requires a system with an electrode area of 64m2. This system has a capital cost of approximately $631,462 with an estimated payback period of 3.86 years.

See more of this Session: Student Poster Session: Catalysis and Reaction Engineering
See more of this Group/Topical: Education

Hydrotreating of C9 Oxygenates Under Supercritical Reaction Conditions

Héctor J. Camareno Ojeda, Chemical Engineering, University of Puerto Rico at Mayagüez, Mayagüez, PR

Jackson W. Ford, Center for Environmentally Beneficial Catalysis and Department of Chemical & Petroleum Engineering, University of Kansas, Lawrence, KS

Bala Subramaniam, Center for Environmentally Beneficial Catalysis, Chemical & Petroleum Engineering Department, University of Kansas, Lawrence, KS

The use of bio-oils as transportation fuel has been limited due to stability and quality issues caused by oxygenated groups. The oxygen content may be reduced by hydrodeoxygenation to produce hydrocarbons and water from bio-oil (Babu, 2008). The water is immiscible with the hydrocarbons and partitions completely into a separate phase. The resulting hydrocarbon layer may then be used for processing to transportation fuels in conventional refineries. To perform such hydrotreating at mild pressures, supercritical hydrogenation of model bio-oil oxygenates was investigated. The $Pt/Al2O3$ - catalyzed hydrogenation of nonanal was investigated in supercritical hexane in a fixed bed reactor at 300°C and pressures of 30-50 bar with a stoichiometric excess of H2 to produce nonane. Other products were liquids nonanol and octane as well as light gases. The product profile at various operating conditions and preliminary conclusions are presented regarding the effectiveness of the proposed hydrotreating approach.

See more of this Session: Student Poster Session: Catalysis and Reaction Engineering
See more of this Group/Topical: Education

Analyzing the Distribution of Water around the Cellobiohydrolase I Linker Peptide

Joseph Mikhail, Department of Chemical and Biomolecular Engineering, Vanderbilt University, Nashville, TN

George Rouvelas, Department of Chemistry, Vanderbilt University, Nashville, TN

Courtney Taylor, Department of Chemical and Biomolecular Engineering, Vanderbilt University, Nashville, TN

Clare McCabe, Department of Chemical and Biomolecular Engineering, Vanderbilt University, Nashville, TN

Cellulose, as the most abundant carbohydrate on earth, holds great potential as an alternative renewable source of transportation fuel through its hydrolysis and subsequent fermentation. Cellobiohydrolase I (CBH 1) from Trichoderma reesei is one of the most active enzymes in the hydrolysis of cellulose. CBH 1 is a multi-domain enzyme, consisting of a large catalytic module containing an active site tunnel and a small cellulose binding module, joined to one another by a 27 residue linker peptide. The goal of this work is to study the density of water around the cellulase enzyme above a cellulose surface via molecular dynamics simulations. Analysis of the water density surrounding the enzyme indicates that the sugar molecules on the linker peptide affects the water structuring around the enzyme and above the cellulose substrate.

See more of this Session: Student Poster Session: Computing and Process Control
See more of this Group/Topical: Education

Multivariate Image Analysis, Histogram Matching, Process Control Chart and Acoustic Signal Assisted Nucleation Detection

Kaoutar Abbou Oucherif, Chemical Engineering, New Mexico Institute of Mining and Technology, Socorro, NM

Levente L. Simon, Institute for Chemical and Bioengineering, Safety and Environmental Technology Group, ETH Zurich, Zurich, Switzerland

Nagy Zoltan, Chemical Engineering, Loughborough University, Loughborough, England

Konrad Hungerbuehler, Chemical Engineering, ETH Zurich, Zurich, Switzerland

The aim of this work is to investigate several image processing and process monitoring technologies to provide systematic, sensitive, robust and low complexity-solutions for nucleation detection using bulk video imaging (BVI). The image analysis techniques are applied for the crystallization of caffeine as a model for active pharmaceutical ingredients, both externally and in-situ. The image frames are analyzed using feature index monitoring where signals for the mean gray intensity, the maximum intensities of the color channels, and the first principal components (PC1) images are calculated. The resulting time series trends are monitored using Shewhart and exponentially weighted moving average (EWMA) statistical process control charts (SPC) where nucleation is detected when the signals exceed the set control limits. A second approach to nucleation detection is histogram matching and computation of distance measures. Histograms of the images are generated and similarity measures are applied to detect a significant change indicating the beginning of nucleation; the current study evaluates the use of Kolmogorov-Smirnof and X2 statistics to determine the statistically significant change between the observed and the reference histograms for a 95% confidence interval. SPC charts were applied to the distance measures. The third approach relies on applying multiway principal component analysis (MPCA) on the color images to obtain the first score, which is also an image. The PCA model residual monitoring performance was also investigated. For acoustic signal based nucleation monitoring, the gray scale images were converted into a 2-channel stereo sound. It was found that this method has less performing nucleation detection capabilities compared to the methods which directly rely on the images. The results demonstrate that MPCA based monitoring is a suited technique for nucleation onset detection. The Statistical control chart assisted monitoring also successfully detect the nucleation onset. Furthermore, the histogram matching methods are not able to provide accurate monitoring due to significant process fluctuations when crystals are not present in solution.

See more of this Session: Student Poster Session: Computing and Process Control
See more of this Group/Topical: Education

Introducing Tissue Engineering Scaffolds to Middle School Teachers and Students

Marcus A. Duffy, School of Chemical Engineering, Oklahoma State University, Stillwater, OK

Karen High, School of Chemical Engineering, Oklahoma State University, Stillwater, OK

Sundararan. V. Madihally, School of Chemical Engineering, Oklahoma State University, Stillwater, OK

Providing K-12 students an earlier exposure to engineering is recognized as a way to prepare them for postsecondary courses. There has been a significant increase in academies and workshops addressing this concept. Since there is a total turnover of students each year, training K-12 teachers has been thought as an efficient approach. Further, teachers can integrate engineering concepts during the academic year which provides a natural transition. Based on this concept, two middle school teachers (teaching Biology and Math) were selected to participate in research dealing with tissue engineering. Teachers worked for six weeks (four days a week) within the research laboratory on formation of porous structures using biodegradable polymers. Teachers were exposed to forming porous structures using chitosan and gelatin solution in various forms using the apparatus available in the laboratory. A low cost freeze drying system that is safe for operation by sixth grade students was developed. The overall cost of performing the experiment is also significantly cheap and less time consuming. An envisioned project the sixth grade students will work on during the academic year is freeze drying chitosan-gelatin solutions. Biology teacher will ask the students to make scaffolds that can support the regeneration of various body parts. Students will survey in the beginning to determine what polymers would be best for use and why. Once they determine what polymer they need, then they will freeze and air dry the premade chitosan-gelatin solution to form 2D and 3D structure. Another math teacher will help them understand the pore size, pore number and void fraction calculation using scanning electron micrographs given by me. Students will be asked to make scaffolds that can mimic a specific body part. We anticipate that the entire hands-on experience will be a positive reinforcement to their creative learning experience and encourages them to biomedical engineering/sciences.

See more of this Session: Student Poster Session: Education
See more of this Group/Topical: Education

Development of Educational Materials Related to Drug Delivery Systems

Kathryn Whitaker, Chemical Engineering, Rowan University, Glassboro, NJ

Eileen Batten, Chemical Engineering, Rowan University, Glassboro, NJ

Vladimir De Delva, Chemical Engineering, Rowan University, Glassboro, NJ

Dr. Stephanie Farrell, Chemical Engineering, Rowan University, Glassboro, NJ

C. Stewart Slater, Chemical Engineering, Rowan University, Glassboro, NJ

Students from Rowan University's chemical engineering department are working with the Center for Structured Organic Particulate Systems (C-SOPS), funded by the National Science Foundation (NSF) and located at Rutgers University, to create educational materials based on drug delivery systems. These educational materials introduce engineering concepts at the high school and college level. The materials developed include an introduction to drug delivery, a web-based experiment on tablet production and drug release testing, photographs of tablet production equipment, and a problem set with solutions. The problems developed vary in difficulty and technical content. The problems focus on drug release systems and pharmaceutical processes to teach introductory engineering concepts such as unit conversions, the Ideal Gas Law, and mass and energy balances. The educational materials will be distributed via the Pharmahub website at https://pharmahub.org.

See more of this Session: Student Poster Session: Education
See more of this Group/Topical: Education

Learning Fugacity, Fugacity Coefficient and the Phase Equilibrium Criteria through 3D Phase Diagrams for Pure Components

Mateo Gómez-Tena, Departamento de Ingeniería Química y Alimentos, Universidad de las Américas, Puebla, Cholula,Puebla, Mexico

José.R. Flores-Tapia, Departamento de Ingeniería Química y Alimentos, Universidad de las Américas, Puebla, Cholula,Puebla, Mexico

Luis G. Ríos-Casas, Departamento de Ingeniería Química y Alimentos, Universidad de las Américas, Puebla, Cholula,Puebla, Mexico

Nelly Ramírez, Departamento de Ingeniería Química y Alimentos, Universidad de las Américas, Puebla, Cholula,Puebla, Mexico

Alejandra Alarcon-Garcia, Projects Department, Ciba, S.A. de C.V. (part of the BASF Group), Puebla, Mexico

In undergraduate Chemical Engineering programs, it is evident the complexity that students have to handle when the professor starts the study of thermodynamic properties and phase equilibrium criteria, but a deeper understanding could be attained by adding a visual approach.This work is related to a previous one regarding the representation of analysis of Gibbs Free Energy in 3D diagrams. The thermodynamic properties data for water were plotted in 3D to represent the resulting equations for G (Gibbs free energy).The database was taken from Haar et. al. (1984). The 3D diagrams obtained were used to generate projections in 2 dimensions: P-G, G-P, T-G, G-T, for homogeneous fluid phases and vapor-liquid equilibrium. This time it is focused on the behavior of concepts like fugacity and fugacity coefficient, as additional equilibrium criteria. These thermodynamic definitions are obtained from the Gibbs Free Energy and are calculated with an equation of state. The 3D diagrams for ln f as a function of P and T, are presented. The mesh is constructed with values at constant temperature and pressure and the curves for constant vapor fraction are included inside the vapor-liquid region. The 3D diagrams obtained can be used to generate projections in 2 dimensions ln f-P.In the last few years, this methodology has been applied in the Thermodynamics course for undergraduate Chemical Engineering students. The professor has perceived that students show good acceptance of this educational material and the evaluations have revealed a better understanding of these concepts.

See more of this Session: Student Poster Session: Education
See more of this Group/Topical: Education

The Repairing of Misconceptions in Thermodynamics through the Use of Inquiry Activities

Jeff Stein, Department of Chemical Engineering, Bucknell University, Lewisburg, PA

Misconceptions exist in all fields of learning and develop through a person's preconception of how the world works. The focus of this research is to repair misconceptions in thermodynamics by using inquiry-based activities. Inquiry-based learning is a method of teaching that involves hands-on learning and self-discovery. Previous work has shown inquiry-based methods result in better conceptual learning by students relative to traditional lectures. The thermodynamics activities were designed to guide students towards the correct conceptual understanding through observing a preconception fail to hold up through an experiment or simulation. The developed activities focus on the following topics in thermodynamics: Internal Energy versus Enthalpy, Equilibrium versus Steady State, and Entropy. For each topic, two activities were designed to clarify the concept and assure it is properly grasped. Each activity is described in a packet containing experimental instructions along with pre- and post- analysis questions. All activities and packets are performed in small groups as hands-on experiments or simulations, require a written prediction and comparison to the observations, and draw the students' interest. Preliminary data were collected this summer and showed an improvement in conceptual understanding for students with misconceptions. Currently, the activities are being tested at a number of universities.

See more of this Session: Student Poster Session: Education
See more of this Group/Topical: Education

Design of Experiments to Optimize the Production of Chemical Products From Waste Crude Glycerol

Jason B. Gish, Chemical and Materials Engineering, University of Kentucky, Paducah, KY

Jeffrey R. Seay, Chemical and Materials Engineering, University of Kentucky, Paducah, KY

Biodiesel is a sustainable and clean source of energy both in the manufacturing process and upon consumption. Additionally, it is produced from renewable resources such as vegetable oils or animal fats. However, the process of manufacturing biodiesel from vegetable oils generates glycerol as a byproduct. Biodiesel production yields 1 kg of glycerol for every 9 kg of biodiesel. Previous research has identified numerous industrially important chemical products that can be manufactured from glycerol, such as the industrially important specialty chemical acrolein. Developing process models for the production of acrolein from glycerol will be useful because it will identify which products are the most economically viable and establish cost targets for switching production from crude oil derived propylene to glycerol.

This contribution is focused on development the process simulation models that will serve as the basis for a flexible economic model for determining the optimal product slate for a multipurpose reactor based on glycerol dehydration chemistry. A design of experiment approach will be utilized in the aid of finding the optimum operating conditions of the system. This analysis will be part of a holistic approach performed by applying process simulation in combination with laboratory experiments for the conceptual design of chemical processes.

See more of this Session: Student Poster Session: Environmental
See more of this Group/Topical: Education

Recycling and Its Importance

Derek A. Smith, Department of Chemical and Biomolecular Engineering, University of South Alabama, Mobile, AL

Dr. Srinivas Palanki, Chemical and Biomolecular Engineering, University of South Alabama, Mobile, AL

Dr. Shafik Sadek, Chemical and Biomolecular Engineering, University of South Alabama, Mobile, AL

Recycling is an important practice that consumers and industry alike can do to help conserve non-renewable resources (fossil fuels) that are used for energy and help the environment in which we live. Recycling efforts include recycling paper, glass, plastics, and metals. This study investigates the efforts that the government and local municipalities take to encourage recycling among consumers and private industries, as well as, the amount of energy and non-renewable resources conserved when products are recycled. Results indicate that local and government programs have been designed to promote recycling and encourage purchase of recycled materials vs. raw materials. As a result of using recycled materials, energy consumption is decreased and raw materials are conserved. Recycling also reduces the destruction of forests and natural habitats due to the mining of raw materials, as well as, water and air pollution via the production process using raw materials. The resulting effect of recycling is only a small part of conserving energy and fossil fuels; recycling must be conjoined with other conservation efforts like renewable energy resources to help rid the total dependence on fossil fuels.

See more of this Session: Student Poster Session: Environmental
See more of this Group/Topical: Education

Energy Conservation in the United States

Caron S. Kost, Department of Chemical and Biomolecular Engineering, University of South Alabama, Mobile, AL

Dr. Srinivas Palanki, Department of Chemical and Biomolecular Engineering, University of South Alabama, Mobile, AL

Dr. Shafik Sadek, Department of Chemical and Biomolecular Engineering, University of South Alabama, Mobile, AL

Energy conservation is an important part of the United States' energy future. Presently, the primary sectors for energy consumption are electricity, transportation, and industrial. Any conservation program must be centered on these three sectors in order to be effective. This study investigates government initiatives to conserve energy in each of these sectors. Results indicate that current government efforts such as the American Recovery and Reinvestment Act, hybrid vehicle tax credits, and the Landfill Methane Outreach Program will do very little to decrease energy usage in the US. However, energy conservation can decrease the rate at which energy consumption increases. In order for the United States to implement a sustainable and cost effective energy policy, conservation must be combined with renewable energy sources in order to phase out the use of fossil fuels as a primary energy source.

See more of this Session: Student Poster Session: Environmental
See more of this Group/Topical: Education

The Use of Residues as New Energy Resources

Daniel Esteban Gómez Iñiguez, Chemical Engineering, SOLBEN A.C. and ITESM, Monterrey, Mexico

As global demand intensify and prices rise, energy companies are going to the ends of the earth to uncover new supplies. However, researchers hadn't considered the use of several simple residues from industrial processes that could be used as an energy source. This presentation is not a usual presentation; the idea of my poster is joining different researches I had done during the last year and present it in a macro project that relates every investigation in a single poster. This poster would not include every detail of all the researches but I would be prepared to answer with facts and proves any question anyone could have about all of them. Everything started with a Biodiesel production efficiency research that involved different kind of oils for Biodiesel production and as a result I concluded that Jatropha Curcas oil haves lot of advantages not only in production but in the whole process of oil extraction and environmental impact to our society. Moreover I was able to make decantation process in minutes by using solar energy in the process instead of waiting for ours and I was able to prevent the water waste in the secondary saponification reactions by using an Ionic exchange resin for purifying biodiesel before using it. After this, I was part of the installation of the Biofuels Quality Testing Labs at the ITESM and I started making several ASTM proves to the Biodiesel I got from every experiment for the future Mexican ASTM Rules of this energy source. Afterward, I was really concerned about the greatest residue of biodiesel production, glycerin. Biodiesel industry producers and consumers are considering expanding flood of crude glycerin. The U.S. biodiesel industry is expected to produce about 1.4 billon pound of glycerin between 2006 and 2015, according to a study of the LECG Inc. director, John Urbanchuk. Actually, around 100 kg of glycerin are produced per each ton of Biodiesel. In addition, sawdust has been a wood residue without many possibilities of utilization. Considering that the annual production of wood in Mexico is of more than 8 million of cubic meters, from which the annual production of sawdust surpasses 2,800,000 cubic meters, sawdust volumes are expected to increase. Its accumulation represents a serious problem of environmental contamination, especially where the soil is deposited. Therefore the development of alternatives for its operation is considered a subject of interest for Worldwide Wood Industries. Making a proportioned mixture between glycerin and sawdust, in an approximate relation 1:3 might be part of the solution for lots of sustainability problems. The result of the mixture would be a viscous blend, and by employing a mechanical compression we might produce sawdust-glycerin logs. These blocks have a greater efficiency than firewood, because glycerin acts as an ignition source for sawdust. Definitely, burning glycerin at high temperatures provokes the discharge of several gases, but therexs a method to control them. For controlling emissions, the project is based on burning these blocks inside a closed chamber where gases would be captured for preventing toxic emissions to our surroundings and by making a geological storage of carbon dioxide we could help the regeneration of feature petroleum reserves and avoid contaminating our world. The energetic result of the blockxs combustion would be used for generating water vapor. This steam would pass through a generator turbine giving as a result electric energy obtained from a green design in a sustainable price, due to the effective waste management and the efficient use of energy. The real value of the wood waste as a fuel source must take into account its available heat content, the operating costs of production needed to handle and convert it to useable energy, before any worthwhile comparative studies can be made. This innovation investigation was awkward by the Mexican Chemistry Nobel Price, Dr. Mario Molina and nowadays this procedure is being applied in the SOLBEN A.C. Company. Last summer another project that consisted in the Biodiesel production with algae was done in Heidelberg, Germany at the IFEU, Institute of Energy. I was working during a month in the oil extraction process and algae growth for lowering its biodiesel production cost. This research was completed at Nove Hrady, Czech Republic with an algae growth investigation where I used the Biogas Fugat (Residue from Biogas production) in a mixture with water to improve the algae growth rate. Impressively, the results were really positive. We were able to grow 47% times faster Scenedesmus Dimorphus algae for its future oil extraction and a pure Biodiesel production. This by just giving a new use to the fugat that was being throw away by some companies and otherxs where using it as a liquid fertilizer. Unlike most other investigators, I founded a company (SOLBEN A.C.) where all this projects and other ones are being applied. Bioenergy producers are using various wastes for creating energy by a viable and sustainable way, supporting directly the energetic problem that is approaching in many nations as well as the problem of sustainability and global warming. At the same time the objective of my company is helping the development of rural areas in my country by reactivating camps with the plantation of Jatropha Curcas and making these towns auto-sustainable. This year we constructed a Biodiesel pilot plant and we are going to start planting 35,000 hectares of Jatropha Curcas for oil exporting to Spain where oil would be converted into Biodiesel. This company is purely Mexican and shows how researchers, no matter their age could launch their projects into the business market to help their environment and at the same time the people of our world. Finally I would like to I recognize the assist of Dr. Miguel Angel Romero Ogawa, Dr. Oliver Probst and Ing. Vítězslav Březina.

See more of this Session: Student Poster Session: Environmental
See more of this Group/Topical: Education

Enhancing Bioremediation by Exploiting Bacterial Chemotaxis: 2-Dimensional Microcosm Studies

Kathryn L. Strobel, Chemical Engineering, University of Virginia, Charlottesville, VA

Christian Griebler, Institute of Groundwater Ecology, Helmholtz Zentrum München, Neuherberg, Germany

Jun Liu, Chemical Engineering, University of Virginia, Charlottesville, VA

Roseanne M. Ford, Chemical Engineering, University of Virginia, Charlottesville, VA

In-situ bioremediation is limited by the inability to deliver the bacteria to the source of the chemical contaminant which they will degrade. Chemotaxis, the ability of some bacteria to sense higher concentrations of chemical attractants and swim toward them, can improve the effectiveness of bioremediation by directing bacteria toward contaminant sources. In this study, a two-dimensional lab-scale model aquifer was used to determine the effect of chemotaxis on the transport of bacteria exposed to a chemoattractant concentration gradient in porous media. The model was packed with sand and an aqueous medium was pumped across horizontally, simulating groundwater flow in a homogenous aquifer. Sodium benzoate was injected continuously to generate a steady concentration gradient transverse to flow and chemotactic *P.putida* F1 or a non-chemotactic mutant of the same species was injected below the attractant. Samples were collected at the end of the microcosm to provide vertical concentration profiles of the bacteria and attractant. The chemotactic bacteria center of mass was found to be an average of 1.24 ± 0.86 cm closer to the attractant injection than the non-chemotactic mutant. These results suggest that chemotaxis can improve the transport of bacteria to contaminated zones, enhancing the effectiveness of *in-situ* bioremediation.

See more of this Session: Student Poster Session: Environmental
See more of this Group/Topical: Education

On the Prevention of Trace Metal Emissions From Coal Combustion

Ariel Atkins, Stanford University, Stanford, CA

Angela Boysen, Stanford University, Stanford, CA

Jennifer Wilcox, Energy Resources Engineering, Stanford University, Stanford, CA

Understanding the oxidation of selenium during the coal combustion processes is crucial to preventing its release into the atmosphere. Coal combustion provides a major source of energy, accounting for 52% of electricity in the US. Due to projected increases in coal use worldwide, attention has turned to the potentially harmful effects such as the well known problem of CO_2 emission and the emissions of various trace elements such as Se, Hg, and Ar. An *ab initio* study of selenium oxidation/reduction in coal combustion was conducted at the CCSD(T) level of theory with the aug-cc-pVTZ basis set. Twenty-nine different reactions involving Se, SeH, SeH_2, SeO, SeO_2, SeCl, $SeCl_2$, and SeOH were considered. For each species, calculated geometries and vibrational frequencies were compared to experimental data. Reaction enthalpies were calculated at a variety of temperatures and compared to experiment at 298.15 K. The Gibbs free energy of reaction and corresponding equilibrium constant were calculated for each reaction and plotted as a function of temperature to find the linear relationship between $\ln(k_{eq})$ and 1/T describing each reaction's thermodynamics. It was found that SeO and SeO_2 are the most thermodynamically favorable products and the formation pathways of SeCl, $SeCl_2$, and SeH_2 are also possible.

See more of this Session: Student Poster Session: Environmental
See more of this Group/Topical: Education

Toxicity of Nanoparticles: Cerium and Hafnium Oxides

Dominique A. Ingato, Chemical and Biomolecular Engineering, University of Pennsylvania, Philadelphia, PA

James A. Field, Chemical and Environmental Engineering, University of Arizona, Tucson, AZ

Reyes Sierra, Chemical and Environmental Engineering, University of Arizona, Tucson, AZ

Antonia Luna-Velasco, Chemical and Environmental Engineering, University of Arizona, Tucson, AZ

Nanoparticles have become an important component of many industrial processes and products over the past decade. In particular to this project, the semiconductor industry uses nanoparticles in CMP (Chemical-Mechanical Polishing) slurries during the planarization process. Currently, there are no effective methods of removing the suspended nanoparticles from the slurries before they are sent to waste-water treatment plants. The objective of this study was to determine the toxicity of the nanoparticulate slurry compoun, cerium oxide, on bacteria from waste-water treatment sludges as well as on aquatic bacteria and human cells. Cerium oxide nanoparticles were not toxic to methanogens but inhibited aerobic bacteria at high concentrations.

See more of this Session: Student Poster Session: Environmental
See more of this Group/Topical: Education

Economically Profitable and Environmentally Sustainable Conceptual Process Design of Biodiesel Byproduct Crude Glycerol

Kandace Y. Ramey, Chemical and Materials Engineering, University of Kentucky, Paducah, KY

Jeffrey R. Seay, Department of Chemical and Materials Engineering, University of Kentucky, Paducah, KY

A sustainable and clean source of energy both in the manufacturing process and upon consumption is biodiesel. Biodiesel is produced from renewable resources such as vegetable oils or animal fats. However, in the biodiesel production process, 1 kg of glycerol is generated for every 9 kg of biodiesel produced. This formulates large amounts of waste materials that could be used constructively. Previous research has identified numerous industrially important chemical products that can be manufactured from glycerol. In particular, many C3 compounds, currently manufactured from propylene can also be manufactured from bio-based glycerol. Developing process models for the production of these, and other products, from glycerol will be useful because it will identify which products are the most economically viable and establish cost targets for switching production from crude oil derived propylene to glycerol. This contribution is focused on developing a conceptual design of the glycerol dehydration process. The process designs produced from this research will be based on minimizing environmental impact while providing industries with economically efficient and technically feasible processes. This analysis will be part of a holistic approach performed by applying process simulation in combination with laboratory experiments for the conceptual design of chemical processes.

See more of this Session: Student Poster Session: Environmental
See more of this Group/Topical: Education

Geobacter Sulfurrenducens Growth in Iron Oxides Media

Karla M. Ortiz Perez, Department of Chemical Engineering, University of Puerto Rico, Mayagüez Campus, Mayagüez, PR

Ruth Helmus, Department of Biochemistry & Molecular Biology, Penn State University, University Park, University Park, PA

Lan Dang, Department of Chemical Engineering, Penn State University, University Park, University Park, PA

Ming Tien, Department of Biochemistry & Molecular Biology, Penn State University, University Park, University Park, PA

Wayne Curtis, Department of Chemical Engineering, Penn State University, University Park, University Park, PA

The Department of Energy (DOE) has become very interested in using microbial systems to remediate radioactive metal contaminants, such as Technetium (Tc) and Uranium (U). These materials are found naturally in their soluble form. Water sources, such as lakes, rivers, and aquifers, contaminated with these and other metal traces represent an emerging issue for environmental and human health. After some studies it is demonstrated that dissimilatory metal reducing (DMR) microorganisms, including different types of Geobacter are able to reduce metals like iron. DOE is interested in a system where Tc and U are present because the reduction of these ones results in their precipitation as insoluble oxides that are easier to take care. To study this system, researchers have used iron reduction as a model system. Geobacter sulfurrenducens is able to reduce iron insoluble ($Fe3+$) into iron soluble ($Fe2+$). Geobacter sulfurrenducens employ carbon sources as electron donors in order to reduce the studied metals (electron acceptors). The electron donor will be sodium acetate and the electron acceptor ferrihydrite ($Fe3+$ source). The mainly approach of this study is the development of a technique, using iron oxide as the electron acceptor, that allows the total reduction of metals. Iron is used to develop this procedure because it is possible to measure the $Fe2+$ present in solutions using a ferrozine assay. The method will be used to reduce metals that are persistent in groundwater but they are insoluble after being reduced. Insoluble metals will be easier to treat in order to have water free of these pollutants. The places where Tc and U are principally found do not have all the nutrients that Geobacter needs to survive. If the interested bacterium is not able to survive it will not be able to perform the reduction of the interested metals, for this reason the substrate usage rate that Geobacter need to survive is also presented.

See more of this Session: Student Poster Session: Environmental
See more of this Group/Topical: Education

"Essential Oil Constituents of Petiveria Alliacea L. by Soxhlet Extraction and GC-MS"

Edwin O. Medina-Santos, Chemical Engineering Department, University of Puerto Rico-Mayagüez Campus, Mayagüez, PR

Maritza De Jesus-Echevarría, Chemistry Department, University of Puerto Rico-Mayagüez Campus, Mayagüez, PR

Abstract

Since the end of the last century numerous studies have been carried out to investigating the insecticidal activity of herbs. Since 1949 the Petiveria alliacea L. plant have been recognized as a prospect source of natural insecticide. To explore the possibility of this field a study of the essential oil constituents of this plant was investigated. Petiveria alliacea L. is an herbaceous perennial that grows up to 1 m in height. It is indigenous to the Amazon rainforest and tropical areas of Central and South America, the Caribbean, and Africa. The soxhlet extraction technique was carried out with two different solvents, dichloromethane, CH2Cl2 and methanol, CH3OH using 0.4g of the plant powder. Gas Chromatography coupled to Mass Spectrometry, GC/MS was used to identify the essential oil constituents. The dichloromethane and methanol extract have several common compounds but methanol have more chemical constituents. The essential oil of the methanol extract was three times more than dichloromethane extract. The most abundance compounds is phytol follow by 4-vinyl-2-methoxy-phenol and phytol and hexadecanoic acid in methanol and dichloromethane extracts respectively. Some of the constituents are classify as insecticide, benzenecarboxylic acid, 5,6,7,7a-Tetrahydro-6-hydroxy-4,4,7a-trimethyl-2(4H)-benzofuranone, benzenemethanethiol, 4-vinyl-2-methoxy-phenol and others.

See more of this Session: Student Poster Session: Environmental
See more of this Group/Topical: Education

Air Pollution and Public Health in Mayagüez, Puerto Rico

Lysmarie Santos-Velazquez, Chemistry, University of Puerto Rico,-Mayagüez Campus, Mayagüez, PR

Elienisse Rodriguez-Medina, Chemistry, University of Puerto Rico,-Mayaguez Campus, Mayagüez, PR

Maritza De Jesús-Echevarría, Chemistry, University of Puerto Rico- Mayagüez Campus, Mayagüez, PR

Air pollution around the world has been a major environmental health issue for people living in metropolitan areas. This health issue arises as a consequence or urban development and also with the atmospheric activities of the Sahara dust and Soufrière Hills volcano in the Caribbean. Epidemiologic studies suggesting airborne particulate matter as one of the principal causes of respiratory illnesses. Ambient air pollution and its health effects have not been studied in detail in Puerto Rico. It research conduct a study to evaluate whether there was an association between the air pollution, Total Suspended Particles PM10 (TSP/PM10) point sources and a risk of Chronic Obstructive Pulmonary Disease (COPD) in the Mayagüez municipally. The instrument used for this research was a High Volume Sampler (HVS) and the COPD data was provided by the ASES "Administración de Seguro de Salud de Puerto Rico". The high volume sampler was employed for twenty four hour periods from 2005 to 2007 years. The highest values of TSP were during February, March and April months of 2005 (μg/m3): 36.24, 44.48 and 37.55, respectively and February, September and November 2006 (μg/m3): 37.82, 34.67 and 33.39, respectively. There a correlation of the highest value of the TSP in the three years during February and March months. The preliminary results shows that there a predisposition when the TSP value increase the number of respiratory diseases hospitalization also increase, specially during the months of February and September. This results point out that a direct proportional correlation exits between the TSP values and the respiratory disease if the data of other Health Insurance Plans were included. Our research cans now awareness about the possible consequences of the air pollution and the effect in the public health. At the present day, no TSP measurements obtained exceed the limits impose by the National Ambient Air Quality Standard (NAAQS).

See more of this Session: Student Poster Session: Environmental
See more of this Group/Topical: Education

Direct Screening Analysis of PAHs in Atmospheric Particulate Matter with HS/SPME and GC/MS

José Roberto Rivera-Negrón, Chemical Engineering, Chemistry Department, University of Puerto Rico - Mayaguez Campus, Mayaguez, PR

Maritza De Jesús-Echevarría, Chemistry Department, University of Puerto Rico - Mayaguez Campus, Mayagüez, PR

Polynuclear aromatic hydrocarbons (PAHs) are by-products of the combustion of organic matter and are carcinogenic compounds. Sources of atmospheric PAHs are mostly anthropogenic (e.g. motor vehicles, industrial processes, domestic heating, waste incineration, and tobacco smoke) and natural processes (e.g. forest fires and volcanic eruptions). Urban measurements of the atmospheric PAHs in several cities have been reported and several analytical methodologies have been employed to evaluate the PAHs contents in the atmosphere. A particularly promising extraction technique is headspace solid-phase microextraction HS/SPME due to its several advantages over other extraction techniques. It is a sample preparation technique that requires no solvents and permits sample transfer and analysis with little or no modifications to existing gas chromatographic mass spectrometer, GC/MS instruments. Methods employing SPME can be easily adaptable to extraction PAHs of airborne particulate matter especially those compounds lower than four congeners. The validation of the HS/SPME technique coupled to GC/MS for the analysis of PAH's in atmospheric particulate matter, collected in fiberglass filters using a High Volume Sampler instrument was explore. The validated methodology can be use to analyze qualitative and quantitative the PHA's compounds in air samples (TSP/PM10).

See more of this Session: Student Poster Session: Environmental
See more of this Group/Topical: Education

Reaction Kinetics of CO_2 Sequestration by $Mg(OH)_2$ in High Ionic Strength Solutions

Caitlin D. Allen, Sandia National Laboratories & New Mexico Institute of Mining and Technology, Soccorro, NM

Martin B. Nemer, Performance Assessment and Decisions Analysis, Sandia National Laboratories, Carlsbad, NM

Haoran Deng, Performance Assessment and Decisions Analysis, Sandia National Laboratories, Carlsbad, NM

Previous studies of the reaction pathways of CO_2 sequestration by magnesium provide little experimental data regarding the behavior of the P_{CO2} versus time in a closed system of high ionic strength solution. This important for emulating post-closure conditions in the Waste Isolation Pilot Plant (WIPP), a transuranic waste repository held in a deep-geologic bedded salt formation. At WIPP, MgO is emplaced as an engineered barrier to limit the actinide solublizing effects of CO_2 formed from possible cellulose, plastic, and rubber (CPR) biodegredation. The objective of this study is to obtain experimental evidence for the functional form of the CO_2 sequestration rate-law by magnesium in concentrated brines. Purified $Mg(OH)_2$ was added to DI water or 5m NaCl in HDPE serum bottles. The headspace was charged with CO_2. Bottles were capped and stirred continuously. The headspace CO_2 concentration was monitored versus time using GC/MS. Two distinct CO_2 sequestration timescales were observed. XRD of solid products indicates the presence of hydromagnesite(5424). An integrated rate equation was developed by modeling a sum of series of first order reactions. Rate constants were determined with k_1 = 0.31 ± 0.05 hours^{-1} for the fast timescale, and k_2 = 0.01 ± 0.005 hours^{-1} for the slow timescale.

See more of this Session: Student Poster Session: Environmental
See more of this Group/Topical: Education

Liposomes Drug Encapsulation and Uptake in Cancer

Caitlin D. Allen, New Mexico Institute of Mining and Technology, Socorro, NM

Lipophilic anticancer agents cannot be intravenously injected in their free form due to their extensive insolubility in water, hindering the development of anticancer activity these compounds possess. The inability to devise a mechanism that would enable effective administration of many potentially potent lipophilic drugs imposes a significant obstacle in the ongoing search for selective anticancer biochemical inhibitors. Progress towards such a mechanism has been achieved through the encapsulation of lipophilic drugs in vesicles of self-assembled phospholipid known as liposomes. Lipophilic drugs reside in the shielded hydrophobic region of the liposome's amphipathic phospholipid bilayer, increasing their aptitude for administration. Investigation of liposomes as a medium for the administration of lipophilic drugs has ensued due to the variety of potential biological applications these structures possess. This research monitors the effects of polyethylene glycol (PEG) and silica coated liposomes, for the encapsulation and delivery of chemically modified anticancer drugs upon jurkat and gila cell strains.

See more of this Session: Student Poster Session: Environmental
See more of this Group/Topical: Education

Development of Region Specific Optimized Integrated Biorefinery Models

Andrea Ramsey, Chemical Engineering, University of Kentucky, Lexington, KY

Jeffrey Seay, Chemical and Materials Engineering, University of Kentucky, Paducah, KY

Environmental impacts of fuel production processes have become a great concern for scientists and environmentalists alike. As a result of the increased interest in sustainability, research has begun to focus on the development of efficient, environmentally responsible energy production methods using both renewable and non-renewable resources. Gasification is a prime method for creating energy in an efficient manner and may be used with both coal and biomass in either separate processes or in a co-fired process. Gasification is a partial combustion reaction that produces synthesis gas, more commonly termed syngas, which may be burned directly or converted into chemical products and liquid fuels such as bioethanol. Biomass gasification is not only an environmentally clean source of energy but also poses a viable disposal means for biomass waste such as field crop residues. Because of the low energy density of most bioresources, it is necessary to have accurate models optimizing the process and supply chains for a specific geographical region. Optimization of environmental impact, cost, energy usage, and energy production is based on the specific feedstocks and product needs of a particular region. This contribution will optimize the environmental and economical factors for the Jackson Purchase region of Western Kentucky, USA.

See more of this Session: Student Poster Session: Environmental
See more of this Group/Topical: Education

Heavy Metals Remediation Via Cyclic Electrowinning/Precipitation (CEP)

Lakshmi Madhavan, Chemical Engineering, Brown University, Providence, RI

J.M. Calo, Chemical Engineering, Brown University, Providence, RI

Pengpeng Yao, Chemical Engineering, Brown University, Providence, RI

Indrek Küloats, Chemical Engineering, Brown University, Providence, RI

This paper describes a Cyclic Electrowinning/Precipitation (CEP) System, that has been developed for the removal heavy metals (e.g., Cu(II), Ni(II), Cd(II), Pb(II)) from aqueous waste streams or leachates originating from contaminated sources such as Brownfields or Superfund sites. The novelty of this automated system lies in its coupling of metal reduction by electrowinning with in-process precipitation/redissolution to produce a programmable cyclic process that can be matched to varying metal mixture compositions and concentrations. The precipitation/redissolution part of the process addresses two important issues: it (1) concentrates the metal ions in solution so that the electrowinning steps operate at high current efficiencies; and (2) reduces metal ion concentrations in the final effluent water to very low concentrations. In the current system a spouted particulate electrode (SPE) is used to circulate conductive particles that serve as cathode surfaces for metal reduction. This process removes and concentrates dilute metal ions onto the conductive particle surfaces, producing an extremely large volume reduction from the original wastewater. Although the main objective of the CEP system is remediation, the potential for recovery of single purified metals for recycling also exists. Specific experimental results concerning the performance of this system are presented.

See more of this Session: Student Poster Session: Environmental
See more of this Group/Topical: Education

Water Ordering On Aluminum Oxide Surfaces

K. Alexandria Cassady, Energy, Environmental and Chemical Engineering, Washington University in St. Louis, St. Louis, MO

Cynthia S. Lo, Energy, Environmental and Chemical Engineering, Washington University in St. Louis, St. Louis, MO

Aluminum oxide is a useful material in engineering applications such as environmental remediation for the removal of heavy metals from water, and advanced materials such as ceramics and coatings. The structure of the clean and hydroxylated aluminum oxide (11-20) surface has been studied using density functional theory. The lowest-energy surface structure has been found to be the stoichiometric surface, which is in stark contrast to the results on other aluminum oxide surfaces (e.g., (0001), (1-102)). The hydroxylated surfaces have also been studied with density functional theory, where four water molecules have been dissociated per unit cell. The results show that the stoichiometric surface termination is favored in aqueous environments as well.

See more of this Session: Student Poster Session: Environmental
See more of this Group/Topical: Education

Investigation of Atmospheric Aerosol Properties Using Environmental AFM

David M. Marchese, Chemical Engineering, Bucknell University, Lewisburg, PA

Timothy M. Raymond, Chemical Engineering, Bucknell University, Lewisburg, PA

Atmospheric aerosols affect both global and regional climate by altering the radiative balance of the atmosphere and acting as cloud condensation nuclei. Despite an increased focus on the research of atmospheric aerosols due to concerns about global climate change, current methods to measure the hygroscopic properties of aerosols are limited in various ways by experimental procedure. The objective of this research is to investigate the utility of environmental AFM for imaging atmospherically relevant aerosols as they respond to changes in relative humidity. Traditional aerosol generation and collection techniques were used in conjunction with environmental AFM to measure the hygroscopic properties of common organic and inorganic aerosols. The results of this research demonstrated the utility of environmental AFM for measuring the hygroscopic properties of aerosols. Further research in this area will lead to an increased understanding of the role of organic and inorganic aerosols in the atmosphere, allowing more accurate climate models to be developed.

See more of this Session: Student Poster Session: Environmental
See more of this Group/Topical: Education

Arsenic Removal From Water through Simple Mucilage-Based Filters

Angel D. Rodriguez, Chemistry, University of Puerto Rico Mayaguez Campus, Mayaguez, PR

Dawn I. Fox, Chemical and Biomedical Engineering, University of South Florida, Tampa, FL

Norma Alcantar, Chemical and Biomedical Engineering, University of South Florida, Tampa, FL

Developing communities that depend on ground water as their main source may be exposed to arsenic and current commercial removal methods are very expensive. Research was conducted in order to develop an effective and affordable filter capable of removing arsenic from water. Mucilage from the Opuntia ficus-indica cactus was chosen as one of the components of the filter because of it's know capability of lowering arsenic concentration. Non-geling extract was used during all experiments. In order to improve the filters capabilities on arsenic removal, factors like solution pH levels and column composition were studied and optimized to obtain the best results. Mucilage concentration was varied in order to determine if there was a relationship between the amount of mucilage present and the amount of arsenic removed with the filter. Other affecting factor, like washing the sand previous to experiments, were also observed during the series of experiments. A sand:clay filter variation, with a 90:10 proportion respectively, demonstrated the best performance regarding arsenic removal, after achieving a 97.68 percent of arsenic removal from an aqueous solution.

See more of this Session: Student Poster Session: Environmental
See more of this Group/Topical: Education

Zero Waste Semiconductor Facility

Omar Leon, Chemical Engineering, New Mexico State University, Las Cruces, NM

A process designed to achieve zero waste from a semiconductor facility has been completed. Moving towards zero waste by means of high temperature and burning in the presence of excess oxygen with potent scrubber towers to obtain Btu value. CO_2 released from the high temperature burning is fed to algae for the production of ethanol. Additionally, a reverse osmosis facility is installed to aid the incinerating plant in its need for water. Research was conducted to evaluate the feasibility of producing methane gas from combined solid/food and agricultural waste. The agricultural waste was feedlot manure acting as a catalyst for the reaction. The results showed that the combined waste can produce as high as 1.82 m^3 of CH_4 per day. The process of recycling the chemical waste was done by diffusion dialysis. The proposed design accounts for solid as well as chemical waste with a 94% recovery of all material discarded from the facility.

See more of this Session: Student Poster Session: Environmental
See more of this Group/Topical: Education

Mechanism of Metal-Reduction with Geobacter Sulfurreducens

Lan Dang, Department of Chemical Engineering, Penn State University, University Park, University Park, PA

Ryan Muthard, Department of Chemical Engineering, Penn State University, University Park, University Park, PA

Ruth Helmus, Biochemistry and Molecular Biology, Penn State University, University Park, University Park, PA

Ming Tien, Biochemistry and Molecular Biology, Penn State University, University Park, University Park, PA

Wayne Curtis, Department of Chemical Engineering, Penn State University, University Park, University Park, PA

Geobacter Sulfurreducens, an iron-reducing microbe, has great potential for remediation of metal-contaminated water. The ability of Geobacter to reduce and precipitate metals provides a means to sequester toxic ions that would otherwise move freely in water systems. This is done by feeding the organism sufficient carbon to maintain respiratory activities. Several studies have been conducted to determine Geobacter's kinetic parameters on soluble terminal acceptors, but very little analysis has been conducted on the difference between the organism's kinetic parameters from a continuous culture and batch culture. This is quantitatively defined by Geobacter's growth yield and maintenance respiration requirements. A chemostat capable of operation under strictly anaerobic conditions was constructed and Geobacter grown on acetate as the terminal electron donor and ferric citrate as the terminal electron acceptor. The value of growth yield of 0.61 gDW/gacetate and maintenance coefficient of 0.356 gacetate/gDW/hr is consistent with previous reports. The dramatically different yield experienced on different electron acceptors has led speculation of divergent proposed mechanisms of metal reduction. Further research is currently aimed toward examining the kinetics of Geobacter on soluble and insoluble electron acceptors to further comprehend the mechanism of extracellular metal reduction.

See more of this Session: Student Poster Session: Environmental
See more of this Group/Topical: Education

Analysis of Biodegradable Bone/Polyurethane Composite Using Raman Spectroscopy

Michael J. Skoumal, Chemical and Biomolecular Engineering, Vanderbilt University, Nashville, TN

Severe bone fractures often require autologous bone grafts that involve the use of invasive surgery, which poses numerous risks to the patient. One possible alternative is the use of injectable polyurethane that enhances healing through the natural tissue remodeling process by providing biodegradable bone scaffolding for the cells. This polyurethane biomaterial has been shown to be biocompatible, biodegradable, non-cytotoxic, and facilitates cell migration and proliferation. Furthermore, the injectable nature of the polyurethane allows for minimally invasive surgery and quicker patient recovery time. Previously, the time-intensive method histomorphometry was used to analyze cell growth in the area of interest. The objective of this study is to investigate the use of Raman spectroscopy as a method of quickly and quantitatively analyzing the bone tissue to determine the degree of tissue remodeling in the polyurethane implant site.

See more of this Session: Student Poster Session: Food, Pharmaceutical, and Biotechnology
See more of this Group/Topical: Education

The Use of Variable Depth Hydrogel Stamps as An Efficient Means for Providing Biocompatible Gradients

Halle A. Senger, Mechanical Engineering, The University of Akron, Akron, OH

Ashley Madon, Mechanical Engineering, The University of Kentucky, Lexington, KY

Dewayne Lee Thompson, Mechanical Engineering, The University of Kentucky, Lexington, KY

Dr. Christine Trinkle, Mechanical Engineering, The University of Kentucky, Lexington, KY

Dr. J. Zach Hilt, Chemical and Materials Engineering, The University of Kentucky, Lexington, KY

The controlled application of proteins and other biomolecules to generate patterned surfaces is a crucial part of many processes, for example creating biosensors and directing cell growth for tissue engineering. Traditionally, poly-dimethylsiloxane (PDMS) elastomeric stamps have successfully been used to transfer proteins with sub-micron resolution. However, the resulting patterns are limited to a single, uniform protein concentration. The technique presented here utilizes variable depth hydrophilic stamps as a diffusion medium, allowing protein gradients to be easily applied and accurately reproduced. Specifically, stamps of a 2-hydroxyethyl acrylate (HEA) hydrogel made with poly(ethylene glycol) diacrylate (PEG 400 DA) as a crosslinker and 2% agarose hydrogel are used to demonstrate variable-concentration pattering of a fluorescently-labeled protein. Both hydrogels were able to provide a wide range of diffusion for varying depths. The HEA had the widest range, about 2.5 times larger than agarose when compared at equal depths. From these results, it can be concluded that it is possible to create protein gradients via variable stamp depth. Ultimately, optimization studies can be undertaken to find the best suited hydrogel for varying proteins.

See more of this Session: Student Poster Session: Food, Pharmaceutical, and Biotechnology
See more of this Group/Topical: Education

A Computational Investigation of Bone Biology

Devin P. Sullivan, Chemical and Biomolecular Engineering, Vanderbilt, Nashville, TN

The bone cycle is a continuous and dynamic system in which old or damaged bone is constantly removed and replaced by new bone. This system consists of two basic cell types, osteoclasts and osteoblasts. The osteoclasts are responsible for the catabolic effect of bone resorption while the osteoblasts form bone, an anabolic process. The coordination between these two cell types is crucial in maintaining appropriate strength in one's bones. This system works in highly coordinated groups called basic multicellular units or BMUs. Utilizing various modeling techniques will lead to new clinical inquiries and treatments of diseases.

This work is a summary of my 3 month research scholarship visiting UWA. During that time I had the opportunity to work on a variety of problems related to bone regulation ranging from in-vitro to in-vivo applications. Despite the obvious and extensive self-regulation of bone tissue, biologists have been able to do little to understand the complexities of these cell-cell interactions. Several computational models attempt to shed light on these interactions, covering a large scope of biological queries related to the bone system. This presentation aims to highlight a few of these models and their potential applications. The first model will highlight the balance in differentiation of mesenchymal stem cells into osteoblasts versus adipocytes. The next model investigates the effects of parathyroid hormone on overall bone activity and resorption. Lastly, a spatial model of a single BMU demonstrates the spatial movement of these cells as they interact. The discussion will focus on the potential therapeutic applications of each model. Of particular interest will be the predictive nature of the models beyond that of traditional biological approaches.

See more of this Session: Student Poster Session: Food, Pharmaceutical, and Biotechnology
See more of this Group/Topical: Education

Tissue Regeneration Using Various Structures

Kenneth J. Walker, School of Chemical Engineering, Oklahoma State University, Stillwater, OK

Kelsy Thompson, School of Chemical Engineering, Oklahoma State University, Stillwater, OK

Sundararan. V. Madihally, School of Chemical Engineering, Oklahoma State University, Stillwater, OK

Objective: The objective of the research was to produce chitosan and chitosan-gelatin porous structures and proceed to colonize these structures with Human Foreskin Fibroblasts (HFF-1). Chitosan and gelatin are both biocompatible and biodegradable polymers, which is a requirement for the scaffolds. The porous structures are two dimensional (2-D), three dimensional (3-D), and hydrogels. The importance of this study is the applications in which the scaffolds can be used. The 2-D membrane is often used for wound healing, while the 3-D porous structures can be fabricated to replace multicellular membranes for organs. Hydrogels can be formed as a substitute for tissues such as cartilage using a minimally invasive procedure. Methods: As previously stated, there are three scaffolds that have been constructed for the purpose of the research. All three matrices are made with a 0.5% w/v chitosan or chitosan-gelatin solution. The two dimensional membrane is fabricated by air drying the solution in a sterile environment. The three dimensional matrix is made by freezing the solution to -20oC and then freeze drying the frozen solution. Hydrogels are formed by incubating the solution at 37oC (98.6oF). HFF-1 cells are cultured in serum containing media for more rapid growth. The cells are switched to serum-free media prior to seeding. Cells are seeded once the 2-D and 3-D scaffolds are constructed whereas the hydrogels are not formed until at the time of seeding. The seeded matrices are incubated four days changing the media every second day. The removed media is tested for collagen and elastin using Sircol Soluable Collagen Assay and Fastin Elastin Assay kits. On the fourth day the cells are stained with Actin and DAPI in order to view the cells in the different layers of the matrices. MMP-2/MMP-9 is also assessed. Results: All matrices were constructed using chitosan and chitosan-gelatin and were successfully seeded with HFF-1 cells. Standard curves for collagen and elastin assays have been achieved to fit the actual Collagen and Elastin data in the curve. Cells show spreading and attachment on chitosan-gelatin structures. Testing for MMP-2/MMP-9 must be accomplished for further analysis. Conclusions: Multiple types of matrices can be constructed using the same solution by altering the processing method without additional chemical reactions necessary.

Funding provided by: KJW was funded by the ARRA NIH REU Program, KT was funded by OK-INBRE program.

See more of this Session: Student Poster Session: Food, Pharmaceutical, and Biotechnology
See more of this Group/Topical: Education

Dynamic Pressure Measurements in Trophoblast Cells

Andrea L. Jones, Division of Engineering, Brown University, Providence, RI

Stephanie E. McCalla, Engineering, Brown University, Providence, RI

Anubhav Tripathi, Division of Engineering and Warren Alpert School of Medicine, Brown University, Providence, RI

Jared Robins, Women and Infants Hospical, Providence, RI

Blastocyst formation is critical for early embryo development and implantation. Trophoblast cell differentiation in the blastocyst can impact several diseases of pregnancy including preeclampsia and intrauterine growth restriction. Unfortunately, few models exist to adequately study these complicated events. In this work, we will explore that mechanisms by which the trophoblast vesicle forms. We hypothesize that the acellular centers of the vesicles contain fluid under pressure. To prove this, we will measure intra-vesicle pressure using a highly sensitive sensor developed and patented by the investigators. Using this technique functional trophoblast vesicles, developed using prototype bioreactors can be tested by altering the shape, size, number of cells and culture conditions. This work will assist the final goal of producing spherical vesicles that preferentially differentiate into extra-villous trophoblast cells, the cells that predominate in blastocysts and are responsible for embryo attachment and implantation.

See more of this Session: Student Poster Session: Food, Pharmaceutical, and Biotechnology
See more of this Group/Topical: Education

Assembly of Bacteria at Oil-Water Interfaces

Danielle Lussier, Chemical Engineering, Rowan University, Glassboro, NJ

Anderson Shum, Harvard University, Cambridge, MA

Connie Wilking, Harvard University, Cambridge, MA

The way bacteria assemble themselves at oil-water interfaces and how they behave when encapsulated in double emulsions is of interest both for determining what the stabilization properties of the bacteria are and for studying gene expression in varying conditions. Previous research has only studied hydrophobic bacteria at the interfaces of single emulsions. The unique aspect of this experimentation is that, along with simple techniques, microfluidic devices were used to produce both single and double emulsion droplets of controlled sizes. Bacillus subtilis, Pseudomonas aeruginosa, and Escherichia coli bacteria were used for the experimentation, each having certain distinct characteristics and providing variation. Single emulsions were made with silicone oil, alone with ten other oils to further determine what the cause of stabilization. Double emulsions with concentrated bacteria solutions were prepared in an attempt to fill the droplets and grow a biofilm. The testing confirmed that E. coli and silicone oil are the most successful combination for producing a stable emulsion. Testing different oils also showed that the stability of the emulsions increased as the number of carbons in the chemical structure increased. Through experimentation some interesting properties of ferrofluids, when emulsified with bacteria were discovered. P. aeruginosa, which sometimes oriented themselves vertically, were the only bacteria that showed any organized or specific arrangement at the oil-water interface.

See more of this Session: Student Poster Session: Food, Pharmaceutical, and Biotechnology
See more of this Group/Topical: Education

Peptide-Functionalized Affinity Hydrogels for Reducing Local Inflammatory Response

Patrick D. Boyer, Department of Chemical and Biomolecular Engineering, Vanderbilt University, Nashville, TN

Chien-Chi Lin, Department of Chemical and Biological Engineering, Howard Hughes Medical Institute and University of Colorado, Boulder, CO

Alex A. Aimetti, Department of Chemical and Biological Engineering, University of Colorado, Boulder, CO

Kristi S. Anseth, Department of Chemical and Biological Engineering, University of Colorado, Howard Hughes Medical Institute, Boulder, CO

Cell delivery using poly(ethylene glycol) (PEG) hydrogels has great potentials in tissue engineering and regenerative medicine applications, including islet encapsulation and transplantation to treat type 1 diabetes. However, one critical limitation of current hydrogel encapsulation technique is that it cannot prevent the diffusion of small antigenic, chemotactic, and cytotoxic molecules. These small molecules may hinder the survival of encapsulated cells in a multifaceted way. Here, we describe the synthesis, characterization, and application of peptide-functionalized affinity hydrogels capable of sequestering a pro-inflammatory chemokine, monocyte chemotactic protein-1 (MCP-1), secreted by the encapsulated pancreatic β-cells. We first demonstrate that extending the affinity peptide sequence away from the crosslinking site enhances MCP-1 binding and decreases release in PEG hydrogels. We then show that the novel affinity hydrogels significantly reduce the release of MCP-1 secreted by encapsulated MIN6 β-cells and isolated mouse islets. This strategy may improve the survival of encapsulated β-cells due to decreased MCP-1 secretion and reduce local inflammatory response.

See more of this Session: Student Poster Session: Food, Pharmaceutical, and Biotechnology
See more of this Group/Topical: Education

Binding and Inactivating Pathogenic *E. Coli* Using Antimicrobial Peptide Cecropin P1-Cys Fragments

Amy M. Manaresi, Chemical Engineering, Rowan University, Glassboro, NJ

Terri A. Camesano, Chemical Engineering, Worcester Polytechnic Institute, Worcester, MA

Escherichia coli O157:H7 is a pathogenic strain of bacteria that commonly contaminates foods such as ground beef and spinach. One potential way to prevent ***E. coli*** contamination is to attach antimicrobial peptides (AMPs) to food preparation surfaces. Previous studies have shown that the AMP cecropin P1-cys (CP1-cys) can effectively bind and kill ***E. coli*** when attached to a gold substrate. The purpose of this study was to determine if fragments of CP1-cys attached to gold could kill ***E. coli*** as effectively as the whole peptide. Quartz Crystal Microbalance with Dissipation monitoring (QCM-D) was used to attach the peptide fragments and bacteria to the gold substrate, and a LIVE/DEAD® BacLight™ Kit quantified the live versus dead cells on the substrate. The performances of fragments H2N-LSKTAKKLENSAKKC-OH and H2N-EGIAIAIC-OH were compared to the performance of the whole peptide, H2N-SWLSKTAKKLENSAKKRISEGIAIAIQGGPRC-OH. Initial studies indicate that although the fragments can bind to the gold substrate and kill ***E. coli***, the thickness of peptide on the gold as well as the amount of cells killed were relatively less than when the entire peptide was used. This could lead to further understanding of AMP behavior with respect to substrate binding and the binding and inactivation of ***E. coli***.

See more of this Session: Student Poster Session: Food, Pharmaceutical, and Biotechnology
See more of this Group/Topical: Education

PEG-Fe_3O_4 Hydrogel Nanocomposites for Combined Chemotherapy and Hyperthermia Treatment of Cancer

Jenna M. Shapiro, Chemical and Materials Engineering, University of Kentucky, Lexington, KY

Samantha A. Meenach, Chemical and Materials Engineering, University of Kentucky, Lexington, KY

J. Zach Hilt, Chemical and Materials Engineering, University of Kentucky, Lexington, KY

Kimberly W. Anderson, Chemical and Materials Engineering, University of Kentucky, Lexington, KY

Chemotherapy is a conventional cancer treatment that is still limiting for some types of cancer. Hyperthermia, the heating of cancerous tissues to 40-45°C, can increase the efficacy of chemotherapy, specifically paclitaxel. We hypothesize that iron oxide hydrogel nanocomposites can be used for a dual-therapy application to treat cancer. Hydrogels composed of the macromer poly(ethylene glycol) (PEG) methyl methacrylate and crosslinker poly(ethylene glycol) dimethacrylate were fabricated with iron oxide nanoparticles incorporated in the hydrogel matrix. These nanoparticles can be remotely heated by an alternating magnetic field (AMF), thus producing the hyperthermia effect from the nanocomposite. When imbibed with paclitaxel, the hydrogels have the potential to provide synergistic heating and chemotherapy in a local area. Swelling analysis indicated an inverse relationship between temperature and volume swelling ratio (Q), and between crosslinking density and Q. AMF heating of the hydrogel nanocomposites indicated higher crosslinked hydrogels exhibit a greater ΔT. Hydrogels with lower crosslinking densities exhibited an overall faster release of paclitaxel due to increased effective diffusivity. Hyperthermia studies have demonstrated that these hydrogel nanocomposites can be remotely heated, causing an increase in cytotoxicity for M059K glioblastoma cells. Current studies are focusing on the cytotoxicity effect of combined hyperthermia and paclitaxel release.

See more of this Session: Student Poster Session: Food, Pharmaceutical, and Biotechnology
See more of this Group/Topical: Education

Alkaloid Extraction and Purification From *Catharanthus roseus* Hairy Root Tissue

Diane C. Brown, Chemical and Biological Engineering, Iowa State University, Ames, IA

Guy W. Sander, Chemical and Biological Engineering, Iowa State University, Ames, IA

Jacqueline V. Shanks, Chemical and Biological Engineering, Iowa State University, Ames, IA

Plant metabolic engineering of ***Catharanthus roseus*** may provide improved ways of producing vincristine (Oncovin®) and vinblastine (Velbe®) which are used in the treatment of cancer. The tabersonine branch point is very important for the synthesis of vindoline which is a key precursor to vincristine and vinblastine. Genetically modified ***C. roseus*** hairy root lines are used to analyze the flux of metabolites at the tabersonine branch point. Genes inserted by recombinant DNA techniques can amplify or suppress flux through existing metabolic pathways and can create new metabolic pathways for desired products that are not expressed in the wild-type line. By suppressing flux to undesired products like hörhammericine, tabersonine flux can be diverted to the vindoline-vinblastine-vincristine pathway. One of the current research challenges is the lack of commercially available standards to quantify low-level metabolites at the tabersonine branch point. A biomass extraction and analytical HPLC protocol was adapted for semi-preparative scale in order to obtain tabersonine, lochnericine, and hörhammericine standards from ***C. roseus*** hairy root tissue. Standards will enable improved study of the metabolic pathways originating at the tabersonine branch point and will allow for the determination of previously unidentified co-eluting tabersonine-like compounds.

See more of this Session: Student Poster Session: Food, Pharmaceutical, and Biotechnology
See more of this Group/Topical: Education

Novel Isotope Tracer Method for Quantifying Glucose Metabolism in Type-2 Diabetes

Lina M. Aboulmouna, Chemical and Biomolecular Engineering, Vanderbilt University, Nashville, TN

Jamey D. Young, Chemical and Biomolecular Engineering, Vanderbilt University, Nashville, TN

Masakazu Shiota, Molecular Physiology and Biophysics, Vanderbilt University Medical Center, Nashville, TN

Accurately assessing in vivo rates of glucose turnover, gluconeogenesis, glycogen breakdown and futile cycling in the liver are crucial to understanding the control of whole-body glucose metabolism and how it becomes dysregulated in conditions such as diabetes and obesity. Currently, these rates are estimated using methods that do not account for the full complexity of intracellular biochemical reaction networks, and therefore often lead to inconsistent results. In this study, we are conducting metabolic flux analysis (MFA) studies using a unique formulation of 13C- and 2H-labeled glucose tracers and Gas Chromatography-Mass Spectrometry (GC-MS) profiling in order to understand the regulation of liver glucose fluxes in both fasted and fed Sprague-Dolly rats. By analyzing GC-MS labeling data of plasma glucose and lactate using our custom MFA software tools, we are developing novel strategies to quantify whole-body glucose metabolism using an integrated experimental and computational approach.

See more of this Session: Student Poster Session: Food, Pharmaceutical, and Biotechnology
See more of this Group/Topical: Education

The Effect of Nitric Oxide On Uncharacterized Genes in Escherichia Coli K12

Jake R. Gillilan, Chemical and Biological Engineering, Iowa State University, Ames, IA

Donovan S. Layton, Chemical and Biological Engineering, Iowa State University, Ames, IA

Ai-Ling Teh, Statistics, Iowa State University, Ames, IA

Derrick K. Rollins Sr., Chemical and Biological Engineering, Iowa State University, Ame, IA

Laura R. Jarboe, Department of Chemical and Biological Engineering, Iowa State University, Ames, IA

Summary

Nitric oxide (NO) is produced by the human immune system. The response of Escherichia Coli (E. coli) to this compound has been studied using genomic and phenotypic analysis. E. coli infects the urinary tract and is most commonly found in food borne illnesses and can cause death. This project investigates the contribution of uncharacterized E. coli genes to the NO response. Genes were selected for analysis by Differential Principal Component Analysis of transcriptome data. We cloned and tested the phenotype of these uncharacterized genes in E. coli, using traditional PCR and cloning techniques. The resulting phenotype was evaluated using and inhibitory dose of DeaNO and a five to eight minute interval of measurement.

See more of this Session: Student Poster Session: Food, Pharmaceutical, and Biotechnology
See more of this Group/Topical: Education

Quantifying Erythrocytes by Solution Resistance Measurement

Amanda L. Mixon, Biological Engineering, Mississippi State University, Mississippi State, MS

Soumya K Srivastava, Chemical Engineering, Mississippi State University, Mississippi State, MS

Adrienne R Minerick, Chemical Engineering, Mississippi State University, Mississippi State, MS

Microfluidics is a rapidly growing field with applications ranging from protein or genomic analysis to inkjet printer technology. In the past, researchers have used many methods in quantifying cells and particles in a microdevice. Some of the methods are based on optical means, where a fluorescence signal is used to track the particle movement or to physically count the particles present. They both could be laborious, and time consuming. The purpose of this research is to relate DC solution resistance to the concentration of red blood cells within microchannels. A microdevice with a single, straight channel of 80μm in width, 1.5cm in length and 50μm in depth was fabricated with soft lithography and cast as a PDMS device. The device connections are shown in Figure 1. Resistance was measured over the channel with the erythrocytes using the Relation (1). Dextrose added buffers were used to quantify the dependence of solution conductivity alongside blood sample dilutions. DC voltages were applied lengthwise down the channel via electrodes and voltage change was measured across the resistor. This method has the potential to be used for quantifying cells at multiple outlet ports simultaneously with minimal analysis required.

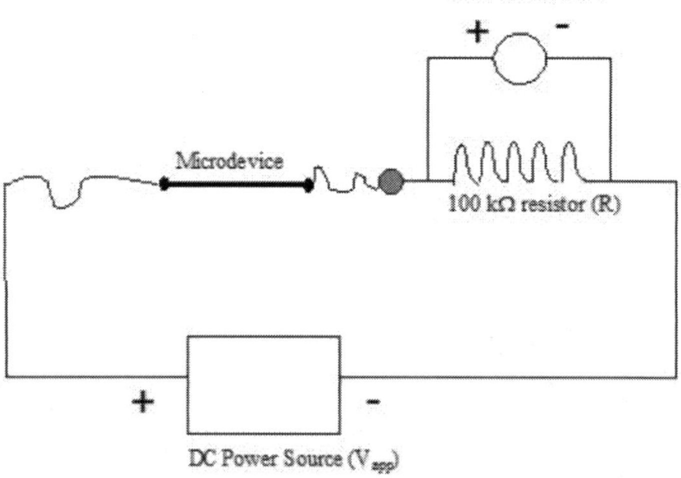

Figure 1: Device Connections for Experimentation

$$R_{channel} = \frac{V_{applied}}{I_{measured}} - R_{resistor} \qquad (1)$$

See more of this Session: Student Poster Session: Food, Pharmaceutical, and Biotechnology
See more of this Group/Topical: Education

Use of Alginate Hydrogels as Non-Immunogenic Reactive Substrates for Neutrophil Adhesion in in Vitro Flow Adhesion Assays

Sara Yacob, Chemical Engineering, University of Michigan, Ann Arbor, MI

Omolola Eniola-Adefeso, Chemical Engineering, University of Michigan, Ann Arbor, MI

In vitro cell-free adhesion assays have previously been useful in elucidating neutrophil receptor-ligand interactions and cell signaling. However, these cell-free assays often involve the use of plastic or glass substrate that can generate a significant amount of non-specific adhesion. Moreover, neutrophils are highly active cells that can undergo significant non-specific activation on plastic and glass substrates - particularly when experiments are conducted at body temperature. Alginate hydrogels are non-immunogenic in nature and thus support cell adhesion only when chemically grafted with active biological molecules. As such alginate hydrogels are attractive for in vitro cell adhesion studies. We show that alginate hydrogels modified with leukocyte adhesion ligands such as the selectins and ICAM-1 can support rolling and firm adhesion of neutrophils in a parallel plate flow chamber. Specifically, neutrophils interacting with P-selectin coupled gels displayed slow rolling in the flow chamber while maintaining their spherical shape at 37oC - suggesting that cells remain unactivated in the absence of an external activating stimulus. When ICAM-1 was coupled to the gel surface, neutrophil firm adhesion was observed. Compared to the traditional glass cover slip, alginate hydro-gels also present a better structural model of the cell wall. Finally, we show that alginate gels were able to withstand vacuum pressure during experimentation for extended periods of time. Overall, the described alginate hydrogel for use in parallel plate flow chamber shear adhesion assays can allow further study of neutrophil rolling and firm adhesion dynamics at more physiological conditions.

See more of this Session: Student Poster Session: Food, Pharmaceutical, and Biotechnology
See more of this Group/Topical: Education

RC Materials Based On Nanocomposite Sol-Gel Systems

Chelsie E. Bottom, Chemical and Materials Engineering, University of Kentucky, Lexington, KY

Ashley Hawkins, Chemical and Materials Engineering, University of Kentucky, Lexington, KY

J. Zach Hilt, Chemical and Materials Engineering, University of Kentucky, Lexington, KY

Remotely controlled (RC) drug release is a growing area of research for its biomedical applications and ability to alter drug releasing therapy after implantation. Thermosensitive block copolymer sol-gel systems are promising drug delivery vehicles for RC release because of their ability to go through temperature dependent sol-gel-sol phase changes. The phase transitions can be tailored to create gelation upon injection of a polymer/drug solution into the body, trapping the drug. Through the incorporation of Fe3O4 nanoparticles that heat when exposed to an alternating magnetic field (AMF), the temperature of the system can be increased remotely to initiate the phase transition and resultant drug release. The release of a model drug, Vitamin B12, from nanocomposite poly(ethylene-oxide-b-propylene-oxide) systems upon AMF exposure, which will raise the temperature and increase drug transport, is to be demonstrated. Phase transitions for this polymer were determined to be in desirable ranges for in vivo purposes, and the nanocomposite proved to heat when exposed to an AMF. Vitamin B12 release from the polymer when placed in water baths of different temperatures was greater for samples in a near solution state than in the gel state. Current work includes studying drug release from nanocomposite polymer systems by AMF exposure.

See more of this Session: Student Poster Session: Food, Pharmaceutical, and Biotechnology
See more of this Group/Topical: Education

Macrophage Responses to Fluid Pressure: A Potential Effector at the Tissue-Implant Interface

Danielle M. Frechette, Dept of Biological and Agricultural Engineering, University of Arkansas, Fayetteville, AR

Lars M. Bjursten, Dept of Clinical Sciences, Lund University, Malmö, Sweden

Geert W. Schmid-Schönbein, Dept of Bioengineering, University of California-San Diego, La Jolla, CA

D.A. Puleo, Center for Biomedical Engineering, University of Kentucky, Lexington, KY

Hainsworth Y. Shin, Center for Biomedical Engineering, University of Kentucky, Lexington, KY

The objective of this study was to assess the sensitivity of macrophages to fluid pressures as a first step towards identifying parameters that modulate the tissue implant interface. We hypothesized that fluid pressure at an implant interface influences macrophage activity. To address this hypothesis, we assessed superoxide production by, and the F-actin cytoskeletal organization of, HL-60 derived macrophages after a 2 h exposure to multiple pressure oscillations by varying either their mean (MP) (5-40 mm Hg), pulse pressure (PP) (0-7.5 mm Hg), or frequency (0-1.5Hz). From these experiments, we observed significantly ($p<0.05$) reduced superoxide production by the cells under all conditions tested. Interestingly, the reductions in superoxide release were independent of the MPs and oscillation frequencies, and instead depended on the PPs. We also observed a reorganization of the cytoskeleton in cells exposed to 20 mm Hg MP at PPs ranging from 0-7.5 mm Hg at 1.5 Hz relative to paired uncompressed controls. These results demonstrate that fluid pressure modulates macrophage activity and function at a cellular level. Pressure may be a potential parameter that plays a role in wound healing, e.g. at the tissue-implant interface.

See more of this Session: Student Poster Session: Food, Pharmaceutical, and Biotechnology
See more of this Group/Topical: Education

The Fabrication of Functionalized Biodegradable Polyesters with Applications in Drug Delivery

Joseph Russell, Chemical Engineering, Bucknell University, Lewisburg, PA

This poster session will detail methodology for the production of functionalized polyesters by synthesis of modified α-hydroxy acid monomers and their subsequent polymerization. Polyesters are among the most frequently utilized biomaterials in the medical industry. Although polyesters are ideal with respect to biocompatibility, their physical properties often hinder their application for medical activities. In order for biodegradable polymers to meet the demands of controlled release pharmaceuticals, they must surmount difficulties with mechanical properties, toxicity and degradation rate. To engineer materials that are capable of hurdling these challenges, structural alterations to current polymers and the chemistry of such structurally augmented polymers must be explored in greater detail. It is believed that through chemical modification these properties can be tailored, making the polymers increasingly valuable to modern health care. The modified polymers will likely possess important implications for improved drug delivery and other medical applications such as controlled release drug therapies, which offer possible solutions to some of the existing drug delivery problems.

See more of this Session: Student Poster Session: Food, Pharmaceutical, and Biotechnology
See more of this Group/Topical: Education

Synthesis and Characterization of Polyaniline-Polymer Acid Complexes for Use in Biosensing Applications

Alper A. Konuk, Chemical Engineering, The University of Texas at Austin, Austin, TX

Carolyn L. Bayer, Biomedical Engineering, The University of Texas at Austin, Austin, TX

Nicholas A. Peppas, Chemical Engineering, The University of Texas at Austin, Austin, TX

Conducting polymers have the potential to be used for biosensor applications because they can be made to exhibit a measurable change in their conductivity in response to the presence of biomolecules. The concentration of the biomolecule determines the extent of the change in conductivity, thereby providing the basis for a sensing mechanism.

We have targeted polyaniline (PANI) because it has been shown to become conductive when doped with poly(2-acrylamido-2-methylpropane sulfonic acid) (PAAMPSA). A method has been developed for the template synthesis of in-situ PANI-PAAMPSA films that are deposited on glass substrates with gold electrodes. The use of a polymer acid means that polyaniline can now be incorporated into aqueous solutions without losing its conductivity, which increases its versatility and the ease with which it can be used in a sensing device. Going hand in hand with this achievement is the development of a method for the deposition of PANI-PAAMPSA films on a substrate, which is of great importance because even if PANI-PAAMPSA maintains its conductivity in aqueous solutions, it cannot be used as a sensor because it will be dissolved into the solution. By mounting the film on a substrate, it can be exposed to an aqueous solution without being dissolved and thus function as an effective biosensor.

PANI-PAAMPSA nanofilms with thicknesses of approximately 500 nm have been formed on glass substrates with thermally evaporated gold electrodes through the use of in situ methods, which have produced thinner films than those made previously by using drop or spin-coating methods, and are more ideal for achieving a fast sensor response. The homogeneity of the films has been increased by treating the substrates with 3-triethoxysilylpropylamine prior to exposure to the PANI-PAAMPSA reaction solution.

The change in conductivity of the PANI-PAAMPSA films after exposure to pH levels ranging from 4.5 to 8.80 has been measured and found to show a linear response. Biomolecular sensing capabilities of PANI-PAAMPSA films have also been evaluated using 10 mg/ml solutions of lysozyme and bovine serum albumin (BSA) in a 150 mM sodium phosphate buffer. The conductivity of the films exposed to only the buffer, lysozyme and BSA was found to be 0.109, 0.362 and 0.788 S/cm respectively. The change in the conductivity of the films has been shown to occur within five minutes. The results of both the pH and biomolecule studies have been supported by UV-VIS spectroscopy, which shows appropriate shifts in absorbance peaks either towards the non-conductive pernigraline form of polyaniline or its conductive emeraldine form. The films from the pH studies have also been shown to re-dope when rinsed in water.

These results support the notion that PANI-PAAMPSA is a promising platform for a biosensing device.

See more of this Session: Student Poster Session: Food, Pharmaceutical, and Biotechnology
See more of this Group/Topical: Education

In Vitro Activation of Murine Dendritic Cells by Combinatorially and Conventionally Fabricated Polyanhydride Nanospheres

Ashley N. Yeager, Department of Chemical and Biological Engineering, Iowa State University, Ames, IA

Latrisha Petersen, Chemical and Biological Engineering, Iowa State University, Ames, IA

Bret Ulery, Chemical and Biological Engineering, Iowa State University, Ames, IA

Michael Wannemuehler, Veterinary Microbiology & Preventive Medicine, Iowa State University, Ames, IA

Balaji Narasimhan, Chemical and Biological Engineering, Iowa State University, Ames, IA

With the rapid evolution of many diseases, the need for a high throughput approach to the discovery and design of drug/vaccine delivery devices is evident. Combinatorial polymer synthesis and nanosphere fabrication methods provide a rapid means of testing multiple variables (i.e., polymer chemistry) in a single experiment that can lead to accelerated discovery and optimization of adjuvants and/or drug delivery devices. In this work, the combinatorial process was used to study the activation of C3H/HeOuJ dendritic cells (DCs) by varying polyanhydride chemistries. Biodegradable polyanhydrides have been studied for use in single dose vaccines. In particular, polyanhydride nanospheres composed of 1,8-bis(p-carboxyphenoxy)-3,6-dioxaoctane (CPTEG) and 1,6-bis(p-carboxyphenoxy)hexane (CPH) have not only been shown to be promising candidates for drug delivery, but also as adjuvants. The CPTEG:CPH copolymer system has been shown to possess unique properties such as tailorable release kinetics and the ability to stabilize fragile proteins, making it an ideal candidate for vaccine delivery. The combinatorial evaluation of DC activation by CPTEG:CPH nanospheres was compared to conventional methods for validation of this novel, high throughput process. Specific cell surface markers and cytokine profiles were evaluated in order to compare the performance of nanospheres of varying chemistries and those made by the different processes. It was observed that nanospheres fabricated combinatorially affected the expression of cell surface markers and the production of cytokines in a similar manner to the conventionally fabricated nanospheres; this was demonstrated by an increase in cell surface marker expression and cytokine production as the molar ratio of CPTEG within the copolymer increased, making it less hydrophobic. These insights are valuable in the rational design of single dose vaccines based on nanospheres.

See more of this Session: Student Poster Session: Food, Pharmaceutical, and Biotechnology
See more of this Group/Topical: Education

Experimental Study of Simulated Weightlessness in Model Bone Systems

Krissie E. Hunt, Animal Sciences, North Carolina A&T State University, Greensboro, NC

Brandon K. Mills, Chemical Engineering, North Carolina A&T State University, Greensboro, NC

Whitney L. McIver, Chemical Engineering, North Carolina A&T State University, Greensboro, NC

Kenneth L. Roberts, Chemical Engineering, North Carolina A&T State University, Greensboro, NC

Past astronauts have lost up to 13% of their bone mass during space missions according to NASA and can lose up to 2% of their bone mass each month. Recent research on bone loss in the literature involves the evaluation and analysis of microgravity and its effects on humans, mice, and other animals. Gravitational effects on earth-based skeletal development influences the size, shape, strength of bone, the vascular supply, and fluid flow to the skeletal system. Gravity also provides the mechanical stimulus necessary for bone growth, development, and maintenance. Past astronauts have lost up to 13% of their bone mass during missions in space according to NASA and can lose up to 2% of their bone mass each month.

Previous work determined the effects of simulated microgravity on weight gain, feed and water intake, urine and fecal output, and bone decomposition in rats in an effort to further understand of the treatment of osteoporosis in human patients. Development of a test model at North Carolina A&T State University was carried out in partnership with NASA-Ames Research Center to study the effects of weightlessness for future manned space missions. An experimental model to study simulated weightlessness in Sprague Dawley Rats was developed to gain further understanding of the physiological phenomena that occur in a microgravity environment. A model replica was made of the NASA Morley-Holton cage for weightlessness referred to as the Micro-Gravitational Metabolic Cage (MGMC). Non-weight bearing versus weight bearing conditions were studied over a 42-day period. Weight gain, urine and fecal output were observed to be moderately affected in the non-weight bearing rats. The water and feed intake were observed to be similar in both groups.

Several environmentally benign and biocompatible model systems such as smectite clays, layered double hydroxides, hydroxyapatite, and swelling synthetic micas will be employed as the inorganic matrix for the synthesis of biopolymer nanocomposite materials. These synthesized bionanocomposites will be exposed to model microgravity conditions for bone demineralization and characterized before and after exposure using XRD, SEM, TEM and elemental analysis. The experimental data will be compared with previously obtained experimental and literature data for bone loss in rats and humans.

See more of this Session: Student Poster Session: Food, Pharmaceutical, and Biotechnology
See more of this Group/Topical: Education

Quantitative Evaluation of Mechanical Algal Cell Rupturing Techniques and Their Effectiveness in the Extraction of Value Added Products

Andrew J. A. Volk, Chemical and Materials Engineering, University of Alberta, Edmonton, AB, Canada

Algae are major sources for high value added products with applications in the energy, pharmaceutical, cosmetic and food industries. The primary difficulty with the use of algae for the production of added value products is the harvesting step. Algae have a rigid cell wall composed of cellulose, glycoproteins and additional polysaccharides. It is this composition that makes extraction of the added value products difficult. To evaluate various extraction techniques we developed a model using the green alga Chlorella protothecoides. This organism is known to produce high quantities of oil when grown in nitrogen deprived environment. Oil is an ideal added value product to study the extraction of as it can be readily quantified in the cell and after extraction. In this study we used this model to compare drying and grinding with a Cell Disruptor. The extent of cell breakage was measured by comparing pictures of the cultures before and after processing. The amount of value added product recovered following separation, solvent extraction and centrifugation was also used as a measure of the extent of cell breakage and efficiency of the processes. We concluded that the Cell Disruptor was more efficient than drying and grinding.

See more of this Session: Student Poster Session: Food, Pharmaceutical, and Biotechnology
See more of this Group/Topical: Education

A Kinetic Study of in Vitro Lysis of Mycobacterium Smegmatis

B.J. Valente, Department of Chemical and Biomolecular Engineering, University of Nebraska-Lincoln, Lincoln, NE

Elsje Pienaar, Department of Chemical and Biomolecular Engineering, University of Nebraska-Lincoln, Lincoln, NE

Alan Fast, Department of Chemical and Biomolecular Engineering, University of Nebraska-Lincoln, Lincoln, NE

Aaron M. Fluitt, Department of Chemical and Biomolecular Engineering, University of Nebraska-Lincoln, Lincoln, NE

Scott E. Whitney, Department of Chemical and Biomolecular Engineering, University of Nebraska-Lincoln, Lincoln, NE

Robert J. Fenton, Department of Veterinary & Biomedical Sciences, University of Nebraska-Lincoln, Lincoln, NE

Raul G. Barletta, Department of Veterinary & Biomedical Sciences, University of Nebraska-Lincoln, Lincoln, NE

Ofelia Chacon, Department of Veterinary & Biomedical Sciences, University of Nebraska-Lincoln, Lincoln, NE

Hendrik J. Viljoen, Department of Chemical and Biomolecular Engineering, University of Nebraska-Lincoln, Lincoln, NE

The emergence of drug-resistant strains of tuberculosis has made nucleic acid amplification the diagnostic test of choice, replacing the traditional acid-fast stain and culture test. The amplification test consists of 1) sputum sample collection, 2) lysis of bacilli to extract DNA, 3) amplification of DNA by PCR, and 4) detection of PCR products. We describe the kinetics of lysis in the DNA extraction step, using *Mycobacterium smegmatis* in a Tris-EDTA buffer. *M. smegmatis*, often used as a surrogate of *M. tuberculosis*, was transformed with green fluorescent protein (GFP) genes and lysed at different temperatures in a cuvette. The extent of lysis was tracked by measuring the increase in the fluorescence signal as the cells lysed. Two theoretical models were formulated to provide a quantitative description of lysis. First, a Monte Carlo simulation modeled the cell wall as layers of blocks, in which blocks exposed to the lysing solution are removed at random. The model produced a time-varying probability distribution function that was analyzed using the Fokker-Planck equation. Second, in a reaction engineering approach, a differential equation describing the change in the number of exposed blocks on each layer was solved. Both models yielded results consistent with the experiments. The most important findings were that 1) the activation energy of lysis is 22 kcal/mol, 2) cells lyse after 14-17% loss in cell wall thickness, 3) the initial distribution of cell wall thicknesses was determined, and 4) near-complete lysis is achieved in 200 s at 80°C or 90 s at 90°C. The results will aid the design of a lysis protocol for nucleic amplification tests.

See more of this Session: Student Poster Session: Food, Pharmaceutical, and Biotechnology
See more of this Group/Topical: Education

The Development of Bacterial Resistance to Silver Dressings

Amanda Chapman, Chemical Engineering, University of Alberta, Edmonton, AB, Canada

Dr. Robert E. Burrell, University of Alberta, Edmonton, AB, Canada

Patricia L. Nadworny, University of Alberta, Edmonton, AB, Canada

Silver, in the form of AgNO3, has been used as an antimicrobial agent for the last 125 years. Initially it was delivered as a 1% AgNO3 solution (6352 mg/L) but over the years this was reduced to a 0.5% AgNO3 solution (3176 mg/L) that was applied up to 12 times per day. In 1967, 1% silver sulfadiazine cream (3025 mg/kg) was introduced, and it became the standard therapeutic treatment for burns world wide. More recently, new silver dressings have entered the medical marketplace. These dressings often release less than 5 mg/L of silver. Nadworny (personal communication) has shown that organisms such as Pseudomonas aeruginosa readily develop resistance to silver when exposed to successively higher doses of silver. They do not readily develop resistance to single high doses of silver (>3000 mg/L). This raises concerns about the appropriateness of using silver dressings which release low doses of silver. In this study, we examined the ability of clinically relevant bacteria to develop tolerance/resistance to silver released from commercially available silver-containing dressings. The results showed that bacteria could develop resistance to low dose silver dressings after repeated exposures. This is a significant concern for patients treated with dressings, which contain inappropriate levels of silver.

See more of this Session: Student Poster Session: Food, Pharmaceutical, and Biotechnology
See more of this Group/Topical: Education

Flow Variability in Continuous Powder Feeders

Daniel Mateo, Chemical Engineering, University of Puerto Rico Mayaguez Campus, San Juan, PR

Continuous processing of powder blends has the ability to control disturbances online, avoiding the loss of valuable processing materials and enabling effective process control. As part of this product development process, research efforts have focused on examining and optimizing the production of homogeneous solid mixtures, which requires consistent feeding of the individual components. Minimizing variability of the individual components in a powder blend product requires the close study of feeding performance in order for optimization. This work seeks to look at the feeding of a Loss-in-Weight feeder to find maximum feed rate as well as the performance at several intermediate federates for different feeder conditions. To obtain a consistent flow rate it is necessary to optimize feeder performance by selection of the proper feeder tooling (screw, screen). In order to evaluate flow variability, automated sampling was used which involves a precision scale with output to a computer. The Schenck AccPro II scale was used to record the weight every 1 s for three materials flowing through a Gericke GLD feeder. The feeder configurations that are examined are open helixes without center rods and closed augers of three different sizes. The data collected from the feeder showed that proper selection of the screw type and size has a significant impact on the capacity and performance of the feeder.

See more of this Session: Student Poster Session: Food, Pharmaceutical, and Biotechnology
See more of this Group/Topical: Education

Discovering Novel Peptide Inhibitors of LpxA and LpxD through Phage Display

Jose A. Burgos, Chemical Engineering, University of Puerto Rico Mayaguez Campus, Mayaguez, PR

Garry D. Dotson, Medicinal Chemistry, University of Michigan, Ann Arbor, MI

Lipid A biosynthesis is essential for membrane integrity and survival of gram negative bacteria. Because of its essentiality, this pathway is a potential target for inhibition in the development of novel antimicrobial agents. LpxA is an enzyme that catalyzes the first step in lipid A biosynthesis, the transferring of an R-3-hydroxyacyl moiety of R-3-hydroxyacyl-ACP to the 3-OH position of UDP-GlcNAc. Likewise LpxD, which catalyzes the third reaction of lipid A Biosynthesis, transfers an R-3-hydroxyacyl moiety of R-3-hydroxyacyl-ACP to the amine of UDP-3-O-(R-3-OHC14)-GlcN. Using phage display, a 12 amino acid peptide library (Ph.D.-12; New England Biolabs) was screened against immobilized LpxA and LpxD from several species of gram negative bacteria. This method was successful in finding peptides that bind to E. coli LpxA and LpxD, N. meningitidis LpxA and LpxD, and P. aeruginosa LpxD. Phage that bound to the targeted proteins were amplified and their DNA sequenced in order to identify the peptides. The binding strength between ecLpxA-binding phage and ecLpxA, nmLpxA, ecLpxD, nmLpxD, and paLpxD were compared by performing HRP-Conjugated Anti-M13 Phage Coat Antibody binding assays. We show that phage # 18 obtained from the ecLpxA phage display screen, binds tightly not only with ecLpxA, but also with ecLpxD and paLpxD. Identifying different peptides that bind to the targeted enzyme(s) and determining their binding strength, will allow focused efforts on a subset of the peptides identified in our phage display screens to further test for inhibition of cell growth in gram-negative bacteria.

See more of this Session: Student Poster Session: Food, Pharmaceutical, and Biotechnology
See more of this Group/Topical: Education

Role of P-Glycoprotein in the Transport of Amyloid-β Protein in Cerebral Amyloid Angiopathy

Emily E. Matherly, Biomedical Engineering Program, University of South Carolina, Columbia, SC

Francisco J. Gonzalez, Department of Chemical Engineering, University of South Carolina, Columbia, SC

Melissa A. Moss, Department of Chemical Engineering, University of South Carolina, Columbia, SC

Cerebral amyloid angiopathy (CAA), a cerebrovascular deposition of the amyloid-B (AB) protein, is a well-known cause of hemorrhagic stroke. Altered AB transport may be responsible for vascular AB accumulation. However, it remains unclear how the transport of AB from the brain to the vasculature is impeded to facilitate AB accumulation. P-glycoprotein (Pgp), an ATPase transporter, interacts with amphipathic molecules and mediates their removal from the brain. It was hypothesized that Pgp is involved in the transport of AB through the BBB. AB assembly state(s) transported from the basolateral to the apical side of confluent monolayers of human brain microvascular endothelial cells were compared using endothelial permeability coefficients (Pe). Both AB monomer and AB aggregates were transported across monolayers. In addition, AB assembly state(s) that interact with Pgp were identified using an in vitro assay that measures ATPase activity of Pgp. These results revealed that Pgp preferentially binds soluble AB aggregates, but not monomer of fibril. Together, these findings suggest a role for Pgp I transporting soluble AB aggregates across the BBB.

See more of this Session: Student Poster Session: Food, Pharmaceutical, and Biotechnology
See more of this Group/Topical: Education

Development of Antioxidant Polymeric Nanoparticles for the Suppression of Doxorubicin Mediated Vascular Chemotoxicity

Kevin Baldridge, Chemical and Materials Engineering, University of Kentucky, Lexington, KY

P. Wattamwar, Chemical Engineering, University of Kentucky, Lexington, KY

K. Anderson, Chemical Engineering, University of Kentucky, Lexington, KY

T. Dziubla, Chemical Engineering, University of Kentucky, Lexington, KY

The maximum tolerable dose of of chemotherapeutics is primarily limited by the toxic side effects caused by non-specific accumulation of drug in healthy tissue. While drug targeting can reduce this nonspecific accumulation, even the best targeting systems still have >80-90% of drug end up outside the tumor. As such, active targeting combined with a method to suppress toxicity in healthy tissue may provide a means of improving treatment outcomes by allowing greater doses of chemotherapeutic to be used. Doxorubicin (DOX) is a chemotherapeutic agent which can generate reactive oxygen species (ROS) that can induce cell death. The hypothesis of this study is that antioxidant polymeric nanoparticles (AoP NPs) delivered concurrently with DOX will reduce apparent DOX toxicity in non-cancerous cells. To evaluate this hypothesis, various antioxidants were evaluated as chemoprotectants against DOX in Human Umbilical Vein Endothelial Cells (HUVEC). Depending on the antioxidant used and its concentration, it was found that it can function as either an antioxidant or pro-oxidant. In the presence of antioxidant, IC50 of DOX was found to be 0.200 &muM with 1mM Trolox, and 1.219 &muM with 1mM vitamin E (liposomes), which were lower and higher, respectively, than that of DOX with no antioxidant, IC50=0.315 &muM.

See more of this Session: Student Poster Session: Food, Pharmaceutical, and Biotechnology
See more of this Group/Topical: Education

Bioreactor Design for Plant Propagation

Sydney E. Shaw, Chemical Engineering, The Pennsylvania State University, University Park, PA

Jeffrey S. Larsen, Chemical Engineering, The Pennsylvania State University, University Park, PA

Wayne R. Curtis, Chemical Engineering, The Pennsylvania State University, University Park, PA

Plant tissue culture provides a means to proliferate plants without the requirement of going through seeds, thus enabling the rapid propagation of superior plants such as sterile hybrids or disease resistant varieties. We are developing this technology for increasing the productivity of cultivated plants including fungal resistant *Theobroma cacao* (Chocolate Tree) via embryogenesis and seedless watermelon varieties via organogenesis. Four untested watermelon varieties were compared to select the variety most amenable to shoot proliferation. Productivity of this variety was compared in temporary immersion bioreactors (TIBs) relative to the standard protocol on solid media. Cotyledon explants were dissected from germinated seedlings and cultured on MS media supplemented with benzylaminopurine (BAP) and indoleacetic acid (IAA), inducing shoot formation. Media was refreshed every three weeks. The cotyledon explants in TIBs developed much slower than their solid media counterparts, with some evidence of budding along the cut edges. The waxy coating on the cotyledon explants may have prevented the absorption of the liquid media. In contrast, secondary shoot proliferation was successful in TIB culture. This work establishes the feasibility of utilizing bioreactors for watermelon propagation while setting the stage for future process scale-up and the analysis of oxygen mass transfer limitations in the developing tissues.

See more of this Session: Student Poster Session: Food, Pharmaceutical, and Biotechnology
See more of this Group/Topical: Education

Metabolic Analysis of Prostate Cancer

Hannah Lois Lintag, Ralph E. Martin Department of Chemical Engineering, University of Arkansas, Fayetteville, AR

Christa N. Hestekin, Ralph E. Martin Department of Chemical Engineering, University of Arkansas, Fayetteville, AR

There is currently great opportunity to expand the knowledge of cancer in areas concerning the effect of altered protein expression on central metabolism. The proteome involved in energy usage for cell proliferation might have developed alterations so that it could accommodate the more rapid growth and development of cancer cells. The purpose of this project is to conduct a proteomic study on the cellular metabolic processes relevant to carcinogenesis and cancer propagation. Understanding the function of protein in metabolic networks will help develop more effective approaches to detection, monitoring, and treatment of cancer. A normal prostate cell line (control) has been be grown along with two prostate cancer cell lines, and from these cell lines, protein were extracted. The protein systems of interest, those that are associated with central metabolism, have been analyzed via two-dimensional gel analysis (2DGE). The results show that analysis with 2DGE promises great potential in diagnostic medicine.

See more of this Session: Student Poster Session: Food, Pharmaceutical, and Biotechnology
See more of this Group/Topical: Education

Understanding Cancer Cell Motility and Interactions with the Extracellular Matrix

Ranjini Krishnamurthy, Chemical and Biomolecular Engineering, Johns Hopkins University, Baltimore, MD

Stephanie Fraley, Chemical and Biomolecular Engineering, Johns Hopkins University, Baltimore, MD

Denis Wirtz, Chemical and Biomolecular Engineering, Johns Hopkins University, Baltimore, MD

Cancer research focuses primarily on biochemical signaling in the tumor microenvironment. While this is one crucial aspect of understanding cancer, the effects of mechanical signals, such as stiffness and dimensionality, are proving to be important as well. However, most experiments performed in this area so far have been done on 2-dimensional (2D), planar substrates, which do not accurately represent the physiological environment in vivo. Prior research shows that cells embedded in 3-dimensional (3D) extracellular matrices display drastically different motility characteristics and, as they do not form focal adhesions in 3D matrices, the differences might be due to the altered role of the focal adhesion proteins. Nonetheless, it has proven difficult to isolate the reason for these differences, as altering the 3D extracellular matrix leads to changes not only in stiffness but also in pore size and ligand density.

The first objective of my research is to develop a method that will provide controlled stiffening of a 3D collagen gel, while maintaining constant pore size and ligand density. For the first time, this will allow for quantitative determination of the effect of stiffness alone on cell motility. There are many methods of altering the stiffness of a 3D collagen gel. Here, we chose to compare the effect of increasing collagen concentration-which alters stiffness, pore size, and ligand density--against using Transglutaminase II enzyme (TGII) as a cross-linker. We show that the pore size and ligand density of Transglutaminase II crosslinked gels are unchanged compared to the uncrosslinked gel. Cell viability and effect of stiffness on 3D motility were assayed by embedding HT1080 Human Fibrosarcoma cells in the matrix and using time-lapse microscopy along with other novel techniques including intracellular microrheology and a 3D matrix traction assay. The results showed wild type cell velocity decreased as collagen concentrations increased as well as when the matrix was stiffened with TGII.

Since focal adhesion proteins are thought to be important in the mechanotrasduction of external signals to the inside of the cell, we next embedded HT1080 cells with various focal adhesion proteins knocked down in uncrosslinked and crosslinked gels. We measured motility characteristics, such as velocity and number of protrusions, of the knockdown cells in a crosslinked matrix and compared them to those measured in uncrosslinked gels and also to wild type cells in both conditions. This allowed us to determine which focal adhesion proteins are important to the cell in sensing stiffness in 3D matrices.

Our findings indicate that stiffness of the extracellular matrix indeed plays a critical role in cell motility by altering the velocity and reducing the number of protrusions. The development of this novel and accessible technique of stiffening 3D matrices using TGII allows for exclusive analysis of an important variable to cancer cell motility-stiffness. Further study will lead to a greater understanding of the molecular mechanisms driving cell migration and may reveal a possible treatment or prevention method for cancer. Isolating and understanding the mechanical transduction pathway in a cancer cells will provide greater insight into the mechanism of metastatic tumors and the role that the surrounding extracellular matrix plays.

See more of this Session: Student Poster Session: Food, Pharmaceutical, and Biotechnology
See more of this Group/Topical: Education

Exploration of the Phylogenetic Relationship Between Ten Escherichia Coli Lab Strains

Donovan S. Layton, Chemical and Biological Engineering, Iowa State University, Ames, IA

Brelan Moritz, Microbiology and Cell Science, University of Florida, Gainesville, FL

Stephen J. Willson, Department of Mathematics, Iowa State University, Ames, IA

Laura R. Jarboe, Chemical and Biological Engineering, Iowa State University, Ames, IA

L.O. Ingram, University of Florida, Gainesville, FL

Despite the extensive characterization and use of Escherichia coli in metabolic engineering, the relationship between the various lab strains remains underexplored. Here we compare ten E. coli lab strains: C36 (ATCC# 8677), C (ATCC# 8739), W (ATCC# 9637), B (ATCC # 11303), NCTC 9001 (ATCC# 11775), W3100 (ATCC # 14948), ML308 (ATCC # 15224), B/r (ATCC# 23227), MG1655 and DH10B. E. coli O157:H7 and Salmonella typhimurium LT2 were used as non-lab strain controls. 10 kb of sequence data was obtained from seven non-essential regions: ptsG, yeiG, eutS, talA, xylAB, atpI and mutL. A phylogenetic tree of these strains was generated, showing that strains W and NCTC 9001 are the most distinct from the other eight lab strains. Surprisingly, O157:H7 was more similar to the other nine lab strains than NCTC 9001. This analysis revealed a truncation in TalA of strain B and a lack of TalA activity was evidenced by increased menadione sensitivity. Similar to previous reports, the MutL region is prone to variability, with half of the lab strains containing a frame shift that results in premature truncation of MutL. We also found that, of the ten strains, only strain W contains the gene encoding penicillin acylase. Strains W and B/r contain the stably inherited pRK2 plasmid and the immigration control region is found in strains B, B/r, W3100 and MG1655.

See more of this Session: Student Poster Session: Food, Pharmaceutical, and Biotechnology
See more of this Group/Topical: Education

Polycatechin Nanoparticles: a Novel Reactive Oxygen Species Scavenging Drug Delivery Vehicle

Sydney Shaffer, Chemical Engineering, Georgia Institute of Technology, Atlanta, GA

D. Scott Wilson, Chemical Engineering, Georgia Institute, Atlanta, GA

Niren Murthy, Biomedical Engineering, Georgia Institute of Technology, Atlanta, GA

Cellular reactive oxygen species (ROS) such as superoxide, hydrogen peroxide, and hydroxyradical play a key role in the progression of inflammatory disease. In healthy cells, biological mechanisms maintain equilibrium between the concentration of ROS and the concentration of antioxidant molecules and enzymes that reduce them to a less destructive form, but in the progression of many diseases this balance is disrupted and an excess of ROS are produced within the cell. The objective of this project was to design a ROS scavenging nanoparticle drug delivery system that makes use of excessive ROS production in inflamed tissue as a targeting signal for the delivery of a therapeutic and that can simultaneously act as therapeutic by reducing cellular ROS levels. In order to carry out this task, we have developed a family of nanoparticles formulated from ROS-scavenging catechin polymers. In this presentation, we demonstrate the formulation of polycatechin into a nanoparticle, characterize the superoxide dismutase equivalent activity of polycatechin nanoparticles, and confirm in cell culture the ROS-scavenging ability and ROS mediated drug release.

See more of this Session: Student Poster Session: Food, Pharmaceutical, and Biotechnology
See more of this Group/Topical: Education

Xenobiotic Metabolism of Myclobutanil in Cultured Hepatocytes

Christopher F. Doe, Department of Chemical and Biochemical Engineering, Rutgers University, Piscataway, NJ

Vidya V. Iyer, Department of Chemical and Biochemical Engineering, Rutgers University, Piscataway, NJ

Charles M. Roth, Department of Chemical and Biochemical Engineering, Department of Biomedical Engineering, Rutgers University, Piscataway, NJ

Marianthi G. Ierapetritou, Department of Chemical and Biochemical Engineering, Rutgers University, Piscataway, NJ

Myclobutanil is a fungicide of the triazole family of conazoles that acts by inhibiting ergosterol formation. Since the liver is the primary location of detoxification, rat hepatocytes were isolated and used for our study. We are able to investigate this through metabolic flux analysis which uses the extracellular metabolites to evaluate the flux of these molecules into and out of the cell. Through the analysis of elementary modes in the well determined metabolic pathways, it can be determined the course of metabolism of myclobutanil. The aim of the experiment is to test the effect of myclobutanil dose on hepatocytes over a three day period. In three six-well plates, hepatocytes were cultured and tested using 4 concentrations ranging from 0.15 mM to 1.2 mM with controls. Initial analysis appears to indicate that the higher concentrations of myclobutanil result in cell death over the three day period with higher concentrations at earlier time points. It also appears that urea production is inhibited by DMSO, and even more so by myclobutanil. The results from this experiment can be used for future research as guidelines for optimal choice of myclobutanil concentrations to maximize concentration without killing most of the hepatocytes.

See more of this Session: Student Poster Session: Food, Pharmaceutical, and Biotechnology
See more of this Group/Topical: Education

Liposomal Microarrays for Toxin Detection

Jude Phillip, Chemical Engineering, The City College of New York, New York, NY

Isabel Sole, Chemical Engineering, The City College of New York, New York, NY

Alexander Couzis, Chemical Engineering, The City College of New York, New York, NY

The technology of protein microarrays (or "protein chips") is a recent innovation used for screening of the binding interactions of proteins (e.g. antigens, cell membrane receptors, enzymes) with potential binding partners or ligands. The ligands can be other proteins (e.g. antibodies or hormones), enzyme substrates, or small synthetic molecules such as oligopeptides, or their analogues (peptoids, etc.) prepared as part of a combinatorial chemistry library. In this work, we have developed a new microarray platform for arraying membrane receptors for multiplexed screening of the binding interactions of these receptors with binding ligands. We used interferometric lithography to make large areas of periodic photoresist structures on a Si wafer for use as templates for metal deposition. Following metal deposition, the photoresist was removed, leaving ordered metal arrays on a flat Si surface. The bottom of the wells was subsequently functionalized with an amine termination by backfilling with an amine silane. The amine functionalized wells were then conjugated to Neutravidin labeled with a FITC dye. As a proof of concept experiments, we used the intact liposome microarray format and known receptor-ligand system to check the applicability of the liposome microarray platform for biodetection.

See more of this Session: Student Poster Session: Food, Pharmaceutical, and Biotechnology
See more of this Group/Topical: Education

Effect of Initial Particle Size On Particle Growth During Wet Granulation in a High Shear Mixer

Barbara Enid Rodriguez Davila, Chemical Engineering, University of Puerto Rico, Mayaguez Campus, Villalba, PR

Carlos Velázquez, Chemical Engineering, University of Puerto Rico, Mayaguez, PR

Effect that initial particle size has on particle growth during wet granulation in a high shear mixer

The main objective of this research is to study the effect that initial particle size has on particle growth during wet granulation in a high shear mixer. Wet granulation in a high shear mixer is primarily a process for particle enlargement that is used in many industries such as pharmaceutical, mineral processing, agriculture, food, and detergent. Particle size enlargement is attained by the addition of a liquid to a powder or a mixture of powders that are to be mixed in a high shear mixer. Wet binders such as povidone are used to improve tablet strength and friability. The process consists of the granulation of different particle sizes of lactose anhydrous and lactose monohydrate with a solution of povidone 2.5% as a binder. To obtain the best results in the granulation process 345g of lactose anhydrous and 90g of lactose monohydrate were used. The povidone solution was added at a 12.5% of humidity. Granule growth in high shear wet granulation is a dynamic process in which granules are continuously forming and breaking down. This type of behavior is named Bimodal Particle Size Distribution and was evident in all the experiments. This granulation process is done in order to prevent segregation of the constituents of the powder mix, to improve the flow properties of the powder mix, to improve the compaction characteristics of the powder mix, and to reduce dust particles.

See more of this Session: Student Poster Session: Food, Pharmaceutical, and Biotechnology
See more of this Group/Topical: Education

7-t-Butyldimethylsilyl-10-Hydroxycamptothecin (DB-67) Precipitation in the Presence of Ferrous Ions: A Liposomal Delivery Strategy

Kimberly P. Doan, Chemical and Materials Engineering, San Jose State University, San Jose, CA

Tian-Xiang Xiang, Pharmaceutical Sciences, The University of Kentucky, Lexington, KY

Thomas Dziubla, Department of Chemical and Materials Engineering, University of Kentucky, Lexington, KY

Brad Anderson, Pharmaceutical Sciences, The University of Kentucky, Lexington, KY

7-t-butyldimethylsilyl-10-hydroxycamptothecin (DB-67/AR-67), a drug that exhibits anti-cancer activity, is currently in clinical development for the treatment of solid tumors. The clinical efficacy of DB-67 may be improved by targeting its delivery to solid tumors. Liposomal encapsulation is a promising formulation approach to improve the delivery of chemotherapeutic agents to tumor sites. However, formulation strategies for liposomal loading and retention of hydrophobic drugs such as the neutral camptothecin DB-67 have had limited success. Therefore, complexation and/or salt formation of DB-67 with iron ions are explored as a potential loading and retention strategy. The extent of drug precipitation due to potential formation of Fe2+-DB-67 salts was investigated at suspension conditions of pH = 7.8, [DB-67]=100 µg/mL, and [Fe2+] = 0 and 0.001M. At one hour after suspension formation, the DB-67 concentration in the supernatant was analyzed by HPLC.In the suspension with iron ions, 52.60% of DB-67 was found in the precipitate whereas the sample without the presence of iron ions had 4.55% of DB-67 in the precipitate. Results suggest that Fe2+ cations may significantly affect the degree of precipitation of DB-67 carboxylate. Future studies will be attempted to load Fe2+-DB-67 salt in liposomes and explore its effect on drug release.

See more of this Session: Student Poster Session: Food, Pharmaceutical, and Biotechnology
See more of this Group/Topical: Education

Metabolically Controlled 3D Cellular Integration Into Lipid Silica Films

Theresa M. Lee, Chemical Engineering, University of New Mexico, Sandia National Labs, Albuquerque, NM

Eric C. Carnes, Chemical Engineering, University of New Mexico, Sandia National Labs, Albuquerque, NM

Robbie Castillo, Chemical Engineering, University of New Mexico, Sandia National Labs, Albuquerque, NM

Carlee Ashley, Chemical Engineering, University of New Mexico, Sandia National Labs, Albuquerque, NM

Deanna Lopez, Chemical Engineering, University of New Mexico, Sandia National Labs, Albuquerque, NM

C. Jeffrey Brinker, Chemical Engineering, University of New Mexico, Sandia National Labs, Albuquerque, NM

We have discovered a unique metabolically and optically controlled lithography approach that allows patterned integration of live cells into nominally solid-state devices and maintains their viability under extreme conditions of desiccation and starvation. We observe that yeast, bacterial, and mammalian cells deposited on lipid/silica thin film mesophases actively reconstruct the surface to create a fully 3D bio/nano interface composed of localized lipid bilayers enveloped by a lipid /silica mesophase. Remarkably this integration process selects exclusively for living cells over the corresponding apoptotic cells. The localized lipid interface maintains fluidity, accessibility, and functionality of the cellular surface and viability of the cell as we show, for example, by exposure of an immobilized mammalian immune cell to a model endotoxin, lipopolysaccharide (LPS) from E coli. Recognition of LPS by cell surface receptors activates the Toll-like receptor 4 (TL-4) signaling pathway, confirming functions including molecular recognition, recruiting of adaptor molecules, activation of kinases, etc. Optical definition of cellular integration is achieved by dose-dependent UV exposure of the lipid/silica mesophase film. Short exposures reduce the film contact angle with water and promote integration. Longer exposures solidify the lipid/silica mesophase, preventing integration. Double exposures define first patterns of cellular integration and second high diffusivity pathways to or between cells via photolytic degradation of the lipid mesophase template and attendant formation of ordered monosized, mesopores. Temperature dependent studies along with those employing sodium azide to interfere with ATP production confirm that integration is metabolically controlled. Biaxial stress measurements reveal that integration is accompanied by a low frequency oscillatory stress that is communicated between and appears to synchronize the behaviors of the integrating cells. Auxiliary nanocomponents added to the lipid/silica mesophase film or introduced with the cell are localized at the cellular surface. Overall this 'lithography with life' approach provides the first demonstration of optically-defined 3D cellular immobilization. It promises a new means to integrate 'bio' with 'nano' into platforms useful to study and manipulate cellular behavior at the individual cell level and to interface living organisms with electronics, photonics, and fluidics.

See more of this Session: Student Poster Session: Food, Pharmaceutical, and Biotechnology
See more of this Group/Topical: Education

Magnetic Resonance Analysis of Physically Crosslinked Biopolymer Gels

Hilary T. Fabich, Chemical and Biological Engineering, Montana State University, Bozeman, MT

Diana Bernin, Chemical and Biological Engineering, Chalmers University of Technology, Göteborg, Sweden

James E. Maneval, Department of Chemical Engineering, Bucknell University, Lewisburg, PA

Joseph D. Seymour, Chemical and Biological Engineering, Montana State University, Bozeman, MT

Sarah L. Codd, Mechanical and Industrial Engineering, Montana State University, Bozeman, MT

Gill G. Geesey, Center for Biofilm Engineering, Montana State University, Bozeman, MT

Using the non-invasive properties of nuclear magnetic resonance (NMR), the mobility of water in a biopolymer gel can be examined by measuring diffusion and magnetic relaxation. Understanding the molecular role of water in physical gelation and water distribution on gel material properties has great potential to increase understanding of biological function [1]. Alginate, and the impact of Ca2+ on gelation has been extensively studied using NMR [2,3]. During formation of the gel under certain conditions, small capillaries are formed inside the gelled structure [4]. These provide molecular transport pathways through the entangled biopolymer network and can control how water and ions are transported through the gel. Another point of interest in this system stems from the role of acetylated and deacetylated alginates in biofilm formation [1] and NMR data on the relaxation and diffusion of water in model alginates provide baseline data for exploring the role of water in biofilm formation.

1. D.E. Nivens et al., J. Bacteriology 183(3) 1047-1057 (2001). 2. N. Nestle and R. Kimmich, Colloids and Surfaces A 115 141-147 (1996). 3. B.P. Hills et al., Mag. Reson. Chem. 38 719-728 (2000). 4. H. Treml, S. Woelki, H.-H. Kohler, Chemical Physics 293 341-353 (2003)

See more of this Session: Student Poster Session: Food, Pharmaceutical, and Biotechnology
See more of this Group/Topical: Education

Peptoid-Based Therapeutics for Alzheimer's Disease

Jennifer Herrera, Chemical Engineering, University of Arkansas, Springdale, AR

Alzheimer's disease (AD) is a slow degenerative disease that causes memory loss and eventually leads to death. AD is caused by the aggregation of the amyloid-beta protein , found on the outside of brain cells; once the amyloid-beta protein begins to aggregate it forms plaques on the extracellular part of the neural cells. So far there is no AD medication available that breaks up the amyloid aggregates. Current research has found that certain molecules bind to the amyloid-beta protein and prevent aggregation. The purpose of this research project is to engineer peptoid-based molecules for the therapeutic treatment of AD. Peptoids form strong helices, have high bioactivity, and are easy and cost-effective to synthesize. Three peptoids designed to have different chemistries at only the 6th side chain have been chosen; the first peptoid is positive, the second peptoid is neutral, and the third peptoid is negative. We have synthesized all three peptoids and are in the process of purification and characterization. The ability of the peptoids to bind amyloid-beta protein will provide important information on the use of peptoids for the therapeutic treatment of AD.

See more of this Session: Student Poster Session: Food, Pharmaceutical, and Biotechnology
See more of this Group/Topical: Education

Effect of Tumor Secreted Soluble Factors On Mesenchymal Stem Cell Microrheology and Heterogeneity of the Cytoskeletal Network

Daniel McGrail, Chemical and Biomolecular Engineering, Georgia Institute of Technology, Atlanta, GA

Michelle Dawson, Chemical and Biomolecular Engineering, Georgia Institute of Technology, Atlanta, GA

Mesenchymal Stem Cells (MSCs) are a pluripotent adult bone marrow (BM) derived stem cell that spontaneously migrate from the BM in response to tumor secreted soluble factors to integrate in tumors [1]. Though actin polymerization is a key step in cell mobility [2], it is unclear how this cytoskeletal reorganization affects the microrheological properties. Previous studies have shown that increased concentrations of actin, leading to increased polymerization, creates more heterogeneity in solution [3]. Using multiple particle tracking, the Brownian motion of 100nm fluorescent particles is tracked to derive the ensemble-averaged mean squared displacement (MSD), which after undergoing a Fourier transformation is used to calculate frequency-dependent complex viscoelastic modulus (G*) and then separated into the elastic (G`) and viscous (G``) moduli [4]. Measurements have shown an increase in distribution of the MSDs after treatment with soluble factors indicating increased heterogeneity. Furthermore, after treatment the cells transition from primarily viscous to primarily elastic rheology, suggesting the reorganization of globular actin to f-actin is modulated by these tumor secreted molecules.

1. Chapel, A., et al., Mesenchymal stem cells home to injured tissues when co-infused with hematopoietic cells to treat a radiation-induced multi-organ failure syndrome. J Gene Med, 2003. 5(12): p. 1028-38. 2. Uribe, R. and D. Jay, A review of actin binding proteins: new perspectives. Molecular Biology Reports, 2009. 36(1): p. 121-125. 3. Apgar, J., et al., Multiple-particle tracking measurements of heterogeneities in solutions of actin filaments and actin bundles. Biophysical Journal, 2000. 79(2): p. 1095-1106. 4. Mason, T.G. and D.A. Weitz, OPTICAL MEASUREMENTS OF FREQUENCY-DEPENDENT LINEAR VISCOELASTIC MODULI OF COMPLEX FLUIDS. Physical Review Letters, 1995. 74(7): p. 1250-1253.

See more of this Session: Student Poster Session: Food, Pharmaceutical, and Biotechnology
See more of this Group/Topical: Education

Microfluidic Chamber Array for High-Throughput Chemical Screening

Emily S. Gong, Chemical and Biomolecular Engineering, Georgia Institute of Technology, Atlanta, GA

Kwanghun Chung, Chemical and Biomolecular Engineering, Georgia Institute of Technology, Atlanta, GA

Jan Krajniak, Chemical and Biomolecular Engineering, Georgia Institute of Technology, Atlanta, GA

Hang Lu, Chemical and Biomolecular Engineering, Georgia Institute of Technology, Atlanta, GA

Microorganisms are used in biological research and drug discovery because of their low complexity, ease of maintenance, and the conservation of fundamental cellular processes between microorganisms and mammals. However, their relatively small size makes it difficult for researchers to physically manipulate these microorganisms. This difficulty limits the experimental throughput and, in some cases, precludes experiments altogether. To address this limitation, we have created a multilayer microfluidic device containing a circular chamber array and a chemical-mixing channel network to perform chemical screenings. In the flow layer, microorganisms are loaded sequentially into rows of circular chambers. Loading regulators in the control layer ensure that a controlled number of organisms is loaded into each chamber. During loading, another set of valves block off portions of the chamber to create a flow channel in the middle to facilitate the loading process. After loading these valves are opened to drastically increase the area of the chamber. Once the chambers are occupied, a chemical gradient is supplied with the use of staged mixing and microorganisms' physiological outputs are analyzed. This microdevice increases the throughput of chemical screenings by allowing a range of chemical concentrations to be tested simultaneously.

See more of this Session: Student Poster Session: Food, Pharmaceutical, and Biotechnology
See more of this Group/Topical: Education

Peptoid-Based Antibody Mimics for Disease Detection Using ELISA Microarray Analysis

James Phillip Turner Jr., Ralph E. Martin Department of Chemical Engineering, University of Arkansas, Fayetteville, AR

Dhaval S. Shah, Ralph E. Martin Department of Chemical Engineering, University of Arkansas, Fayetteville, AR

Shannon L. Servoss, Chemical Engineering, University of Arkansas, Fayetteville, AR

A number of potential protein markers of disease have recently been discovered using techniques such as mass spectroscopy. In order to validate the effectiveness of these protein markers for disease detection, they must be tested by a secondary technique such as ELISA microarray. This technique requires two affinity reagents with high sensitivity and specificity. We are developing peptoid-based antibody mimics that have the potential to be inexpensive, high affinity and specificity, and able to be rapidly selected from combinatorial libraries. The antibody mimics consist of two regions: the constant region to provide stability when immobilized to a solid surface and the variable region for protein binding. The two regions are synthesized individually and attached using click chemistry to produce the peptoid-based antibody mimic. The antibody mimics will be characterized using ELISA microarray analysis, considering both sensitivity and specificity. The long term goal is to develop an accurate method for detecting diseases such as breast cancer from a blood or urine sample.

See more of this Session: Student Poster Session: Food, Pharmaceutical, and Biotechnology
See more of this Group/Topical: Education

Modeling Small Molecule Elution From a Hydrogel Using a Microfluidic Technique

Stephanie M. Evans, Chemical Engineering, Bucknell University, York, PA

Dye elution from hydrogels using a glass on glass microfluidic technique can be used to begin modeling drug release from a bio-contacting material. Photolithography is used to cure a poly(ethylene glycol) diacrylate (PEGDA), dye, and a photoinitiator solution inside a microfluidic device with a channel through the middle of the gel. Pumping water through the channel causes dye to diffuse out of the hydrogel and into the water stream based on a concentration gradient. Channel sizes within the devices range from 300-1000µm to simulate vessels within the body. Digital image analysis of the dye diffusion captured by a stereomicroscope is modeled to determine diffusion coefficients and the dye release profile with respect to time and position in the hydrogel. Analysis of the eluted fluid is performed through ultraviolet-visible spectroscopy to determine a short time diffusion coefficient. The properties of the hydrogel can characterized by the extent of PEGDA crosslinking, which is related to the UV exposure dosage, the initial water and dye content in the PEGDA solution, and the swelling ratio in the gel.

See more of this Session: Student Poster Session: Food, Pharmaceutical, and Biotechnology
See more of this Group/Topical: Education

Calcium-Alginate Capsule Formation and Separation in Microfluidic Devices

Renee N. Perry, Chemical Engineering, Bucknell University, Lewisburg, PA

Janet Tesfai, Chemical Engineering, Bucknell University, Lewisburg, PA

Erin L. Jablonski, Chemical Engineering, Bucknell University, Lewisburg, PA

Microfluidic devices can be used for many applications, including the formation of well-controlled emulsions. In this study, the capability to continuously create mono-disperse droplets in a microfluidic device is used to form calcium-alginate capsules through a suspension polymerization in an oil solution from aqueous droplets of calcium chloride and sodium alginate. Calcium-alginate capsules have many potential uses, such as immunoisolation of cells, microencapsulation of active drug ingredients, and encapsulation of bitter agents in food or beverage products. Capsule formation is accomplished through fusion of a sodium alginate droplet and a calcium chloride droplet. The high surface tension between the droplet of calcium chloride and sodium alginate necessitates the use of the surfactant sodium dodecyl sulfate (SDS) and a soft lithography device with a judiciously designed geometry. After creating the capsules, it is necessary to separate them out of the oil solution and into an aqueous solution. A common method of separation is centrifugation, which can damage both the capsules and the cells inside. The use of a microfluidic device with channel walls of disparate hydrophobicity is used to stabilize co-laminar flow of an oil phase and an aqueous phase. The disparity of hydrophobicity is accomplished by defining one side of the microfluidic device with the hydrogel poly (ethylglycene) (PEG), which adheres to the glass surface through the use of 3-(trichlorosilyl)-propyl methacrylate (TPM). Due to the difference in surface energy within the channel, the aqueous stream is stabilized near PEG-DA and the oil stream is stabilized near a hydrophobic resin (optical adhesive). The technique of passive separation using co-laminar flow has shown promising results in separating the calcium-alginate capsules from the oil phase and into the aqueous phase. The separation device has also shown potential for applications in extraction, specifically extracting a dye from a hydrophobic fluid into a hydrophilic fluid during co-laminar flow within the channel.

See more of this Session: Student Poster Session: Food, Pharmaceutical, and Biotechnology
See more of this Group/Topical: Education

Elucidation of the Mechanism of Action of Antifungal ß-Peptides

Suehelay Acevedo, Industrial Biotechnology Program, University of Puerto Rico, Mayaguez, PR

Juan C. Nieves, Industrial Biotechnology Program, University of Puerto Rico, Mayaguez, PR

Patricia Ortiz-Bermúdez, Department of Chemical Engineering, University of Puerto Rico, Mayaguez, PR

Sean P. Palecek, Chemical Engineering, University of Wisconsin - Madison, Madison, WI

Antimicrobial ß-peptides are oligomers made from artificial amino acids, which have been designed to mimic natural host defense peptides (HDPs). These ß-peptides have shown potent antibacterial and antifungal activities and key structural features have been identified that determine these activities. Despite the amount of literary information found on mechanisms of action for other antifungal agents, the mechanisms of action of the ß-peptides used in this research are not completely understood. It is believed they kill target cells by interacting with and disrupting the integrity of cellular membranes. In this study we have devised a biomolecular strategy to elucidate the genes involved in the sensitivity of yeast to well-characterized antifungal ß-peptides. In order to do this we chose two strains of *Saccharomyces cerevisiae* with the same background where one of them (SNYDER 760) was mutagenized with a mini-Tn3::LEU2 transposon that contained a lacZ gene lacking a promoter and initiator codon. We hypothesize that an increased resistance of this strain to the ß-peptides might indicate a disruption of the genes involved in the sensitivity to these ß-peptides. Therefore, as a control an unmutagenized strain (SKY 760) was submitted to the same treatment. Two different ß-peptides were used in this study, $(ACHC\text{-}V\text{-}K)_3$ and $Y\text{-}(ACHC\text{-}V\text{-}K)_3$. First the Minimum Inhibitory Concentration (MIC) of the two *S. cerevisiae* strains when exposed to the ß-peptides was determined. To gain further insight into the mechanism of action of these ß-peptides, Molecular Biology techniques were employed in order to determine the genes that were active upon exposure of the yeast to the ß-peptides. In the MIC Assays for the $Y\text{-}(ACHC\text{-}V\text{-}K)_3$ ß-peptide, both strains of *S. cerevisiae* grew to the same concentration indicating equal resistance to the ß-peptide. In the MIC Assays for the $(ACHC\text{-}V\text{-}K)_3$ ß-peptide, the strain SNYDER 760 proved more resistant to the ß-peptide by a two-fold dilution. Therefore, a genetic analysis was performed on that strain in order to identify if indeed the cause for resistance was the transposon inserted into this strain's genome. After obtaining these results, we will determine by means of a PCR Vectorette Assay, the genes that were interrupted by this transposon and knowing these interrupted genes might give us some insight into the mechanism of action of these ß-peptides.
ace="Symbol"face="Symbol"face="Symbol"face="Symbol"face="Symbol"face="Symbol"face="Symbol"face="Symbol"face="Symbol"face="Symbol"face="Symbol"face="Symbol"face="Symbol"face="Symbol"face="Symbol"face="Symbol"face="Symbol"face="Symbol"

See more of this Session: Student Poster Session: Food, Pharmaceutical, and Biotechnology
See more of this Group/Topical: Education

Biosensors for the Detection of Glucose in Human Interstitial Fluid

Courtney H. Fox, Department of Chemical and Biomolecular Engineering, North Carolina State University, Raleigh, NC

Biosensors are devices comprising of a selective interface in close proximity or integrated with a transducer, which relays an interaction between the surface and analyte. Additionally, electrochemical biosensors are capable of operating in turbid media, much like the interstitial fluid of the human body. Enzyme-based biosensors are highly effective tools for detecting the presence of a biological molecule in vivo that can be utilized in medical devices and diagnostics. This research project focused on the development of a robust glucose biosensor system that can detect glucose in subcutaneous human interstitial fluid. Large biological molecules, like enzymes or antibodies, can be immobilized to the surface of microelectrodes by forming a stable acyl amino ester on the electrode surface. The co-addition of N'-(3-dimethyaminopropyl)-N'-ethylcarbodiimide (EDC) and N-hydroxysuccinimide (NHS) forms a stable amino ester. Large biological molecules can be easily intercalated into this network, and will be immobilized in the immediate proximity of the electrode. Glucose oxidase (GOx) was densely substituted into this network by replacing the active NHS esters with the primary amines of the enzyme. When tested and calibrated in a solution designed to mimic the composition of human interstitial fluid, the glucose biosensors accurately measured the concentration glucose over an eight-hour operational period, particularly at glucose concentrations within the hypoglycemic regime.

See more of this Session: Student Poster Session: Food, Pharmaceutical, and Biotechnology
See more of this Group/Topical: Education

In Vitro Testing of Cellular Response On Bone Biomaterials

David C. Harris, Department of Chemical and Biomolecular Engineering, Vanderbilt University, Nashville, TN

Edna M. Prieto, Department of Chemical and Biomolecular Engineering, Vanderbilt University, Nashville, TN

Jerald Dumas, Department of Chemical and Biomolecular Engineering, Vanderbilt University, Nashville, TN

Katarzyna Zienkiewicz, Department of Chemical and Biomolecular Engineering, Vanderbilt University, Nashville, TN

Scott A. Guelcher, Department of Chemical and Biomolecular Engineering, Vanderbilt University, Nashville, TN

We are working to address a critical problem in the orthopedic field by developing new, resorbable synthetic bone biomaterials that can be used to more effectively repair problematic fractures by integrating into host tissue and supporting tissue regeneration. The materials are remodeled *in vivo* through the process of osteoclast-mediated resorption, which migrate in from the host bone, and osteoblasts, which subsequently form new bone that repairs the fracture. It follows that any viable biomaterial must demonstrate favorable interactions with these cell types. We are currently performing *in vitro* testing of various allograft bone/polyurethane composite formulations to verify that they support cell viability, proliferation, and differentiation of both osteoclast and osteoblast cells of murine origin. Future work will be directed toward developing a co-culture system to run tests with both cell types concurrently and more accurately model the actual in-vivo environment. We anticipate that such a capability would allow additional fine-tuning of the composite biomaterial to be done during the *in vitro* testing phase, thereby saving time and money later on during *in vivo* testing.

See more of this Session: Student Poster Session: Food, Pharmaceutical, and Biotechnology
See more of this Group/Topical: Education

Protoplast Fusion Between Escherichia Coli and Lactobacillus Brevis Toward the Engineering of Butanol-Tolerant Strains

Matthew S. Rehmann, Chemical and Biomolecular Engineering, University of Pennsylvania, Philadelphia, PA

Katy Kao, Texas A&M University, College Station, TX

One obstacle to the production of butanol from microorganisms is butanol's toxicity: organisms that produce butanol are unable to grow in high butanol concentrations. *Escherichia coli* can be engineered to produce butanol but experiences severely hindered growth in butanol concentrations above 1%. In comparison, *Lactobacillus brevis* is more butanol-tolerant, but *L. brevis* requires more nutritious media to sustain growth than *E. coli* requires. Protoplast fusion between *E. coli* and *L. brevis* was used to engineer bacteria that are more butanol-tolerant than *E. coli* but are able to grow in less nutritious media than that required by *L. brevis*. Five fusants were isolated that could sustain growth on both LB and MRS medium over long periods of time. Three of these fusants contained the 16S rRNA sequence of *E. coli*, and two contained the 16S rRNA sequence of *L. brevis*. Three of the fusants were able to grow in M9 minimum media. Finally, three of them had higher relative growth rates in the presence of butanol than wild-type *E. coli*, and one of the other fusants was able to grow to a higher cell density in 2% butanol than *E. coli*.

See more of this Session: Student Poster Session: Food, Pharmaceutical, and Biotechnology
See more of this Group/Topical: Education

Mechanical Properties of Escherichia Coli F-Pili

Maria L. Vera, Chemical Engineering, Oklahoma State University, Stillwater, OK

Philip M. Silverman, Ph.D., Genetic Models of Disease, Oklahoma Medical Research Foundation, Oklahoma City, OK

Robin Harris, Genetic Models of Disease, Oklahoma Medical Research Foundation, Oklahoma City, OK

Lucinda Maddera, Genetic Models of Disease, Oklahoma Medical Research Foundation, Oklahoma City, OK

The dissemination of antibiotic resistance in bacterial populations is an increasing public health threat to both domestic animals and humans. Bacteria that have acquired resistance to one or more antibiotics create a challenge for the treatment of disease because antibiotics that were once effective become less so. There are several means by which bacteria acquire genes for antibiotic resistance, but the interest of our studies pertains to bacterial conjugation. Bacterial conjugation involves the transfer of genetic material directly from one bacterial cell to another. In the common gram negative bacterium Escherichia coli, a specialized extracellular filament known as the F-pilus mediates the initial contact between two bacteria during conjugation. Currently, the structure, function, and mechanical properties of the F-pilus are unknown. Our study hypothesized that F-pilus flexibility is an important functional property, allowing the filaments to sample a large interstitial volume for potential DNA recipients. Using images of F-pili derived from electron micrographs, we determined the persistence length of F-pili, which is one measure of flexural rigidity. Our results show that F-pili are among the most flexible protein filaments normally produced by cells. These and related ongoing studies are pertinent for understanding the dissemination of antibiotic resistance in bacterial populations.

See more of this Session: Student Poster Session: Food, Pharmaceutical, and Biotechnology
See more of this Group/Topical: Education

Mechanical and Cell Adhesion Properties of Alkylated Chitosan

Kevin B. Miles, Chemical Engineering and Materials Science, Wayne State University, Detroit, MI

Howard W.T. Matthew, Chemical Engineering and Material Science, Wayne State University, Detroit, MI

Chitosan is a biomaterial of interest due to high availability, biocompatibility, and biodegradability. However, improvements in tensile strength and cell adhesion may broaden its use as a scaffold in many tissue engineering applications. The primary amino groups on chitosan provide a site for many grafting reactions that can increase the tensile strength and cell adhesion properties of the polymer. In this study, the effects of N-alkylation on the mechanical and cell adhesion properties of chitosan films were investigated. Chitosan was alkylated with the fatty acids hexadecanoic acid (C16), and decanoic acid (C10), using the carbodiimide EDC to achieve degrees of substitution (DS) ranging from 5-30%. Uniaxial tensile testing of the films showed that 5% DS hexadecanoic acid and 5% DS decanoic acid both improved the maximum tensile strength of chitosan by 12.5% and 3.8%, respectively. For both fatty acids, higher DS values resulted in reduced tensile strength. Preliminary cell culture studies suggested that valvular interstitial cells seeded on cast films exhibited reductions in growth rate with increased DS. Results to date indicate that derivatization with long chain fatty acids improved chitosan mechanical properties at low DS. Cell adhesion and proliferation studies are under way and results will be reported.

See more of this Session: Student Poster Session: Food, Pharmaceutical, and Biotechnology
See more of this Group/Topical: Education

Pediatric Feeding Tube Occlusions

Ashley Baxter-Baines, Chemical Engineering, Rowan University, Glassboro, NJ

Christina Tortu, Chemical Engineering, Rowan University, Glassboro, NJ

Jason Selvaggio, Chemical Engineering, Rowan University, Glassboro, NJ

Dr. Stephanie Farrell, Chemical Engineering, Rowan University, Glassboro, NJ

Many pediatric medical conditions require an enteral feeding tube to provide necessary nutrients when oral feeding is not possible. Although nutritional support is a necessity, tube occlusions can occur frequently, and the replacement of tubing is discomforting to patients. As a result, research and studies are being done to determine the factors that cause feeding tube occlusion. Previously, tube occlusions were studied at a flow rate of 30 ml per hour and a pH of 1.2. The enteral formulas tested were PediaSure® and Nutren Junior® with Fiber, which were delivered by gravity flow feed bags and a simulation of a pulse flow pump. Results showed that PediaSure® coagulated 3 hours after being added to simulated gastric juice, where as Nutren Junior® with Fiber did not. The current work focuses on the investigation of multiple parameters including flow rate, tube material and size, pH and formula type to observe their affects on tube occlusions. Experiments are being conducted using a peristaltic pump with a pressure sensor that indicates the presence of downstream occlusions. The formula is pumped through a feeding tube into simulated gastric juice with an initial pH of 2.5 which represents fasting conditions. Tube diameters from 5-9 French and flow rates of 0.1 - 30 ml/h are being investigated. Elemental, milk-based, and soy-based formulas will be investigated.

See more of this Session: Student Poster Session: Food, Pharmaceutical, and Biotechnology
See more of this Group/Topical: Education

Functional Genomic Study of Exogenous n-Butanol Stress in Escherichia Coli

Becky J. Rutherford, Chemical Engineering, University of California, Berkeley, Berkeley, CA

Robert Dahl, Chemical Engineering, University of California, Berkeley, Berkeley, CA

Richard Price, Biological Engineering, University of California, Berkeley, Berkeley, CA

Peter Benke, Joint BioEnergy Institute, Emeryville, CA

Heather Szmidt, Joint BioEnergy Institute, Emeryville, CA

Aindrila Mukhopadhyay, Joint Bioenergy Institute, Emeryville, CA

Jay D. Keasling, Joint BioEnergy Institute, Emeryville, CA

n-Butanol has been proposed as an alternative biofuel to ethanol, and both Escherichia coli and Saccharomyces cerevisiae have been engineered to produce it. Unfortunately, n-butanol is more toxic than ethanol to these organisms. To understand the basis for its toxicity, cell wide studies were conducted at the transcript, protein and metabolite levels to obtain a global view of the n-butanol stress response. Analysis of the data indicates that n-butanol stress has components common to other stress responses including perturbation of respiratory functions (nuo, cyo operons), oxidative stress (sodA, sodC, yqhD), heat shock and cell envelope stress (rpoE, clpB, htpG, cpxR, cpxP, degP), and metabolite transport and biosynthesis (malE, opp operon). Florescence assays revealed a large increase in reactive oxygen species during n-butanol stress, confirming observations from the microarray and proteomics data. Strains mutant in several genes whose products changed most dramatically during butanol stress were examined for increased sensitivity to n-butanol. Results from these analyses allow identification of key genes that were recruited to alleviate oxidative stress, protein misfolding and other causes of growth defect. Chassis engineering based on these cues may be required in a high titer, butanol-producing host.

See more of this Session: Student Poster Session: Food, Pharmaceutical, and Biotechnology
See more of this Group/Topical: Education

Optimization of Fluorogenic Thrombin Generation Assay

Dominique T. Monteil Jr., Department of Chemical and Biochemical Engineering, University of Maryland, Baltimore County, Baltimore, MD

Mikhail V. Ovanesov, Office of Blood Research and Review, C.B.E.R., U. S. Food and Drug Administration, Bethesda, MD

Blood coagulation is a cascade of enzymatic reactions leading to the formation of thrombin and eventually clotting plasma. Thrombin is the key coagulation enzyme with many functions: thrombin clots blood plasma by cleaving protein fibrinogen; activates platelets and vascular cells; controls its own production by activating Factors V, VIII, XI and Protein C; and controls fibrinolysis via activation of Factor XIII and TAFI. Traditional assays for blood coagulation function (e.g. APTT and PT) are based on the recording of the end-point clotting time in a blood plasma sample mixed with artificial activators. The clotting tests have limited predictive value and cannot be used for control of many pro- and anti-coagulation treatments. The novel microplate thrombin generation (TG) assay was specifically developed to utilize a physiological activator (low amount of tissue factor) and real-time monitoring of thrombin enzymatic activity via a small fluorogenic peptide substrate for thrombin. The TG assay holds promise as a physiologically relevant and sensitive test of global hemostasis, but as yet is has not been validated. In this study, using Michaelis-Menten kinetics, rate limiting reactions were determined and correlations were made between fluorescent intensely and the amount of active enzyme thrombin generated. We then investigated the artifacts that are introduced by the sub-optimal selection of excitation and emission wavelengths and fluorophore concentrations in a microplate reader experiment. These novel experimental approaches provide useful models for the analysis of blood coagulation kinetics in health and disease.

See more of this Session: Student Poster Session: Food, Pharmaceutical, and Biotechnology
See more of this Group/Topical: Education

Evaluation of Intradermal Drug Delivery Devices

Sohyun S. Park, School of Chemical and Biomolecular Engineering, Georgia Institute of Technology, Atlanta, GA

Jyoti Gupta, School of Chemical and Biomolecular Engineering, Georgia Institute of Technology, Atlanta, GA

Samirkumar R. Patel, School of Chemical and Biomolecular Engineering, Georgia Institute of Technology, Atlanta, GA

Mark R. Prausnitz, School of Chemical and Biomolecular Engineering, Georgia Institute of Technology, Atlanta, GA

Background: Traditional vaccine injections use the Mantoux method which involves inserting a hypodermic needle at an angle into the skin such that it delivers the drug intradermally. However, the reliability of this method in terms of localizing drug within the skin requires special training and is variable from person to person.

Objective: The objective of this study was to evaluate three intradermal injection devices to quantify the amount of drug they deliver into skin and to determine the reliability of drug distribution within the dermis.

Methods: The three devices evaluated in this study were: traditional Mantoux injection using a 28-gauge hypodermic needle administered by an experienced user, a custom intradermal injection adapter prototype with a 28-gauge hypodermic needle and an 850 µm long hollow microneedle. Ten male feeder pigs were used in this study with IACUC approval. To quantify the amount of drug delivered into skin, 100 µl of sulforhodamine B was injected into the skin of anesthetized pigs using the different injection devices. The dye left on the skin surface and in the device was collected and quantified using a spectrofluorometer. A mass balance was then used to determine the amount of drug delivered into skin. To determine the distribution of dye within the skin, 100 µl of trypan blue was injected into the skin of freshly euthanized pigs. A skin biopsy was immediately taken followed by freezing the sample. Histology sections of the samples were taken and imaged under brightfield microscopy.

Results: The spectrofluorometer quantification results indicated that the traditional Mantoux method administered by an expert professional, intradermal adapter and hollow microneedle all delivered > 95% of sulforhodamine into skin. The histology results indicated that all injection devices were able to localize dye in the dermis with similar reliability with means ranging from 84-92% of dye in the dermis for the three delivery methods.

Conclusions: The results of this study show that the intradermal adapter and hollow microneedles are as effective as traditional Mantoux method in delivering drug into skin. In this study, all three delivery methods reliably delivered drug to the dermis. However, it must be noted that the reliability of Mantoux injection depends on the user whereas the other two devices are designed to reduce user variability. The results of this study will be used to further optimize the prototype devices to increase reliability and improve injection efficacy.

See more of this Session: Student Poster Session: Food, Pharmaceutical, and Biotechnology
See more of this Group/Topical: Education

Predicting Metabolic Engineering Knockout Strategies for Chemical Production: Accounting for Competing Pathways

Naama Tepper, Computer Science, Technion IIT, Haifa, Israel

Tomer Shlomi, Computer Science, Technion IIT, Haifa, Israel

Metabolic engineering aims to use microbes as factories that can produce and degrade organic molecules for industrial and biomedical purposes. In recent years metabolic engineering was successfully employed to produce various fuels and chemicals, and significant efforts are made for over-producing additional chemicals. Various computational methods are used to design genetic manipulations that can achieve metabolic engineering goals. These methods aim to anticipate the effect of genetic manipulations on cellular metabolism via a metabolic network, searching for specific manipulations that would lead to maximized production rate of chemicals of interest. However, a thorough examination of these methods reveals that they do not account for the presence of alternative pathways in a metabolic network, leading to over-optimistic predictions regarding achievable chemical production rates. In this paper we describe a novel computational method called RobustKnock that predicts genetic manipulations that maximize the guaranteed production rate of chemicals of interest, accounting for the presence of alternative pathways in the network. We show that RobustKnock predicts genetic manipulations that lead to guaranteed production rates of various chemicals in E.coli (e.g. ethanol that can be used as a bio-fuel), while strategies predicted by previous methods lead to zero guaranteed production rates.

See more of this Session: Student Poster Session: Food, Pharmaceutical, and Biotechnology
See more of this Group/Topical: Education

Investigation of the Jak-STAT and MEK-C/EBPâ Signaling Pathways in Response to Cytokine Stimulation in HepG2 Cells

Colby Moya, Department of Chemical Engineering, Texas A& M University, College station, TX

Zuyi (Jacky) Huang, Department of Chemical Engineering, Texas A& M University, College station, TX

Peng Cheng, Department of Chemical Engineering, Texas A& M University, College station, TX

Juergen Hahn, Department of Chemical Engineering, Texas A& M University, College station, TX

Arul Jayaraman, Department of Chemical Engineering, Texas A& M University, College station, TX

Non-alcoholic fatty liver disease (NAFLD) is a metabolic liver disease that encompasses a range of conditions from hepatic steatosis (or simple accumulation of fat in hepatocytes) to end-stage liver diseases such as fibrosis/cirrhosis. It is the most common liver disease worldwide and has been estimated to affect more than 30 million people in the U.S and is strongly associated with different aspects of Metabolic Syndrome. The accumulation of fat in hepatocytes (or steatosis) is the initial step that is common to all aspects of the disease, including steatohepatitis. Even though steatosis is defined as a clinically benign condition as it is reversible and does not lead to liver damage by itself, its progression to steatohepatitis results in strongly adverse effects to liver health. It is becoming evident that pro-inflamamtory cytokines such as TNF-α and IL-6 contribute to disease progression. IL-6 signaling is of specific interest, as it can occur through two pathways that are associated with markedly different outcomes in steatosis. Activation of STAT3 through the Jak-STAT pathway has been shown to decrease steatosis through decreased lipid accumulation and increased fat oxidation. On the other hand, activation of C/EBPβ through the MEK-C/EBPβ pathway increases steatosis by promoting lipid accumulation. Since the two IL-6 activated pathways can potentially influence signaling through the other pathway, the dynamics of IL-6 signaling in terms of pathways used is likely to be an important determinant of the extent of steatosis, as which pathway is utilized affects the relative levels of STAT3 and C/EBPβ in hepatocytes. In order to investigate this hypothesis, we have developed a comprehensive IL-6 signaling model and combined it with green fluorescent protein (EGFP) based transcription factor HepG2 reporter cell lines (STAT3 and C/EBPβ) for profiling relative changes in STAT3 and C/EBPβ expression. Our data shows that the expression of EGFP (and hence, the activation of transcription factors) correlates with the amount and specificity of the cytokine used. Different cytokines (IL-1, IL-6, and OSM) show different levels of GFP expression, suggesting that the signaling pathways of interest are controlled differently by each cytokine.

See more of this Session: Student Poster Session: Food, Pharmaceutical, and Biotechnology
See more of this Group/Topical: Education

Development of a Microfluidic Assay for in Vitro Transcription of Immobilized RNA Template

Allison M. Yorita, Brown University Division of Engineering, Providence, RI

Stephanie McCalla, Brown University Division of Engineering, Providence, RI

Anubhav Tripathi, Brown University Division of Engineering, Providence, RI

This research focuses on the development of a microfluidic assay for in vitro transcription of an RNA template immobilized on microbeads. The detection of RNA requires amplification, due to the small amount present in collected samples. With RNA transcription and amplification of the sample, disease detection can be carried out within a microchip. The microfluidic assay was optimized based on bovine serum albumin (BSA) concentration, microbead concentration, and reaction kinetics. Bovine serum albumin is used as a means to prevent glass surface interactions with the magnetic microbeads. It is expected that increased concentrations of bovine serum albumin will not affect in vitro transcription of RNA, as it is already present as a component in the RNA transcription reaction. In addition, increased concentrations of BSA should decrease the degree of glass surface interactions with DNA coated magnetic microbeads. In using BSA as a dynamic component of the RNA transcription reaction, it is predicted that the RNA from samples can be transcribed and amplified using magnetic beads with immobilized RNA template within a microchip. These beads can then be easily removed with the removal of a magnet on the microchip. The concentration of beads needed to create a magnetized monolayer bead bed within the microchannels will also be optimized.

See more of this Session: Student Poster Session: Food, Pharmaceutical, and Biotechnology
See more of this Group/Topical: Education

Predicting Activation Enthalpies of Cytochrome-P450-Mediated Hydrogen Abstractions: Comparison of Semiempirical PM3, SAM1, and AM1 with a Density Functional Theory Method

Arthur N. Mayeno, Environmental and Radiological Health Sciences, Colorado State University, Fort Collins, CO

Jonathan L. Robinson, Chemical & Biological Engineering, Colorado State University, Fort Collins, CO

Raymond S. H. Yang, Environmental and Radiological Health Sciences, Colorado State University, Fort Collins, CO

Brad Reisfeld, Chemical & Biological Engineering, Colorado State University, Fort Collins, CO

The progression of computational chemistry in conjunction with recent advancements in computer technology and power has made accessible new territory in the study of enzyme kinetics. Here, we focus on predicting activation enthalpies of the hydrogen abstraction step of the cytochrome P450 enzyme-mediated hydroxylation reaction - a key process in the biotransformation of xenobiotics. A rapid methodology was previously developed by Korzekwa et al. (J. Am. Chem. Soc., 112, 7042-7046, 1990) to predict the activation enthalpy of the hydrogen abstraction reaction, using the p-nitrosophenoxy (PNPO) radical as a simple surrogate for the CYP active oxygen species. The activation enthalpy is approximated with a linear regression model using reaction enthalpy and ionization potential (of the substrate radical) as predictor variables, calculated by semi-empirical (SE) method AM1. We have applied this methodology to a set of 24 substrates, using PM3 and SAM1 calculation methods in addition to AM1. Our regression models showed strong predictive capabilities, with R2 values of 0.89, 0.90, and 0.93 for AM1, PM3, and SAM1 methods, respectively. In an effort to evaluate the applicability of the SE results, our calculated activation enthalpies were compared with those calculated by a hybrid density functional theory (DFT) method, B3LYP, using a more realistic iron-oxo-porphine model (FeO), and the results revealed limitations of the PNPO radical model. Thus, predictive models developed using SE predictors provide rapid and generally internally-consistent results, but should be interpreted and used cautiously.

See more of this Session: Student Poster Session: Food, Pharmaceutical, and Biotechnology
See more of this Group/Topical: Education

Acyl-CoA Carboxylases: Domains and Their Organization

Armando Elizondo-Noriega, Department of Chemical en Biological Engineering, Iowa State University of Science and Technology, Ames, IA

Yingfei Chen, Department of Chemical and Biological Engineering, Iowa State University of Science and Technology, Ames, IA

David Cantu, Department of Chemical and Biological Engineering, Iowa State University of Science and Technology, Ames, IA

Peter J. Reilly, Department of Chemical and Biological Engineering, Iowa State University of Science and Technology, Ames, IA

We have gathered all amino acid sequences and three-dimensional structures available digitally of the multi-domain carbon-carbon bond-forming ligases (EC 6.4.1.-). EC 6.4.1.1 (pyruvate carboxylase), EC 6.4.1.6 (acetone carboxylase, of which no known amino acid sequences exist with evidence at protein level digitally), and EC 6.4.1.7 (2-oxoglutarate carboxylase) are fundamentally different from the acyl-CoA carboxylases (ACCs, EC 6.4.1.2 to EC 6.4.1.5) in not being active on substrates containing thioester groups linked to CoA. We are incorporating ACCs, which are important components of the fatty acid synthesis cycle, into the ThYme (Thioester-active enzYme) database. This will aid researchers worldwide who investigate these enzymes. We have found that all of these ligases share very similar biotin carboxylase and biotin carboxylase carrier protein domains. However, their carboxyl transferase domains (CTs) and the order in which their domains are located in the polypeptide chain, or even whether they form one chain, differ. For instance, there are two different CT domains, one found in EC 6.4.1.1 and EC 6.4.1.7, and the other found in EC 6.4.1.2 through 6.4.1.5. The second kind of CT domain exists in two different forms, CT-alpha and CT-Beta. The combination of these two forms or the absence of CT-alpha defines the CT domains in the ACCs.

See more of this Session: Student Poster Session: Food, Pharmaceutical, and Biotechnology
See more of this Group/Topical: Education

A Novel Microfluidic Device That Passively Generates Chemical Gradients and Detects Oxygen Concentrations Under Bacterial Biofilms

Adit Dhummakupt, Biomedical Engineering, Vanderbilt University, Nashville, TN

Philip Samson, Biomedical Engineering, Vanderbilt University, Nashville, TN

Dmitry Markov, Biomedical Engineering, Vanderbilt University, Nashville, TN

John Wikswo, Biomedical Engineering, Vanderbilt University, Nashville, TN

Leslie M. Shor, Department of Chemical, Materials and Biomolecular Engineering, University of Connecticut, Storrs, CT

This paper describes the use of micro-patterned stamps to create biofilm dot arrays for use in a microfluidic device. Staphylococcus aureus was grown on an agar plate and transferred to a glass slide coated with a platinum-porphyrin based oxygen sensing film via a PDMS stamp. After incubation, the resulting biofilms are robust and can be patterned to any shape and size ranging from 500μm to several mm. The technique is simple, reproducible and allows the biofilms to be observed without the need to fix the bacteria for observation. The oxygen film is able to detect aerobic metabolism in the biofilms quickly and easily, although the accuracy of the data is dependent on the maturity of the biofilms. Used in conjunction with a microfluidic device capable of generating a concentration gradient, this method allows for the real-time monitoring of biofilm structure and oxygen usage against a gradient of nutrients, toxins or antibiotics.

See more of this Session: Student Poster Session: Food, Pharmaceutical, and Biotechnology
See more of this Group/Topical: Education

A Study of Novel Degradable Imprinted Networks

Daniel Pulliam, Biomimetic & Biohybrid Materials, Biomedical Devices, and Drug Delivery Laboratories, Department of Chemical Engineering, Auburn University, Auburn, AL

Esmaiel Jabbari, University of South Carolina, Columbia, SC

Mark E. Byrne, Biomimetic & Biohybrid Materials, Biomedical Devices, and Drug Delivery Laboratories, Department of Chemical Engineering, Auburn University, Auburn, AL

In recent years, there has been increased study of molecular imprinting within hydrogels, a technique which creates macromolecular memory for template molecules in polymer networks. In this work, imprinting phenomena is studied within a degradable network structure. Significant application potential exists in medicine and includes controlled drug-releasing scaffolds, matrices, and films. The degradable, imprinted hydrogels created in this study are composed of the macromer, poly(lactide-ethylene oxide-fumarate) (PLEOF), along with the template molecule, diclofenac sodium (DS), and the monomers 2-hydroxyethyl methacrylate (HEMA), diethylaminoethyl methacrylate (DEAEM), and polyethylene glycol 200 dimethacrylate (PEG200DMA). The quality of the memory (i.e., template loading, affinity, and release) will be discussed in relation to the controlled degradation of the networks.

See more of this Session: Student Poster Session: Food, Pharmaceutical, and Biotechnology
See more of this Group/Topical: Education

Enzyme Loading and Activity in Nanoporous Metal Oxides

Kala Bean, Chemical Engineering Department, Tuskegee University, Tuskegee, AL

The activity of the antioxidant enzyme catalase loaded onto the surface or into the pores of silica-based nanoporous materials was studied. Four different types of mesoporous materials were created using published procedures from the Rankin group: amorphous non-porous silica micro particles, nanoporous silica with small radially oriented pores, hollow spherical silica with parallel nanopores (T3), and hollow spherical silica with large, interconnected bimodal nanopores (HSSP-I). The materials were characterized by nitrogen adsorption and dynamic light scattering to confirm that they have characteristics similar to those reported in the literature. Once the materials were synthesized, they were loaded with the enzyme catalase. To test protection of catalase by the silica supports, samples of enzyme-loaded materials were exposed to the proteolytic enzyme pronase and then the temperature was reduce to stop pronase activity. The catalase activity was tested by exposing the either free enzyme or enzyme-loaded materials to peroxide and measuring the rate of degradation using UV-Vis spectrophotometry. Consistent with expectations, the materials with the largest, most accessible pores (HSSP-I) showed the greatest degree of catalase loading and protection from pronase. However, the material with parallel, smaller channels showed some degree of protection from pronase, suggesting either that the pores were partially accessible, or that the enzymes were protected to some degree by adsorbing to the external particle surface.

See more of this Session: Student Poster Session: Food, Pharmaceutical, and Biotechnology
See more of this Group/Topical: Education

A Kinetic Study of the Production of Biodiesel From Algae

Robert W. Dacus III, Chemical Engineering, University of Tennessee at Chattanooga, Lookout Mountain, TN

Dr. Robert C. Mebane, Chemistry, University of Tennessee at Chattanooga, Chattanooga, TN

Dr. Francis J. Jones, Chemical Engineering, University of Tennessee at Chattanooga, Chattanooga, TN

During a time of foreign fuel dependency and high green house gas emissions, it is imperative for the US to develop a biofuel that can alleviate economic and environmental pressures. Biodiesel, a fuel made from plant and animal oils, uses domestic feedstock for production and recycles CO¬2 previously released during combustion. Currently biodiesel is produced from food oils using homogenous base catalysts in a batch reaction process. The creation of significant byproducts, the displacement of a food source, and the batch nature of this process make biodiesel an expensive fuel alternative. The objective of this project is to identify and test pure metal and metal oxide catalysts in order to develop a heterogeneous, continuous flow process. Algae was also investigated as a possible feedstock since its use in production would not displace food sources or arable land. It was found that algae and other vegetable oils can be successfully converted to biodiesel using these metal and metal oxides at low temperatures. Use of the novel catalysts did not create unwanted byproducts that require purification. Successful catalysts also create biodiesel at low temperatures for a variety of vegetable oils containing traditional impurities such as free fatty acids.

See more of this Session: Student Poster Session: Fuels, Petrochemicals, and Energy
See more of this Group/Topical: Education

Performance and Emissions of Petro-Diesel and Biodiesel

Peter J. Coate, Lafayette College, Allegany, NY

The emissions and performance of a variety of plant oils against each other and petro-diesel is tested and compared. Plant oils used in the experiment include sunflower oil, peanut oil, canola oil, and soybean oil. Experiments are conducted using a Yanmar L70EE single cylinder, four cycle, air cooled diesel engine. Each fuel's performance is determined at controlled RPM. Data on torque, horsepower, and emissions are collected and compared. Carbon dioxide, carbon monoxide, and nitrogen oxides are measured using ECOM-AC electrochemical sensor based emission-analyzer. The data collected at each RPM value is analyzed and presented.

See more of this Session: Student Poster Session: Fuels, Petrochemicals, and Energy
See more of this Group/Topical: Education

Proton Conducting Polymer Films within Nanoporous Gold Leaf Electrodes

Alexander R. Trzeciak, Vanderbilt University, Nashville, TN

Kane G. Jennings, Chemical Engineering, Vanderbilt University, Nashville, TN

A novel approach to the design of a fuel cell cathode that contains a minimal amount of platinum incorporates an ionomer-filled nanoporous gold leaf coated with a platinum monolayer. Because of pore size-related limitations and the delicacy of gold leaf, a traditional ionomer substance such as Nafion cannot be simply applied to the surface. Instead, to achieve conformal coating throughout the pores, the ionomer film must be polymerized on-site using surface initiated ring-opening metathesis polymerization of norbornene monomer equipped with a fluorocarbon side chain. The conditions for this reaction must be finely tuned to ensure that monomer has access to all surfaces of the gold, that there is a sufficient amount of monomer to achieve conformal coating, and that the monomer is not in such an excess that would lead to excessive overgrowth or pore plugging. The goal of this research was to determine both the appropriate variables and the optimal reaction conditions for the desired coating. It was found that the two main factors that affect polymerization quality are the gold pore size and the monomer solution concentration. Working ranges in each of these variables were established.

See more of this Session: Student Poster Session: Fuels, Petrochemicals, and Energy
See more of this Group/Topical: Education

Reversible and Irreversible Phase Transitions in Athabasca Bitumen Asphaltenes

Jennifer M. Morrison, Chemical and Materials Engineering, University of Alberta, Edmonton, AB, Canada

John M. Shaw, Chemical and Materials Engineering, University of Alberta, Edmonton, AB, Canada

The precipitation of asphaltenes causes severe fouling problems throughout the production, transport and refining operations of oil companies worldwide. The economic impact is significant. Examples include plugging and wettability reversal in reservoirs, to catalyst and heat exchanger fouling in refineries leading to production failure or production loss in the first example and to more frequent refinery shutdowns in the second example. Asphaltenes are a poorly understood petroleum fraction and there is a critical need of quantitative prediction tools and data related to there thermophysical properties. In this work, the phase behaviour of Athabasca asphaltenes separated from their native hydrocarbon resource using normal alkanes was studied using a combination of differential scanning calorimetry (DSC), temperature modulated differential scanning calorimetry (TMDSC) and rheology measurements in the temperature interval 200 K to 520 K. For pentane through dodecane Athabasca asphaltenes, there is a broad reversible endothermic phase transition, from solid to liquid, overlapping an irreversible exothermic phase transition in the temperature range 330 K and 520 K. Thus, in general, asphaltene thermal and physical properties and hence phase behaviour differ depending on their thermal history in this temperature range. This result may account for the diversity of asphaltene behaviors in apparently similar industrial environments.

See more of this Session: Student Poster Session: Fuels, Petrochemicals, and Energy
See more of this Group/Topical: Education

Solid-Phase Synthesis of Bioinsipried Oligomers for Charge Transfer Investigation

Robert Bonderer, Chemical and Environmental Engineering, University of California, Riverside, Riverside, CA

Valentine I. Vullev, Department of Bioengineering, University of California, Riverside, Riverside, CA

Facile control of the directionality of photoinduced charge transfer is of great importance for solar-energy conversion applications. A broad range of donor-bridge-acceptor and donor-acceptor systems designed and prepared in our group, allow us to investigate various aspects of electron transduction and of the photogeneration of long-lived charge-transfer species. Recently, we theoretically demonstrated the feasibility of bioinspired polypeptides (composed of non-traditional aromatic residues) as potential materials for controlling charge transfer and charge transport. We have used established solution-phase procedures for the preparation of these bioinpsired conjugates. Solution-phase synthesis, however, is challenging and unfeasible for achieving such oligomers with relatively large molecular weight. Automated solid-phase peptide synthesis offers an alternative. Utilizing various Fmoc protocols (used in such automated procedures), however, did not yield the desired macromolecular conjugates. We analyzed our findings and concluded that steric hindrence and the lack of sufficient electrophilicity of the activated carboxyls are the underlying reasons for the failed coupling steps. The purpose of this project is to assay the kinetics of a series of solution phase reaction that lead to the same non-native oligopeptides. We will implement optimal reaction conditions, to avoid the aforementioned issues of steric hinderence and lack of electrophicility. This reaction scheme will then be adapted for solid-phase protocols and the synthesizer programmed accordingly for automating a broad range of bioinspired conjugates. Incorporating these oligopeptides as bridges in donor-bridge-acceptor systems will allow us to investigate their charge-transduction properties. The developed automated solid-phase synthetic protocols will prove useful for any polypeptide sequence involving residues requiring challenging coupling steps.

See more of this Session: Student Poster Session: Fuels, Petrochemicals, and Energy
See more of this Group/Topical: Education

Improving Soluble Fiber Yield From Whole Stillage through Pretreatment and Ion Exchange

Linda Beckwith, University of Nebraska-Lincoln, Lincoln, NE

See more of this Session: Student Poster Session: Fuels, Petrochemicals, and Energy
See more of this Group/Topical: Education

Biodiesel Synthesis, Characterizaton and Emissions Testing Using Chromatographic Methods

Jonathan P. Pena, Chemical Engineering, North Carolina A&T State University, Greensboro, NC

Chawanne T. Rutues, Chemical Engineering, North Carolina A&T State University, Greensboro, NC

Jasmine C. Taylor, Chemical Engineering, North Carolina A&T State University, Greensboro, NC

Kimberly A.G. Moore, Chemical Engineering, North Carolina A&T State University, Greensboro, NC

Stephanie L. Luster-Teasley, Chemical Engineering, North Carolina A&T State University, Greensboro, NC

Biodiesel, also known as fatty acid alkyl esters, which is derived from triglycerides by transesterification with an alcohol, has attracted considerable attention during the past decade as a biodegradable, non-toxic fuel. Because of its environmental benefits and the fact that it can be used in any diesel engine without modification, biodiesel is technically competitive and offers advantages compared to conventional petroleum diesel fuel. Besides being a renewable resource, biodiesel reduces polycyclic aromatic hydrocarbons while engine performance and fuel economy are nearly identical to those of conventional fuels. Several processes have been developed, among which transesterification using alkali-catalysis gave high levels of conversion of triglycerides to their corresponding methyl esters in a short reaction time.

Transesterification of four commercial grade vegetable oils: canola oil, corn oil, peanut oil, and soybean oil, was carried out using methanol and three catalysts: sodium hydroxide (NaOH (s) pellets), liquid sodium hydroxide (1 N NaOH (l) and sodium methoxide (NaOCH3 (s)) as catalysts. The methyl esters of the corresponding oils were separated from the glycerol by-product, purified, and characterized by different methods to evaluate their densities, yields, and composition. From the qualitative analyses performed, it was determined that based on density, yield, and emissions testing, corn biodiesel had most potential for being an alternative for diesel fuel.

See more of this Session: Student Poster Session: Fuels, Petrochemicals, and Energy
See more of this Group/Topical: Education

Determination of a Critical Length Scale for Clustering Instability in Granular Systems

Michael Pacella, Chemical and Biochemical Engineering, UMBC, Baltimore, MD

Daniel Cromer, Chemical and Biological Engineering, University of Colorado at Boulder, Boulder, CO

Christine M. Hrenya, Chemical and Biological Engineering, University of Colorado at Boulder, Boulder, CO

In sharp contrast to familiar molecular fluids, granular fluids composed of inelastic particles are unstable to the formation of macroscopic density inhomogeneities known as clusters. Theses instabilities are of particular engineering interest in fluidized beds, such as those found in coal and biomass gasifiers, where they decrease gas-solid contact area and, subsequently, impact conversion. Clusters are characterized by a critical length scale which defines the system size at which clusters will first begin to appear. In order to quantify a critical length scale for clustering, molecular dynamics simulations are used. These simulations treat the granular particles as hard spheres enclosed in a cube with periodic, adiabatic boundaries. Preliminary simulation results are in good agreement with theoretical results generated by linear stability analyses of the hydrodynamic equations for granular flow. The long term application for the research discussed in this poster is the improvement of the gasification process through an enhanced knowledge of the clustering phenomenon. Although not truly representative of a gasfier, the simulations discussed in this poster are useful in obtaining a "basic principles" understanding of cluster formation and the system length scales required for their formation.

See more of this Session: Student Poster Session: Fuels, Petrochemicals, and Energy
See more of this Group/Topical: Education

Microalgae Derived C10-C14 Biofuels

Christine S. Miranda, Chemical Engineering, University of Puerto Rico Mayaguez Campus, Mayaguez, PR

A renewed interest in sustainable energy raw material sources and technologies emerged strongly because of environmental, geopolitical and national security concerns. The growing awareness of the contribution of traditional fuel usage to climate changes has prompted sustainable energy commercial developments especially in wind, photovoltaic, and biofuels markets. This project addresses the need for renewable energy by developing technologies for controlled growth of Microalgae as a cost effective source of various oils that serve as starting materials for biodiesel and biojet fuels. The integrated Biorefinery technology platform utilizes carbon dioxide produced during bioethanol fermentation as the carbon source for microalgae growth and capitalizes on the long photoperiods in Puerto Rico by using fiber optic delivery sunlight. Molecular genetics and analytical expertise are being used to characterize unique indigenous microalgae and to optimize the production or different oils desired for manufacturing different Biofuels. At the Birorefinery Research Center and affiliated laboratories the emphasis is to develop and implement microalgae based products. Why use microalgae? Using microalgae to produce biodiesel will not compromise production of food, fodder and other products derived from crops. Algae can produce 20 times more oil per acre, compared to the current crops now utilized for bio-fuels production. Algae bio-fuel is eco-friendly also in the sense that it contains no sulfur. It is non-toxic and highly biodegradable. Between 1 and 3% of the total U.S. cropping area would be sufficient for producing algal biomass that satisfies 50% of the transport fuel needs. Algae are the best source for the production of oils and biodiesel. It is only renewable biodiesel that can potentially completely displace liquid fuels derived from petroleum.

See more of this Session: Student Poster Session: Fuels, Petrochemicals, and Energy
See more of this Group/Topical: Education

Effects of Char Particles On the Aging of Bio-Oil Produced From Timber Biomass Pyrolysis

Caitlin Naske, Dave C. Swalm School of Chemical Engineering, Mississippi State University, Mississippi State, MS

Jason Speed, Dave C. Swalm School of Chemical Engineering, Mississippi State University, Mississippi State, MS

Zach Wynne, Dave C. Swalm School of Chemical Engineering, Mississippi State University, Mississippi State, MS

Sarah E. Crosby, Dave C. Swalm School of Chemical Engineering, Mississippi State University, Mississippi State, MS

Keisha B. Walters, Dave C. Swalm School of Chemical Engineering, Mississippi State University, Mississippi State, MS

The viscosity of bio-oil created from biomass pyrolysis has been found to generally increase as the bio-oil ages. This viscosity increase is accelerated by the addition of heat. It is theorized that a major mechanism for this viscosity increase is a polymerization reaction catalyzed by char particles. To test this hypothesis, bio-oil samples created from pine whole tree biomass were aged at 80C for 3 weeks. Prior to aging, a subset of these samples were vacuum filtered (20-25 micrometer pore size) followed by centrifugal filtration (0.2 micrometer pore size). The filtered solids and bio-oil were analyzed using Fourier transform infrared (FTIR) spectroscopy, diffuse reflectance infrared Fourier-transform spectroscopy (DRIFTS), gel permeation chromatography (GPC), and combined gas chromatography-mass spectroscopy (GC/MS). Particle size distribution data was obtained using optical microscopy and Image-Pro Plus software on both the removed solids and unfiltered bio-oil. Measured particle size distributions confirmed a significant decrease in particle density and size after filtering. Removed solids showed very similar FTIR spectra as the unfiltered bio-oil. It was theorized that this was due to a thin coating of bio-oil remaining on the particles. A methanol and acetone washing procedure was developed; this method allowed for the true FTIR spectra of the removed particles to be collected. The filtered solids FTIR spectra showed marked similarity to the char produced during pyrolysis.

See more of this Session: Student Poster Session: Fuels, Petrochemicals, and Energy
See more of this Group/Topical: Education

Supercooled Water in PEM Fuel Cells

Antonio Pistono, Chemical Engineering, Tennessee Technological University, Cookeville, TN

Cynthia Ann Rice-York, Department of Chemical Engineering/Center for Manufacturing Research, Tennessee Technological University, Cookeville, TN

Polymer Electrolyte Membrane Fuel Cells (PEMFCs) face three hurdles for widespread automotive commercialization: cost, cathode catalyst durability and freeze tolerance. This research seeks to understand water mobility during and after freeze of a hydrogen PEMFC. During the process of freezing in a fuel cell, the water inherently present in the membrane electrode assembly undergoes phase transition from liquid to solid. The acidic nature of Nafion® and its presence in the catalyst layer (more specifically the agglomerate) causes supercooled water to appear during the freezing process which is mobile within the membrane, catalyst layer and agglomerate. The cells were conditioned with varying %RH under load and then frozen. Electrochemical Impedance Spectroscopy (EIS) and Cyclic Voltammetry (CV) were performed after freeze. For cells under no load, EIS shows a high frequency intercept (HFI), and generally both high and low frequency arcs (HFA,LFA). Varying %RH did not change the LFA but did change HFI and HFA. A thermal gradient on the cell fixture was used to simulate the end of a stack during freeze. Regression was used to fit circles to the HFA and LFA.

See more of this Session: Student Poster Session: Fuels, Petrochemicals, and Energy
See more of this Group/Topical: Education

Identification of Dominant Mechanisms for Capacity Fade of Lithium-Ion Batteries

Nancy A. Burns, Department of Chemical Engineering, Tennessee Technological University, Cookeville, TN

Ruthvik Basavaraj, Department of Chemical and Biomolecular Engineering, University of Illinois at Urbana-Champaign, Urbana, IL

Venkatasailanathan Ramadesigan, Energy, Environmental and Chemical Engineering, Washington University in St. Louis, St. Louis, MO

Folarin Latinwo, Department of Chemical and Biomolecular Engineering, University of Illinois at Urbana-Champaign, Urbana, IL

Ravi N. Methekar, Energy, Environmental and Chemical Engineering, Washington University in St. Louis, St. Louis, MO

Richard D. Braatz, Department of Chemical and Biomolecular Engineering, University of Illinois at Urbana-Champaign, Urbana, IL

Venkat Subramanian, Energy, Environmental and Chemical Engineering, Washington University in St. Louis, St. Louis, MO

The life-cycle cost of a battery depends on lifetime. The literature reports numerous possible mechanisms for capacity fade in lithium-ion batteries. Mathematical models including these phenomena are few and do not include all postulated mechanisms. Developing such a mathematical model is challenging due to (1) unknown model parameters, (2) cumulative non-separable effects of individual mechanisms occurring simultaneously, and (3) lack of computationally efficient solver schemes. Electrochemical engineering models will be reviewed with a step-by-step description of what each possible mechanism implies in terms of model formulation and numerical simulation. Due to the combination of multiple mechanisms possible for capacity fade, the data available (discharge curves at different cycles) may not be sufficiently informative to identify the dominant mechanisms in a particular operating scenario just by comparing fits of data to simulation results. To address this, Bayesian estimation, uncertainty quantification, and model discrimination techniques will be used to analyze mechanisms for capacity fade. The use of reformulated models for discharge curves facilitates the application of these systems techniques, which employ Markov Chain Monte Carlo simulation and polynomial chaos expansions. The computational cost and the identifiability of the model parameters will be compared for the inclusion of each mechanism.

See more of this Session: Student Poster Session: Fuels, Petrochemicals, and Energy
See more of this Group/Topical: Education

The Chemistry behind the Conversion of Vegetable Oils to Biodiesel

Daniel Denby, Engineering, Hope College, Holland, MI

Abigail Phillips, Engineering, Hope College, Holland, MI

Peter N., Larsen, Science, Holland High School, Holland, MI

Sameer Chavda, Chemistry, Hope College, Holland, MI

Michael J. Misovich, Engineering, Hope College, Holland, MI

Moses N.F. Lee, Chemistry, Hope College, Holland, MI

With the cost of energy rising dramatically in the last few years, scientists and engineers have intensified their efforts to find alternative sources of energy. The use of biologically produced oils, especially used vegetable oils to produce biodiesel, has been heavily investigated as a source of energy. Biodiesel is a mixture of methyl or ethyl esters of fatty acids, and it is generated from alkanolysis of fatty acid triglycerides extracted from vegetable oils. This project has several objectives. 1. Prepare pure sample of methyl esters of palmitic and lauric acid that can be used as internal standards for analysis. 2. Convert vegetable oil to biodiesel using methanolysis and ethanolysis, and develop appropriate analytical techniques to analyze the efficiency of the conversation as well as to examine the composition and purity of the biodiesel. These techniques include thin layer chromatography, gas chromatography-mass spectrometry, infrared spectrophotometry, and nuclear magnetic resonance spectrometry. 3. Develop a chemical process to convert one of the byproducts of biodiesel formation, namely glycerol, to triacetin. This product has multiple uses, including the use as an additive in biodiesel. Initial studies were conducted using pure glycerol. The goal is to use "waste" glycerol derived from biodiesel production processes. Appropriate analytical methods to analyze triacetin also were developed. Results from the aforementioned studies will be presented.

See more of this Session: Student Poster Session: Fuels, Petrochemicals, and Energy
See more of this Group/Topical: Education

Waste Heat Recovery

Pavlo Kostetskyy Sr., Chemical Engineering, Rowan University, Glassboro, NJ

Derek Becht, Chemical Engineering, Rowan University, Glassboro, NJ

Bolaji Adigun, Chemical Engineering Dept., Rowan University, Glassboro, NJ

Rowan engineering students working with Johnson Matthey have identified a green engineering opportunity for reducing their natural gas usage utilizing the heat contained in an exhaust gas stream. A heat exchange system was designed to recover energy from exhaust gases and use it to preheat combustion air fed to this process. Installing this new heat exchanger will increase the pressure drop on the system as well as increase the NOx produced as a result of higher combustion air temperature. To reduce the NOx produced, new ultra-low NOx burners have been specified. In addition, new fans are required to maintain the current gas flowrates in the system. Total savings based primarily on reduced natural gas consumption are estimated at $302,500/yr. The capital equipment proposed is a gas to gas heat exchanger, a combustion air fan, and a scrubber fan. In addition to reducing the carbon footprint of the furnace system, the project is economically viable as the internal rate of return is 20.7% and the payback period is 49 months. This project will reduce the CO2 emissions of the furnace system by 163 tons/yr and decrease the nonrenewable fossil fuel usage of the furnace system by recovering the waste heat generated.

See more of this Session: Student Poster Session: Fuels, Petrochemicals, and Energy
See more of this Group/Topical: Education

Wastewater Converted to Biofuel Via Oleaginous Microorganism Consortium

Rosemary Weathersby, Dave C. Swalm Schoold of Chemical Engineering, Mississippi State University, Mississippi State, MS

Jacqueline Hall, Mississippi State University, Mississippi State, MS

W. Todd French, Dave C. Swalm School of Chemical Engineering, Mississippi State University, Mississippi State University, MS

Rafael Hernandez, Dave C. Swalm School of Chemical Engineering, Mississippi State University, Mississippi State, MS

William E. Holmes, Mississippi State Chemical Laboratory, Mississippi State University, Mississippi State, MS

Andro H. Mondala, Chemical Engineering, Mississippi State University, Mississippi State, MS

Oleaginous microorganisms are able to produce up to 70% of their weight in oil as lipids and their use as a biofuel feedstock is partially limited by the cost to cultivate them. Municipal wastewater has the potential to be a growth medium for these microorganisms due to the fact that wastewater contains carbon and other trace nutrients that a microorganism requires. With this new growth medium, these microorganisms have the potential to treat the wastewater as well as accumulate oil for biofuels. This study is focused on enhancing the consortium's growth through the addition of xylose and glucose into autoclaved municipal wastewater. Preliminary results show that the microorganisms reach stationary phase within 24 hours of cultivation on autoclaved municipal wastewater amended with lignocellulosic sugars. The maximum cell mass was 0.8g/L and a reduction in the chemical oxygen demand (COD) to a minimum within 24 hours of cultivation. The oleaginous microorganism consortium's doubling time on municipal wastewater amended with lignocellulosic sugars is approximately 2.5 hours. These results show that these microorganisms can thrive by utilizing lignocellulosic sugars added to municipal wastewater to produce oil. Using wastewater treatment facilities will enable millions of gallons of oil to be produced across the nation.

See more of this Session: Student Poster Session: Fuels, Petrochemicals, and Energy
See more of this Group/Topical: Education

Oil Production for Biofuels Via Wastewater Treatment

Taylor Raiford, Dave C. Swalm Schoold of Chemical Engineering, Mississippi State University, Mississippi State, MS

Jacqueline Hall, Mississippi State University, Mississippi State, MS

W. Todd French, Dave C. Swalm School of Chemical Engineering, Mississippi State University, Mississippi State, MS

Rafael Hernandez, Swalm School of Chemical Engineering, Mississippi State University, Mississippi State, MS

William E. Holmes, Mississippi State Chemical Laboratory, Mississippi State University, Mississippi State, MS

Andro H. Mondala, Chemical Engineering, Mississippi State University, Mississippi State, MS

Darrell L. Sparks, Chemical Engineering, Mississippi State University, Mississippi State University, MS

In the U.S., biodiesel consumption increased from 450 to 690 million gallons in 2008. Oleaginous microorganisms that produce more than 70% of their weight in oil as lipids could be a potential source of oil to aid in this growing demand. By cultivating these microorganisms on municipal wastewater, the wastewater is treated while simultaneously accumulating oil. This study is focused on comparing the cultivation of an oleaginous microorganism consortium on synthetic primary effluent wastewater to autoclaved primary effluent wastewater amended with synthetic lignocellulosic sugars. Results show that the microorganisms reach stationary phase within 24 hours of cultivation on both the synthetic as well as the municipal wastewater. The growth of the consortium on the synthetic wastewater showed a maximum growth of 0.4g/L compared to 0.66g/L, the maximum growth when cultivated on autoclaved wastewater. Also, these microorganisms consumed all of the glucose within 24 hours of cultivation. The consortium's doubling time on municipal wastewater and the synthetic wastewater is approximately 4 hours and 8 hours, respectively. These results demonstrate that the municipal wastewater could be a viable medium that these microorganisms can thrive on for oil production. This modification could potentially generate billions of gallons of oil for producing biofuels.

See more of this Session: Student Poster Session: Fuels, Petrochemicals, and Energy
See more of this Group/Topical: Education

Municipal Wastewater: Source of Oil Production

Mathew J. Mistilis, Dave C. Swalm Schoold of Chemical Engineering, Mississippi State University, Mississippi State, MS

Jacqueline Hall, Mississippi State University, Mississippi State, MS

W. Todd French, Dave C. Swalm School of Chemical Engineering, Mississippi State University, Mississippi State University, MS

Rafael Hernandez, Dave C. Swalm School of Chemical Engineering, Mississippi State University, Mississippi State, MS

William E. Holmes, Mississippi State Chemical Laboratory, Mississippi State University, Mississippi State, MS

Andro H. Mondala, Chemical Engineering, Mississippi State University, Mississippi State, MS

Oleaginous microorganisms can produce up to 70 percent of their body weight in oil, thus a potential source of oil for biofuel production. A growth medium that has not been utilized for cultivating oleaginous microorganisms is municipal wastewater. However, the wastewater influent contains indigenous microorganisms that could inhibit the growth and oil production of these oleaginous microorganisms. This study is designed to determine the survivability of the oleaginous microorganism consortium cultivated on primary effluent wastewater amended with synthetic lignocellulosic sugars. Preliminary results have shown that these microorganisms reach stationary phase as well as a maximum cell concentration of 0.6g/L within 24 hours of cultivation. The added amount of glucose was consumed within 8 hours of cultivation. Since indigenous microorganisms were present, a plate count method was utilized to enumerate yeast, which makes up 90% of the consortium. The plate count yielded a significant amount of yeast colony forming units, which is significantly higher than the number of yeast colony forming units in raw wastewater. Future works will entail creating an upscale system to cultivate these microorganisms on municipal wastewater in 10L bioreactors. With this upscale, the full potential of producing oil while treating the wastewater can be determined.

See more of this Session: Student Poster Session: Fuels, Petrochemicals, and Energy
See more of this Group/Topical: Education

Influence of Temperature, Gas Flow Rate, and Steam/Biomass Ratio On Gasification of Algae Biomass

Clay Beevers, Chemical Engineering, New Mexico Institute of Mining and Technology, Socorro, NM

R. Bryan Woodruff, Chemical Engineering, University of Colorado at Boulder, Boulder, CO

Amy Oberlin, University of Michigan, Ann Arbor, MI

Alan Weimer, Chemical and Biological Engineering, University of Colorado, Boulder, CO

Biomass gasification is one of the most promising avenues in the field of alternative fuels. It is being looked at both as a method for production of syngas which will be used in catalytic reforming processes such as Fischer-Tropsch, and as a source of pure H_2 for fuel cell applications, just to name a few. The objective of this research was to determine the effects of temperature, gas flow rate, and steam/biomass ratio on the conversion of algae biomass to syngas during steam gasification. The temperature range used in these experiments was 1000°C -1300°C which exceeds the range of 400°C-900°C that is used in most gasification research. The motivation for exceeding the traditional gasification temperature range was to obtain gasification data that can be used to predict gasification results in solarthermal reactors which can easily achieve temperatures in excess of 1000°C. The algae gasification results were also compared with switchgrass gasification data that was obtained using the same experimental methods to determine which might be a better choice for syngas production. The data yielded the results that conversion and H_2 concentration have a strong dependence on increasing temperature. It was also shown that steam to biomass ratio increases and carrier gas flow rate decreases conversion, but both have more negligible effects than temperature. Algae and switchgrass had maximum conversions of 66% and 95% respectively at 1300°C. It was also found that temperature increases the selectivity of H_2 and decreases CO selectivity. The surprising result was that switchgrass had much higher conversions under the same operating conditions even though it contains lignin. Research is currently in progress to explain this result.

See more of this Session: Student Poster Session: Fuels, Petrochemicals, and Energy
See more of this Group/Topical: Education

Design of a Lab-Scale Mini-Plant for Optimization of a Fixed Bed Catalytic Reactor for the Production of Chemicals From Sustainable Feedstocks

Jacob Thomas, Chemical and Materials Engineering, University of Kentucky, Paducah, KY

Sean Hansrote, Chemical and Materials Engineering, University of Kentucky, Paducah, KY

Jeffrey R. Seay, Department of Chemical and Materials Engineering, University of Kentucky, Paducah, KY

The production of biodiesel from vegetable oils yields large quantities of waste crude glycerol as a by-product. This glycerol has the effect of reducing the overall carbon utilization of the biodiesel production process. However, if the crude glycerol could be used to create an industrially important and sustainable product, the carbon utilization and the economics of the overall biodiesel production process could be improved.

It is well known that a range of valuable chemical products can be produced from crude glycerol. Based on the established literature data, a lab-scale miniplant was designed and built to produce an array of C3 compounds from waste crude glycerol. The reactor has the ability to produce compounds such as acrolein, 1,2-propanediol, 1,3-propanediol, hydroxyacetone, and acrylic acid. The purpose of this research is to conduct a series of experiments to determine the optimum operating conditions for this reactor.

Experiments were also run to calibrate the gas chromatograph used in analyzing product samples from the lab reactor. Calibration allowed for determination of the effects of flow rate, composition, temperature, and catalyst concentration on the reaction. These parameters were optimized by identifying trends in the laboratory data.

See more of this Session: Student Poster Session: Fuels, Petrochemicals, and Energy
See more of this Group/Topical: Education

Density Functional Theory Analysis of Methane Dehydrogenation On Platinum Nanoclusters for Liquid Fuels Production

Nathan A. Fine, 1 Brookings Drive, Washington University in St. Louis, St. Louis, MO

Cynthia S. Lo, Energy, Environmental and Chemical Engineering, Washington University in St. Louis, St. Louis, MO

Methane has proven itself to be a useful precursor for the production of liquid fuels and other value-added chemicals through the Fischer-Tropsch process, but currently its potential is limited since it appears to be too energetically stable to undergo direct conversion to higher hydrocarbons and other liquid fuels. It is believed that more technologically advanced nanoscale catalysts may facilitate more economical and direct methods of production. In this study, the physisorption of methane on a 20-atom tetrahedral platinum nanocluster, and the chemisorption of dehydrogenated methane derivatives have been modeled using density functional theory. These calculations provide a strong base for computing the reaction pathway, using the nudged elastic band and related methods, for catalytic methane dehydrogenation on Pt nanoclusters. Furthermore, the nanoparticle structure, composition and placement on a metal oxide support may be varied to design catalysts with improved yield, selectivity, and stability for the direct synthesis of liquid fuels from methane.

See more of this Session: Student Poster Session: Fuels, Petrochemicals, and Energy
See more of this Group/Topical: Education

Photosystem I-Based Biohybrid Photoelectrochemical Cell

Frederick Hijazi, Chemical and Biomolecular Engineering, Vanderbilt University, Nashville, TN

Photosynthesis is the process by which Nature coordinates a tandem of protein complexes of impressive complexity that function to harness roughly 90TW of solar energy on a global scale. Advances in biochemistry and nanotechnology have provided tools to isolate and extract individual components of this process. Here we show how one of these components, Photosystem I (PSI), is incorporated into an electrochemical system to yield a stand-alone biohybrid photoelectrochemical cell that converts light energy into electrical energy. The fabrication of the cell, in which PSI complexes are assembled into a dense multilayer on the surface of the cell's cathode, is described. The cells were irradiated with various intensities of white light and the corresponding responses of a typical cell's chronoamperometric behavior and open circuit potential are reported. The relationship between these responses and the photocatalytic behavior of PSI complexes is described. In addition, we report surface analysis of Photosystem I monolayers by scanning electrochemical microscopy (SECM) in light and dark conditions. SECM allows the direct detection of the electrochemical current measured through the tip electrode produced by the reducing side of PSI by the tip electrode.

See more of this Session: Student Poster Session: Fuels, Petrochemicals, and Energy
See more of this Group/Topical: Education

An Analysis of Biodiesel Production by Continuous Supercritical Methanol Reaction

Nicole Lorenz, University of Arkansas, Fayetteville, AR

Jamie Hestekin, University of Arkansas, Fayetteville, AR

Robert Babcock, University of Arkansas, Fayetteville, AR

Kris White, University of Arkansas, Fayetteville, AR

Biodiesel is a renewable fuel that can be produced from vegetable oils, animal fats, and other feedstocks. It is traditionally produced using either acid or base catalysis, depending on the type of feedstock used. However, these traditional methods produce a great deal of waste and require long conversion times. Supercritical methanol reaction has fast reaction times and low waste production. This method has been explored for biodiesel production, but few studies have investigated the kinetics for a continuous reactor. In this study, a continuous supercritical reaction with methanol has been used to produce biodiesel from tall oil fatty acids, a paper byproduct. The reactor used in these experiments is unique in that it contains static mixers to provide contact between the reactants in case complete supercritical conditions are not reached. This project investigated the kinetics of a continuous supercritical reactor and will test the effect of statics mixers within the reactor.

See more of this Session: Student Poster Session: Fuels, Petrochemicals, and Energy
See more of this Group/Topical: Education

Production of 2,3-Butanediol and Its Derivatives From Biomass

Karen Tieszen, Chemical Engineering, Kansas State University, Manhattan, KS

2,3-Butanediol is a useful chemical feedstock that can be produced through fermentation of sugars from biomass. The need for renewable chemical feedstocks from biomass has increased in recent years due to increases in the price of petroleum-derived chemicals, and the finite supply of petroleum. The most useful product that can be obtained from 2,3-butanediol is methyl ethyl ketone, which can be used as a fuel additive to increase the octane rating of gasoline. 2,3-Butanediol is produced through the fermentation of sugars from biomass by the microorganism klebsiella oxytoca. Furthermore, 2,3-butanediol can be converted through dehydration to methyl ethyl ketone in the presence of catalyst ZSM-5, an aluminosilicate zeolite. The concentration of 2,3-butanediol produced through shake flask experimentation was 0.87 g/L. ZSM-5 was found to convert 2,3-butanediol to methyl ethyl ketone through dehydration. Further optimization of the fermentation and dehydration reactions will allow 2,3-butanediol to be a viable biomass-derived chemical feedstock.

See more of this Session: Student Poster Session: Fuels, Petrochemicals, and Energy
See more of this Group/Topical: Education

Power Recovery in Combined-Cycle Plant Utilizing Organic Power Cycle

Jennifer Guilfoyle, Chemical Engineering, Illinois Institute of Technology, Glen Ellyn, IL

As new technology in the power industry has evolved over the last few decades, new ideas for power generation have emerged. Because combined cycle plants are able to provide two commodities for their customers - heat and power - from one location, they have newfound popularity. With the aforementioned recent technological advances, new types of combined plants can be formed. Research into one such cycle is being conducted, in which energy from a wind farm is utilized to liquefy oxygen that is to be sent to an oxy combustion plant. The use of wind power in condensing oxygen greatly reduces oxygen transportation costs. To make this model more economical, power would need to be produced from the heating of oxygen to offset the power consumed from the condensing process. For this purpose, a low temperature power cycle would be employed, utilizing the waste heat from the power plant as the heat source. Simulations were conducted with various organic fluids in a wide range of temperatures and pressures.

See more of this Session: Student Poster Session: Fuels, Petrochemicals, and Energy
See more of this Group/Topical: Education

A Biodiesel Feedstock of Microbial Triglycerides From Seafood Processing Waste and Glycerol

Bonnie O'Neal, Dave C. Swalm School of Chemical Engineering, Mississippi State University, Mississippi State, MS

Guochang Zhang, Swalm School of Chemical Engineering, Mississippi State University, Mississippi State, MS

W. Todd French, Dave C. Swalm School of Chemical Engineering, Mississippi State University, Mississippi State University, MS

Oils derived from byproducts such as glycerol and shrimp shells with no value could potentially be a cheap source of biodiesel. These byproducts could also be used as carbon and energy source for oleaginous microorganism for oil production. The objective of this investigation is to determine the feasibility of using oleaginous microorganism to convert these byproducts to triglycerides as a feedstock for biodiesel production. Previous studies showed that Cryptococcus curvatus could grow using n-acetyl glucosamine, which is the major sugar product from the hydrolysis of Seafood Processing Waste. Therefore, C. curvatus was first grown in a media with n-acetyl glucosamine as sole carbon and energy source. Then, the cells of C. curvatus were transferred to a fresh nitrogen-free media with glycerol as sole carbon. In the media with a 50g/L of glycerol, the lipids in cells of C. curvatus was accumulated from initial 1.3% to 26.1% of the dry weight; cell mass also increased from 5.5 g/l to 10.4 g/l. The fatty acid profiles of these lipids were determined using GC-FID, which determine the properties of biodiesel. Experimental results indicate that it is feasible that these byproducts can be converted by C. curvatus to triglycerides as a biodiesel feedstock.

See more of this Session: Student Poster Session: Fuels, Petrochemicals, and Energy
See more of this Group/Topical: Education

Optimizing Pilot Plant Biodiesel Production

Abigail R. Phillips, Engineering, Hope College, Holland, MI

Peter N. Larsen, Science, Holland High School, Holland, MI

Daniel Denby, Engineering, Hope College, Holland, MI

Sameer Chavda, Chemistry, Hope College, Holland, MI

Moses N.F. Lee, Chemistry, Hope College, Holland, MI

Michael J. Misovich, Engineering, Hope College, Holland, MI

Conventional biodiesel is a fuel made by converting triglycerides in oils and fats to methyl esters. To improve campus sustainability, Hope College is implementing a small scale pilot plant to convert waste vegetable oil (WVO) from the dining halls to biodiesel. This research investigated alternative raw materials, processing steps, and operating conditions for the pilot plant. Substitution of ethanol for methanol resulted in successful conversion of WVO to ethyl esters; however, conversions varied from 80.0 to 88.1 percent compared to 91.8 to 94.6 percent for methyl ester biodiesel. Furthermore, the raw material cost of ethanol was 45 percent higher than methanol per gallon of biodiesel product. Molar ratios of 5:1 and 6:1 methanol to oil resulted in costs per gallon of biodiesel between $1.61 and $1.80. Using a 4:1 ratio produced a high viscosity waste stream that limited practical implementation. Partial recycling of the waste obtained from the production of biodiesel enabled approximately 51% of the initial alcohol and 88% of the initial catalyst to be reused, decreasing cost per gallon by $0.52. This process was proved to be a feasible option for the college.

See more of this Session: Student Poster Session: Fuels, Petrochemicals, and Energy
See more of this Group/Topical: Education

Microbial Fuel Cells (MFCs)

Caspar Yi, Chemistry and Life Science, United States Military, West Point, NY

Robert Bozic, Chemistry and Life Science, United States Military Academy, West Point, NY

James Sumner, Bio-Inspired Devices and Sensors (BIDS), U.S. Army Research Laboratory, Adelphi, MD

A Microbial Fuel Cell (MFC) is a device that converts chemical energy into electrical energy. MFCs provide waste remediation and most importantly, electricity generation. How the electrons are harnessed and utilized is what interests the Army. Certain dyes and bacteria are capable of anaerobic respiration by an electron acceptor other than oxygen. Effects of certain mediators [methylene blue and DSSNI (UCSB)] were studied using electrochemical, staining techniques, and inexpensive materials. Knowing the mechanisms by which electron transfer occurs will enable different possibilities with different mediators and materials to surface. While the MFC is fairly intermediary in its methods, we are able to use this system to study the effects that certain materials have on the growth of bacteria.

See more of this Session: Student Poster Session: Fuels, Petrochemicals, and Energy
See more of this Group/Topical: Education

Porphyrin Functionalized Carbon Nanotube Doped Polypyrrole for Photovoltaic Applications

Wesley A. Chen, Chemcial and Environmental Engineering, University of California, Riverside, Riverside, CA

Ashok Mulchandani, Chemcial and Environmental Engineering, University of California, Riverside, Riverside, CA

Mangesh Bangar, Chemcial and Environmental Engineering, University of California, Riverside, Riverside, CA

Priti Mulchandani, Chemcial and Environmental Engineering, University of California, Riverside, Riverside, CA

Porphyrins are functional dyes that are known to have photophysical properties. This property makes porphyrin commonly studied for its potential for becoming part of a dye-sensitized solar cell. TSPP was used in this experiment to act as dopant for the electropolymerization of polypyrrole and to functionalize single walled carbon nanotubes. The molecular geometry of the porphyrin, its four planar rings, makes the bonding between SWCNT and the porphyrin possible through π-interactions. Single walled carbon nanotubes are known for their high electrical conductivity. The carbon nanotubes were used to facilitate voltage transfer through the polypyrrole film. Polypyrrole was electrochemically polymerized with the TSPP functionalized SWCNT as the dopant. The morphology of phorphyrin doped polypyrrole and TSPP + SWCNT doped polypyrrole were studied with SEM. The photocurrents of the two films were compared with a polypyrrole doped with a common dopant, sodium chloride.

See more of this Session: Student Poster Session: Fuels, Petrochemicals, and Energy
See more of this Group/Topical: Education

Optimization of $Eu:TiO_2$ Nanoparticles for Aqueous Systems Applications

Francisco J. Chaparro-Carrasquillo, Chemical Engineering, University of Puerto Rico at Mayaguez, Mayaguez, PR

Kacie Louis, Chemistry, University of Wisconsin at Madison, Madison, WI

Robert Hamers, Chemistry, University of Wisconsin at Madison, Madison, WI

Novel $Eu:TiO_2$ nanoparticles, which are able to fluoresce, can be used to understand fundamental processes in biology. They will be introduced in young zebrafish to test the toxicity. The fluorescence previously achieved is not as high as expected. It is ideal to have higher detectable fluorescence from the nanoparticles so they can be tracked through the zebrafish which can be related to toxicity studies. Dibenzoylmethane (DBM) and Benzoic acid (Ba) were added to the europium complex at basic conditions to synthesize $Eu(DBM)_3Ba:TiO_2$ and $Eu(DBM)_3:TiO_2$. Shperical particles were synthesized using a pressure bomb at 120°C and different times intervals. IR studies showed a shift in the C=O peak, from 1,650 to 1500cm^{-1}, indicating that the ligands are interacting with europium through this bond. The SEM images showed that particle size range from 1 to 3µm for particles that were treated at 300 and 180 minutes in the pressure bomb. Nevertheless, solutions that were heated for 30 minutes gave a mixture of micro and nano-scale sized particles. The fluorescence of these particles is being quenched, which is due to the size of the particles. The next step is to improve synthesis, since nano-scale particles, with high fluorescence, are desired.

See more of this Session: Student Poster Session: General Papers
See more of this Group/Topical: Education

Isolating Feline β-Hexosaminidase

Nick Cochran, Scott-Ritchey Research Center, Auburn University, Auburn University, AL

Dr. Doug Martin, Scott-Ritchey Research Center, Auburn University, Auburn University, AL

The GM2 Gangliosidoses are a class of lysosomal storage diseases caused by a deficiency of β-hexosaminidase. This enzyme is made up of two subunits, the α subunit and the β subunit. The gene HEXA codes for the α subunit, while HEXB codes for the β subunit. A human example of GM2 Gangliosidosis is Tay-Sach's disease, caused by mutations in the HEXA gene. This disease currently has no cure, and is always fatal by about 4 to 5 years of age. Other GM2 Gangliosidoses have effectively the same symptoms and outcomes. The principal investigator has treated cats with the human genes that code for the subunits of this enzyme. The treatment did reduce storage material, but the cats showed a significant immune response. The scope of the project for the research assistant was to isolate the two feline HEXA and HEXB genes in plasmids and verify enzyme expression in mammalian cells. The two genes both had unverified expressive sequences, but homology between species provided confidence that the suspected sequences were correct. The HEXA gene was isolated from RNA. The HEXB gene was isolated by primer overlap extension of a shorter sequence that was previously thought to be expressive to obtain the longer expressive gene. Both genes were loaded into plasmids so that the sequence could be verified by sequencing. A collaborator loaded the genes into adeno-associated viral vectors that will be used for treatment. These vectors were used to verify expression in mammalian cells before beginning treatments.

See more of this Session: Student Poster Session: General Papers
See more of this Group/Topical: Education

Surface Enhanced Raman Spectroscopy of Yeast as Model for Possible Biological Detection

Roxannie Gonzalez, Industrial Biotechnology, University of Puerto Rico - Mayagüez Campus, Mayaguez, PR

William Ortiz Rivera, Chemistry, University of Puerto Rico, Mayaguez Campus, Mayaguez, PR

Samuel P. Hernandez-Rivera, Chemistry, University of Puerto Rico, Mayaguez Campus, Mayaguez, PR

Surface Enhanced Raman Spectroscopy (SERS) is considered as a fast, in situ alternative identification method for microorganisms. This technique provides important information about the spectroscopy signatures of cellular components within cells. SERS has significant benefits for industrial microbiology, food microbiology and biological warfare agents detection. The proposed method for the detection of microorganisms employs a fast and simple synthesis of silver colloids based on reduction of silver nitrate with hydroxylamine hydrochloride at alkaline pH. Colloids were characterized using UV-VIS spectroscopy. Yeast was used as the model microorganism since its cellular components and spectroscopic signatures are well known. This process allows for the detection of microorganisms such as yeast using an excitation wavelength of 785 nm. This detection system is an attractive identification method for other biological agents.

See more of this Session: Student Poster Session: General Papers
See more of this Group/Topical: Education

An Examination of Sessile and Pendant Drop Retention Forces

Rajiv R. Jaini, Chemical Engineering, Lamar University, Baytown, TX

Drop adhesion force, the force required to slide a drop along a surface, is an important parameter used in fluid dynamics, agriculture, and environmental science. The tilt plate method is the most conventional way to study drop lateral adhesion forces. However, there are two obvious limitations associated with this method. Firstly, the range of forces that can be applied to the drop are limited by the experimental setup, as the maximum force that can be applied is the drop's weight. Secondly, the tilt stage experimental setup requires that both the normal and lateral (retention) forces, which are applied to the drop, change simultaneously. Accordingly, our group has assembled the first model of a novel instrument that negates these limitations. This instrument, the Centrifugal Adhesion Balance (CAB), allows drops to slide on a surface by the application of centrifugal force. Thus, the drops are no longer limited in terms of the range of forces that can be applied. Additionally, because of the experimental equipment, the normal and lateral force components can be decoupled, effectively eliminating the second limitation. In these experiments, our group has compared truly sessile and truly pendant drop retention forces. We have also examined the relationship between drop resting time and drop retention force. Effectively, we are interested in not only determining the quantitative values for the retention forces but also determining the effect of surface deformation and the parameters that cause surface deformation on these forces.

See more of this Session: Student Poster Session: General Papers
See more of this Group/Topical: Education

Phase Behavior of Complex Systems Using the GC-SAFT-VR Equation of State

Jessica D. Haley, Department of Chemical and Biomolecular Engineering, Vanderbilt University, Nashville, TN

M. Carolina Dos Ramos, Department of Chemical and Biomolecular Engineering, Vanderbilt University, Nashville, TN

Clare McCabe, Department of Chemical and Biomolecular Engineering, Vanderbilt University, Nashville, TN

The statistical associating fluid theory [1] is a commonly used molecular-based equation of state that has been successfully applied to study a wide range of fluid systems. In recent work, the GC-SAFT-VR [2] equation that combines the SAFT equation for potentials of variable range (VR) with a group contribution like approach was proposed, and has been shown to provide an excellent description of the phase behavior of pure non-associating fluids and their mixtures. The parameters for key functional groups (such as CH_3, CH_2, CH, CH_2=CH, C=O, C_6H_5, ether and ester, OH, NH_2, CH=O, COOH) were obtained by fitting to experimental vapor pressure and saturated liquid density data for selected low molecular weight fluids. In this work, the transferability of these parameters is tested by comparing the theoretical predictions with experimental data for pure fluids not included in the fitting process and their binary mixtures. The GC-SAFT-VR approach is found to accurately predict the phase behavior of the systems studied without adjusting the group parameters to binary mixture data.

[1] Chapman et al. *Fluid Phase Equilibria* 52, 31-38 (1989); Chapman et al. *Industrial & Engineering Chemistry Research* 29(8), 1709-1721 (1990). [2] Peng et al. *Fluid Phase Equilibria* 277(2), 131 (2009).

See more of this Session: Student Poster Session: General Papers
See more of this Group/Topical: Education

A Possible Role for Uncoupling Protein 2 in the Development of Insulin Resistance

David A. Goodwin Jr., Artie McFerrin Department of Chemical Engineering, Texas A&M University, College Station, TX

Colby Moya, Artie McFerrin Department of Chemical Engineering, Texas A&M University, College Station, TX

Arul Jayaraman, Artie McFerrin Department of Chemical Engineering, Texas A&M University, College Station, TX

Type 2 Diabetes is becoming a major health concern throughout the world. Recent clinical evidence points to restoring overall insulin sensitivity as the major objective in the management of Type 2 Diabetes. Current research indicates Uncoupling Protein 2 may play a role in the development of insulin resistance and the onset of Type 2 Diabetes. UCP2 over expression has been linked to increases in reactive oxygen species concentration. The presence of UCP2 has been linked to the ability of the cell to perform insulin-stimulated uptake of glucose. Tumor Necrosis Factor α has also been linked to increases in ROS concentration and the insulin sensitivity of the cell. This evidence suggests a link between UCP2, ROS, TNF-α, and insulin resistance. The overall objective of our research is to establish the relationship between UCP2 and ROS in the context of insulin resistance in 3T3-L1 adipocytes with the specific aim of determining the effect of UCP2 expression on ROS concentration. To date we have succeeded in cloning the UCP2 gene into an inducible plasmid, pRev-TRE, inducible by the addition of Doxycycline. The development of a lentiviral vector, able to integrate the pRev-TRE UCP2 plasmid into the genome of 3T3-L1 adipocytes, is currently underway. Our future work will include experiments quantifying the concentration of ROS in the presence of various UCP2 expression conditions including: baseline physiological, blank, and over expression.

See more of this Session: Student Poster Session: General Papers
See more of this Group/Topical: Education

Calculating Vapor Pressure and Concentration for Binary Mixtures through Equations of State

Laine M. Klopfenstein, Engineering, Hope College, Holland, MI

Michael Misovich, Engineering, Hope College, Holland, MI

Equations of state (EOS) relating the variables pressure, volume, and temperature are often used in chemical engineering to predict vapor-liquid equilibrium (VLE) properties. Using cubic EOS, vapor pressures of pure substances can be found as a function of temperature and substance specific parameters using an algorithm. The purpose of this research was to extend previously existing programs for predicting pure component vapor pressure to predict vapor pressures and concentrations of binary mixtures. The Soave-Redlich-Kwong (SRK) cubic equation of state was used to generate data sets in order to better understand the relationships among the variables and parameters characterizing VLE in a binary mixture. These included critical temperature ratio, critical pressure ratio, vapor pressure ratio, and acentric factors. Standard plots such as Px, xy, and relative volatility illustrated the results. Comparisons were made among predictions of Raoult's Law, the Lewis and Randall Rule, and EOS calculations with an ideal solution mixing rule, which show different values for ideal solutions. These data sets cover a range of critical temperature ratios from 0.8 to 1.2, critical pressure ratios from 0.8 to 1.2, and acentric factors from -0.4 to 1.0, and are expected to be useful in future research applying cubic EOS to mixtures.

See more of this Session: Student Poster Session: General Papers
See more of this Group/Topical: Education

Truncation Error of PRSV Vapor Pressures Calculated by a Series Method

Whitney Askew, Engineering, Hope College, Holland, MI

Amy Miller, Engineering, Hope College, Holland, MI

Michael Misovich, Engineering, Hope College, Holland, MI

Equations of state (EOS) describe relationships among pressure, molar volume, and temperature. Cubic EOS can be solved for liquid and vapor volume as well as equilibrium properties such as vapor pressure. A specific cubic EOS, Peng-Robinson-Stryjek-Vera (PRSV), was studied in this work. An existing technique expressed vapor pressure as a polynomial series in temperature. The series coefficients depended upon two substance-specific parameters, the acentric factor ω and a parameter labeled κ_1. Vapor pressures were calculated when the series was truncated after the first, fifth, eighth, and tenth power in temperature and were compared to exact vapor pressures calculated by a fugacity algorithm. Results were determined for values of the acentric factor ω between -0.4 and +0.4 and values of the κ_1 parameter between -0.2 and +0.3. Plots of truncation error versus reduced temperature were made by fixing the value of one of the two parameters and allowing the other to vary. In some cases, varying one of the parameters caused the sign of the truncation error to change. Plots showing 10% error contours were also produced for a fixed value of one parameter while reduced temperature and the other parameter were varied.

See more of this Session: Student Poster Session: General Papers
See more of this Group/Topical: Education

Purification and Analysis of Bacteria Contaminated Water Using Slow Sand Filtration and Metal Ions

Joseph J. Brandonisio, Engineering, Hope College, Holland, MI

Victor M. Kanyi, Engineering, Hope College, Holland, MI

Michael J. Pikaart, Chemistry, Hope College, Holland, MI

The purpose of our research was to improve the efficiency of slow sand water filters. This technology is being used in undeveloped countries to provide cheap, clean drinking water to those who normally would not have it. Slow sand water filters were constructed out of PVC pipes and filled with varying types of layered sand. Each filter was testing the efficiency of one of four different types of sand or two forms of metal ions at killing bacteria. The efficiency was measured by comparing the concentration of fecal coliforms in the output water to that of the input water. These experiments were carried out at room temperature and 4°C in order to observe the effect of temperature on slow sand water filtration. Most traditional water filters rely on a biofilm forming on top of the sand and a lack of oxygen in the column of sand to kill coliforms in the water. Our research added to this system by introducing copper as a germicide to the columns. We added copper sulfate or ACX (copper-zinc alloy) to some of the tubes and observed their bacteria killing efficiency compared to copper-less filters with the same sand type and temperature and used particle acceleration to determine which sands retained the most copper. We also tested bacteria concentrations and pH at varying heights on the column to determine where most of the cell death was taking place.

See more of this Session: Student Poster Session: General Papers
See more of this Group/Topical: Education

The Role of Structure Function Relationships in the Contractility Outputof Vascular Smooth Muscle

Alexander Peyton Nesmith, School of Engineering Applied Sciences, Harvard University, Cambridge, MA

Patrick Alford, School of Engineering and Applied Sciences, Harvard University, Cambridge, MA

Kit Parker, School of Engineering and Applied Sciences, Harvard University, Cambridge, MA

The primary function of vascular smooth muscle cells (VSMCs) is to control the diameter of blood vessels and thus blood pressure via contraction and dilation. However, VSMCs also play a vital role in wound healing. VSMCs are able to perform both of these functions by readily switching phenotype, between contractile and synthetic, in response to environmental cues. Incorrect phenotype function is associated with a number of pathologies, such as atherosclerosis. For VSMCs, boundary conditions affect cell shape, orientation and coupling, all of which affect the cells' ability to contract. This may be via phenotypic switching or by structure-function modulation. The goal of this study is to determine the specific roles of cell geometry and multi-cellular organization in the contractile output of VSMCs. Using microcontact printing and image analysis, different cellular geometries and multi-cellular organizations were quantified in terms of actin alignment, nuclear shape and alignment, and tissue thickness. This can then be compared to the found contractile output of the cells. Thus far, it has been found that the actin alignment of the lines of different width was similar, thinner lines had better nuclear alignment, and thinner lines also altered the aspect ratio of the nuclei. The future work will consist of correlating this data with the data obtained from the contractility experiments.

See more of this Session: Student Poster Session: General Papers
See more of this Group/Topical: Education

Metal/Metal Oxide Nanoparticles Decorated Single Walled Carbon Nanotubes: Towards Room Temperature Gas Sensors

Danny F. Martinez, Chemical and Environmental Engineering and Center for Nanoscale Science and Engineering, University of California Riverside, Riverside, CA

Syed Mubeen, Chemical and Environmental Engineering and Center for Nanoscale Science and Engineering, University of California Riverside, Riverside, CA

Nosang V. Myung, Chemical and Environmental Engineering and Center for Nanoscale Science and Engineering, University of California Riverside, Riverside, CA

The purpose of this study is to develop an efficient method to functionalize single-walled carbon nanotubes (SWNTs) with metal/metal oxide nanostructures that can be used for room temperature gas sensors. This work is motivated by the fact that SWNTs showed enhanced sensitivity towards gas molecules when it is functionalized with certain metals, metal oxides and conducting polymers. Electrochemical functionalization technique was utilized to decorate SWNTs with different metals (e.g. Ag, Cr, Fe) followed by electrochemical oxidation to convert them to metal oxides. Unlike other deposition techniques, electrodeposition allows precise tuning of sensor properties by controlling the size and number of nanoparticles deposited on SWNTs. For instance this study demonstrates that by changing deposition potential and time SWNTs can be used as templates to grow silver nanoparticles (at -1V, 0.5 seconds) or nanowires (at -1 V, 25 seconds). These devices were subjected to different analytes to determine sensing performance of metal/metal oxide gas sensors. Ag nanoparticles demonstrated sensitivity of $\Delta R/R$ of 26% to NH_3 at 100 ppm. Oxidized Ag-nanowire demonstrated sensitivity of $\Delta R/R$ of 32% to NH_3 at 100ppm.

See more of this Session: Student Poster Session: Materials Engineering and Sciences
See more of this Group/Topical: Education

Carbon Nanotube Based Tunable Microbattery

Philippe A. Lacasse, Electro-Optical Systems Lab, Georgia Tech Research Institute, Atlanta, GA

W.J. Ready, Eosl, Georgia Tech Research Institute, Atlanta, GA

Batteries are essential to the functioning of many modern technologies and have been the subject of much research, both in academia and industry. Much of this research has been devoted to decreasing the size and weight of the battery, while simultaneously increasing the energy density and current output. With increasing skill researchers are developing nanofabrication and microelectronics techniques. Some researchers have been turning their attention to using nanoscale patterns for use as electrodes. It is suggested that increased surface area would allow for increased reaction rate (due to larger reaction surface) and better battery performance would be achieved. The work for the first incarnation of the carbon nanotube (henceforth CNT) tunable microbattery project was inspired by work done by Madou and Wang of UC-Irvine (Biosensors and Bioelectronics, 2005), whose design was able to produce 78% more power than a comparable stack-plate battery. These results were possible through the application of a "bed of nails" design. The anodes and cathodes were in alternating rows by the use of an interdigitated array as charge collectors. It is believed that the vertically aligned carbon nanotube arrays grown by the Ready lab would be superior electrodes when compared to the amorphous carbon of the Madou and Wang design. The first semester of research on the tunable CNT based microbattery focused on the evolution of process flows for lab scale testing of the microbatteries. The main variables examined were the material of the interdigitated array, creation and comparison of several fabrication protocols as well as selection of the photoresist. Madou and Wang used a gold thin film in their work, but the Ready lab has experienced poor growth using such a thin film as a substrate in the past. TiN and TiW were chosen as alternative substrates. TiN was chosen for its good resistance values, while TiW was also chosen for its ease in fabrication. Three different fabrication protocols were developed for use; two using Microchem's SU8-2002 (negative) photoresist, and one using Microchem's S-1818 (positive) photoresist.

See more of this Session: Student Poster Session: Materials Engineering and Sciences
See more of this Group/Topical: Education

Environmentally Friendly Levan-Montmorillonite Films

Jennifer Guerrero, Chemical Engineering, University of Puerto Rico at Mayaguez, Mayaguez, PR

Xiaoming Chen, Chemical Engineering, University of South Carolina, Columbia, SC

Dr. Harry Ploehn, Chemical Engineering, University of South Carolina, Columbia, SC

Levan is a naturally occurring homopolysaccharide of D-fructose. In a typical levan structure, the monomers are joined by â(2->6) linkage, while the side chains are joined to the backbone via â(2->1) linkage. The levan produced by microorganisms usually have densely packed globular structure with molecular weights greater than 0.5 million1. Levan is water soluble and has an usually low intrinsic viscosity2. It is a film former under certain conditions. Therefore levan has a potential to be made into 'green' products that are suitable for applications such as barrier coating, membrane separation and flexible packaging. However, levan films exhibit very poor mechanical properties, which would prevent any of their applications from becoming practical. It is the challenge of maintaining levan film integrity that motivated this study. Two different approaches have been tried in our study in order to improve the levan film properties. The first approach involves the use of an inorganic filler to reinforce the levan films. The second approach involves the use of a cross-linking agent to crosslink the levan polymer itself. The current report is largely comprised of the results obtained using the first approach. The levan solutions were characterized using XRD, DSC and TGA and the results indicates that levan intercalates into montmorillonite (MMT) galleries and that the interactions between MMT and levan results favorable in a change in levan thermal properties.

See more of this Session: Student Poster Session: Materials Engineering and Sciences
See more of this Group/Topical: Education

Controlling the Dewetting of Liquid Crystals Using Two-Component Self-Assembled Monolayers

Amit Shavit, Chemical Engineering, University of Massachusetts Amherst, Amherst, MA

Stephanie Malone, Chemical and Biological Engineering, University of Colorado at Boulder, Boulder, CO

Daniel K. Schwartz, University of Colorado at Boulder, Boulder, CO

We have developed new liquid crystal (LC) alignment layers that promote homeotropic LC anchoring and exhibit a low contact angle of the nematic LC phase. These experiments were motivated by the requirements of LC-based biosensors, where thin LC layers are in contact with aqueous solution. In these systems, the LC layer is prone to de-wetting because of the relatively high LC contact angle on a standard homeotropic alignment layer composed of an octadecyltriethoxysilane (C18) self-assembled monolayer (SAM). In contrast, SAMs of ethyltriethoxysilane (C2) exhibit a low LC contact angle, but promote planar anchoring. Therefore, we prepared two-component SAMs of C18 and C2 in different proportions. We characterized our surface using three techniques: LC anchoring/birefringence, contact angle goniometry, and ellipsometry. A two-component mixed SAM, comprised of ~10% C18 and 90% C2, successfully reduced the LC contact angle by over 50% while maintaining homeotropic alignment of the liquid crystal.

See more of this Session: Student Poster Session: Materials Engineering and Sciences
See more of this Group/Topical: Education

Studying the Effect of Pegylation On Fibrinogen Adsorption with Single-Molecule Resolution

Erin L. Chang, Chemical and Biological Engineering, University of Colorado, Boulder, CO

Mark J. Kastantin, Chemical and Biological Engineering, University of Colorado, Boulder, CO

Daniel K. Schwartz, Chemical and Biological Engineering, University of Colorado, Boulder, CO

Poly(ethylene) glycol (PEG) is widely accepted as an effective biocompatible surface due to its ability to inhibit protein adsorption. Total internal fluorescence microscopy (TIRFM) was used to study fibrinogen adsorption and desorption on pegylated fused silica on a single-molecule level. The results of this study show that pegylation does not strongly inhibit fibrinogen adsorption; however, there is also fibrinogen desorption from PEG. Conversely, there is no fibrinogen desorption from unmodified fused silica. The increased ability for adsorbed fibrinogen to later desorb from PEG is a possible explanation for why pegylated surfaces are more resistant to protein accumulation than unmodified fused silica. Moreover, on PEG there is an inverse relationship between fibrinogen fluorescence intensity and desorption rate. It is possible that brighter fibrinogen correspond to fibrinogen aggregates, which suggests that these aggregates do not desorb from pegylated surfaces as easily as single fibrinogen molecules.

See more of this Session: Student Poster Session: Materials Engineering and Sciences
See more of this Group/Topical: Education

Nanocomposites of An Engineering Plastic, Poly(ether ether ketone) with Carbon Nanofibers

Kimberly Dikovics, Stevens Institute of Technology, Hoboken, NJ

Dilhan Kalyon, Chemical Engineering, Stevens Institute of Technology, Hoboken, NJ

Engineering plastics must provide excellent mechanical properties at elevated temperatures for myriad applications, especially in automotive and aerospace industries. PEEK is an engineering polymer which is widely used because of its superior mechanical properties, high melting temperature (380 °C) and excellent resistance to chemical corrosion. Carbon nanofibers (CNFs) are novel three-dimensional reinforcements, formed when graphene sheets are assembled into cone or tube-like structures. Polymer composites with small concentrations of CNFs generally exhibit significant increases in mechanical properties, as well as electrical and thermal conductivity. Incorporating CNFs into PEEK thus has the potential to generate nanocomposites with improved mechanical properties while conserving high melting point and chemical resistance. This improvement is due to a variety of potential factors, including the reinforcement effect of the CNFs forming a network structure, increased crystallinity caused by CNFs acting as a nucleating agent, or by thermal crosslinks forming within the polymer structure. In this investigation, PEEK CNF nanocomposites were prepared by two separate methods (melt mixing and dissolution processes), and at varying CNF concentrations (0.5 to 2% by weight). Linear viscoelastic properties of the molded PEEK/CNF nanocomposite specimens were characterized using both tensile and torsional oscillatory deformations. Furthermore, differential scanning calorimetry and X-ray diffraction were employed to determine the percent crystallinity and crystal size of the PEEK as functions of CNF concentration and processing method. The results suggest that improvements in mechanical properties can only occur under favorable dispersion conditions for the CNFs and in the absence of the thermal degradation of the PEEK.

See more of this Session: Student Poster Session: Materials Engineering and Sciences
See more of this Group/Topical: Education

Grafted Block Copolymer Nanoassembly Drug Carrier

Daniel Lindsay, Chemical Engineering, University of Arkansas, Fayetteville, AR

Andrei Ponta, Pharmaceutical Sciences, University of Kentucky, Lexington, KY

J. Zach Hilt, Chemical and Materials Engineering, University of Kentucky, Lexington, KY

Younsoo Bae, Pharmaceutical Sciences, University of Kentucky, Lexington, KY

This project aims to develop a novel drug carrier, in the form of a grafted block copolymer nanoassembly (GNA) created from functional poly(ethylene glycol)-poly(amino acids). The GNAs contain a hydrophilic exterior and a hydrophobic interior, allowing tumor specific delivery of poorly-water soluble anticancer drugs through the enhanced permeability and retention (EPR) effect. The GNAs were prepared by conjugating multiple amphiphilic polymer 'grafts' bound to a fluorescein-tagged polymer 'scaffold.' Conjugation reactions of grafts with scaffolds were tested under various conditions with different reaction times, ratios, and concentrations. UV/VIS spectroscopic analysis revealed average grafting yields of 2-3 grafts (260 nm) per scaffold (494 nm) with 53 binding sites. Physiochemical properties of GNAs were evaluated by taking particle size with dynamic light scattering measurements at pH 5.0 and 7.4. Results suggested that GNAs have greater stability than conventional self-assembling nanoassemblies (SNAs). Drug loading experiments were subsequently conducted with GNAs using a model drug (17-allylaminogeldanamycin). Colorimetric quantification showed drug loading of 18% and 48% for GNAs and SNAs, respectively.

GNAs with robust globular structures are envisioned to provide a more facile and efficient delivery paradigm for hydrophobic anticancer drugs than other existing injection vehicles, taking advantage of the EPR effect.

See more of this Session: Student Poster Session: Materials Engineering and Sciences
See more of this Group/Topical: Education

Characterization and Activation Temperature Modification of Flexible Nanopourous Sorbents

Meghan E. Clardy, Chemical Engineering, University of Puerto Rico - Mayaguez, Mayaguez, PR

Jose N. Primera-Pedrozo, Chemical Engineering, University of Puerto Rico - Mayaguez, Mayaguez, PR

Omar Garcia-Ricard, Chemical Engineering, University of Puerto Rico - Mayaguez, Mayaguez, PR

Arturo J. Hernandez-Maldonado, Chemical Engineering, University of Puerto Rico - Mayaguez, Mayaguez, PR

Flexible nanoporous adsorbents can be used for gas purification and storage. CPL-2 and UPRM-4m are two of these materials synthesized in the UPRM NSSAL lab which are selective towards CO2 adsorption. Flexible materials such as these have been shown to be affected by the pre-activation (degas) temperature during post synthesis procedure. In order to further modify the synthesis of these materials for improved efficiency, the optimal degas temperature for each material was determined. CPL-2 has the highest adsorption capacity and surface area while maintaining selectivity after degassing at 1000°C. Sr-UPRM-4m had the highest surface area after degassing at 900°C and continued higher adsorption potential over previous, non ion exchanged stages. By determining these optimal degas temperatures, the post-synthesis modification has been improved to maximize the adsorption capacity of each material and therefore improve the potential for applications in purification and storage.

See more of this Session: Student Poster Session: Materials Engineering and Sciences
See more of this Group/Topical: Education

Mimicking of Iron-Sulfur Cluster Proteins: Synthesis of Half-Amphiphilic Dendrimer

Jodie Greene, North Carolina State University, Raleigh, NC

Christopher Gorman, Chemistry, North Carolina State University, Raleigh, NC

Iron sulfur proteins are used in the body for electron transport, control of gene expression, substrate binding, and enzyme activity. The behavior of the iron-sulfur cluster depends greatly on the protein structure surrounding it. In particular, the redox potential of the cluster and the rate of electron transfer to and from the cluster can vary over a wide range depending on the protein environment in which the cluster is found. To understand how structure affects these parameters, an artificial but controlled set of structures are being synthesized and studied. These structures are dendrimers of varying hydrophobicity with the iron sulfur cluster at their topological center. The dendrimers encapsulate the iron sulfur cluster and allow us to explore the electronic properties of the cluster while eliminating the solvent environment of the body. The half hydrophilic/half hydrophobic dendrimer is being explored for its encapsulating properties. The half hydrophobic is predicted to act like a lipid and wrap around the cluster to protect it. The half hydrophilic is predicted to act allow the dendrimer to be dissolved in polar solvents, like water. This structure is predicted to allow the dendrimer to have a dual role of core protection and water solubility. Dendrimers are being prepared with one, two and three generations (layers of hyperbranching) through a series of organic synthesis steps. Subsequently, electrochemistry measurements are being performed to determine their redox potentials and rates of electron transfer.

See more of this Session: Student Poster Session: Materials Engineering and Sciences
See more of this Group/Topical: Education

Synthesizing Iron and Iron Oxide Nanoparticles in the Membrane Domain for Reactions at Interfaces

Eric L. Coker, Chemical & Materials Engineering, University of Kentucky, Lexington, KY

Dibakar Bhattacharyya, Chemical & materials Eng., University of Kentucky, Lexington

Scott Lewis, Dept. of Chemical & Materials Engineering, University of Kentucky, Lexington, KY

J. Zach Hilt, Chemical and Materials Engineering, University of Kentucky, Lexington, KY

Iron in various oxide forms has proven to have wide applicability, with possibilities ranging from the inactivation of viruses to the remote controlled heating of desired areas. Zero valent iron (ZVI), ferrous, ferric, ferryl, as well as Fe2O3 and Fe3O4 all contain unique and desirable properties as nanoparticles. In combining multiple disciplines, we are looking at immobilizing Fe0/Fe2O3 and Fe3O4 nanoparticles in the polymer domain. The autocatalytic reaction that occurs at the interface of Fe0 nanoparticles in the presence of oxygen to produce hydrogen peroxide and the hydroxyl radical has large implications that are being investigated. Much work is also being done to develop a technique to immobilize Fe3O4 nanoparticles in the polymer membrane domain. If this can be accomplished, the objective will then be to look at the remote controlled heating of the magnetite nanoparticles made possible by magnetite's superparmagnetic properties. An efficient technique has already been developed to immobilize ZVI nanoparticles in the polymer membrane domain. This is done through the ion exchange of ferrous ions onto a PVDF-PAA membrane (polyvinyladine fluoride membrane complexed with Polyacrylic Acid), followed by reduction with Sodium Borohydride. In addition to this, magnetite (Fe3O4) nanoparticles have been synthesized in solution by the following reaction: Fe^(2+) + 2Fe^(3+) + 8OH^-→ Fe_3 O_4 + 4H_2 O So the objective is now to combine these two techniques and synthesize a membrane containing Fe3O4 nanoparticles.

See more of this Session: Student Poster Session: Materials Engineering and Sciences
See more of this Group/Topical: Education

Polymer-Based Amplification of Cellular Protein Detection

Katerina R. Voigt, Department of Chemical Engineering, Kansas State University, Manhattan, KS

Heather J. Avens, Department of Chemical and Biological Engineering, University of Colorado, Boulder, CO

Brad J. Berron, Department of Chemical and Biological Engineering, University of Colorado, Boulder, CO

Vivek Balasubramaniam, Univeristy of Colorado at Denver School of Medicine, Aurora, CO

Gregory J. Seedorf, Univeristy of Colorado at Denver School of Medicine, Aurora, CO

Chistopher N. Bowman, Department of Chemical and Biological Engineering, University of Colorado, Boulder, CO

Fluorescent staining of cells and tissues is critical in the field of histology because it enables scientists to visualize the locations of cellular molecules such as proteins. Current methods of staining require expensive materials, are time-consuming, and may require high-quality microscopes with which to view the results. A novel polymerization-based amplification approach uses visible light to non-covalently attach highly-fluorescent nanoparticles to targeted proteins, increasing sensitivity and signal intensity, as well as allowing multiple targets to be stained in a single cell.

See more of this Session: Student Poster Session: Materials Engineering and Sciences
See more of this Group/Topical: Education

Evaluation of a Molecular Spring Nanothermometer

Hector E. Lopez Carretero, Artie McFerrin Department of Chemical Engineering, Texas A&M University, College Station, TX

Diego A. Gómez-Gualdrón, Artie McFerrin Department of Chemical Engineering, Texas A&M University, College Station, TX

Perla B Balbuena, Artie McFerrin Department of Chemical Engineering, Texas A&M University, College Station, TX

Classical molecular dynamics (MD) simulations are used to evaluate the feasibility of a molecular spring nanothermometer. A system composed of a gold slab of six layers and four thiolated polyethylene glycol (PEG) molecules, was located in a central simulation box and repeated periodically to represent the self-assembly of PEG molecules on the gold surface. The interactions between gold and PEG were obtained from density functional theory (DFT) simulations and correlated with a Morse effective force field model. It is known that the (111) surface of a gold crystal is the lowest in energy, for this reason our system has this surface exposed to PEG. The simulations were performed in DLPOLY 2.20 at temperatures between 293 K and 333 K. This was done to obtain information that would help us evaluate the ability of the system as a thermometer in biologically favorable temperatures. The velocity autocorrelation function (VACF) for each simulation was obtained and Fourier transformed to find vibrational spectra at different temperatures. Analysis of these calculated spectra may permit to get some insights into the design of practical devices where the variation of vibrational modes produces different luminescence at each temperature, which is the principal characteristic of this nanothermometer.

See more of this Session: Student Poster Session: Materials Engineering and Sciences
See more of this Group/Topical: Education

Crystal Growth of Li[Ni1/3Co1/3Mn1/3]O2 as a Cathode Material for High-Performance Lithium Ion Batteries

Thinh Vo, Chemical and Environmental Engineering, UC Riverside, Riverside, CA

Jianxin Zhu, Chemical and Environmental Engineering, UC Riverside, Riverside, CA

Matthew Woodley, Electrical Engineering, UC Riverside, Riverside, CA

James Weaver, Chemical and Environmental Engineering, UC Riverside, Riverside, CA

David Kisailus, Chemical and Environmental Engineering, UC Riverside, Riverside, CA

Recently, there has been a tremendous effort to solve our energy problem by developing multiple alternative routes to energy conversion. Many of these conversion technologies operate cyclically (e.g., photovoltaics generate energy during daytime hours) and require some type of energy storage. Although batteries technologies exist for this purpose, there are multiple issues limiting their performance and lifetime. In addition, current materials (e.g., cathodes such as LiCoO2) suffer from poor energy densities, degradation and are associated with high costs.

Li[Ni1/3Co1/3Mn1/3]O2 was recently discovered to be a novel cathode material for lithium ion batteries. This material was introduced by Makimura and Ohzuku (2001) and has since received a lot of attention due to its high reversible capacity, environmentally friendly nature, long life cycle and light weight. In addition, the substitution of Mn and Ni cations render it a less costly than LiCoO2. Since its initial discovery, some research has been done to investigate the performance of this promising material.

One limiting factor of this and other cathode materials is the lack of control of crystal size and shape. Inhomogeneous particle sizes lead to reduced performance and lifetime. A thorough investigation of the crystal growth behavior of these materials would provide critical insight to controlling the material size and shape and lead to longer lasting, higher performance cathodes.

In this study, we have investigated the crystal growth behavior and cation arrangement of the Li[Ni1/3Co1/3Mn1/3]O2 material. Samples were heated between 700-900 0C at for set durations. Scanning Electron Microscopy and X-Ray Diffraction analysis enabled the determination of activation energy and thus, the mode of diffusion during calcinations. We believe that by understanding the mechanism of diffusion during calcinations, we can completely control the size and morphology to provide an optimal cathode material.

See more of this Session: Student Poster Session: Materials Engineering and Sciences
See more of this Group/Topical: Education

Biomaterials for Drug Delivery to the Injured Spinal Cord

Raymond Seekell, Chemical Engineering, Villanova Univesity, Villanova, PA

Nathan Grantier, Chemical Engineering, Villanova Univesity, Villanova, PA

Noelle K. Comolli, Chemical Engineering, Villanova Univesity, Villanova, PA

Spinal cord injury (SCI) affects nearly 250,000 Americans with an estimated additional 10,000 cases per year (NIH, 2001). The vast majority of these cases are young adults in their early twenties, who face enormous physical challenges with no treatment currently available. In addition, long-term care of these patients creates an overwhelming financial burden on the healthcare system. In order to successfully repair the lost and damaged tissue following SCI and promote functional recovery, a number of problems need to be solved: (1) Survival of nervous tissue needs to be increased and lost cells need to be replaced; (2) The glial scar which is created after the injury and serves as a barrier to axonal growth must be dissolved; (3) The immune reaction that follows the initial injury and leads to inflammation and secondary injury has to be modulated and finally, (4) Regenerating axons must be guided to their target cells to form functional synapses.

Most current designs for spinal cord repair after an injury focus on replacement of cells using either a polymeric or cellular based scaffold design. The goal of this research is to expand the use of biomaterials to address the prevention of inflammation and secondary injury as well as to provide axonal guidance. The first part can be achieved using thin polymeric films that release anti-inflammatory drugs over a 3-5 day period locally to the injury site. These films can be made from either a non-degradable material that would need to be later removed (poly vinyl alcohol - poly vinyl pyrollidone), or ideally a degrading poly lactic glycolic acid (PLGA) that would not require a second surgery.

Research has shown that axons will respond to specific neurotrophins (NT-3, BDNF and GDNF) and grow towards increasing concentrations of them (Taylor et al, 2006). It is therefore believed that if increasing gradients of BDNF could be created beyond the cellular or scaffold replacement at the injury site, then the axons could be guided back into the uninjured tissue to remake the necessary connections for functional recovery. The goal of this project is to create increasing gradients of BDNF that can be sustained over time in vivo after a spinal cord injury. The BDNF is encapsulated in a degradable polymeric (poly-lactic-acid) microparticle. These microparticles degrade slowly over 6-8 weeks when injected into tissue, allowing a controlled release of the BDNF over time. Injections at increasing concentrations moving away from the injury site can create BDNF gradients in the tissue that can be sustained for 6-8 weeks. This should allow for the regenerating or replaced axons to reconnect with uninjured tissue leading to functional recovery after a spinal cord injury. These microparticles can then be implemented along with any promising axonal replacement or regeneration design for the injury site allowing increased chances for recovery.

Taylor L, L. Jones, M. Tuszynski, A. Blesch. "Neurotrophin-3 Gradients Established by Lentiviral Gene Delivery Promote Short-Distance Axonal Bridging beyond" Cellular Grafts in the Injured Spinal Cord J Neurosci 2006;26(38):9713-9721.

See more of this Session: Student Poster Session: Materials Engineering and Sciences
See more of this Group/Topical: Education

Sonolytic Synthesis of Biocompatible, Magnetic Metal Oxide Nanomaterials for Therapeutic Systems

Jayson P. Wicker, Chemical Engineering, North Carolina A&T State University, Greensboro, NC

Kristen A. Uitenham, Chemical Engineering, North Carolina A&T State University, Greensboro, NC

Timothy H. McArthur, Chemical Engineering, North Carolina A&T State University, Greensboro, NC

Gerard L. Moore, Chemical Engineering, North Carolina A&T State University, Greensboro, NC

Kenneth L. Roberts, Chemical Engineering, North Carolina A&T State University, Greensboro, NC

In this project iron oxide and iron oxide-based magnetic nanomaterials were manufactured using sonolytic & chemical synthesis techniques for biomedical applications. Technology challenges involved with dimensional consistency (or the control of size and morphology), biocompatibility, and the scale up of bench-scale syntheses to produce large batch sizes of nanomaterials. When specifically considering the application of bimetallic nanomaterials, additional design considerations will include the control of relative metal compositions and requirements of production consistency regarding the layering of active metal on the surfaces of the nanomaterials.

Magnetic nanomaterials were produced using wet chemistry / reactive processing and sonolytic methods. Superparamagnetic iron oxide (SPIO) nanoparticles were synthesized with iron chlorides using sonochemical methods. All magnetic nanomaterials will be characterized using x-ray diffraction (XRD), elemental analysis, BET surface area analysis, scanning and transmission electron microscopies (SEM and TEM) and superconducting quantum interference device (SQUID) magnetometric analysis. The effects of binary metal oxide composition on the magnetic and structural properties will be discussed.

See more of this Session: Student Poster Session: Materials Engineering and Sciences
See more of this Group/Topical: Education

Microwave Synthesis of Biocompatible Quantum Dots

Jeweliet A. Yost, Chemical Engineering, Bucknell University, Lewisburg, PA

Sarah Wood, Chemical Engineering, Bucknell University, Lewisburg, PA

Brandon Vogel, Chemical Engineering, Bucknell University, Lewisburg, PA

Fluorescent quantum dots (QD) are collections of semiconductor atoms on the order of a few nanometers. Their emission properties are dictated by the surface state and traps of the nanoparticles, and the size of the particle. Quantum dots are an emerging alternative to traditional dyes for solar cells and bio-imaging because of the ability to tailor the chemistry of the QD surface, their high quantum efficiency and their stable emission. One of the draw backs of QDs is the requirement of oxygen and water free synthesis conditions, and difficulty producing high quantum efficiency water soluble particles. We present a general method of synthesizing water soluble core-shell cadmium selenide (CdSe) QDs with quantum efficiencies of greater than 20 percent; the highest reported to date for water soluble quantum dots. The water soluble core-shell QDs are synthesized by microwave heating from commercially available precursors in the presence of water and oxygen in a matter of minutes. This methodology allows for the controlled synthesis of an array of different QDs, and the ability to tailor the chemical functionality of the QD surface with ligands. We also present a method of encapsulating the CdSe QD cores in a zinc sulfide shell to fill in electronic traps and increase the particles' quantum efficiency. We have further been able to increase the quantum efficiency by altering the surface chemistry of the cores. To date we have successfully synthesized QD cores of the yellow-red spectrum. Our current research focuses on revising previously developed procedures to synthesize CdSe QD cores of the full color spectrum as well as cores of other metals.

See more of this Session: Student Poster Session: Materials Engineering and Sciences
See more of this Group/Topical: Education

Improving Dielectric Properties by Grafting P(VDF-CTFE) to TiO_2

Matthew T. Irwin, Deptartment of Chemical and Biomolecular Engineering, Vanderbilt University, Nashville, TN

Qing Wang, Deparment of Materials Science and Engineering, The Pennsylvania State University, University Park, PA

The development of dielectric materials that can exhibit large energy densities is critical for the creation of more efficient technologies. In order to fulfill this need, the ferroelectric copolymer poly(vinylidene fluoride-chlorotrifluoroethylene) is currently being investigated due to its large dielectric constant and high breakdown strength. Traditionally, in order to improve dielectric properties, polymers have been physically blended with nanoparticles that exhibit desirable dielectric behavior. However, physical blends are often laden with polymer and nanoparticle agglomerates, resulting in large dielectric loss. In this study, we have investigated the dielectric properties of a nanocomposite synthesized using a novel grafting technique that involves chemically bonding the random copolymer P(VDF-CTFE) (91/9 mol%) to the nanoparticle TiO_2. Grafting significantly increases nanocomposite dispersion, which results in a larger dielectric constant, smaller dielectric loss, and, ultimately, a larger energy density.

See more of this Session: Student Poster Session: Materials Engineering and Sciences
See more of this Group/Topical: Education

Infrared Spectroscopic Characterization of Bifunctional Organic Linker Molecules

Julius A. Edson, Chemical Engineering, The City College of New York, City University of New York, New York, NY

Ilona Kretzschmar, Chemical Engineering, The City College of New York, City University of New York, New York, NY

The overall focus of this project is the self-assembly of three-dimensional "colloidal" molecules using micrometer and submicrometer-sized particles. The assembly of particles is controlled by (i) the modification of particle surfaces with reactive patches, creating so-called patchy particles, and (ii) the use of organic bifunctional linkers to connect the colloidal building blocks. In the project presented here, we are exploring molecules that may be used as bifunctional linkers between patchy particles. These linker molecules have functional groups that are reactive towards a specific patch material on the particles. In order to determine whether the linker molecules have successfully reacted with the patches, their presence must be detected. The most convenient way to characterize the molecules is infrared (IR) spectroscopy, which is very sensitive towards the functional groups available in the molecules. In order to understand what change the molecules undergo during reaction, a comprehensive analysis of the linker molecules with IR absorption spectroscopy is needed. To this end, we measured the IR spectra of the free molecules in an appropriate IR set up. The molecules characterized with IR in this report are: 1,8-octanedithiol, 1,8-diaminooctane, 1,8-octanediol, and 1,8-octanedioic acid. Using the best spectra collected, peaks are analyzed to determine peak position and interpreted using available literature values. More importantly, peaks that are unique to each linker and functional group are identified. Two of these linker molecules, 1,8-octanedithiol, and 1,8-diaminooctane, are then reacted with gold surfaces, to form self-assembled monolayers. Subsequently, the IR spectra of these surfaces, using Diffuse Reflectance setup, are taken and compared to the spectra of the free molecules in solution. We find that the bifunctional molecules form organized monolayers using the difunctional molecules. For the monofuntional molecules, 1-Octanethiol and 1-Octylamine, the spectra are weak indicating a disorganized monolayer formation.

See more of this Session: Student Poster Session: Materials Engineering and Sciences
See more of this Group/Topical: Education

De Novo Design of Peptide-Calcite Biomineralization Systems

Sarah Schrier, Chemical and Biomolecular Engineering, Johns Hopkins University, Baltimore, MD

Elizabeth A. Specht, Chemical and Biomolecular Engineering, Johns Hopkins University, Baltimore, MD

David L. Masica, Program in Molecular and Computational Biophysics, Johns Hopkins University, Baltimore, MD

Jeffrey J. Gray, Chemical and Biomolecular Engineering, Johns Hopkins University, Baltimore, MD

Many organisms produce complex, hierarchically structured, inorganic materials via protein-influenced crystal growth-a process known as biomineralization. Understanding this process would shed light on hard-tissue formation and guide efforts to develop biomaterials. We created and tested a computational method to design protein-biomineralization systems. The algorithm folds a protein from a fully extended structure and simultaneously optimizes the fold, orientation, and sequence of the protein adsorbed to a crystal surface. We used the algorithm to design peptides (16 residues) to modify calcite (CaC03) crystallization. We chemically synthesized six peptides that were predicted to bind different states of a calcite growth plane. All six peptides dramatically affected calcite crystal growth (as observed by scanning electron microscopy), and the effects were dependent on the targeted state of the {001} growth plane. Additionally, we synthesized and assayed scrambled variants of all six designer peptides to distinguish cases where sequence composition determines the interactions versus cases where sequence order (and presumably structure) plays a role. Scrambled variants of negatively charged peptides also had dramatic effects on calcite crystallization; in contrast, scrambled variants of positively charged peptides had a variable effect on crystallization, ranging from dramatic to mild. Special emphasis is often placed on acidic protein residues in calcified tissue mineralization; the work presented here suggests an important role for basic residues as well. In particular, this work implicates a potential role for basic residues in sequence-order specificity for peptide-mineral interactions.

See more of this Session: Student Poster Session: Materials Engineering and Sciences
See more of this Group/Topical: Education

Micro-Scale Effects On the Wettability of Electrospun PVP-Titania Fiber Mats On ITO Substrates in Aqueous Media

Kevin L. Lyon, Department of Chemical and Materials Engineering, University of Idaho, Moscow, ID

D. Eric Aston, Department of Chemical and Materials Engineering, University of Idaho, Moscow, ID

Nanowires have been produced in dense mats by electrospinning a solution of dispersed polyvinylpyrrolidone (PVP) and a titania precursor onto indium tin oxide (ITO) plates. Parameters affecting the wettability of these mats are under investigation, including contact angle dynamics, mat segment density, spinning and curing conditions, and micro-scale fluid interactions with the wire mats. Mats that were spun at a short distance for contact times of 30 seconds, one minute, and three minutes indicate that the hydrophilic properties of the mat increase with mat thickness. Contact angles as high as 150° were measured, but there was not a distinct pattern in contact angles within a single specimen set. Dynamics of the contact angle measurement indicate that placement of a water droplet onto samples must be done carefully to retain high contact angles. If the droplet is dropped onto the mat surface from a height of 1.5 cm, the hydrophilic properties were shown to decrease the contact angles as much as 38° due to micro-scale wetting of the wire network. Future work will include counting the number of wire intersections within a segment of a sample to determine a segment density and determine a relationship between microscopic configuration and wettability characteristics.

See more of this Session: Student Poster Session: Materials Engineering and Sciences
See more of this Group/Topical: Education

Preparation of Superhydrophobic Polymethylene Films

Nabijan Nizamidin, Chemical Engineering, Vanderbilt University, Nahville, TN

G. Kane Jennings, Chemical Engineering, Vanderbilt University, Nashville, TN

Juan C. Tuberquia, Chemical Engineering, Vanderbilt University, Nahville, TN

Some plants such as Lotus (Nelumbo nucifera) have been known to exhibit special self cleaning properties termed "Lotus Effect", in which dirt is picked up from the surface when the surface leave comes in contact with water droplets. Previous studies have demonstrated that the combination of low surface energy materials with nano- and micrometer scale topography is responsible for this behavior. This superhydrophobic effect has applications in fields such as micro fluidics and water repellent paints and fabrics. Here we described the preparation of low energy surface energy materials with nano- and micrometer scale features using a combination of soft lithographic patterning and surface-initiated polymerization. Polymethylene films were synthesized in the experiment using no specific topographical design. For this method, necessary surface topography is produced after the films are allowed to polymerize past thickness of 2500-3000 nm. Soft lithography patterning method and no topographical design method were compared to improve the superhydrophobicity of the films. Particular emphasis was placed on soft lithography patterning of polymethylene film to achieve the optimal surface roughness to produce the best superhydrophobic films. It has been proven that with soft lithography patterning, superhydrophobic polymethylene films can be produced with film thickness of 1500-2000 nm.

See more of this Session: Student Poster Session: Materials Engineering and Sciences
See more of this Group/Topical: Education

Redox Stable Conductivity of B-Site Doped SrTiO3 for SOFC Anodes

William C. Holler, Chemical Engineering, Bucknell University, Lewisburg, PA

Brandon Smith, Chemical Engineering, Bucknell University, Lewisburg, PA

Michael D Gross, Chemical Engineering, Bucknell University, Lewisburg, PA

The stability of solid oxide fuel cell (SOFC) anodes during cycles of reducing and oxidizing environments (redox cycling) is a critical parameter of durability, particularly for portable applications. The standard Ni-YSZ anode is not stable during redox cycles because the volume of Ni substantially increases upon oxidation. This expansion ultimately results in mechanical failure of the fuel cell. To address this limitation, $SrTa_xTi_{1-x}O_3$ (STT) has been investigated as a potential material for redox-stable anodes. While Ni is both an e^- conductor and an oxidation catalyst, STT can only provide e^- conduction and a separate catalyst must be added. Dense STT and porous STT-YSZ composites, for x = 0.01, 0.05, and 0.10, were sintered in both air at 1873 K and hydrogen at 1773 K. X-ray diffraction (XRD) indicated a single perovskite phase for STT and that STT and YSZ are chemically compatible. Conductivities greater than 1 $S \cdot cm^{-1}$ were achieved for porous STT-YSZ with x = 0.01 and 0.05 when initially reduced at 1773 K. Without this high temperature reduction, conductivity was too low for SOFC anode applications. STT with x = 0.01 demonstrated the highest redox stable conductivity with a porous conductivity of 0.9 $S \cdot cm^{-1}$ after oxidation at 1273 K.

See more of this Session: Student Poster Session: Materials Engineering and Sciences
See more of this Group/Topical: Education

Atomic Layer Deposition of Zinc Oxide in a Fluidized Bed Reactor for Novel Sunscreens and Cosmetics

Samantha I. Johnson, Department of Chemical and Biological Engineering, University of Colorado, Boulder, CO

David M. King, Department of Chemical and Biological Engineering, University of Colorado, Boulder, CO

Jianhua Li, Department of Chemical and Biological Engineering, University of Colorado, Boulder, CO

Xinhua Liang, Department of Chemical and Biological Engineering, University of Colorado, Boulder, CO

Alan W. Weimer, Department of Chemical and Biological Engineering, University of Colorado, Boulder, CO

Thin film deposition techniques have many applications in electronic, semiconductor, and ceramic processes. One such technique is atomic layer deposition (ALD), which can be used to create pinhole-free, thickness-controlled films of metal oxides. While ALD on flats has been extensively studied, the deposition on nanoparticles is fairly novel and merits exploration. A fluidized bed reactor with in situ mass spectroscopy was used to deposit metal oxides onto the surfaces of primary particles. Zinc oxide was deposited by ALD on spherical silicon dioxide particles and high surface area titanium dioxide nanoparticles in order to create novel UV absorbing fillers for sunscreens and cosmetics products. In situ mass spectroscopy allowed for real-time study and control of the ALD process; diffuse transmittance analysis was used to measure the optical properties of the nanocomposite particles. Analysis showed that the bandgap of the deposited zinc oxide shifted nonlinearly with film thickness up to about 5 nm, after which the bulk bandgap of zinc oxide was observed. Mass spectroscopy analysis of the exiting gas stream showed when the reaction product generation terminated, and when unreacted precursor escaped from the bed. This also validated a process control strategy for monitoring particle ALD processes at any scale.

See more of this Session: Student Poster Session: Materials Engineering and Sciences
See more of this Group/Topical: Education

Development of Target-Specific Antifungal Beta-Peptides

Jeffrey Aquino, Chemical Engineering, University of Puerto Rico-Mayaguez Campus, Mayaguez, PR

Patricia Ortiz, Chemical Engineering, University of Puerto Rico-Mayaguez Campus, Mayaguez, PR

Samuel H. Gellman, Department of Chemistry, University of Wisconsin-Madison, Madision, WI

Our project is focused on the design and characterization of engineered materials with intrinsic antifungal activity. We have chosen beta-peptides as a platform for the synthesis of target-specific antifungal agents. The already studied molecular interactions of antimicrobial synthesis peptides with cell membranes of pathogenic fungi and bacteria provide an excellent start point from which to optimize our designs. We propose the utilization of chitin-binding domains (CBDs) as models to create target-specific beta-peptides. CBDs are short amino acid sequences that bind specifically to the N-acetyl glucosamine subunits of the structural biopolymer chitin. We hypothesize that since chitin is a specific structural component of fungal cell walls, a CBD-based beta-peptide will exhibit an enhanced antifungal activity and decrease cytotoxicity particularly in mammalian cells, which lack chitin as a structural component. We will synthesize and characterize CBD-based beta-peptides and assess their activity against fungal cells in vitro using Candida albicans and Cryptococcus neoformans, important human pathogens. Furthermore, we will characterize the in vivo activity of CBD-based beta-peptides using murine pathogenesis models.

See more of this Session: Student Poster Session: Materials Engineering and Sciences
See more of this Group/Topical: Education

Development of Metal Nanowires for Sensor Applications

Felix Delgado, Chemical Engineering, University of Puerto Rico Mayaguez Campus, Las Piedras, PR

Nanomaterials, SiC semiconductor, metal oxides and microfabrication technology is used to develop H2, CxHy, NOx, CO, O2, and CO2 sensors for some application. Silicon-based processing techniques (commonly referred to as microelectromechanical systems or MEMS technology) and their use in the fabrication of chemical sensing microsystems have enabled the fabrication of new sensor structures and sensor arrays with high capabilities but limited size, weight, and power consumption. Work has been on-going to develop an integrated smart leak detection system based on miniaturized sensors to detect hydrogen, hydrocarbons, or hydrazine, and oxygen. The approach is to implement MEMS based sensors incorporated with signal conditioning electronics, power, data storage, and telemetry enabling intelligent systems. The final sensor system will be self-contained with a surface area comparable to a postage stamp.

See more of this Session: Student Poster Session: Materials Engineering and Sciences
See more of this Group/Topical: Education

Synthesis and Characterization of Poly(beta-amino ester) Hydrogel Systems as Orthopaedic Implants for Growth Plate Regeneration

Juan Carlos Cordova, Chemical and Materials Engineering, University of Kentucky, Lexington, KY

Ashley Hawkins, Chemical and Materials Engineering, University of Kentucky, Lexington, KY

D.A. Puleo, Center for Biomedical Engineering, University of Kentucky, Lexington, KY

Todd A. Milbrandt, Department of Orthopaedic Surgery, University of Kentucky, Lexington, KY

Dr. J. Zach Hilt, Chemical and Materials Engineering, The University of Kentucky, Lexington, KY

Biodegradable hydrogels are widely researched biomaterials for many applications, such asin drug delivery and tissue engineering. This research focuses on the synthesis and characterization of biodegradable hydrogel scaffolds as orthopaedic implants for the regeneration of native growth plate tissue in patients suffering from growth plate injuries. Poly(beta-amino ester) hydrogel systems have been tailored to have the necessary degradation rates, mechanical properties, porosity, and cytotoxicity that can be used to invoke the regeneration of the native growth plate. Initial work included the synthesis and characterization of the macromer precursors. Macromers were characterized using gel permeation chromatography (GPC) and FTIR analysis. Hydrogels were then polymerized through a chemical initiated free radical polymerization and conversion was measured using FTIR analysis. The mechanical properties of the gels were determined through compression testing. Porosity was characterized using scanning electron microscopy (SEM) imaging. The cellular interactions with D1 pluripotent cells were also studied by fluorescent tagging of cells which were then seeded onto the hydrogel scaffold and subsequent cell viability analysis using the live/dead assay.

See more of this Session: Student Poster Session: Materials Engineering and Sciences
See more of this Group/Topical: Education

Quantum Chemical Characterization of Candidate Insensitive High Energetic Materials

Rebecca K. Lindsey, Chemical Engineering, Wayne State University, Detroit, MI

In recent years, much effort has been put into the development of high performance insensitive munitions (IM). Such materials exhibit high thermal stability, resistance to impact, and high energetic performance [1]. Development of these materials has offered a challenge due to the tendency for IM compounds to exhibit poorer energetic performance in comparison to the traditionally used, higher sensitivity materials such as RDX and TNT [2,3].

In earlier times, when looking for novel energetic materials to meet these criteria, chemicals would be synthesized in a trial-and-error fashion under the hope that the product would meet the newly set standards [4]. This approach has proved to be expensive in both time and resources, and thus the need for a new method arose.

In this work, we perform quantum chemical analysis on candidate IM compounds Dinitropyrazole (DNP), Nitrotriazolone (NTO), 1-methyl-2,4,5-trinitroimidazole (MTNI) to determine heats of formation in the solid, liquid, and vapor phase, heat of sublimation, heat of vaporization, and torsional barriers. From this information we are able to quantify the strength of intramolecular interactions and gain insight into the impetus of these materials. Calculations for heats of formation, vaporization, and sublimation are performed using two different atom equivalent methods described by B. M. Rice and E. F. C. Byrd et. al. [5,6]. Torsional Barriers are calculated at HF, Density functional, and MP2 levels of theory using the 6-31+G(d,p) basis set.

[1] Davies, P. K.; Provatas, A. Weapons Systems Division, DTSO Defense and Science Technology Oranization 2006 Australia

[2] Huynh, M. H. V. ; Hiskey, M. A. ; Chavez, D. E. ; Naud, D. L. ; Gilardi, R. D. JACS 2005, 127, 12537.

[3]Wallace, L ; Cronin, M. P. ; Day, A. I. ; Buck, D. O. Environ, Sci. Technol. 2009, 43, 1993

[4]Rice, B. M. ; Hare, J. J. J. Phys. Chem. A 2002 106, 1770

[5]Rice, B. M. ; Pai, S. V. ; Hare, J. J. Comb. Flame 1999 118, 445

[6]Byrd, E. F. C. ; Rice, B. M. J. Phys. Chem A 2006 110, 1005

See more of this Session: Student Poster Session: Materials Engineering and Sciences
See more of this Group/Topical: Education

Polymers From Furfural

Olga O. Jennings, Chemical & Biochemical Engineering, University of Iowa, Iowa City, IA

Christopher M. Comer, Chemical & Biochemical Engineering, University of Iowa, Iowa City, IA

Julie L. P. Jessop, Chemical & Biochemical Engineering, University of Iowa, Iowa City, IA

Furfural is a promising starting material for renewable chemicals and plastics. Furfural is derived from variety of agricultural byproducts, mainly corn stover. When furfural derivatives such as furan and 5-bromo-2-furfural are irradiated together with ultraviolet light, difuran compounds are formed through a photochemical reaction. These materials are the potential starting blocks used to synthesize biodegradable polymers with excellent mechanical properties and thermal resistance. A protected-difurfural compound has been synthesized in 50-60% yield. However, the initial reactants and solvent can be recovered and reused in a secondary reactor to increase the overall yield of the reaction. Continued research has focused on the deprotection of the compound, its conversion to an alcohol, and polymer synthesis and characterization.

See more of this Session: Student Poster Session: Materials Engineering and Sciences
See more of this Group/Topical: Education

Epoxide/Acrylate Hybrid Photopolymerizations: Synergy of the Dual Photoinitiator System

Na Yeon Kang, Chemical & Biochemical Engineering, University of Iowa, Iowa City, IA

Hoseop Eom, Chemical & Biochemical Engineering, University of Iowa, Iowa City, IA

Julie L. P. Jessop, Chemical & Biochemical Engineering, University of Iowa, Iowa City, IA

Photopolymerization uses light to form long chains of molecules. This method has been widely applied due to its energy savings, solvent-free systems, and ease of reaction control. Hybrid systems contain acrylates and epoxides, each reacting separately to form polyacrylate and polyepoxide chains in the presence of radical and cationic photoinitiators, respectively. In this research, using Raman spectroscopy, we examine how the interplay between the two photoinitiators affects the kinetics of radical and cationic photopolymerizations in the hybrid systems. Understanding this synergy enables mitigation of sensitivity to oxygen and water during the hybrid photopolymerizations.

See more of this Session: Student Poster Session: Materials Engineering and Sciences
See more of this Group/Topical: Education

Development of Nanocomposite Shape Memory Polymers

Zhengwang He, Department of Chemical and Materials Engineering, University of Kentucky, Lexington, KY

Nitin S. Satarkar, University of Kentucky, Lexington, KY

J. Zach Hilt, Chemical and Materials Engineering, University of Kentucky, Lexington, KY

Shape memory polymers (SMP) are a class of polymers that have the capabilities of actuating systems during their transition from a temporary shape to permanent shape. In many systems, the shape memory effect (SME) can be stimulated by a change in temperature. Since direct heating may not always be applicable, nanocomposite systems have been developed that allow for the remote control of the heating. Specifically, the SMP that has been studied is an epoxy-based system. The nanoparticles studied included both iron oxide (Fe_3O_4) nanoparticles and multiwall carbon nanotubes (MWCNT), which can be remotely heated in 300 kHz and 13.65 MHz radiofrequency (RF) fields, respectively. Different loadings of nanoparticles were incorporated into the SMPs to tailor the remote heating property. DSC and FT-IR characterizations were done to determine the glass transition temperatures and polymer conversions. In addition, a couple potential applications of these systems were demonstrated.

See more of this Session: Student Poster Session: Materials Engineering and Sciences
See more of this Group/Topical: Education

Solubility of Organic Solvents in Poly(butadiene): A Capillary-Column Inverse Gas Chromatography Study

See more of this Session: Student Poster Session: Materials Engineering and Sciences
See more of this Group/Topical: Education

Optimization of Atom Transfer Radical Polymerization From Silica Nanoparticles

Yang Yang, Chemical Engineering, University of Virginia, Charlottesville, VA

Maura McEwan, Chemical Engineering, University of Virginia, Charlottesville, VA

David Green, Chemical Engineering, University of Virginia, Charlottesville, VA

Polymer nanocomposites (PNCs) are used widely in coatings, optics, catalysis, and plastics. Uniform dispersion of nanoparticles within the polymer melt is desired for control of material properties including strength, hardness and thermal and electrical conductivity. Grafting polymer chains from particle surfaces is a practiced technique to prevent aggregation of particles, and thus, promote particle dispersion. Atom transfer radical polymerization (ATRP) is a specialized reaction scheme to obtain high graft densities of polymer brushes. Improved engineering control of ATRP is needed to create well-defined, dense polymer brushes from nanoparticles to ultimately improve nanoparticle dispersion within PNCs.

Our goal is to optimize ATRP reaction conditions of methyl methacyrlate (MMA) from silica nanoparticles to achieve polymer brushes of high graft density, low polydispersity, and well-controlled molecular weight of the polymer brush. To this end, the effect of ligand, metal catalyst, particle concentration, and temperature were investigated. Characterization techniques of gas chromatography (GC), gel permeation chromatography (GPC), and thermal gravimetric analysis (TGA) were used to quantify the reaction kinetics, molecular weight, and graft density. Ultimately, optimizing reaction conditions of ATRP to form dense, uniform polymer brushes on nanoparticle surfaces enables control of particle interactions and fine-tuning of material properties of PNCs.

See more of this Session: Student Poster Session: Materials Engineering and Sciences
See more of this Group/Topical: Education

Phase Equilibrium of Solvents in Polybutadiene: A Capillary-Column Inverse Gas Chromatography Study

Romny, Ronald P. Garcia Danner Sr., Chemical Engineering, The Penn State University, Pennsylvania, PA

Ronald P. Danner Sr., Chemical Engineering, The Penn State University, Pennsylvania, PA

See more of this Session: Student Poster Session: Materials Engineering and Sciences
See more of this Group/Topical: Education

Investigation of Dynamic Relaxation Properties of Poly(methyl methacrylate)/Silica Nanocomposites

Alexander L. Heilman, Chemical and Materials Engineering, University of Kentucky, Lexington, KY

Anthony C. Comer, Chemical and Materials Engineering, University of Kentucky, Lexington, KY

Andrew Placido, Chemical and Materials Engineering, University of Kentucky, Lexington, KY

Douglass S. Kalika, Chemical and Materials Engineering, University of Kentucky, Lexington, KY

Understanding the relaxation characteristics of polymers and polymer nanocomposites can offer insight into the nature of the chain motion and, in the case of nanocomposites, potential chemical and physical interactions at the particle-polymer interface. Using dynamic mechanical analysis and broadband dielectric spectroscopy, we studied the sub-glass and glass transitions of poly(methyl methacrylate) nanocomposites containing up to 30 wt% native and surface-modified silica nanoparticles. In systems containing the latter, favorable surface interactions between the polymer and the nanoparticles were observed, and at sufficiently high concentrations a second glass transition event was detected, suggesting a restriction of local chain motion in the vicinity of the nanoparticle surface. Dielectric spectroscopy was used to probe the influence of increasing nanoparticle concentration on the sub-glass and glass-rubber transitions, as well as the polarizability of the nanocomposite system, namely the interfacial polarization encountered at high temperatures and low frequencies. Although an expected increase in interfacial polarization was observed with increased nanoparticle loading, the particles did not have a discernable effect on either the sub-glass or glass transitions of the nanocomposite systems.

See more of this Session: Student Poster Session: Materials Engineering and Sciences
See more of this Group/Topical: Education

Solid-State Fabrication of Polymer Blends: Cryogenic Milling and Solid-State Shear Pulverization

Marc F. Henry, Chemical Engineering, Bucknell University, Lewisburg, PA

Katsuyuki Wakabayashi, Department of Chemical Engineering, Bucknell University, Lewisburg, PA

Blending of two or more polymers can allow for tailored materials with highly tunable and customizable attributes, and are of interest in both industry and research. However, most polymer combinations are thermodynamically immiscible and current methods of polymer blend production have significant limitations. In immiscible blends, reactive extrusion or introducing a third "compatibilizer" species can "compatibilize" the resulting systems, but are only applicable to specific polymer families or chemically-modified polymers.

Studies have been conducted using mechanochemical reactions to prompt polymer chain scission, which can generate compatibilizing block copolymers from the component species in situ via free radical formation and recombination. Two techniques which utilize such mechanochemical principles are cryogenic milling and solid-state shear pulverization (SSSP). Cryogenic milling is a batch-scale process in which polymer pellets are contained within a processing vial and ground via a magnetically-driven tungsten carbide impactor. SSSP utilizes a continuous twin-screw extruder operating at below ambient temperatures.

This study focuses on comparing polymer blends made via common melt-state processing techniques to those fabricated through both batch cryogenic milling and continuous SSSP. Current results indicate phase stabilization and compatibilization via both methods.

See more of this Session: Student Poster Session: Materials Engineering and Sciences
See more of this Group/Topical: Education

Synthesis and Characterization of Sulfonated Poly(styrene-isobutylene-styrene) for Chemical and Biological Protective Clothing

Ariangelis Ortiz, Chemical Engineering, University of Puerto Rico, Mayaguez Campus, Mayaguez, PR

David Suleiman, Chemical Engineering, University of Puerto Rico, Mayaguez Campus, Mayaguez, PR

Poly(styrene-isobutylene-styrene) (SIBS) of 15% weight fraction polystyrene and 100k number average molecular weight (Mn) was sulfonated and functionalized to create ion selective polymer membranes for chemical and biological protective clothing among other applications. Highly selective polymer-metal nanocomposite membranes were created by ion exchange substitution with +2 cations, Mg^{+2}, Ca^{+2} and Ba^{+2}. Selectivity experiments were performed using Dimethyl Methyl Phosphonate (DMMP), a molecule similar to the chemical warfare agent Sarin. The highest Water/DMMP selectivity in the gas phase and the least water absorption percentage was obtained with Ba^{+2}. Polymer membrane properties were evaluated by different characterization techniques including Fourier transform infrared spectroscopy, thermogravimetric analysis, and small angle X-ray scattering. Results from these experiments show significant differences between the sulfonated and ion substituted polymers that can explain the variations in selectivity and water swelling behavior.

See more of this Session: Student Poster Session: Materials Engineering and Sciences
See more of this Group/Topical: Education

Does Pipe Size Affect Slow Crack Growth Resistance?

Monica M. Frey, Chemical Engineering, Kansas State University, Manhattan, KS

Dr. Z. Jimmy Zhou, Technical Service & Development, The Dow Chemical Company, Freeport, TX

The objective of this research was to investigate the effect of pipe geometry and stress intensity on slow crack growth (SCG) failure time of polyethylene (PE) pressure pipe. PE pressure pipe has an outstanding service record over the last 50 years. As PE is increasingly utilized to extrude larger diameter pipes, the stress intensity effect becomes critical. At the same design stress, the stress intensity increases with pipe size, especially for a scratched pipe. Although it is known that increased stress intensity will accelerate the SCG failure process, little work has been done to investigate the effect of pipe size on SCG resistance. In this research study, samples with different thicknesses and notch depths were tested under various stresses in order to create a model for failure time based on overall stress intensity. Preliminary data shows that stress intensity significantly decreases SCG failure time. More specifically the failure time decreases by up to 5.6 times for a hypothetical 36 inch pipe when compared to a 4 inch pipe. This model can be used to estimate the expected lifetime of various size pipes against SCG failure.

See more of this Session: Student Poster Session: Materials Engineering and Sciences
See more of this Group/Topical: Education

Improving the Electrical Properties of Doped ZnO Nanowires Via Post-Synthesis Processing for Solar Applications

Yangluo Jim Wang, Chemical and Biomolecular Engineering, Computational and Applied Mathematics, Rice University, Houston, TX

Rodrigo Noriega, Applied Physics, Stanford University, Stanford, CA

Sujay Phadke, Mechanical Engineering, Stanford University, Stanford, CA

Saahil Mehra, Materials Science and Engineering, Stanford University, Stanford, CA

Alberto Salleo, Materials Science and Engineering, Stanford University, Stanford, CA

Zinc oxide (ZnO) is a wide band-gap material that can potentially replace indium tin oxide (ITO) as a cheaper alternative for transparent conducting electrodes. A study is undertaken to reduce the sheet resistance of low-temperature, solution-grown Ga-doped ZnO (Ga:ZnO). 1% Ga:ZnO nanowires were synthesized via a decomposition of metal salts in trioctylamine at 300°C. Post-deposition annealing in nitrogen reduced the sheet resistance by several orders of magnitude. Other methods such as the fabrication of Ga:ZnO-Ag composites and Ga:ZnO/PEDOT:PSS bilayers were investigated. The composites had a sheet resistance on the same order as unannealed Ga:ZnO, whereas Ga:ZnO/PEDOT:PSS bilayers had a comparable resistivity to the best annealed Ga:ZnO film.

See more of this Session: Student Poster Session: Materials Engineering and Sciences
See more of this Group/Topical: Education

Effect of Pressure On a Budesonide-Ethanol Spray During the Supercritical Anti-Solvent Precipitation Process

Andrew P. Klinger, Chemical Engineering, Auburn University, Auburn, AL

Rajeshwar Chinnawar, Chemical Engineering, Auburn University, Auburn, AL

Christopher B. Roberts, Chemical Engineering, Auburn University, Auburn, AL

Steve R. Duke, Chemical Engineering, Auburn University, Auburn, AL

This study seeks to understand the effects of pressure on the creation of micron sized steroid particles through a supercritical anti-solvent precipitation process. A 1 weight % budesonide in ethanol solution was sprayed through a 100 μm inner diameter capillary nozzle into a high pressure vessel containing supercritical CO_2. The spray characteristics were visualized by a high magnification video system, which captured images through viewing windows on the sides of the vessel. The supercritical CO_2 temperature was held constant at 323 K, and particles were produced for operating pressures of 84, 89, and 94 bar. Images were taken at three distances from the nozzle. Images taken from the nozzle tip were used to measure the jet break up length, which decreased with increasing operating pressure. The average droplet diameters were measured at 3 and 14 mm distances away from the nozzle at each operating pressure. The droplet diameters increased from 3 to 14 mm and decreased with increased operating pressure. The collected budesonide particles were analyzed under a scanning electron microscope. The particles were spherical and about 1 μm in diameter for each operating pressure. The visualization methods allow particle properties to be related to process and jet properties.

See more of this Session: Student Poster Session: Materials Engineering and Sciences
See more of this Group/Topical: Education

A Study of Diffusion and Partitioning of Organic Compounds Into Polydimethylsiloxane (PDMS)

Jennifer L. Greene, Systems Biology and Bioengineering Undergraduate Research Experience (SyBBURE), Vanderbilt University, Hendersonville, TN

Leslie M. Shor, Department of Chemical, Materials and Biomolecular Engineering, University of Connecticut, Storrs, CT

Polydimethylsiloxane (PDMS) has many advantages over other materials commonly used in the fabrication of microfluidic devices. PDMS is inexpensive, and also has excellent biocompatibility and optical transparency which makes it particularly well-suited for many biological applications. However, one disadvantage of PDMS is the potential for solvents to swell the material, and for hydrophobic organic compounds to partition from the aqueous phase into the PMDS walls of microfluidic devices. Very high surface area to volume ratios common in microfluidic devices exacerbates this problem. The goal of this study was to measure diffusion and partitioning of commonly used organic compounds including antibiotics as a function of PDMS formulation and treatments including monomer extraction and plasma activation of cured surfaces. This research provides the data necessary to more accurately model and predict partitioning and diffusion effects into and out of PDMS. Better prediction of micro-scale chemical conditions within microfluidic devices will enhance the utility of microfluidic approaches in the basic biological sciences.

See more of this Session: Student Poster Session: Materials Engineering and Sciences
See more of this Group/Topical: Education

Phase Equilibrium of Organic Solvents in Polybutadiene: A Capillary-Column Inverse Gas Chromatography Study

Romny Garcia Sr., Chemical Engineering, University of Puerto Rico at Mayaguez, Coamo, PR

The capillary-column inverse gas chromatography (CCIGC) at infinite dilution of solvent has been used to determine diffusivity and partition coefficients of polymer-solvent systems in polybutadiene (PBD, 98% cis). Experiments were performed at 45, 60 and 80°C. Partition coefficients were used to determine the following thermodynamics parameters: infinitely dilute weight fraction activity coefficients and Flory-Huggins interaction parameters. The results indicated that ethylbenzene is the most suitable solvent for polybutadiene. The calculated weight-fraction activity coefficient data were compared to the prediction of the group contribution, lattice-fluid equation of state GCLF-EoS and the Uniquac Functional-group Activity Coefficient (UNIFAC) free volume models.

See more of this Session: Student Poster Session: Materials Engineering and Sciences
See more of this Group/Topical: Education

Poly(ethylene glycol)-Based Hydrogel Nanocomposites with Iron Oxide and Single-Walled Carbon Nanotubes for Hyperthermia Cancer Therapy

Christopher R. Barton, Chemical & Materials Engineering, University of Kentucky, Lexington, KY

Nitin S. Satarkar, University of Kentucky, Lexington, KY

Samantha A. Meenach, Chemical & Materials Engineering, University of Kentucky, Lexington, KY

J. Zach Hilt, Chemical and Materials Engineering, University of Kentucky, Lexington, KY

Kimberly W. Anderson, Chemical & Materials Engineering, University of Kentucky, Lexington, KY

Since nanocomposites have been shown to have remote heating capabilities and have demonstrated biocompatibility, a variety of applications in medicine are possible, particularly in hyperthermia cancer therapy. Hyperthermia is a type of treatment in which body tissue is exposed to moderate temperatures (up to 45°C) for a period of time. In this research, hydrogel nanocomposites were demonstrated as potential candidates for remote heating using electromagnetic fields. Specifically, hydrogel nanocomposites based on poly(ethylene glycol) (PEG) and loaded with either iron oxide particles or single-walled carbon nanotubes, were fabricated via free-radical polymerization. Polymer-coated iron oxide nanoparticles were selected because of remote heating properties and biocompatibility within the body, while carbon nanotubes are of interest due to their ability to enhance the mechanical properties of composites. Heating studies were done to investigate the effects of composition of PEG and nanoparticles on heating properties. The nanocomposites were remotely heated via various electromagnetic fields to achieve surface temperatures in the hyperthermia range. Different particle loadings, combined with different electromagnetic field strengths, caused different heating results.

See more of this Session: Student Poster Session: Materials Engineering and Sciences
See more of this Group/Topical: Education

Hyperthermia Cancer Therapy Using Poly(ethylene glycol)-Based Magnetic Hydrogel Nanocomposites

Lynndsey M. Klenk, Chemical and Materials Engineering, University of Kentucky, Lexington, KY

Kimberly W. Anderson, Chemical and Materials Engineering, University of Kentucky, Lexington, KY

James Z. Hilt, Chemical and Materials Engineering, University of Kentucky, Lexington, KY

Samantha A. Meenach, Chemical and Materials Engineering, University of Kentucky, Lexington, KY

This research studies the use of hydrogel nanocomposites consisting of poly(ethylene glycol) methyl ether methacrylate and poly(ethylene glycol) dimethacrylate (PEGMMA/PEGDMA) containing iron oxide nanoparticles for thermal cancer therapy applications. Hyperthermia is defined as the heating of cancerous tissues to between 40 and 45°C, and when used in conjunction with other cancer therapeutics such as radiation and chemotherapy, hyperthermia has been shown to improve the treatment success rate. The magnetic nanoparticles in these hydrogel nanocomposites give the system the ability to be remotely heated using an external alternating magnetic field. In addition, the ability to control the swelling properties of these systems provides control over drug delivery properties. Previous swelling analysis studies indicate that systems with lower crosslinking density have a higher volume swelling ratio. Heating studies using an alternating magnetic field show that the heating of these nanocomposites is dependent on both the amount of iron oxide loading in the gel and the strength of the alternating magnetic field. This research demonstrates that these hydrogel nanocomposite systems can be a successful thermal cancer therapeutic through showing that M059K glioblastoma cells can be killed selectively in vitro using hydrogel nanocomposite systems.

See more of this Session: Student Poster Session: Materials Engineering and Sciences
See more of this Group/Topical: Education

Continuous Solid-State Fabrication of Polymer Nanocomposites

Benjamin D. Aldrich, Chemical Engineering, Bucknell University, Lewisburg, PA

Paul J. Hubert, Bucknell University, Lewisburg, PA

Marc F. Henry, Bucknell University, Lewisburg, PA

Joshua T. Clark, Bucknell University, Lewisburg, PA

Cara D. Ziegler, Bucknell University, Lewisburg, PA

Polymer nanocomposites are advanced materials that contain small amounts of nanoscale fillers in the polymer matrix. The physical property improvements in the resulting nanocomposites lead to a wide range of high-performance applications from packaging to automotive parts.

There are several widely-used techniques to fabricate polymer nanocomposites, each with distinct advantages and weaknesses. In situ polymerization and solution mixing require the use of external chemicals in batches, while the high temperature operation in melt mixing can lead to practical issues like polymer degradation and filler re-agglomeration.

Solid-state processing of polymer nanocomposites is a relatively novel method which applies high shear and compressive forces at temperatures below the melt and/or glass transition temperature of the polymer. Solid-state shear pulverization (SSSP) is an industrially-applicable, continuous solid-state process in which polymer/nanofiller blends are pulverized within a chilled twin-screw extruder.

This poster discusses the features of Bucknell's new solid-state pulverizing system, focusing on the screw setup and feeder design. Mechanical properties of pure polymers are investigated to gain insight into the mechanochemistry occurring in SSSP. Model polypropylene-based nanocomposites made via SSSP are compared with equivalent samples made via conventional melt-mixing processes. Various physical property measurements were conducted to evaluate the processing-structure-property relationships.

See more of this Session: Student Poster Session: Materials Engineering and Sciences
See more of this Group/Topical: Education

Encapsulation of Hewl Crystals IN Polymeric Networks

Jose A. Jimenez, Chemical Engineering, University of Puerto Rico-Mayagüez Campus, Orocovis, PR

Madeline Torres, Ph.D., Chemical Engineering, University of Puerto Rico-Mayagüez Campus, Orocovis, PR

Alexandra Martinez, Chemical Engineering, University of Puerto Rico-Mayagüez Campus, Orocovis, PR

Protein encapsulation and crystallization are under much investigation due to the advantages they possess, such as the amount of applications that can result like the use in biosensors, drug delivery purposes, filter membrane creation, etc. Crystallization and encapsulation can also permit us to learn more about the structure and properties of the proteins. There are many advantages but the more important ones are the comfort and stability these methods offer proteins, which is very important since proteins are extremely sensitive when outside of their natural environments and the ability of manipulate the protein crystals. However, there are many other things that must be taken in consideration like pH, temperature, solvents, release, ect. All the experiments where produced using the protein lysozyme from hen egg white (HEWL), this due to the property of been more resistant to harsh conditions than other proteins. We are using two methods for protein crystallization: Batch method and the Hanging Drop method. Varying the concentrations of Methacrylic acid (anionic, MAA) and poly(ethylene glycol) dimethacrylate n=1000 (cross-linker, PEGDMA1000), adding the protein crystals to the solution and using UV Polymerization the membranes are obtained with the crystals inside the polymeric networks. The qualitative studied morphologies were: 88% MAA - 40%Solvent, 88% MAA - 50%Solvent, 88% MAA - 60%Solvent, 90% MAA - 40%Solvent, 92% MAA - 40%Solvent, 95% MAA - 40%Solvent, 98% MAA - 40%Solvent and 99% MAA - 40%Solvent. The best result were obtained at 88% MAA - 40%Solvent (ethanol 30%), where the crystals lasts months without suffering any damage, in other words remaining intact and active or stable and their structure remain almost intact compare with the others concentration, where at time cero the crystalline structure looks very damage.

See more of this Session: Student Poster Session: Materials Engineering and Sciences
See more of this Group/Topical: Education

Evaluation of the Selectivity of An Insulator-Based Dielectrophoretic Microdevice: Theoretical and Experimental Results

Ana V. Chávez-Santoscoy, Biotechnology and Food Engineering and Biotechnology Center, Tecnologico de Monterrey, Monterrey, Mexico

Javier Baylon-Cardiel, Electrical and Computing Engineering, Tecnologico de Monterrey, Monterrey, Mexico

Blanca H. Lapizco-Encinas, Biotechnology and Food Engineering, Tecnologico de Monterrey, Monterrey, Mexico

Microscale separations is a novel and rapid growing field; microdevices offer significant advantages over traditional bench scale equipments. When working on the microscale, only small quantities of samples and reagents are required, higher resolution and sensitivity are obtained, and microdevices are low-cost and require shorter processing time [1]. Due to the important advantages that microfluidics there is an increasing interest in the development of separations techniques that can be utilized on the microscale. Dielectrophoresis (DEP) has become one of the most important micro analytical techniques, since it has great potential for the manipulation of a wide array of bioparticle [2-4]. Insulator-based DEP (iDEP) is a dielectrophoretic mode where non uniform electric field are created employing insulating structures. In this work we present an evaluation of the selectivity of a microdevice by employing mixtures of polystyrene nano and microparticles. Experimental and mathematical work was carried out with the objective of measuring how iDEP can selectively immobilize certain types of particles in the presence of other types of particles. It is expected that the results from this study will provide with guidelines for the design and operation of iDEP microdevices.

References [1] Whitesides, G. M., Nature 2006, 442, 368-373. [2] Lapizco-Encinas, B. H., Rito-Palomares, M., Electrophoresis 2007, 28, 4521-4538. [3]Gascoyne, P. R. C., Vykoukal, J., Electrophoresis 2002, 23, 1973-1983. [4]Voldman, J., Ann. Rev. Biomed. Eng. 2006, 8, 425-454.

See more of this Session: Student Poster Session: Separations
See more of this Group/Topical: Education

Surfactant-Assisted Aqueous Extraction of Soybean Oil From Soybean and Recovery of Oil From Skim

Jordan R. Milligan, Chemical Engineering, Rose-Hulman Institute of Technology, Terre Haute, IN

Kerry A. Campbell, Chemical and Biological Engineering, Iowa State University, Ames, IA

Charles E. Glatz, Chemical and Biological Engineering, Iowa State University, Ames, IA

Aqueous extraction processing (AEP) is a safer and more environmentally friendly oil extraction method than conventional solvent extraction. SDS and phospholipid were added as emulsifiers to the extraction mixture to test their effect on extraction yield of protein and oil during AEP. It was found that the addition of SDS to the extraction increased the amount of oil released from the insoluble fraction of the soybean; however, this was not true for phospholipid. A protein-rich skim fraction resulting from AEP was isolated and analyzed by differential scanning calorimetry to determine the effect of the emulsifiers on the native protein conformation. The addition of phospholipid did not affect the native conformation of either glycinin or β-conglycinin; however, SDS-assisted extraction resulted in the denaturation of β-conglycinin. Efforts to break the skim emulsion and induce aggregation of oil droplets to form a cream via electrostatic shielding with electrolyte, exclusion aggregation with a polyelectrolyte, a secondary extraction performed on the skim fraction were unsuccessful.

See more of this Session: Student Poster Session: Separations
See more of this Group/Topical: Education

Separations From Pharmaceuticals to Fuels Using Grafted Calixarenes as Designed Adsorbents

Theodore D. Swift, Department of Chemical and Biological Engineering, Northwestern University, Evanston, IL

Justin M. Notestein, Department of Chemical and Biological Engineering, Northwestern University, Evanston, IL

Calixarenes are cavity-containing small molecules often suggested to be useful for separations. When functionalized with solubilizing groups or when attached to solid surfaces, these compounds can be applied in selective extractions or adsorption. Here, results are presented along two directions. In the first, semi-empirical modeling studies are used to advance the understanding of extraction of positively changed, amphipathic amide amine drugs using strongly negatively charged water-soluble calixarenes. In the second area, calixarenes are examined for the uptake of dipolar molecules such alcohols from aqueous solution. Insight from semi-emperical modeling is compared to materials synthesized by covalent attachment of calixarenes to solid supports. Calixarenes of different ring substituents, bridging groups, and attachment points are examined. The attachment of the calixarene is described and the resulting hybrid organic-inorganic materials are characterized primarily by TGA and UV-visible. Uptake onto solid adsorbents is measured by depletion of bulk solution followed by gas chromatography or spectrophotometric titration.

See more of this Session: Student Poster Session: Separations
See more of this Group/Topical: Education

Utilization of Peptoids in Membrane-Based Chiral Separations

Charles E. Starbird, Chemical Engineering, University of Arkansas, Fayetteville, AR

Jamie A. Hestekin, Chemical Engineering, University of Arkansas, Fayetteville, AR

Shannon L. Servoss, Chemical Engineering, University of Arkansas, Fayetteville, AR

Chirality is a common property of organic chemicals that has become an increasingly important aspect of chemical production, particularly in the production of pharmaceuticals, agrochemicals, and fragrances. Chiral compounds come in two versions, called enantiomers, which share most of the same physical properties, have identical NMR and IR spectrums, and react similarly in non-chiral systems. Currently techniques for separation of enantiomers include: crystallization, kinetic resolution, chromatographic separation, and membrane-based separations. The low material and operating costs make membrane-based separations a popular choice in industry. One disadvantage of membranes is that they become less enantioselective with multiple uses. Our research intends to increase selectivity by including peptoids in the membrane to add structural stability. Peptoids can be synthesized easily using an automated peptide synthesizer and can be created to arbitrary length with a large variety of side-chains. Properly substituted, they can form robust and stable helical structures which can be used to provide greater rigidity to the membrane. We are exploring the inclusion of peptoids within the polymer matrix in conjunction with the molecular imprinting technique often used in membrane-based chiral separations. This should be a relatively simple but effective way of increasing the long-term selectivity of membranes.

See more of this Session: Student Poster Session: Separations
See more of this Group/Topical: Education

Revolutionary Membranes to Slash Energy Consumption in Kraft Pulping Processes

Gina M. Bunster, Chemical & Biomolecular Engineering, Georgia Institute of Technology, Atlanta, GA

Ryan P. Lively, Chemical & Biomolecular Engineering, Georgia Institute of Technology, Atlanta, GA

William J. Koros, Chemical & Biomolecular Engineering, Georgia Institute of Technology, Atlanta, GA

A membrane separation process is considered to replace two thermally-driven units, which consume 50% of evaporation energy, in the Kraft pulping process. In particular, pressure-driven nanofiltration followed by reverse osmosis has the potential to reduce this energy consumption by 8-10 fold. Nanofiltration can remove larger organic and inorganic species while additional reverse osmosis processing can produce a purified water permeate and an inorganic-rich retentate for recycle. The first stage of this project is focused on developing an economically viable nanofiltration membrane that exhibits chemical resistance to harsh weak black liquor streams. A nanofiltration membrane comprising an ultra-thin Nafion® layer on a robust, relatively inexpensive microporous fiber support is being pursued. Nafion® is a unique material that can withstand conditions of high alkalinity and elevated temperatures while maintaining high permeability to water. If Nafion® shows adequate rejection of organics and larger ions, it would provide the basis for this nanofiltration unit resulting ultimately in significant energy savings. Currently, the forest products industry consumes 2.74 quadrillion BTU annually, which is equal to 12% of the total U.S. manufacturing energy demand. Implementation of hollow fiber modules containing defect-free Nafion® sheath layers of 0.3-1 micron thicknesses has the potential to save the U.S. pulp and paper industry about 100 trillion BTU annually. UV-Vis spectroscopy is used to monitor the breakthrough and rejection of various ionic species as a function of time in order to probe the nanofiltration capabilities of Nafion®. Preliminary experiments using commercially available Nafion® (EW = 1100) membranes have shown reasonable water flux and rejection properties.

See more of this Session: Student Poster Session: Separations
See more of this Group/Topical: Education

Novel Pervaporation Membranes for Separation of Solvent/Water Mixtures

Jason Wu, Chemical and Materials Engineering, University of Kentucky, Lexington, KY

V. Smuleac, Chemical and Materials Engineering, University of Kentucky, Lexington, KY

D. Bhattacharyya, Chemical and Materials Engineering, University of Kentucky, Lexington, KY

Traditionally, the pervaporation of miscible solvent/water where the solvent is the major component is performed using hydrophilic membranes (such as PVA or zeolites). In the present paper, a new type of pervaporation membrane (primarily amorphous perfluoropolymers, hydrophobic) was studied for separation of solvent/water mixtures. This membrane has high free volume and is inert for all solvents, and has a remarkable mechanical, chemical and thermal stability. The water is transported by solution diffusion model and the separation in solvent is primarily based on molecular sieving (size exclusion) principles. The membrane shows a high stability for operation over a broad range of feed concentration without swelling; the operating temperature does not have a significant effect on membrane selectivity. Water selectivities as high as 349 and 500 for ethanol/water and IPA/water mixtures (98/2 %wt solvent/water) and fluxes of 0.15 and 0.05 kg/m^{2}h, respectively, were obtained at 22°C. The pervaporation of more complex (ternary) mixtures of ethanol-ethyl acetate-water showed that this system could be successfully applied for solute separation based on complete size exclusion of ethyl acetate. This research has been funded by a NIH-SBIR program subcontract to the University of Kentucky.

See more of this Session: Student Poster Session: Separations
See more of this Group/Topical: Education

Phase Equilibria and Selectivity of Sulfonated Poly (styrene-isobutylene-styrene) for Specialty Separations

Alexandra Gonzalez, Chemical Engineering, University of Puerto Rico, Mayaguez Campus, Mayaguez, PR

Cristina Diaz, Chemical Engineering, University of Puerto Rico, Mayaguez Campus, Mayaguez, PR

David Suleiman, Chemical Engineering, University of Puerto Rico, Mayaguez Campus, Mayaguez, PR

Poly (styrene-isobutylene-styrene) (SIBS) of 22.5% polystyrene weight fraction and 65K number-average molecular weight (Mn) was sulfonated to very high ion exchange capacity (IEC). The sulfonated polymers were cross-linked with several +2 cations (Mg^{+2}, Ca^{+2}, Ba^{+2}). The phase equilibria and selectivity of the resulting membranes was investigated for various specialty separation applications (e.g., water purification). Water swelling experiments reflect significant differences (30%-300%) for the different membranes. Acetone/water selectivity results are also presented. FT-IR results show the interaction of the metal with the asymmetric oxygen of the sulfonic group, which explains the different phase-equilibria and selectivity results.

See more of this Session: Student Poster Session: Separations
See more of this Group/Topical: Education

Molecular Imprinted Silica for Separation of Sugars From Cellulosic Biomass

Elizabeth A. Conklin, Chemical Engineering, University of Louisville, Louisville, KY

Stephen E. Rankin, Chemical and Materials Engineering, University of Kentucky, Lexington, KY

Barbara L. Knutson, Department of Chemical and Materials Engineering, University of Kentucky, Lexington, KY

Improvement of sugar yields in the hydrolysis step of cellulosic biomass conversion to biofuels partly depends on the separation of the inhibition products (e.g glucose, xylose and cellobiose). The separation of these saccharides will allow a more efficient conversion to ethanol as well as new revenue streams. Molecular imprinted silica is synthesized by nanocasting in a sol-gel procedure using glucose as template. After surfactant extraction the silica materials may possess sites that can recognize glucose molecules. FTIR analysis on washed and unwashed samples of silica confirmed the removal of the sugar template. The imprinted silica materials possess large surface area which correlates with the amount of sugar template used. This presentation will compare adsorption of glucose in glucose-imprinted and non-imprinted silica materials, and determine cellulase enzyme adsorption by the glucose-imprinted silica material and its effects on enzyme activity.

See more of this Session: Student Poster Session: Separations
See more of this Group/Topical: Education

Palladium Catalyst Removal From Reaction Mixtures by Fixed Bed Adsorption

Eileen Batten, Chemical Engineering, Rowan University, Glassboro, NJ

Daniel O'Connell, Chemical Engineering, Rowan University, Glassboro, NJ

Pamela L. Kubinski, Rowan University, Glassboro, NJ

Megerle L. Scherholz, Rowan University, Glassboro, NJ

Caitlin A. Boyd, Chemical Engineering, Rowan University, Glassboro, NJ

The removal of palladium catalyst from a Heck coupling reaction mixture was investigated using an external fixed bed adsorption column containing Quadrapure TU resin. The objective was to develop mathematical models that accurately predict Pd breakthrough for the Heck coupling-Quadrapure TU system. Local equilibrium, mass transfer limited, and reaction limited models were investigated. The local equilibrium model predicts breakthrough assuming the resin is fully utilized. However, experimental breakthrough occurs before the bed is completely saturated. As a result, more complex models were developed that account for mass transfer resistance and slow surface reactions that reduce adsorption efficiency. The mass transfer models require a mass transfer coefficient; therefore, a mass transfer coefficient was estimated using the linear mass transfer model to replicate experimental data from Spring 2008 Column Run 7 (CR7-S08). The resulting mass transfer coefficient was 7.82 x 10-3 cm/hr. Mass transfer resistance did not accurately represent the Heck coupling-Quadrapure TU system due to inconsistencies between predicted and experimental breakthrough times. Therefore, two additional models were investigated that assume the surface reaction between the Pd containing species and the adsorbent is the rate limiting step. The first model represents the decay of the resin adsorbent as it becomes saturated throughout the packed bed, in which the reaction rate is time dependent. This model is characterized by two parameters, the Damköhler Number and Activity Number, which may be calculated using two different methods. In the first method, Activity Number is dependent on the Damköhler Number. The parameters are calculated simultaneously in the second method. The resin decay model was fit to CR1-F08 and CR2-F08. For both column runs, Method 2 predicted Damköhler Numbers greater than Method 1. Activity Numbers determined using Method 1 were greater than the values obtained using Method 2 for both runs. Reasons for inconsistencies between methods have not been determined. The second reaction limited model is a reaction-deactivation model, in which the rate is dependent on the concentration of Pd containing species. However, a solution to this model has not yet been developed. Currently, surface reaction models assume a first order reaction between Pd and Quadrapure TU. However, a reaction rate order of 2.5 was estimated using both Integral and Differential methods. While the surface reaction may be the limiting step in the Heck coupling-Quadrapure TU system, a solution to the model is needed before the accuracy of the reaction-deactivation model can be determined.

See more of this Session: Student Poster Session: Separations
See more of this Group/Topical: Education

Gas Separation Using Oscillatory Flow in Wavy-Walled Tubes

Theodore J. Warner, Chemical Engineering, University of Idaho, Moscow, ID

Aaron M. Thomas, Chemical Engineering, University of Idaho

This research examines the use of oscillatory flow in wavy-walled tubes in order to separate carbon dioxide and hydrogen in a non-dilute mixture. This application is commonly used on an industrial scale in coal fired power plants. Studies analyzed the separation behavior using a bank of tubes and how they compared to theoretical work as well as previous research. Using nitrogen as a carrier gas, flow oscillation and amplitude of oscillation were adjusted to change separation behavior. Results followed previous convention and suggested the possibility of oscillatory flow separation in future industrial operations.

See more of this Session: Student Poster Session: Separations
See more of this Group/Topical: Education

Electrochemical Detection of 2,4,6-Trinitro-Toluene

Jeffrey Chin, Chemistry and Life Science, United States Military Academy, West Point, NY

Robert G. Bozic, Chemistry and Life Science, United States Military Academy, West Point, NY

Scott A. Banta, Columbia University, New York, NY

Alan C. West, Chemical Engineering, Columbia University, New York, NY

Long term monitoring of waste disposal sites for Munitions Constituents, MCs, using the Environmental Protection Agency Method 8330A, high performance liquid chromatography, is a costly analytical technique for which there is not a current comparable alternative. This has inspired research and development of fast, low cost techniques to detect parts per billion concentrations of MCs such as such as 2,4,6-TNT as well as other MCs in ground water. In collaboration with the U.S. Army Engineer Research and Development Center in Vicksburg, MS, Columbia University Department of Chemical Engineering and the Department of Chemistry and Life Science at the United States Military Academy have on-going research efforts combining electrochemical engineering, microfluidics, and protein engineering with a goal of developing engineering fundamentals to enable in-situ detection of MCs. Recent efforts included examination of adsorptive effects of MCs to a gold micro electrode surface endowed with an alkanethiol self-assembled monolayer. The current focus is on development of methods that will selectively detect concentrations of an MC from roughly nanomolar quantities to reliably detectable levels via electrochemical methodologies using a peptide separation unit in line with an electrochemical sensor. An optimization study on the electrode size and instrumentation for electrochemical detection of TNT was conducted during this Advanced Individual Development Program. It was determined that the same surface treatment used for a rotating disc electrode (diameter = 5 mm) experiment can be replicated on an electrode 4400 times smaller (d = 0.075 mm).

See more of this Session: Student Poster Session: Separations
See more of this Group/Topical: Education

Biodiesel Dilute n-Alcohol Mixture Characterization by Physicochemical Gas Chromatography

N. S. Bobbitt, Chemical Engineering, University of Arkansas, Fayetteville, AR

J. W. King, Chemical Engineering, University of Arkansas, Fayetteville, AR

The use of gas chromatography for the measurement of the physicochemical properties of solutions and surfaces is a versatile technique and is frequently referred to as "inverse gas chromatography." In this study, we have used a semi-volatile solvent, a commercial biodiesel sample, as the coated stationary phase and characterized its interaction with injected solutes consisting of C1 - C4 alcohols at finite vapor concentrations corresponding to p/po of 0.1 - 0.5 over the temperature range of 55 - 85 °C. Modifications were made to a Gow MAC Model 550 gas chromatograph (GC) to allow the precise measurement of partition or absorption isotherms of alcoholic solutes and their reduction from skewed chromatographic profiles, with subsequent recording on a laptop computer. Solutes were injected onto columns consisting of 16, 20, and 25 weight percent biodiesel on a high surface area support, while column temperature variation was tracked by data logging the response of various J-type thermocouples placed in the chromatographic oven. Measurements were performed in triplicate for a representative portion of solutes, temperatures, and column loadings and single injections were used for the remainder of the measurements. The resultant chromatographic profiles were relatively symmetrical in the case of methanol and ethanol at lower temperatures (55-65 °C) and changed to "tailing" profiles at temperatures (75-85 °C) corresponding to Langmuir-type sorption isotherms. By contrast, n-propanol and n-butanol peak profiles exhibited "fronting" peaks over the entire temperature range, commensurate with anti- Langmuir sorption behavior. In general as temperature increased, the sorption isotherm shifted to lower uptake values at given pressure and retention times shifted toward shorter times, indicating a reduced level of interaction between the alcoholic solute and the biodiesel substrate. Isotherms consisting of vapor mole or weight fraction uptake in biodiesel versus relative partial pressure (a measure of solute fugacity) were developed by deconvoluting that portion of the peak profile attributed to absorption from the contribution to a Gaussian symmetrical profile contribution due to diffusion of the solute in the chromatographic column. Alcohol sorption isotherm data was measured during the passage of n-alcohol peaks through the calibrated thermal conductivity detector of the GC, and the isotherms were fitted to polynomial equations of appropriate order. The chromatographic peak shapes and resultant isotherm arise from both positive and negative deviations from Raoult's Law depending on the alcohol-biodiesel pair, the temperature, and the vapor state concentration of the alcoholic solute. These data trends will be discussed along with heat of solution data and Henry's Law constants at various p/po values to provide a picture of the intermolecular interactions responsible for the solution behavior of these biodiesel/n-alcohol systems. Such isotherms and derivative thermodynamic data can be used in the design of V-L separation processes for alcohol devolatilization from biodiesel as well as the use of biodiesel (vegetable oil methyl esters) as a "green" sustainable solvent system.

See more of this Session: Student Poster Session: Separations
See more of this Group/Topical: Education

A Peptide Based Separator for 2,4,6- Trinitro-Toluene

Seth D. Johnson, Chemistry and Life Science, United States Military Academy, West Point, NY

Robert G. Bozic, Chemistry and Life Science, United States Military Academy, West Point, NY

Scott A. Banta, Columbia University, New York, NY

Alan C. West, Chemical Engineering, Columbia University, New York, NY

Long term monitoring of waste disposal sites for Munitions Constituents, MCs, using the Environmental Protection Agency Method 8330A, high performance liquid chromatography, is a costly analytical technique for which there is not a current comparable alternative. This has inspired research and development of fast, low cost techniques to detect parts per billion concentrations of MCs such as such as 2,4,6-TNT, 2,4-DNT, 2,6-DNT, 1,3,5-TNB, 1,3-DNB and Tetryl as well as other MCs in ground water. In collaboration with the U.S. Army Engineer Research and Development Center in Vicksburg, MS, Columbia University Department of Chemical Engineering and the Department of Chemistry and Life Science at the United States Military Academy have on-going research efforts combining electrochemical engineering, microfluidics, and protein engineering with a goal of developing engineering fundamentals to enable in-situ detection of MCs. Recent efforts included examination of a separation stage based on a peptide. The current focus is on development of methods that will selectively detect concentrations of an MC from roughly nanomolar quantities to reliably detectable levels via electrochemical methodologies using a peptide separation unit in line with a electrochemical sensor. During this AIAD, progress was made toward the ultimate goal of a Peptide Separator Electrochemical Sensor (PSES) designed to detect TNT. A technique was developed for using a peptide to separate and concentrate TNT from salt water solution. Using a peptide, a resin, and a separation column, the effect of the loading concentration of TNT in salt water solution was studied. Optical measurements of the concentration of the loading solution, supernatant after reaction in batch system, and the eluent were taken using a spectrophotometer and compared as part of an initial kinetics study.

See more of this Session: Student Poster Session: Separations
See more of this Group/Topical: Education

ENERGY BALANCES - Tools for Structured Problem Solving and Visual Learners

Inci Ayranci, Chemical and Materials Engineering, University of Alberta, Edmonton, AB, Canada

Suzanne Kresta, Chemical and Materials Engineering, University of Alberta, Edmonton, AB, Canada

The material and energy balances course is a crucial course in the chemical engineering curriculum. In the energy balances part of the course, definition of the reference state and calculation of specific enthalpies relative to the reference state are confusing concepts for students. Our aim in this talk is to provide some tools to clarify the role played by the reference state and the importance of applying a reference state consistently. The first tool uses maps to develop the reference state concept, targeting visual learners. The second step is to use enthalpy-temperature diagrams to illustrate the path independence of changes in enthalpy, and the impact of choosing different reference states. The specific enthalpy data is organized in material balance tables for calculations, reinforcing the importance of structured problem solving practices. Finally, at several points in the unit, summary checklists and decision trees are used to reinforce the concepts.

See more of this Session: Free Forum On Chemical Engineering Education I
See more of this Group/Topical: Education

Non-Traditional Methods to Enhance Creativity for Engineering Innovation

Peter Ludovice, School of Chemical & Biomolecular Engineering, Georgia Institute of Technology, Atlanta, GA

Lew Lefton, School of Mathematics, Georgia Institute of Technology, Atlanta, GA

Richard Catrambone, School of Psychology, Georgia Institute of Technology, Atlanta, GA

Enhanced creativity among U.S. engineers and scientists is required in the face of strategic needs for innovation in numerous areas including: energy, the environment and health. NSF's third generation Engineering Research Centers explicitly require an educational component to enhance creativity to improve innovation. We have applied an approach to improving creativity that has traditionally not been used in technical innovation. The approach uses concepts that have been successfully applied to innovation in non-technical fields, such as business and marketing, by overcoming the two basic challenges that have prevented success in science and engineering. The first challenge is overcoming the cultural aversion in scientists and engineers to exploiting a natural origin of creative energy: humor. The theoretical equivalence of humor and innovation is well documented, and improvisational humor is nothing more than a stochastic simulation algorithm that traverses idea space. However, scientists and engineers have difficulty excepting this concept. Even successful applications such as the "Humor Room" at the Kodak Corporation find resistance among technical managers because they believe this approach is inappropriate. We have modified improvisational humor exercises to traverse technical idea space and to overcome the limitations that have prevented success of this approach in the technical arena. The results from preliminary application of this protocol to engineering design will be discussed. These results indicate the potential for this approach to positively impact technical innovation.

See more of this Session: Free Forum On Chemical Engineering Education I
See more of this Group/Topical: Education

The Teaching of the Freshman Chemical Engineering Course

David L. Silverstein, University of Kentucky, Paducah, KY

Donald P. Visco, Chemical Engineering, Tennessee Technological University, Cookeville, TN

Margot Vigeant, Chemical Engineering, Bucknell University, Lewisburg, PA

Donald R. Woods, McMaster University, Hamilton, ON, Canada

The authors will present the results of the first survey in the resumption of the Chemical Engineering Education Project Committee series of annual surveys on how chemical engineering courses are taught. The survey is now conducted by the AIChE Education Division. This year's survey focuses on the freshman engineering courses and will detail how freshmen are introduced to engineering, what topics are taught, in what environment they are taught, and how they are assessed. ABET criteria, teaching methods, and novel approaches are focus elements of this survey. While the survey focuses on courses specific to chemical engineers, innovative teaching methods in general introduction to engineering courses are included as well.

See more of this Session: Free Forum On Chemical Engineering Education I
See more of this Group/Topical: Education

Teaching Chemical Engineers about Teaching

Daniel Heath, Chemical and Biomolecular Engineering, The Ohio State University, Columbus, OH

James Rathman, The Ohio State University, Columbus, OH

Stephanie Rohdieck, Faculty & TA Development, The Ohio State University, Columbus, OH

Knowledge and techniques which enhance student learning have been developed through rigorous educational research. However, chemical engineers are not often exposed to this work. Furthermore, the task of plunging into this body of knowledge in order to become a better teacher is both time consuming and daunting. However, the Chemical and Biomolecular Engineering Department at the Ohio State University believes that incorporating this information into the graduate student curriculum and providing opportunities to take this knowledge into the classroom is beneficial in preparing future faculty members. Therefore, a graduate student, the department's curriculum committee, and Faculty & TA Development worked together in order to develop the Chemical Engineering Teaching Mentorship Experience.

The Chemical Engineering Teaching Mentorship Experience is a two-quarter graduate-level elective course which organizes and highlights educational knowledge and presents the material in a way which is relevant to the chemical engineering classroom. The first quarter of the Teaching Mentorship Experience is designed to introduce participating graduate students to ideas and techniques to enhance teaching. Some of the topics covered in the course are listed below.

OSU policies and procedures. Students are exposed to topics such as academic misconduct, FERPA, sexual harassment, and disabilities awareness.

Teaching portfolios. The advantages of maintaining a teaching portfolio is introduced to students, and the participants are required to draft a philosophy of teaching statement.

Student learning. Students are exposed to basic ideas in cognitive theory, different levels of learning (foundational knowledge, structural knowledge, knowledge transfer), and learning styles.

Designing lessons. The importance of course planning, writing specific learning objectives, student-centered versus a teacher-centered classroom, and active learning activities are discussed.

Assessing student learning. Ideas on how to assess student learning are presented. Assessment is divided into formal and informal assessment methods. Formal assessments include graded assignment where the use of primary trait analyses to maximize the impact of graded assignments is highlighted. Informal assessment (or classroom assessment techniques) are also introduces as a method of assessing student learning on a more ongoing basis.

Assessing teaching effectiveness. One of the goals of this program is to make future educators reflective about their teaching and to understand that teaching is an iterative process which can continuously be modified and improved. This last topic introduces students to the idea of collecting feedback from their students on their own teaching and making necessary changes to their courses and philosophy of teaching.

Also during the first quarter of the experience, participants are paired with a faculty mentor. The participants use their new educational knowledge and also the experiences of their faculty mentor to plan and develop 2 - 3 lessons along with the associated homeworks, test questions, active learning activities, and assessment devices. During the second quarter, the students take their lessons into their mentor's classroom, assess their students learning, and their own teaching effectiveness. To date, 4 graduate students have participated in the mentorship program.

See more of this Session: Free Forum On Chemical Engineering Education I
See more of this Group/Topical: Education

Production of Biodiesel From Algae: An Experiment-Based Course for Introducing Students to Research Methods

Kevin N. West, Department of Chemical and Biomolecular Engineering, University of South Alabama, Mobile, AL

Silas J. Leavesley, Chemical and Biomolecular Engineering, University of South Alabama, Mobile, AL

Graduate coursework in chemical engineering is typically focused on high level, theoretical treatment of the core disciplines or special topics. While both of these areas are necessary, these courses are usually theoretical and do not expose students to hands-on experimental work that many of them will be asked to perform for the research component of their degree. In addition, related topics such as a literature search for experimental methods and analysis of experimental data are often not covered in a formal manner. Even students that work on computational research projects would strongly benefit from exposure to experimental research methods. We have implemented an experiment-based special topics course that is designed to provide masters students with a strong foundation for experimental work and aid them in performing quality research.

Like many masters programs throughout the country, the graduate program in the Chemical and Biomolecular Engineering Department at the University of South Alabama has three core components: completion of required course work (5 graduate chemical engineering core courses), demonstration of a mastery of the fundamental chemical engineering disciplines (a qualifying exam), and the successful defense of a thesis based on original research. Though this process is theoretically rigorous, the only place for developing competencies in experimental work is in the research component, and if a student's work is computational, this component is completely absent.

To provide this experience, we created an experimental-based course for first-year graduate students focusing on production of biodiesel from algae, an opened ended challenge that allowed them to focus on specific experimental research aspects. The class was divided into three groups corresponding to the three essential operations required for the process: (1) growth of the algae (bioreactor design and testing), (2) harvesting the algae and extraction of the triacylglycerols, and (3) transesterification of the triacylglycerols with methanol to produce biodiesel (fatty acid methyl esters) and glycerol. This created the opportunity for student teams to be exposed to three separate facets of experimental research: biological reactors, separations technologies, and reactor design and kinetics.

Students in this course were initially charged with performing a literature search on their particular topic, including fundamental background information, safety concerns and basic experimental details. Student teams summarized their work in bi-weekly oral presentations to the class. After students had demonstrated sufficient understanding of the process and potential solutions, they were required to develop a detailed research plan outlining their proposed experiments, timeline, and the information they expected to collect. After their research plan was approved, they were allowed to begin experiments.

The algae growth team designed a phyto-bio-reactor and used the reactor to measure algae growth kinetics and to generate a sufficient mass of dried algae needed to test lipid extraction processes. Independent parameters, such a light intensity, CO_2/air feed ratio, and temperature were varied. The extraction team measured the lipid content of algae samples using Soxhlet extraction to determine if chemical extraction alone was sufficient to remove the triacylglycerols from the biological matrix. The reaction group examined the transesterification reaction kinetics with specific attention paid to the effects of temperature, catalyst type, and alcohol concentration. Additionally, each group was also given the task of performing either modeling or scale-up calculations based on their experimental data.

The final product from the class was the production of technical reports detailing the experimental process - from initial planning to final experimental results - along with final oral presentations by each team. We present the results from this class from an instructional perspective on how effectively the students were able to take on a foreign subject, internalize the background information, develop and execute a research plan, and conduct experiments that can lead to publication of quality experimental data. Student work from the course will be presented as well as modifications to the instructional process that we plan to implement in future courses and suggestions for other institutions interested in offering such a course.

See more of this Session: Free Forum On Chemical Engineering Education I
See more of this Group/Topical: Education

Incorporating Soft Skills Into Graduate Seminar Series

Sundararajan. V. Madihally, School of Chemical Engineering, Oklahoma State University, Stillwater, OK

Graduate programs in various institutions are developed to advance the technical competency of the graduate students. Many schools have the seminar series as a mandatory course for all the graduate students. This seminar is primary used as a method of advancing their technical competencies. They are introduced to cutting edge research topics by inviting faculty members from their own institution, other institutions or industrial personnel. Chemical Engineering discipline has expanded to include diverse and broad range of research topics. and many research topics could address only a portion of the students. Further, conducting seminars for the entire duration of the semester may be cost prohibitive, leading to reduction in number of seminars. Interestingly, there are very few seminars that introduce graduate students to non-technical content that could be paramount to their future success. For example, basics of grant writing, the importance of maintaining a laboratory notebook, writing a technical report, ethical behavior, chemical safety demonstrations or academic integrity are not addressed.

In this regard, we have incorporated few topics into the seminar series for last three semesters at Oklahoma State University (OSU). Most of the speakers are from different departments within the University. Each seminar also included a critique (homework) submission to get the feedback on their like and dislike about the presentations. These responses have been very positive and encouraging. Summary of the survey (administered after two semesters) results indicate that graduate students at OSU prefer seven to eight technical seminars and four to five seminars addressing soft skills. Student survey indicated the possibility of many other topics such as time management, technical writing, financial management and interpersonal management. In summary, the seminar series can be used as a possibility of incorporating some of the soft skills into the graduate program.

See more of this Session: Free Forum On Chemical Engineering Education I
See more of this Group/Topical: Education

Discovery, Development and Design; a 3D Approach to Teaching Transport Phenomena

Katherine S. Ziemer, Chemical Engineering, Northeastern University, Boston, MA

Al Sacco Jr., Department of Chemical Engineering, Northeastern University, Boston, MA

Elizabeth J. Podlaha, Chemical Engineering, Northeastern University, Boston, MA

Rebecca L. Carrier, Department of Chemical Engineering, Northeastern University, Boston, MA

Sumru Erkut, Wellesley Centers for Women,, Wellesley College, Wellesley, MA

Jennifer Grossman, Wellesley Centers for Women, Wellesley College, Wellesley, MA

Corinne L. McKamey, Wellesley Centers for Women, Wellesley College, Wellesley, MA

Frank Marealle, Chemical Engineering, Northeastern University, Boston, MA

Numerous studies in the educational literature indicate that pedagogical methods featuring student interaction are more effective than strict lecture-style teaching [1]. Clearly, students more actively engaged in the learning process through group work, problem solving, and open-ended questions, are better prepared for how they will be expected to perform in their professional engineering careers. Prince and Felder [2] noted in 2006 that most traditional engineering education is deductive, where students are taught fundamentals and theories associated with a specific topic, and then progress to applying those learned skills to problem solving. An alternative to this traditional approach is inductive instruction, a term that encompasses a wide variety of possibilities, including inquiry learning, problem-based learning, project-based learning, case-based teaching, discovery learning, and just-in-time-teaching. Northeastern University is combining both deductive and inductive educational approaches to develop an experiential, open-ended, problem-based curriculum which requires hands-on experimentation to understand and develop key chemical engineering principles and concepts. The intent is to integrate all laboratory experiments as part of an associated class with each core chemical engineering topic, rather than having laboratory experiments reside in a single course in the junior year. Towards this end, the Transport/Separations curriculum sequence has been redesigned to utilize laboratory and demonstration modules with an experiential strategy that introduces different modes of learning through discovery, development and design; (3D approach). Externally evaluated assessment tools compare the attitudes and knowledge of students in the standard curriculum (the control group) and students in the modified curriculum (the experimental group). Through the 3D experiential approach, students explore the natural engineering world and discover the underlying concepts (discovery). They are engaged through improving and innovating an experiment (development), and then use their newly acquired knowledge to design an experiment or unit operation (design). Newly created discovery, development and design experiments for fluid flow will be shared along with preliminary assessment results of student learning and attitudes.

1. Felder, R. M. & Brent, R. "Learning by Doing". Chemical Engineering Education 37, 282-283 (2003); Felder, R. M. & Brent, R. "Understanding Student Differences". Journal of Engineering Education 94, 57-72 (2005); Felder, R. M. & Silverman, L. K. "Learning and Teaching Styles in Engineering Education." Engineering Education 78, 674-681 (1988)

2. Prince, M. J. & Felder, R. M. "Inductive Teaching and Learning Methods: Definitions, Comparisons, and Research Bases". Journal of Engineering Education 95, 123-138 (2006).

See more of this Session: Free Forum On Chemical Engineering Education I
See more of this Group/Topical: Education

The Effect of Peer Instruction On Students' Construction of Conceptual Understanding in Thermodynamics

Milo D. Koretsky, Chemical, Biological, and Environmental Engineering, Oregon State University, Corvallis, OR

Bill J. Brooks, Chemical, Biological, and Environmental Engineering, Oregon State University, Corvallis, OR

This study uses the Web-based Interactive Science and Engineering (WISE) Learning Tool as a platform to investigate the effectiveness of Peer Instruction on the explicit understanding of undergraduate students in chemical engineering thermodynamics. WISE is designed to utilize the college's Wireless Laptop Initiative so that every student in a class is simultaneously engaged. This tool facilitates active learning in the classroom, metacognition, and formative assessment of student learning. Peer Instruction is a technique where an instructor presents a multiple-choice conceptual question to the class. Students answer individually at first and next are shown a "poll" of the class responses. They then form groups and discuss the problem with peers, and finally answer again individually. The cohort in this study was 69 students in the second term of a junior level chemical engineering thermodynamics class. No rationalizations were provided for answer choices; instead, each time students answered, they were asked to reflect on their choice and provide a short written explanation. They were also asked to indicate their confidence in the answer chosen. The student explanations, both before and after group consultation, were coded based on completeness and correctness. Four questions were analyzed, two in which a majority of the class initially reported the correct multiple choice answer and two in which the minority had the correct answer. On exercises in which the popular answer is also the correct answer, analysis of code gains compared to code losses indicate a statistically significant increase in apparent understanding. However, on those exercises in which the majority of the class chose incorrectly, there was no statistically significant improvement of understanding for the class as whole. Furthermore, in these cases there were a significant number of students that changed their answer to the popular, incorrect answer. These results lend support to the theory of social constructivism and the consensuality principle, and can help guide the use of Peer Instruction in the classroom.

See more of this Session: Free Forum On Chemical Engineering Education I
See more of this Group/Topical: Education

Chemical Engineering Curriculum Development in Russia and Kazakhstan

Stefaan J.R. Simons, Department of Chemical Engineering, University College London, London, United Kingdom

Sergey Mikhalovsky, School of Pharmacy and Biomolecular Sciences, University of Brighton, Brightin, United Kingdom

Gennady Kuvshinov, Chemical Engineering, Novosibirsk State Technical University, Novosibirsk, Russia

Gulzhakhan Yeligbayeva, Department of Oil, Gas and Polymer Refining Chemical Technology, Kazakh National Technical University, Almaty, Kazakhstan

The collapse of the former Soviet Union and the transition of the new independent states to market economies has highlighted the need for a radical overhaul of the higher education system, particularly in engineering sciences. Due to national requirements, in regards to human development and economic growth, chemical engineering has been highlighted as a key discipline that must be modernised to provide qualified home-grown talent to run the highly technical facilities required to bring about the goals of energy security, increased life expectancy and sustainability. Several universities and organisations across Europe and the CIS are now collaborating to develop modern chemical engineering degree programmes at key institutions in Kazakhstan and Russia. The ultimate goal is for these new programmes to gain international recognition for the quality of the education and for the discipline as a whole.

See more of this Session: Free Forum On Chemical Engineering Education I
See more of this Group/Topical: Education

Introduction and Background

You don't know what you don't know. Take the sciences, for example. Many high school students think they are boring, not the least bit interesting. What they don't know is how creative the sciences are; that the "how" of their iPods, cell phones and video games is even more fascinating than the tools themselves. To breach that gap between perception and reality, a team of Louisiana Tech researchers developed three complementary and linked programs based on three certainties: 1) students of all ages are fixated on technology that gets smaller on a seemingly daily basis; 2) 'hands-on' is the best way to capture and keep their interest; and 3) the most efficient way to reach the greatest number of students is through their teachers.

The three-fold result is 1) *Nanoscience Education and Research Outreach (NERO)*, a six-week National Science Foundation (NSF) RET summer program that teaches middle and high school science and math teachers how to involve their students in hands-on scientific research, 2) *Creating Connections*, a Graduate Teaching Fellows in K-12 Education (GK-12*)* program that provides science, technology, engineering and math (STEM) graduate students with opportunities to acquire key career enhancing skills including communicating STEM subjects to technical and non-technical audiences, leadership, team building, and teaching while enriching STEM learning and instruction in K-12 schools, and 3) a *Shell/Louisiana Regional Collaboratives (LRC) nanoSCIENCE* program whose goals are to prepare teachers with the content and pedagogical knowledge, infrastructure and support that will allow them to engage their students in challenging hands-on activities and investigations.

NERO and Creating Connections are three-year programs funded by the National Science Foundation (NSF); both are the result of a Louisiana EPSCoR *Planning Grant for Major Initiatives* awarded to the Dr. David Mills. Funding of the Shell/LRC project was awarded with a key objective of building synergies with existing K-12 outreach projects.

This presentation will address three critical areas in the development of K-12 engineering outreach grants. These are project development and the grant proposal, design of effective and engaging outreach programs, and the successful implementation of outreach programs. The PIs have over eight years in leading K-12 outreach programs. We will present our 'best practices' based on our own 'hard lessons learned' and invite others to share in a later open panel discussion on writing competitive grant proposals, program development, and implementation.

Louisiana Tech University's K-12 Outreach Programs

Louisiana Tech University's K-12 nanoSCIENCE Outreach program builds on partnerships established though Louisiana Tech's CATALyST (Center for Applied Teaching and Learning to Yield Scientific Thinking) Center and local parish schools. The targeted partner schools display similar patterns: poor performance in STEM fields, students from low socioeconomic backgrounds, and high percentages of minority students.

NERO, an NSF RET Site Program

NERO assists teachers in learning how to replace teaching 'at' students with involving them in hands-on scientific research projects so they can learn for themselves how fascinating it can be to ask and answer scientific questions by observing and doing experiments. NERO creates an environment in which the teachers themselves conduct bionano- and nanotechnology research using state-of-the-art equipment under the mentorship of Louisiana Tech University engineering and science research faculty and the assistance of GK-12 Teaching Fellows.

Teachers create learning modules and curricular activities for their students which they and their research mentors test and retest on four Saturday workshops during the academic year. NERO has seen teachers embrace research and gain confidence in their knowledge base, laboratory skills, and science communication abilities. They return to their classrooms with experiences to truly engage their students in meaningful scientific investigations and to prepare them for the highly technical workforce.

NSF GK-12 Creating Connections program

Creating Connections has developed Teaching-Learning Labs in three north Louisiana schools—Simsboro High School and Monroe's Carroll Jr. High School and Martin Luther King Jr. Middle School. Staffed by Louisiana Tech graduate engineering and science students, each lab focuses on creating instructional modules and hands-on student engagement in science and engineering. Louisiana Tech University STEM graduate students (GK12 Teaching Fellows) contribute to Louisiana's effort to advance scientific knowledge through partnerships with K-12 communities. As they gain the communication skills that enable them to explain science to people of all ages, GK12 Teaching Fellows also inspire transformation in the K-12 formal and information learning environments and stimulate interest in science and engineering among students and teachers.

Shell/LRC nanoSCIENCE Program

The Shell nanoSCIENCE project is part of the Louisiana Regional Collaboratives Program, a dissemination project of the Texas Regional Collaboratives Program that is funded by the Shell and other corporate partners. The nanoSCIENCE program conducts summer and academic year teacher professional development workshops for north Louisiana math and science teachers, emphasizing the theme of "molecular science and nanotechnology," a combination of biology and engineering at the nanoscale. Special emphasis is given to developing teacher professional development programs that emphasize energy research, nanotechnology, and their societal applications and implications. The project has as its long-term objective to develop strong, lasting partnerships and cultural and scientific awareness among K-12 teachers, STEM researchers and graduate students.

Discussion

By capitalizing on the synergies provided though directly linking three K-12 outreach programs; Louisiana Teach has had a significant impact on mathematics and science teaching in our parish partner schools. This linkage enabled us to train Master Teachers, who now assist in all programs to develop and deliver K-12 workshops, pilot new activities and ideas, decrease development of outreach activities through cross program use, effectively engage research faculty in academic engagement. We have successfully engaged middle and high school teachers in conducting research, translating this experience into classroom learning experiences, improved communication of science and the scientific research process to K-12 schools and the community, enhanced mathematics and science teaching in targeted 6-12 classrooms, and developed a collaborative learning environment. K-12 students have enjoyed experiences that have developed their knowledge and skills necessary to succeed in the current and future technology-driven workplace and received encouragement to extend their education and pursue STEM careers. STEM graduate students have increased their confidence in their ability to work with teachers and students, developed a toolkit of strategies for use in working with K-12 schools, and obtained insights into K-12 classrooms. Participating research faculty have gained an understanding of appropriate and effective outreach methods and received training in communicating their research.

Undergraduate Research: REU Sites, Summer Fellowships, and Academic Year Experiences

Steve R. Duke, Chemical Engineering, Auburn University, Auburn University, AL

This talk will describe practices that result in successful multifellow-multimentor REU sites, individual summer undergraduate fellowships, and academic year part-time research experiences.

See more of this Session: The Directorate for Education and Human Resources at NSF: Opportunities and Awards
See more of this Group/Topical: Education

ITEST: Green Energy Technologies in the City

Angela Calabrese Barton, Teacher Education, Michigan State University, East Lansing, MI

Scott Calabrese Baraton, Chemical Engineering and Materials Science, Michigan State University, East Lansing, MI

Edna Tan, Teacher Education, Michigan State University, East Lansing, MI

Shari Levine Rose, Teacher Education, Michigan State University, East Lansing, MI

Green Energy Technologies in the City (GET City) is a year round program for youth focused on the science and engineering of energy sustainability using advanced information technologies. The program provides youth opportunities to develop scientific research skills and conceptual understanding related to energy technologies, production and sustainability. During the 2007-2008 year, participants investigated the concept of urban heat islands and to what extent it applied to their home town of Lansing, MI. They created a digital video documentaries showcasing their findings, which were presented to an audience at Michigan State University. They also created public service announcements which aired on local network television that focused on key findings from their investigations into energy supply and demand issues (oil production and use and its environmental impact in terms of carbon emissions) comparing contemporary practices to those of previous generations and projecting what might happen two generations into the future. Recently, we have focused on a proposed new coal-based power plant for Lansing, the economic and environmental aspects of the decision-making process, and the alternatives available, including alternative energy sources and conservation. This talk will provide an overview of the GET City program, describing how information technology provides participating youth with the tools to communicate a coherent and engaging story-line to broader audiences, situating the youth as "community science experts".

See more of this Session: The Directorate for Education and Human Resources at NSF: Opportunities and Awards
See more of this Group/Topical: Education

Thoughts On Developing and Directing a Successful NSF IGERT Program

Kenneth F. Reardon, Chemical and Biological Engineering, Colorado State University, Fort Collins, CO

As stated by the National Science Foundation, IGERT is the NSF's "flagship interdisciplinary training program, educating U.S. Ph.D.

scientists and engineers by building on the foundations of their disciplinary knowledge with interdisciplinary training." What are the elements of a successful IGERT proposal and program? Some thoughts on this question will be provided in the context of the new IGERT program in Multidisciplinary Approaches to Sustainable Bioenergy at Colorado State University.

See more of this Session: The Directorate for Education and Human Resources at NSF: Opportunities and Awards
See more of this Group/Topical: Education

The Scholarship of STEM Education

Daina Briedis, Chemical Engineering and Materials Science, Michigan State University, East Lansing, MI

Education and teaching scholarship is growing in importance as engineering programs seek to prepare students for the workforce of the future. It is encouraging to know that faculty with dedication to teaching and pedagogy can find scholarly outlets and financial support for their interests. This presentation will describe the path that a small kernel group of faculty took to building a well-funded intercollegiate, interdisciplinary research program in STEM education.

See more of this Session: The Directorate for Education and Human Resources at NSF: Opportunities and Awards
See more of this Group/Topical: Education

A Panel Discussion

Scott Calabrese Barton, Michigan State University, East Lansing, MI

Kenneth F. Reardon, Chemical and Biological Engineering, Colorado State University, Fort Collins, CO

Steve R. Duke, Department of Chemical Engineering, Auburn University, Auburn, AL

Daina Briedis, Chemical Engineering and Materials Science, Michigan State University, East Lansing, MI

David K. MIlls, Institute for Micromanufacturing and Biological Sciences, Louisiana Tech University, Ruston, LA

This is a panel discussion amongst all of the presenters in this session.

See more of this Session: The Directorate for Education and Human Resources at NSF: Opportunities and Awards
See more of this Group/Topical: Education

SOME VERY EASY AND IMPORTANT LABORATORY PRACTICES FOR TRANSPORT PHENOMENA.

Benito Serrano Rosales[1*], Víctor Javier Cruz Delgado, Juan José Mejía Montoya, Jesús Moreira del Rio.

Programa de Ingeniería Química, Facultad de Ciencias Químicas, Universidad Autónoma de Zacatecas, Campus Siglo XXI, Carretera a Guadalajara, Km 6, Ejido La Escondida, Zacatecas Zac, México 98160, Tel 52-492-925-6690, Ext 6137, Fax 52-492-921-3990, beniserra@prodigy.net.mx

In this material it is intended to share the experience gained during more than ten years of proposing and teaching some laboratory practices of transport phenomena to undergraduate students. These practices were designed with economical constrains but with a high degree of motivation, due the lack of them.

All the practices of laboratory should require observation, measurement and calculations, but due their nature, some of them ask additionally to the student: 1) practical abilities (for example to close or to open a valve in the correct time, to perform the measurement of a property, etc), 2) to analyze the theory and to obtain mathematical models and to solve the resulting equations, in order to compare the analytical solution with the experimental data, 3) Some combinations of this two points. The practices described below require the execution of the points previously described.

1.- Four practices of heat transfer.

a) Qualitative evidence of the different capacity for several materials to conduct heat.

In this practice it is shown trough a very simple experiment how the aluminum, bronze and iron have different thermal conductivities. Several small candles (as those used in the birthday cakes) are placed along the length of a cylindrical bar for each material. A burner is placed at one end of the bar and it is observed that as the heat advances through the bar, the candles melt and fall at different times in each material, because of the difference in conductivity for each bar.

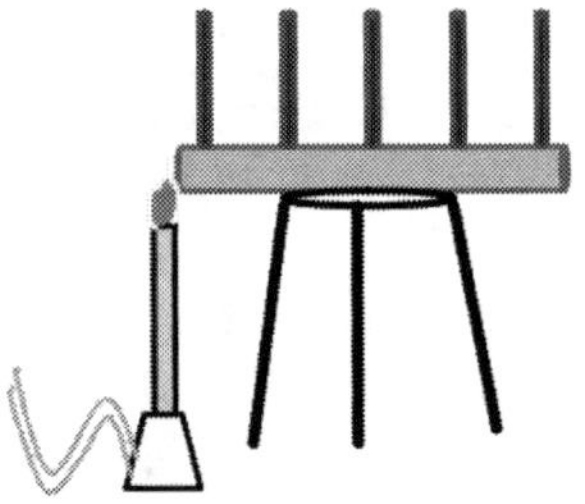

Figure 1. Scheme for heat conduction experiment.

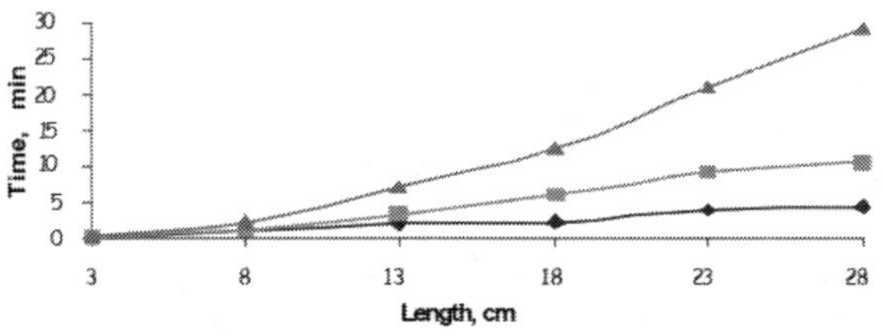

Figure 2.- Candles drop-times at different lengths for (♦) Al, (■) Br, and (▲) Fe.

Figure 2 shows the drop - times of the candles during the experiment. It is observed that for iron it takes longer for the candles to fall. Then the steel has the biggest resistance for heat transfer.

b) Determination of the transient axial temperature profiles.

A burner is placed at one end of a bronze bar equipped with holes; thermocouples are placed in different positions in the length of the bar in order to measure the axial temperature profiles at different times (Figure 3). In this way, curiosity is generated in the students about looking for a mathematical model that describes the experimental temperature profiles. This practice intends to be more quantitative.

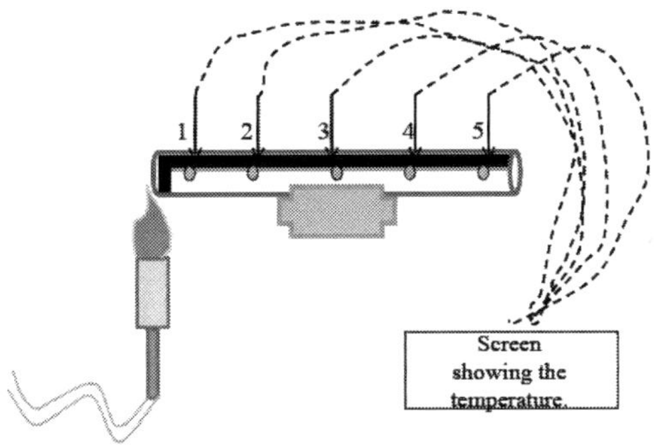

Figure 3.- Diagram for the second experiment

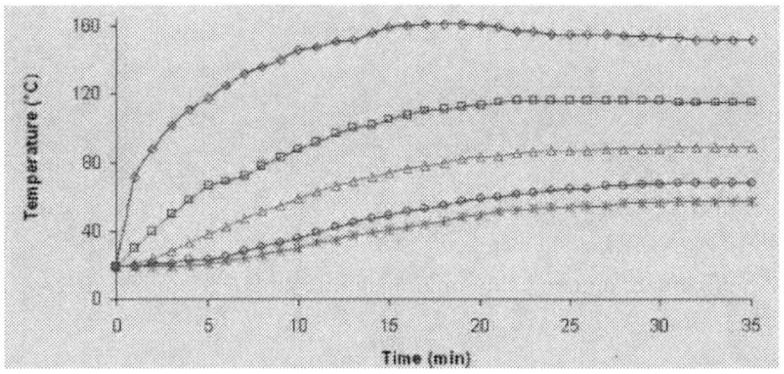

Figure 4.- . Temperature profiles for a bronze bar. Thermocouples (◊) 1, (□) 2, (Δ) 3, (○) 4 and (*) 5.

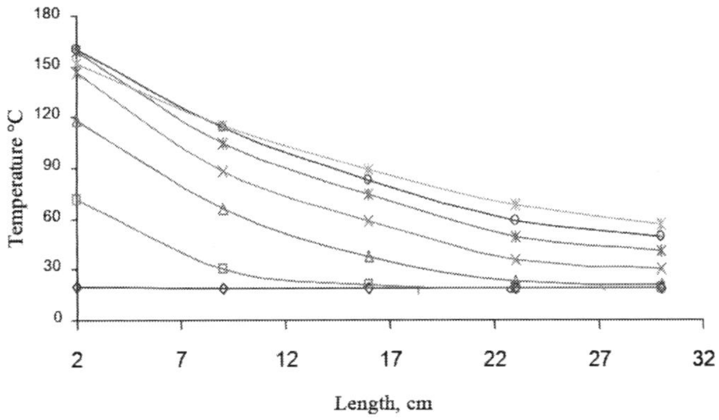

Figure 5.- Axial temperature gradients for the bronze bar. (◊) 0 min, (□) 1 min, (Δ) 5 min and (×) 10 min, (×) 15 min, (0) 20 min, (*) 35 min.

The experimental results are shown in figures 4 and 5, indicating that when the time is infinite (after 25 minutes), a constant temperature gradient is achieved.

c) Determination of the provided heat and measurement of the axial transient temperature profiles.

In the former practices, the amount of heat added to the bars was unknown. In this experiment, it is provided using a piece of equipment with an electrical resistance and it is calculated knowing the Ampers and Volts (Figure 6). The axial temperature profiles are measured at different times and the experiment is prolonged until eventually the steady state is reached, and the Fourier Law is illustrated. This practice is performed with and without insulation to study its effect on the axial temperature profiles. Figure 7 shows that with insulation, the temperature profiles are higher. The different possible boundary conditions to be used in this system are also indicated, and besides, the student is encouraged to find the solution of the partial differential equation contained in the mathematical model describing this process.

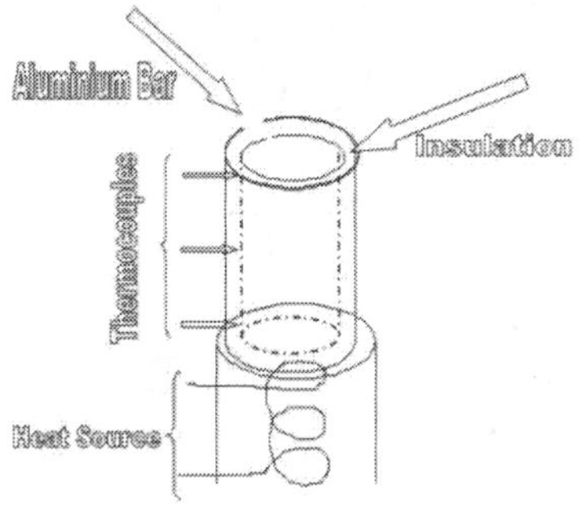

Figure 6.- Prototype of the experiment to measure the axial profiles and provided heat.

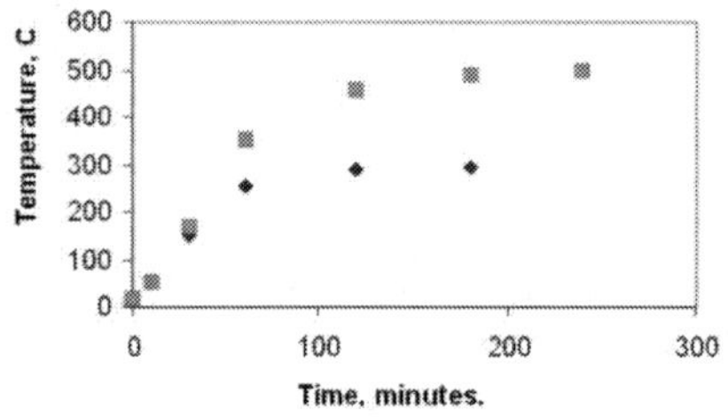

Figure 7.- Temperature profiles (■)with insulation and (♦)without insulation.

Mathematical Modeling.

The vertical metallic bar is exposed to a source of heat in the bottom, in the cylindrical side it is insulated with material of glass fiber, and in the top, it is exposed to the environment. The axial temperature in the center of the bar is measured at different times. Knowing that the temperature U is the dependent variable and that there are two independent variables, these being the time t and length z, heat transfer within the bar is only due to conduction, therefore, the differential equation that describes this phenomenon is:

$$\alpha \frac{\partial^2 U}{\partial z^2} = \frac{\partial U}{\partial t}$$

The Boundary conditions are:

In the bottom the heat flow is constant, therefore, z = 0, $-k\frac{\partial U}{\partial z} = q_0$

On the top exposed to the environment: At z = L, $-k\frac{\partial U}{\partial z} = h(U - U_{environment})$

The initial condition is that before the application of heat, the bar is at constant temperature, $U_0 = 25°$ C.

For $t < 0$, at $0 < z < L$, $U_0 = 25°$.

The objective is to solve this model in analytical form using the techniques of variable separation and the theory of Sturm Liouville. Also the technique of orthogonal collocation will be used, to obtain an approximation to the analytical solution. These solutions will be compared with the experimental profiles obtained with the prototype. Due the insulation of the bar on the cylindrical side, it is considered there are not radial temperature gradients, and because of that, they were not included in the mathematical model.

Determination of the provided heat.

In order to measure the given heat by the iron (source of heat), three methods were used.

1.- Method of Fourier to evaluate the heat flow q.

Once it was considered the steady state was reached, and the temperature gradient was constant, the axial temperature was measured, and using the reported values in the literature about the thermal conductivity of the material, the provided heat flow was determined using the Fourier law.

$$\frac{q_x}{A} = -k\frac{dT}{dx}$$

2.- Method using the voltmeter and amperimeter.

During the realization of the experiments with the proposed set - up, an external heat source was used, which was controlled to provide a constant heat flow through a rheostat. A voltmeter and an amperimeter were connected to this system, in order to know the power of the calefactory element in contact with the bars of the different materials, using the relation:

Power = Volts * Amperes = (Joule /Coulomb) *(Coulomb /Sec)

Power = Joule / Sec.

Now it is also necessary to divide this value for the heat transfer area, with a diameter of 5.1 cm, to obtain the heat flow per unit area.

3.- Calorimeter.

A calorimeter was used to measure the temperature rise in regular time intervals of a known water volume contained in a wicker, placed on the top of the bar once the steady state was reached. The heat losses due to glass heating were evaluated, knowing the heat capacity of the glass. These data were substituted in the equation to measure the heat flow in the following form:

$Q = \mathrm{m\ Cp}\ \Delta T.$

d) Determination of the continuity boundary conditions for the temperature profile.

Two vertical bars of different materials are placed together in series (as a composite wall) and several thermocouples are placed in both bars (Figure 8). In the bottom, heat is provided and the transient temperature profiles are measured until reaching the steady state. In the position where the bars are joined, a continuity in the temperature profile should be observed and also the change in the slopes from one material to the other. This practice also illustrates technical problems that normally happen, and inasmuch as they are solved, the student reinforces his knowledge and strengthens his character to do scientific research.

The measured profiles are illustrated in figure 9, but the continuity in the temperature is not clearly illustrated, maybe for technical problems, not solved yet.

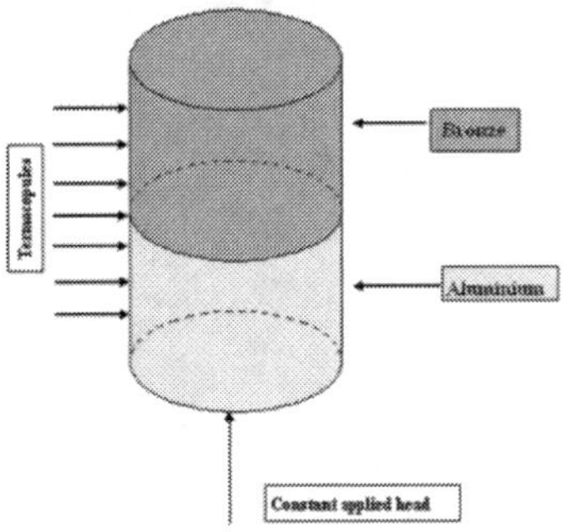

Figure 8.- Bars in series to illustrate the continuity in temperature between two materials See thermocouples.

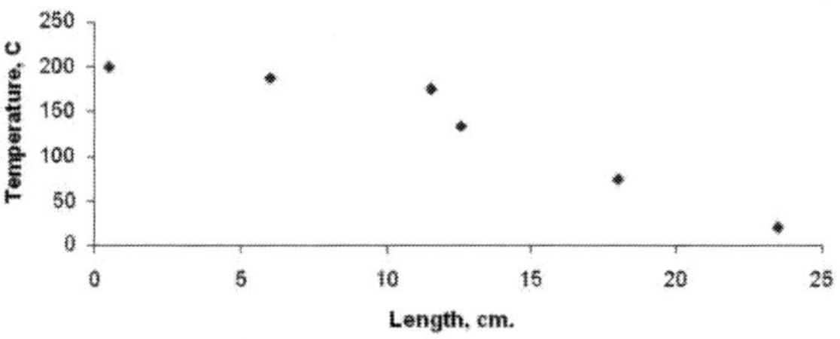

Figure 9.- Axial temperature profiles of temperature in two materials in series.

2.- Two practices of momentum transport.

b) Determination of the height profile for the liquid surface in a rotating container.

Water is placed in a cylindrical container and after that it is put in rotation at different angular velocities (Figure 10). Several photographs are taken (Figure 11) and they are compared with the solution of the differential equation resulting from the momentum analysis and the boundary conditions according to the problem.

Equation of continuity.

$$\frac{\partial \rho}{\partial t} = -\left(\frac{\partial}{\partial x} \rho u_x + \frac{\partial}{\partial y} \rho u_y + \frac{\partial}{\partial z} \rho u_z \right)$$

Momentum equation (Navier – Stokes).

$$\rho \frac{Du}{Dt} = -\nabla p - [\nabla \cdot \tau] + \rho g$$

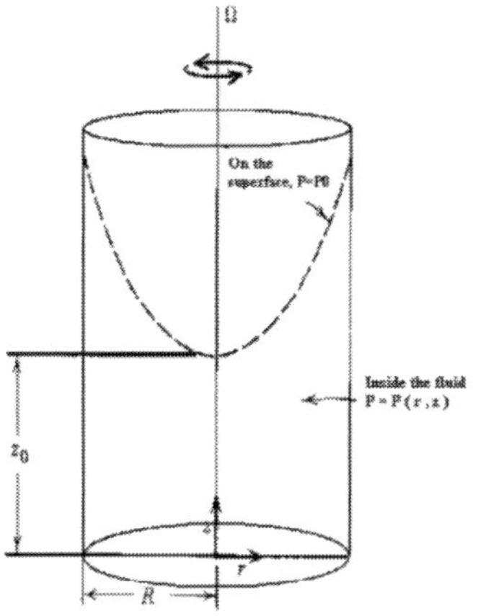

Figure 10.- Sketch of the rotating device

Figure 11.- Photography of the container in rotation.

Radial Component.

$$\rho\left(\frac{\partial u_r}{\partial t} + u_r \frac{\partial u_r}{\partial r} + \frac{u_\theta}{r}\frac{\partial u_r}{\partial \theta} - \frac{u_\theta^2}{r} + u_z \frac{\partial u_r}{\partial z}\right) =$$

$$= -\frac{\partial p}{\partial r} + \mu\left(\frac{\partial}{\partial r}\left[\frac{1}{r}\frac{\partial}{\partial r}(r u_r)\right] + \frac{1}{r^2}\frac{\partial^2 u_r}{\partial \theta^2} - \frac{2}{r^2}\frac{\partial u_\theta}{\partial \theta} + \frac{\partial^2 u_r}{\partial z^2}\right) + \rho g_r$$

After the analysis of the significant and irrelevant terms:

$$\rho \frac{u_\theta^2}{r} = \frac{dp}{dr}$$

Angular component.

$$\rho\left(\frac{\partial u_\theta}{\partial t} + u_r \frac{\partial u_\theta}{\partial r} + \frac{u_\theta}{r}\frac{\partial u_\theta}{\partial \theta} + \frac{u_r u_\theta}{r} + u_z \frac{\partial u_\theta}{\partial z}\right) =$$

$$= -\frac{1}{r}\frac{\partial p}{\partial \theta} + \mu\left(\frac{\partial}{\partial r}\left[\frac{1}{r}\frac{\partial}{\partial r}(r u_\theta)\right] + \frac{1}{r^2}\frac{\partial^2 u_\theta}{\partial \theta^2} + \frac{2}{r^2}\frac{\partial u_r}{\partial \theta} + \frac{\partial^2 u_\theta}{\partial z^2}\right) + \rho g_\theta$$

After the analysis of the significant and irrelevant terms:

$$0 = \mu \frac{d}{dr}\left(\frac{1}{r}\frac{d}{dr}(ru_\theta)\right)$$

Solving this differential equation

$$u_\theta = \frac{1}{2}C_1 r + \frac{C_2}{r}\Big\} \textit{ general solution.}$$

at $\quad r = 0 \quad u_\theta = \text{finite}$

For $r = R$, u_θ is $R\Omega$.

$u_\theta = \Omega r$, this is the particular solution.

Axial component

$$\rho\left(\frac{\partial u_z}{\partial t} + u_r\frac{\partial u_z}{\partial r} + \frac{u_\theta}{r}\frac{\partial u_z}{\partial \theta} + u_z\frac{\partial u_z}{\partial z}\right) = -\frac{\partial p}{\partial z} + \mu\left(\frac{1}{r}\frac{\partial}{\partial r}\left[r\frac{\partial u_z}{\partial r}\right] + \frac{1}{r^2}\frac{\partial^2 u_z}{\partial \theta^2} + \frac{\partial^2 u_z}{\partial z^2}\right) + \rho g_z$$

After the analysis of the significant and irrelevant terms:

$$0 = -\frac{dp}{dz} - \rho g$$

The purpose here is to find the height surface profile using simultaneously the former equations resulting from the physical analysis of the several components.

$$\rho\frac{u_\theta^2}{r} = \frac{dp}{dr}$$

$$\frac{\partial p}{\partial r} = \rho\Omega^2 r$$

$$\text{Now,}\quad dp = \frac{\partial p}{\partial r}dr + \frac{\partial p}{\partial z}dz$$

$$dp = (\rho\Omega^2 r)dr + (-\rho g)dz$$

According with the former procedure, the last equation is an exact differential equation.

$$p(r,z) = \rho\Omega^2\frac{r^2}{2} - \rho g z + C_1$$

$$p = p_0 \quad a \quad r = 0 \quad y \quad z = z_0$$

$$z - z_0 = \left(\frac{\Omega^2}{2g}\right) r^2$$

The comparison between the theoretical and experimental data is shown in figure 12:

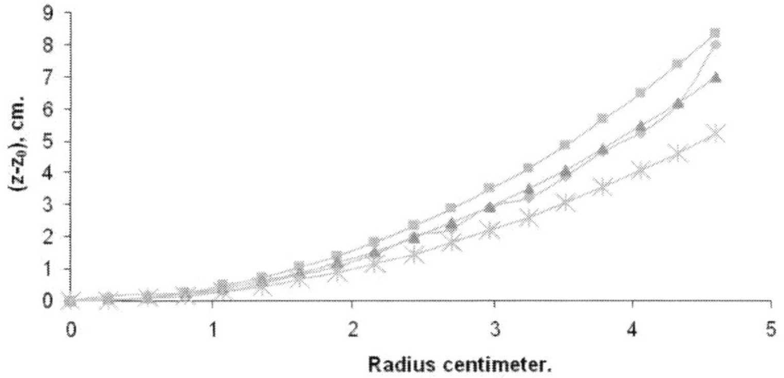

Figure 12.- Height profiles for the container in rotation. 22 rps *, 27.8 rps ■, 25 rps ▲, Experimental ◆

a) *Determination of the height profile of the liquid surface in a fixed cylindrical vessel with a spinning central bar.*

Water is placed in a blender modified with a central bar instead of the blends (Figures 13 and 14). The blender is turned on and the effect of the spinning velocity on the form of the liquid surface is studied. Pictures are taken and they are compared with the solution of the differential equation obtained from the momentum analysis.

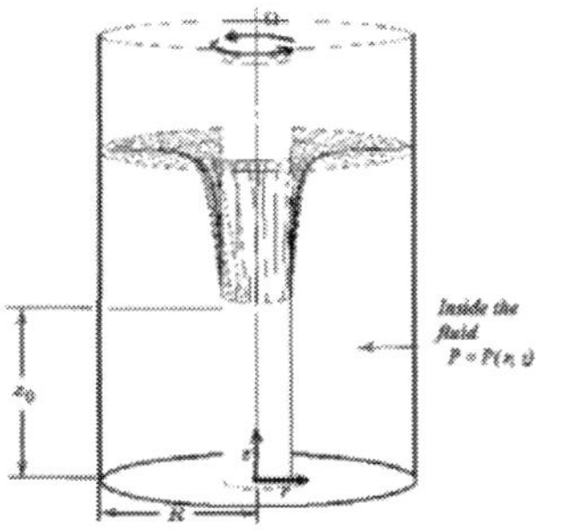

Figure 13.- Sketch of the height profiles.

Figure 14.- Photography of the stirrer.

In this case, the differential equation is the same but the boundary conditions are different.

$$r = R \quad u_\theta = 0,$$

$$u_\theta = \kappa R\Omega \quad para \quad r = \kappa R$$

$$u_\theta = \frac{1}{2}C_1 r + \frac{C_2}{r}\Big\} \textit{General Solution}$$

$$\text{Angular velocity} = \left(\frac{\kappa\Omega}{\kappa - 1/\kappa}\right)\left(r - \frac{R^2}{r}\right) \Rightarrow \textit{Particular Solution}$$

$$dp = \frac{\partial p}{\partial r}dr + \frac{\partial p}{\partial z}dz$$

$$z - z_0 = \frac{A}{\rho g}\left[\frac{r^2}{2} - 2R^2\ln\left(\frac{r}{\kappa R}\right) - \frac{R^4}{2r^2} + B'\right]$$

The comparison between the theoretical and experimental data is shown in figure 15:

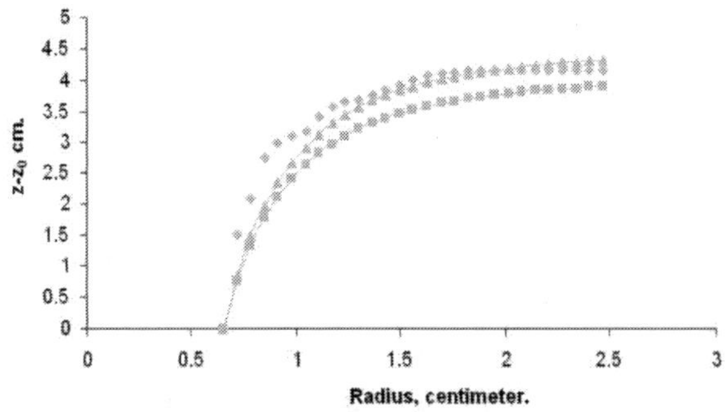

Figure 15.- Comparison between the experimental profile and several theoretical profiles.

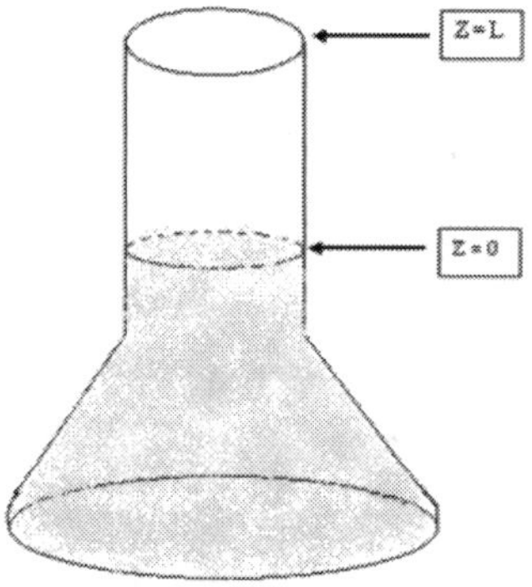

Figure 16.- Evaporation of ethanol.

Model 142 rps ■ , Model 150 rps ▲, Experimental ◆

3.- Mass transport Experiments.

Consider ethanol is evaporating from a flask with air flowing on top of the beaker (see Figure 16). In this process, diffusion and convection take place.

It is desired to evaluate the amount of ethanol evaporated in a period of time. Fick law is used.

$$N_{Az} = -cD_{AB}\frac{dx_A}{dz} + x_A(N_{Az} + N_{Bz})$$

$$N_{Az} = \frac{cD_{AB}}{z_2 - z_1} \ln\left(\frac{1 - x_2}{1 - x_1}\right)$$

The fraction mol of ethanol in the surface is 0.057. During the evaporation process, the system was weighted and in 90 minutes 0.12 gr of ethanol were evaporated, against 0.01399 gr predicted by the theory. The challenge for the student here is to explain this difference.

In summary, facing the lack of economical resources and given the necessity of having laboratory practices, some time it is possible to work using very elemental and cheap pieces of equipment to illustrate the concepts and to face the students to solve the resulting equations from the theoretical analysis, to perform measurements, and to propose solutions for the technical problems.

Conclusions

Not always the lack of economical resources is the main obstacle to propose and to do laboratory practices in transport phenomena as it was shown in the present material. Some times, it is possible to design very easy and cheap laboratory practices.

With very simple prototypes it is possible to illustrate concepts and to get mathematical models that face the students to solve them and to compare the solutions with the experimental data.

Of course, there will be always technical problems.

References.

1. Juan Jose Mejia Montoya, "Determinacion de la Superficie de Liquidos en Rotacion" Tesis de licienciatura en ingeniería química, Facultad de Ciencias Quimicas, Universidad Autonoma de Zacatecas, UAZ, Abril 2007.

2. Victor Javier Cruz Delgado, "Conduccion de Calor en Barras Cilindricas" Tesis de licienciatura en ingeniería química, Facultad de Ciencias Quimicas, Universidad Autonoma de Zacatecas, UAZ, Octubre 2002.

Unsteady State Heat Transfer Modeling in the Undergraduate Engineering Laboratory to Estimate Time of Death

Jimmy L. Smart, Chemical & Materials Engineering, University of Kentucky, Paducah, KY

Historically, estimation of time of death has always been a valuable forensic tool. In England, 925 AD, St. John of Beverly was appointed "keeper of the pleaser of the crown", later to be shortened to "crowner", then finally to "coroner". At that time, only crude "forensic" tools of rigor mortis and corpse warmth were used to gage time of death. It was not until 1600, that the first criminal autopsy was conducted by Ambrose Pare (French physician) [1,2].

By 1811, an English physician named John Davey had formulated his famous law based upon the principle of algor mortis (slow cooling of a warm-blooded corpse). Davey's Law states that 1.5 oF is lost per hour as measured in the armpit. Therefore, at least in theory, it should be possible to record the temperature of a corpse and having knowledge of the heat loss rate curve, be able to estimate time of death. In reality, the rate curve is complicated by many factors and it is often difficult to pinpoint time of death.

Over the years, there have been numerous models proposed to estimate time of death [3]. Results of these models can be useful to crime scene investigators, but are not admissible into courts of law. There is too much error and variability associated with these current models.

Students in the undergraduate engineering laboratory explored use of an unsteady state heat transfer model to estimate time of death. A large diameter wood cylinder was constructed to approximate the trunk of a human body. Interestingly enough, wood has approximately the same thermal diffusivity as the human body. A hole was drilled and a thermocouple was placed at the approximate relative location of the anus. The cylinder was heated to 37 deg C (98.6 deg F) in an oven. At time zero, the cylinder was removed from the oven and experimental data of temperature decay was collected until the cylinder cooled to room temperature (about 21 deg C).

As a laboratory exercise, students are told a crime scene investigator (CSI) had examined an apparent murder victim. The CSI had recorded the victim's rectal temperature between a two hour interval. Also, noted was the ambient temperature within the room. Students are asked to estimate time of death to aid in solving this crime. Students use a Fourier series [4] to model experimental data collected from unsteady state heat transfer from the wooden cylinder. They calibrate their model and test the sensitivity of the model to variations in crime scene conditions. Based upon their test results, students are able to estimate time of death.

1. Sachs, Jessica S., Corpse: Nature, Forensics, and the Struggle to Pinpoint Time of Death, Perseus Books Group: Basic Books, NY, October 2002.

2. Miller, Hugh, What the Corpse Revealed: Murder and the Science of Forensic Detection, St. Martin's Press, NY, June 1999.

3. Henssge, Claus, The Estimation of the Time Since Death in the Early Postmortem Period, 2nd Ed., London: Arnold Publishing, 2002.

4. Incropera, F.P. and D.P. DeWitt, Fundamentals of Heat and Mass Transfer, J. Wiley & Sons, New York, 2002.

See more of this Session: Free Forum On Chemical Engineering Education II
See more of this Group/Topical: Education

A Simple Setup for Experiments in Reactor Dynamics, Control and Transport Phenomena

Silas J. Leavesley, Chemical and Biomolecular Engineering, University of South Alabama, Mobile, AL

Srinivas Palanki, Chemical and Biomolecular Engineering, University of South Alabama, Mobile, AL

In this paper, a simple experimental setup is described that can be used in multiple courses in the undergraduate chemical engineering curriculum. The motivation for developing this setup comes from the high cost of purchasing "canned" experiments that are suitable for the undergraduate curriculum. The experiment consists of a glass reactor with a cooling coil made of copper tubing. The cooling coil can be connected to either air or water. The flow rate of fluid in the coil is adjusted via an electronic control valve. LABVIEW software is used to acquire thermocouple measurements and control the valve actuation. The LABVIEW process controls toolbox is used, which gives a wide variety of PID controller configurations and options. This allows students to experiment with different control configurations and PID settings, while monitoring the response of the reactor temperature. The following experiments can be conducted in this simple apparatus:

1. Implementation of the PID controller to maintain a desired set-point temperature in the reactor.

2. Calculation and comparison of heat transfer coefficients when the fluid in the cooling coil is water or air.

3. Experimental simulation of various batch reactor dynamic conditions, such as reactor start-up.

4. Comparison of flow rates for air and water, and estimation of the pressure drop across a length of coiled copper piping.

This experimental setup costs less than $1000 to build and operate (not including the cost of the computer and LABVIEW). We will describe our experience in using this setup in the process control class and the heat transfer class. Possible extensions of these experiments to the reaction engineering class as well as the fluids class will be discussed.

See more of this Session: Free Forum On Chemical Engineering Education II
See more of this Group/Topical: Education

A Learning Lesson From Thermodynamics: Ideal Fluids. Not True in Transport!

Jennifer Anne Pascal, Chemical Engineering, Tennessee Tech University, Cookeville, TN

Pedro Arce, Chemical Engineering, Tennessee Technological University, Cookeville, TN

Many of us would most likely remember that Thermodynamics was not a trivial subject from a learning point of view. However, the instructor began the course by first introducing the basic concepts and behaviors of ideal fluids and mixtures. Therefore, the ideal gas law was an important learning tool to describe the system behaviors in many topics. Likewise, ideal mixtures were very useful when the focus switched to systems with more than one component. Only after students have mastered a good level of understanding of this subject, real gases and fluids as well as real mixtures were introduced.

The picture described above is entirely different when one analyzes the textbooks related to the subject of transport phenomena, with the exception of the analysis of hydrostatics. Here, the nature of the fluid from the point of view of the ideal vs. real behavior is irrelevant since the fluid velocity is zero. As soon as the focus of the subject switches to fluid motion, the textbook introduces the not so very inviting tetrahedral domain analysis of tensors (?) and stresses and incorporates them into the momentum conservation equation. This is done at a point where students do not even understand motion. Moreover, ideal fluids are considered in many textbooks as a graduate level subject!

In this contribution, the authors introduce a very systematic analysis of the momentum conservation equation by using ideal fluids. The efforts will show the enhancing effect on learning by following the analysis of different cases without the complications of viscous flows and the associated viscous flow theory. This contribution also introduces an effective way of studying the cases of viscous flows after many of the basic cases of motion have been studied with ideal fluids. Illustrative examples and discussion will be offered in the presentation.

See more of this Session: Free Forum On Chemical Engineering Education II
See more of this Group/Topical: Education

Virtual Experiments and Graphical Insights Into Thermodynamics

David A. Gallagher, CAChe Research, Beaverton, OR

The latest software tools enable students to quickly and easily perform virtual experiments in thermodynamics on their own laptop computers. Software tools that incorporate quantum chemistry and statistical thermodynamics methods do not require empirical data and so, can release the students from the constraints of their own limited experience and available chemicals. They allow a wide range of physical properties of compounds and mixtures, including vapor pressure, boiling points, activity coefficients, azeotropes, partition coefficients, liquid-liquid equilibria, etc., to be predicted for even hypothetical compounds. However, more than just numbers, the associated graphical visualization offers an intuitive understanding of the fundamental phenomena that influence physical properties and fluid thermodynamics. For example, visualization of electrostatic potentials on the solvent-accessible surface, and sigma profiles, etc., illustrate the potential for hydrogen bonding, dipole-dipole, dipole-induced-dipole and Van der Waals interactions, providing clear insight into the molecular interactions that determine the physical properties. Finally, these new tools enable the student to explore their own creativity and experiment with off-the-wall ideas, with the safety and convenience of a laptop computer, in the lab, the classroom, and at home. This presentation will be illustrated with examples of experiments performed with the COSMOtherm software package, along with a brief overview of its underlying computational methodologies.

See more of this Session: Free Forum On Chemical Engineering Education II
See more of this Group/Topical: Education

Explosion at T2 Laboratories

Robert J. Hall, P.E.; Mary E. Nikityn; Lucy A. Sciallo; Jeffrey J. Wanko, P.E., C.S.P.

U.S. Chemical Safety and Hazard Investigation Board

Abstract

At 1:33 pm on December 19, 2007, a powerful explosion and subsequent chemical fire killed four employees and destroyed T2 Laboratories, Inc. (T2), a chemical manufacturer in Jacksonville, Florida. It injured 32, including four employees and 28 members of the public who were working in surrounding businesses. Debris from the reactor was found up to one mile away, and the explosion damaged buildings within one quarter mile of the facility.

On December 19, T2 was producing its 175th batch of methylcyclopentadienyl manganese tricarbonyl (MCMT). At 1:23 pm, the process operator had an outside operator call the owners to report a cooling problem and request they return to the site. Upon their return, one of the two owners went to the control room to assist. A few minutes later, at 1:33 pm, the reactor burst and its contents exploded, killing the owner and process operator who were in the control room and two outside operators who were exiting the reactor area.

The CSB found that a runaway exothermic reaction occurred during the first (metalation) step of the MCMT process. The CSB tested the T2 batch recipe to determine the most likely failure scenario. A loss of sufficient cooling during the process likely resulted in the runaway reaction, leading to an uncontrollable pressure and temperature rise in the reactor. The pressure burst the reactor and the reactor's contents ignited creating an explosion equivalent to 1,400 pounds of TNT.

1.0 Introduction

1.1 Summary

At 1:33 pm on December 19, 2007, an explosion and fire destroyed T2 Laboratories, Inc. (T2), a chemical manufacturer on the north side of Jacksonville, Florida (Figure 1). The explosion, which was felt and heard 15 miles away in downtown Jacksonville, killed four T2 employees, including a co-owner. It injured 32, including four T2 employees and 28 members of the public at surrounding businesses.

Figure 1. Aerial photograph of T2 taken December 20, 2007.

Debris from the explosion was found up to one mile away, and the blast damaged buildings within one quarter mile of the facility. The City of Jacksonville subsequently condemned buildings used by four of the businesses surrounding T2. Three of these businesses relocated operations while their buildings were repaired; the remaining business, a trucking company adjacent to T2, permanently closed due to lost business.

On December 19, T2 was producing methylcyclopentadienyl manganese tricarbonyl (MCMT) in a 2,450-gallon chemical batch reactor. At 1:23 pm, the process operator directed an outside operator to call the two owners, who were off site, to report a cooling problem and request they return. Upon their return, one of the owners went to the control room to assist. A few minutes later, at 1:33 pm, the reactor burst and its contents exploded, killing the owner and process operator in the control room and two outside operators who were exiting the reactor area.

1.2 Investigative Process

Due to the multiple deaths, many injuries, and extensive community damage, the CSB launched an investigation on December 19, 2007. The CSB investigation team interviewed employees of T2, emergency responders, eye witnesses, and officials from the City of Jacksonville. The team documented off-site damage, located and photographed reactor debris, and interviewed injured parties and eyewitnesses. JSO assisted the CSB in debris documentation and collection.

The CSB conducted laboratory testing of the chemical reaction used in the T2 process, recreated the most likely scenario, and calculated the pressure relief capacity necessary to vent the uncontrolled reaction. The CSB also conducted a community damage survey of the 32 structures damaged by the blast, using the data collected to estimate the amount of TNT necessary to produce a blast of equivalent magnitude.

1.3 T2 Laboratories, Inc.

T2 Laboratories, Inc. (T2)[1] was a small privately-owned corporation located in Jacksonville, Florida, that began operations in 1996. A chemical engineer and a chemist[2] founded T2 as a solvent blending business and co-owned it until the incident. From 1996 to 2001, T2 operated from a warehouse located in a mixed-used industrial and residential area of downtown Jacksonville. T2 blended and sold printing-industry solvents; it also blended pre-manufactured MCMT to specified concentrations for Advanced Fuel Development Technologies, Inc. (AFD), a third-party distributor.

In 2001, T2 leased a 5-acre site in a north Jacksonville industrial area and began constructing an MCMT process line. In January 2004, T2 began producing MCMT in a batch reactor.[3] By December of 2007, MCMT production was the primary business operation. On the day of the incident, T2 employed 12 people and was producing its 175th MCMT batch (Batch 175).

1.4 Methylcyclopentadienyl Manganese Tricarbonyl

Methylcyclopentadienyl manganese tricarbonyl (MCMT) is an organomettalic compound used as an gasoline additive to increase octane rating. The Ethyl Corporation[4] originally developed MCMT in the late 1950s. T2 manufactured and sold MCMT under the trade name Ecotane.

MCMT is a combustible liquid and is very toxic by inhalation or skin contact. Both the National Institute for Occupational Safety and Health (NIOSH) and OSHA set exposure limits for MCMT.[5] Although MCMT decomposes quickly when exposed to light, the EPA designates MCMT as an extremely hazardous substance (EHS)

---

[1] In the course of designing and constructing the MCMT plant, the T2 owners formed two other business enterprises, which were eventually subsumed into T2. These ventures involved the same personnel; for simplicity this report refers to the company as T2 throughout. Following the incident, T2 has ceased production operations.

[2] The T2 owners are herein referred to as "owner/chemical engineer" and "owner/chemist."

[3] In batch reactions, process operators feed fixed amounts of raw materials into a reactor at specified times, according to a chemical recipe. The entire process occurs within the reactor, which is emptied between batches.

[4] The Ethyl Corporation is now known as Afton Chemical, which is the only domestic producer of MCMT.

[5] NIOSH's recommended exposure limit is 0.2 milligrams per cubic meter average concentration over 10 hours. OSHA's permissible exposure limit is 5 milligrams per cubic meter at any one time.

2.0 Incident Description

2.1 December 19, 2007 Incident

On the evening of December 18, 2007, a night shift process operator cleaned and dried the reactor in preparation for a new MCMT batch. At about 7:30 am on December 19, the day shift process operator began manufacturing Batch 175 from the control room adjacent to the process line. He likely followed the routine batch procedures, loading the reactor with the specified quantities of raw materials using an automated process control system (Section 3.2). An outside operator hand-loaded the reactor with blocks of sodium metal, then sealed the reactor. At about 11:00 am, the process operator began heating the batch to melt the sodium and initiate the chemical reaction, while monitoring the temperature and pressure on the process control screen.

Once the sodium melted, at 210°F (98.9°C), the process operator likely started the mixer (agitator). Heat from the reaction and the heating system continued raising the temperature in the reactor. At a reaction temperature of about 300°F (148.9°C), the process operator likely turned off the heating system as specified in the procedure, but heat from the reaction continued increasing the mixture temperature.

At a temperature of about 360°F (182.2°C), the process operator likely started cooling, using the process control system, as specified in the procedure. However, the mixture temperature in the reactor continued to increase.

At 1:23 pm, the process operator had an outside operator call the owners to report a cooling problem[6] and request they return to the site. Upon their return, the owner/chemical engineer went to the control room to assist and the owner/chemist searched for the plant mechanic. Employees indicated that after visiting the control room, the owner/chemical engineer went to the reactor. He told an outside operator—who was coming to the control room to investigate the audible alarms—that he thought there would be a fire and motioned employees to get away. The owner/chemical engineer then returned to the control room.

By 1:33 pm, the reactor's relief system could no longer control the rapidly increasing temperature and pressure of the runaway reaction. Eyewitnesses from nearby businesses reported seeing venting from the top of the reactor and hearing a loud jet engine-like sound immediately before the reactor violently ruptured, its contents exploding. The explosion killed the owner/chemical engineer and process operator who were in the control room (50 feet from the reactor) and two outside operators who were leaving the reactor area (Figure 2). Another outside operator and the plant mechanic were injured. The owner/chemist was sheltered from the force of the explosion by a shipping container, but suffered a non-fatal heart attack during the incident.

---

[6] Survivors' accounts indicate that the process operator (deceased) reported a cooling system problem.

Figure 2. Control room.

2.2 Injuries and Community Damage

The explosion killed four T2 employees and injured 32 workers at T2 and surrounding businesses. All of the people at T2 during the incident—eight T2 employees and one truck driver making a delivery—were injured or killed. The four fatally injured employees died of trauma as a result of the explosion. Another T2 employee was critically injured and hospitalized for several months.

The CSB conducted a community survey of the surrounding businesses to characterize injuries and structure damage (Figure 4). At the nine businesses within 1,900 feet of the reactor, the explosion injured 27 workers. Of those, 11 suffered lacerations and contusions, seven reported hearing loss, and five fell or were thrown by the force of the blast. The CSB survey team photographed and catalogued 32 damaged structures, some as far as 1,700 feet from T2.

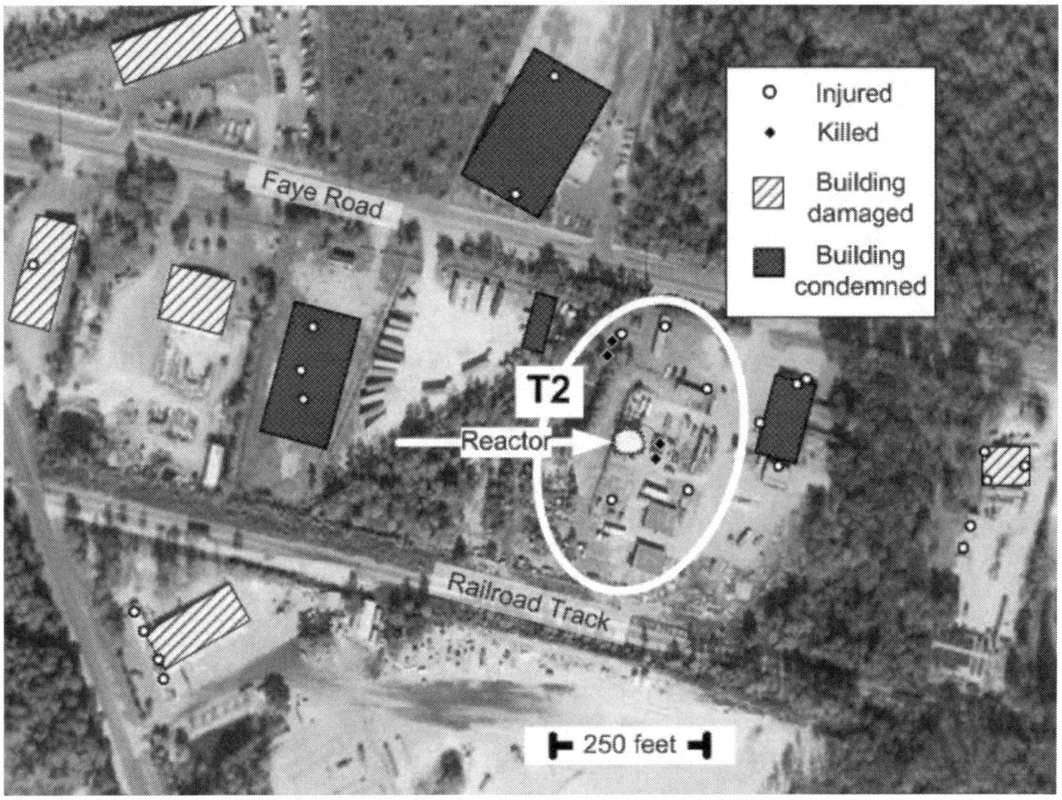

Figure 3. Injury and business locations.

The explosion leveled the plant, propelling debris in all directions. Two large steel support columns from the reactor structure traveled about 1,000 feet along Faye Road in both directions. A 2,000-pound section of the 3-inch-thick reactor head (Figure 4) impacted railroad tracks adjacent to T2, pushing a rail out of place, before impacting and damaging a building about 400 feet from the reactor. The explosion threw piping from inside the reactor hundreds of feet onto the other businesses and wooded areas surrounding T2. The 4-inch diameter agitator shaft from the reactor was thrown about 350 feet across Faye Road in two large pieces that imbedded in a sidewalk and the ground (Figure 5).

Figure 4. Portion of the 3-inch-thick reactor.

Figure 5. Agitator shaft pieces.

Businesses near T2 sustained heavy damage. Structures sustained window damage from the overpressure up to 1,700 feet away from T2. The explosion destroyed a trucking company office trailer located 250 feet from the reactor (Figure 6). If trucking company employees had been in the trailer at the time of the incident, it is likely that they would have been seriously injured or killed.[7] Two warehouses located about 400 to 500 feet from the reactor both sustained heavy damage; nine of the 25 employees at these two businesses were injured.

Figure 6. Trucking company trailer adjacent to T2.

[7] Due to the upcoming holiday, trucking company employees had dismissed for the day shortly before the incident.

3.0 T2 MCMT Process

3.1 Process Development

Working primarily from patents T2's owner/chemist duplicated and tested a three-step process for making MCMT in the laboratory. Between 2000 and 2001, the owner/chemist ran about 110 test batches of MCMT in a one-liter reactor.

In 2001, T2 leased its Faye Road site, in an area zoned for heavy industry.[8] With financial support from a number of investors, T2 designed and constructed a full-scale MCMT production plant. T2 hired consulting engineers to assist in the process design, control system engineering, and project management.

On January 9, 2004, T2 began manufacturing its first full-scale MCMT batch (Batch 1) in the new process line. Batch 1 produced an unanticipated exothermic reaction in the first step; T2 noted the anomaly, adjusted the batch recipe and production procedures to include reactor cooling in the first step,[9] and began a new production batch. Between February and May 2004, T2 manufactured nine more MCMT batches, adjusting the recipe and procedures between batches. Yields varied from zero to about 70 percent.[10] Batch 5 resulted in an uncontrolled (runaway) exothermic reaction in the first step. In Batch 10, the temperature also increased beyond expectations—though not as severely as in Batch 5—due to the exothermic reaction.

On May 24, 2004, following Batch 11, T2's owner/chemical engineer sent a memo to investors declaring successful plant startup and full-scale MCMT production. Although all of the investors had ended their involvement by late 2004, T2 continued producing and selling MCMT on an irregular basis, relying upon sales to pay for more raw materials. On July 28, 2005, producing Batch 42, T2 increased batch size by one-third. T2 manufactured batches from 2005 until the incident (Batch 175) at this larger size.

A process operator ran each step of the batch reaction from a control room adjacent to the process line. The T2 plant manager, a chemical engineer, began working as a process operator in 2004. In 2006, T2 hired two additional chemical engineers to run the process throughout multiple weekday shifts. Each batch required about 48 hours to manufacture. By December 2007, client demand had increased, and T2 produced three batches a week.

---

[8] Although the site was properly zoned, T2 did not obtain all necessary permits prior to plant construction.

[9] The process design already included reactor cooling, which was required in later production steps.

[10] A 70 percent MCMT process yield was a good outcome based on the owner/chemist's research.

3.2 Manufacturing Process

T2 manufactured MCMT in three steps that occurred sequentially within a single process reactor (Figure 7). The National Annealing Box Company of Washington, Pennsylvania, originally constructed the reactor in 1962 for an internal pressure of 1,200 psig.[11] T2 purchased the reactor in 2001 and contracted a firm specializing in pressure vessels to refurbish, modify, and test the reactor. The modifications included replacing and adding of piping nozzles and reducing the maximum allowable working pressure from 1,200 psig to 600 psig.[12]

A 4-inch[13] vent pipe that made two 90-degree pipe bends before connecting to a 4-inch rupture disk provided overpressure protection for the reactor. Employees stated that the rupture disk was set at 400 psig. A pressure control valve installed in a 1-inch vent pipe, which branched off of the 4-inch vent pipe below the rupture disk, controlled reactor pressure.

The MCMT process required both heating and cooling. A heating system circulated hot oil through 3-inch piping installed around the inside of the reactor. A cooling jacket covered the lower three quarters of the reactor. A pipe from the city water system connected to the bottom of the jacket through a control valve and a common supply/drain connection. Water was injected into the jacket and allowed to boil; steam from the boiling water vented to atmosphere through an open pipe connected to the top of the jacket.

---

[11] National Annealing designed and fabricated the reactor in accordance with the 1959 version of the American Society of Mechanical Engineers (ASME) Boiler and Pressure Vessel Code.

[12] Although the nameplate pressure was reduced this had little effect on the inherent pressure-retaining ability of the reactor.

[13] All piping sizes presented in this report are nominal pipe size (NPS).

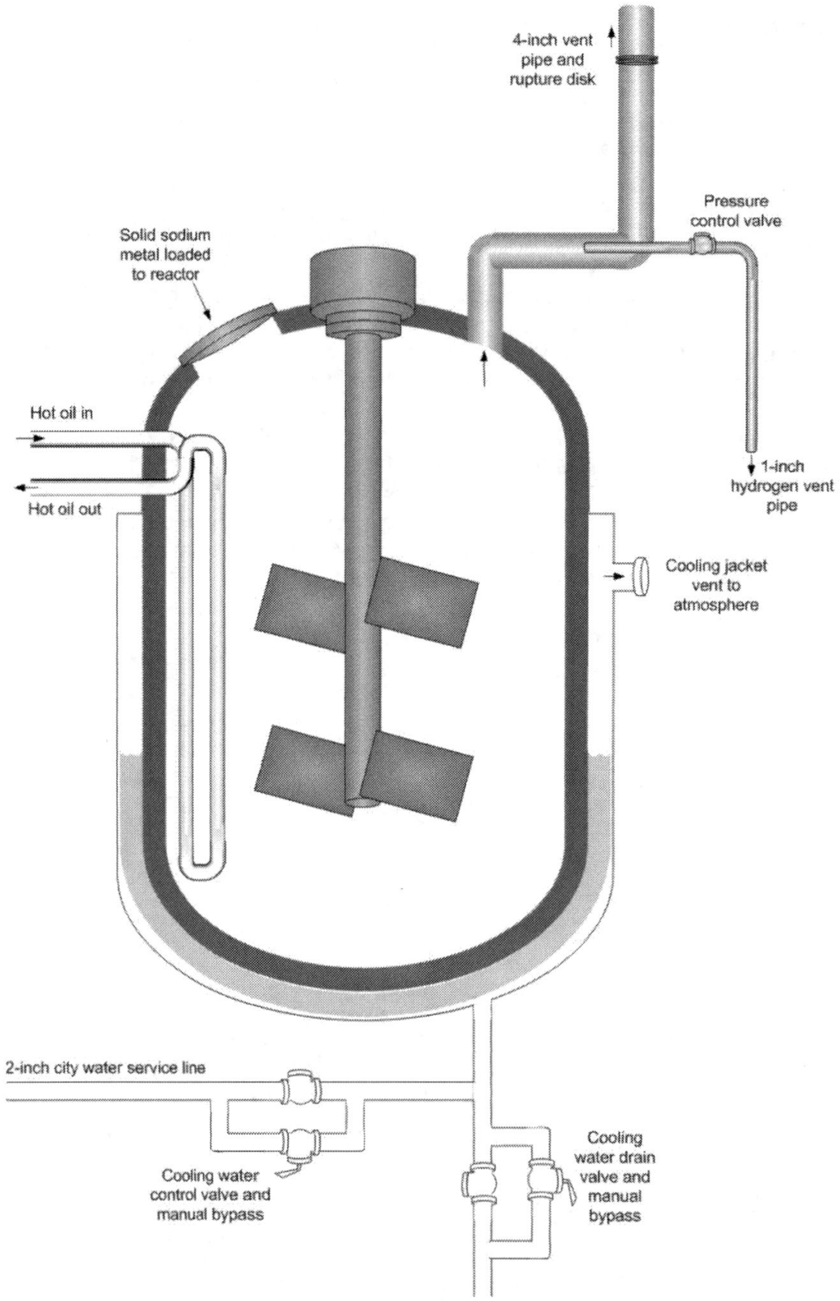

Figure 7. Reactor cross-section.

Each MCMT production step required the process operator to add raw materials and adjust heating, cooling, and pressure using a computerized process control system (Figure 9). On December 19, 2007, the incident occurred during the first (metalation) step of the reaction. The second and third steps of the reaction, which did not occur on December 19, 2007, were also conducted within the reactor. Appendix A discusses the complete reaction chemistry.

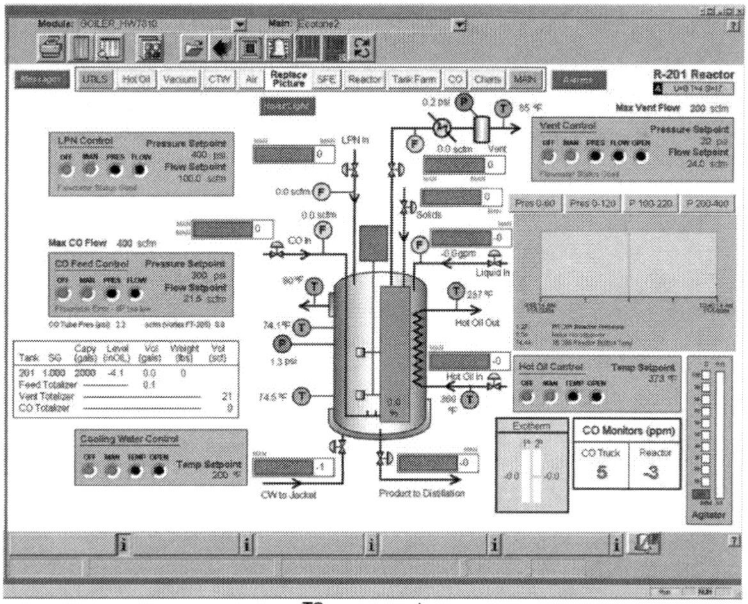

T2 screen capture

Figure 8. Process control system screen.

In the first reaction step (called metalation), the process operator fed a mixture of methylcyclopentadiene (MCPD) dimer[14] and diethylene glycol dimethyl ether (diglyme) into the reactor. An outside operator then hand-loaded blocks of sodium metal through a 6-inch gate valve on top of the reactor, closing the valve when complete. The process operator began heating the mixture with the hot oil piping system, setting reactor pressure control at 50 psig (3.45 bar) and hot oil temperature control at 360°F (182.2°C).

Heating the mixture began the metalation reaction by melting the sodium and splitting each MCPD dimer molecule into two MCPD molecules. The melted sodium then reacted with the MCPD to form sodium methylcyclopentadiene, hydrogen gas, and heat. The hydrogen gas vented to the atmosphere through the pressure control valve and 1-inch vent line.

[14] MCPD dimer is a molecule consisting of two identical MCPD molecules.

Once the mixture temperature reached 210°F (98.9°C), the process operator started the agitator. At a reaction temperature of about 300°F (148.9°C), the process operator turned off the hot oil system; heat generated by the metalation reaction continued to raise the mixture temperature.

At a temperature of about 360°F (182.2°C), the process operator initiated the control system cooling program, which intermittently injected water into the jacket based on the rate of reaction temperature increase. The operating procedures used at the time of the incident included no emergency instructions for loss of cooling. However, earlier procedures—which included emergency instructions—directed the operators to fully open the water supply valve and the manual bypass valve. A secondary (backup) source of water stored on site was not immediately available to the process operator in an emergency.

4.0 Incident Analysis

The incident at T2 was one of the most energetic explosions investigated by the CSB. The CSB estimated the energy release of this explosion to be equivalent to 1,400 pounds of TNT. The blast occurred due to a runaway chemical reaction that generated high temperature and pressure in the reactor. Runaway chemical reactions are extremely dangerous, but their causes are well documented.[15] The CSB evaluated possible causes of the runaway reaction, including

- cross-contamination of the reactor,
- contamination of raw materials,
- wrong concentration of raw materials,
- local concentration of chemical within the reactor,
- application of excessive heat, and
- insufficient cooling.

The CSB determined contamination scenarios to be unlikely. The T2 owner/chemist stated that batch-to-batch contamination of the reactor had previously occurred, resulting in low yields or batch polymerization. Additionally, T2 had experienced a runaway chemical reaction on its first production batch in a clean reactor. Contamination of the raw materials was similarly unlikely, as all of the raw materials used on the day of the incident came from shipments that had been used in previous successful batches.

Based on the reaction chemistry, raw material concentration abnormalities were not credible failure scenarios. Sodium was hand-loaded by operators in the form of one four-drum pallet per batch, making amount variation extremely unlikely. Varying local concentrations of chemicals within the reactor would reduce rather than accelerate reaction rate, since uniform distribution of the three metalation reaction raw materials results in the maximum reaction rate.

Heat was applied to the mixture using the hot oil system; had heating continued beyond 300°F (148.9°C), the CSB calculated that the cooling system would have easily overwhelmed it. The capacity of the cooling system was more than 10 times greater than the maximum capacity of the hot oil system.

The CSB determined insufficient cooling to be the only credible cause for this incident, which is consistent with witness statements that the process operator reported a cooling problem shortly before the explosion. The T2 cooling water system lacked design redundancy, making it susceptible to single-point failures including

---

[15] *Lees' Loss Prevention in the Process Industries* (3rd edition)

- water supply valve failing closed or partially closed.
- water drain valve failing open or partially open.
- failure of the pneumatic system used to open and close the water valves.
- blockage or partial blockage in the water supply piping.
- faulty temperature indication.
- mineral scale buildup in the cooling system.

Interviews with employees indicated that T2 ran cooling system components to failure and did not perform preventive maintenance. On at least one prior occasion since 2006, the reactor cooling drain valve had failed during operations and required repair. Additional credible cooling system failures include formation of mineral scale inside the jacket that could interfere with system heat removal capacity or loose scale blocking the inlet/drain pipe and causing it to stick open. Although a control system malfunction or operator error might also contribute to insufficient reactor cooling, there is no evidence to indicate that either of these occurred.

The CSB contracted laboratory testing of the batch recipe used by T2 on the day of the incident (Appendix B). The results of this testing showed that the standard T2 chemical recipe, without sufficient cooling, was capable of producing the extreme temperature and pressure necessary for the violent reactor failure on December 19, 2007.

When the CSB conducted tests in a closed (sealed) test cell, two exothermic reactions were observed using the T2 recipe. The first exothermic reaction occurred at about 350°F and was the desired reaction between the sodium and the MCPD. A second and more energetic exothermic reaction occurred when the temperature exceeded 390°F (198.9°C); this reaction was between the sodium and the diglyme solvent. The pressure and temperature rise during the second exothermic reaction were about 32,000 psig per minute (2,206 bar per minute) and 2340°F per minute (1300°C per minute), respectively, and burst the test cells.

Using the data obtained from these tests, the CSB determined that it is unlikely that an overpressure relief device of any size set at 400 psig could have prevented failure of the reactor once the second exothermic reaction began. Failure could only be prevented by relieving at a lower pressure during the first exothermic reaction and allowing the MCPD and diglyme solvent to boil and vent, removing both heat and reactants. Had T2 set its 4-inch reactor rupture disk at 75 psig, rather than the 400 psig used, the runaway reaction likely would have been relieved during the first exothermic reaction, precluding the second exothermic reaction. This could have prevented the catastrophic reactor failure that occurred.

5.0 Reactive Hazard Analysis

Companies developing chemistry for commercial production, like T2, must fully research the hazards involved. Hazards may be identified in each phase of commercial development, from laboratory testing to plant operations, and actions taken to assess risk and mitigate or eliminate potential consequences. Many guidelines, including free resources, address these topics (Section 6.0).

5.1 T2 Research and Development

T2 developed its MCMT chemistry based primarily on patents granted in the late 1950s and early 1960s (Appendix D). While patents document process chemistry, they generally do not document process hazards. All three steps of the MCMT process involved toxicity, flammability, or reactivity hazards. Hazardous chemicals used or generated during production steps included sodium metal, carbon monoxide, hydrogen, and organometallic compounds. A literature search performed by the CSB found little published information on the production of MCMT other than the patents, and no published information specific to its reactivity hazards.

A lack of available process hazard information makes laboratory testing especially important. The T2 owner/chemist performed laboratory testing in a 1-liter glass reactor to establish the MCMT process chemistry and determine maximum product yield. He reported that he never observed extreme exothermic behavior during testing and that test temperatures never exceeded 380°F (193°C). By not investigating the reaction's behavior at higher temperatures, the owner/chemist did not observe evidence of exothermic runaway potential.

Cooling requirements in the one-liter laboratory reactor did not accurately indicate the amount of cooling needed in the full-scale T2 reactor. Although the laboratory reactor required occasional heating and did not require cooling, T2 employees reported that additional cooling was determined to be necessary during multiple process upsets in early production batches.

Other CSB investigations (Section 5.4) have found that reliance on laboratory testing can lead to dangerous underestimation of full-scale batch temperatures and has been a principle factor in accidents involving reactive chemistries. *Designing and Operating Safe Chemical Reaction Processes* (HSE, 2000) discusses the importance of proper scale-up design of process equipment and the potential for incidents to occur in full-scale processes that have appeared uneventful in laboratory testing.

5.2 Process Hazard Analysis

Process hazard analysis (PHA) in the development phase helps establish operating limits and identify operating strategies to prevent runaway reactions. PHAs for batch reactor systems should evaluate potential process deviations and equipment malfunctions, including agitator failure, loss of cooling, contamination, and mischarging feed stocks, all of which are common causes of runaway reactions.

One of T2's design consultants identified the need to perform a hazard and operability study (HAZOP, a type of PHA) during scale-up. A comprehensive HAZOP likely would have identified the need for testing to determine the thermodynamic and kinetic nature of the reaction, as well as the limitations of the cooling and pressure relief systems. CSB found no evidence that T2 ever performed the HAZOP.

Similarly, T2 sized the reactor relief devices based on anticipated normal operations, without considering potential emergency conditions. T2 employees stated that the owner/chemical engineer sized the rupture disk based on the maximum expected hydrogen gas generation during normal operation. T2 could not provide any documents related to the sizing and set point of the rupture disk. However, there was no evidence that T2 evaluated a runaway reaction as a possible overpressure source.

5.3 T2 Incident History

The runaway reaction on December 19, 2007, was not the first unexpected exothermic reaction that T2 experienced; three of the first 10 MCMT batches resulted in unexpected exotherms.[16] Each of these occurred during metalation, and in each instance the batch recipe was slightly different. T2 did not repeat batch recipes to isolate the problem, instead changing recipes in each of the first 10 batches.[17] T2 announced successful commercial operation to its stakeholders after Batch 11.

In 2005, at Batch 42, T2 increased the batch size by one third. There are no records of additional chemical or process analysis conducted as part of this recipe change, which may have introduced significant new risks. A greater volume of reactants increased the energy that the reaction could produce, and likely altered cooling and pressure relief requirements.

When the MCMT process yielded unexpected results in early batches, T2 did not halt production, investigate causes, and redesign the process. Instead, T2 attempted to control unexpected reaction results on-line through operator controls or minor alterations to continue running the process as it was already constructed. As demand grew, T2 increased batch size and frequency with no additional documented hazard analysis.

5.4 Reactive Hazard Recognition

Chemists and chemical engineers involved in developing and operating the T2 MCMT process were unaware of the need to perform runaway reaction testing, address emergency relief, and identify and evaluate the causes of process upsets. Unexpected exothermic reactions were managed as they occurred, and T2 employees expected that the owner/chemical engineer could control any future incidents, as evidenced by the call to request his assistance shortly before the explosion. Although the owner/chemical engineer told employees he thought a fire would occur, none of the T2 employees appreciated the potential for a catastrophic explosion.

---

[16] These exothermic behaviors were either stronger than T2 expected or occurred at unexpected steps in the reaction.

[17] The fire following the December 19, 2007, explosion destroyed documentation on later batches and batch recipes.

The CSB has investigated four previous runaway reaction incidents caused in part by not recognizing the reactive hazard. The incidents were

- an explosion at Morton International, Inc. in Paterson, New Jersey on April 8, 1998;
- an explosion at Concept Sciences, Inc. in Hanover Township, Pennsylvania on February 19, 1999;
- a toxic vapor cloud release at MFG Chemical, Inc. in Dalton, Georgia on April 12, 2004; and
- a flammable vapor release and explosion at Synthron, LLC in Morganton, North Carolina on January 31, 2006.

These incidents, summarized in Appendix E, and the one at T2 are responsible for a combined 10 deaths and over 200 injuries.

5.5 Hazard Education

The T2 owner/chemical engineer had a bachelor's degree in chemical engineering and was active in his university's chemical engineering curriculum advisory board. However, reactivity hazard awareness is not currently a fundamental component of chemical engineering curricula. Although both the owner/chemical engineer and owner/chemist held bachelor's degrees and had prior chemical industry experience, neither had previously worked with reactive chemical processes. Hence, they were ill-prepared to appreciate and recognize the reactivity hazards of the MCMT process.

In 2006, the Mary Kay O'Connor Process Safety Center[18] surveyed 180 chemical engineering departments at U.S. universities to determine whether process safety was part of their chemical engineering curricula. Of the universities surveyed, only 11 percent required process safety education in the core baccalaureate curriculum. An additional 13 percent offered an elective process safety course. The survey did not specifically address reactivity hazard awareness education; however, a sampling of curricula indicates that the inclusion of reactivity hazards in chemical engineering education is limited.

Accreditation of U.S. engineering programs, including chemical engineering, is a voluntary, peer-reviewed process to assure consistent curricular quality. Baccalaureate programs are accredited by the Accreditation Board for Engineering and Technology, Inc. (ABET) based on recommendations from profession-specific industry groups. ABET reviews and approves chemical engineering programs based on criteria recommended by the American Institute of Chemical Engineers (AIChE). There is no requirement for accredited baccalaureate chemical engineering programs to include process safety or reactive hazard awareness in their curricula.

---

[18] The Mary Kay O'Connor Process Safety Center is part of the Texas Engineering Experimental Station at Texas A&M University.

This page intentionally left blank.

6.0 Reactive Hazard Resources

A number of resources are available to chemical manufacturers designing reactive processes. In 2002 the CSB published a comprehensive report, *Improving Reactive Hazard Management*, documenting the rate, types, and causes of reactive incidents and providing recommendations to improve regulatory coverage. The American Institute of Chemical Engineers (AIChE) has actively promoted reactivity hazard education, operating groups specifically tailored to provide reactivity safety and design information to students and process engineers

6.1 The CSB Reactives Study

The CSB found that reactive incidents are a serious problem in the U.S., and that both management system and regulatory improvements are needed to help facilities control reactive hazards. The report studied reactive incidents, causal factors, and preventive measures, and outlined the screening, hazard identification, hazard review, operating procedures, and training needed to prevent reactive incidents. [19]

The report documented 167 serious reactive incidents in the United States between January 1980 and June 2001 that resulted in 108 deaths, hundreds of injuries, and significant public impacts. Ongoing CSB monitoring indicates that reactive incidents—such as the T2 explosion—continue to occur.

The report also found that 70 percent of reactive incidents occurred in the chemical manufacturing industry, with 35 percent due to runaway reactions like the one that occurred at T2.[20] While 42 percent of reactive incidents resulted in fires and explosions, another 37 percent caused toxic emissions. Many reactive incidents occurred at small manufacturing sites like T2.

More than 50 percent of the 167 incidents documented in the CSB report involved chemicals not covered by existing OSHA Process Safety Management (PSM) or EPA Risk Management Program regulations (Section 7). The CSB recommended that EPA and OSHA expand reactive hazard coverage under these regulations. EPA now requires reporting of reactive chemical incidents under RMP reporting rules and OSHA has taken steps to increase industry awareness of reactive hazards, but neither has fully implemented the CSB recommendations.

6.2 Reactive Hazard Reference Materials

Numerous publications document safe practices for reactive hazard management. The reference section of this report lists resources that are widely available to plant designers and

---

[19] Available for download at http://www.csb.gov.

[20] Of reactive incidents, 25 percent originate in reactors, with the rest occurring in a wide range of equipment.

operators. Both OSHA and EPA maintain web pages dedicated to reactive chemical hazards.[21] These pages provide information on recognizing and safely managing reactivity hazards as well as links to additional information, case studies, and screening tools. In addition, an AIChE-published reference book, *Essential Practices for Managing Chemical Reactivity Hazards*, was developed by the Center for Chemical Process Safety (CCPS, discussed below) with support from EPA, OSHA, the American Chemistry Council, and the Society of Chemical Manufacturers and Affiliates (formerly the Synthetic Organic Chemical Manufacturers Association). This publication is available online free of charge at http://info.knovel.com/ccps/; the online version includes the full text and interactive support software.

6.3 American Institute of Chemical Engineers

AIChE is a professional organization that provides chemical process safety guidance through education, training, and outreach. Two AIChE groups, the Design Institute for Emergency Relief Systems (DIERS) and the Center for Chemical Process Safety (CCPS), are active in improving reactivity hazard management and education.

6.3.1 Design Institute for Emergency Relief Systems

DIERS is a non-profit organization formed by AIChE in 1976 to promote improved emergency relief design. Presently more than 230 domestic and international member companies interested in design, use, or manufacture of emergency relief systems comprise the DIERS Users Group, which meets semiannually.

DIERS methodology has earned broad industry acceptance and recognition. OSHA's PSM regulation recognizes AIChE/DIERS technical reports as "generally recognized and accepted good engineering practice." OSHA has also recommended using DIERS methodology for reevaluation of relief sizing following a runaway reaction incident.

DIERS Users Group publications address screening and management of chemical reactivity hazards, reactivity test apparatus design and evaluation, scale-up, and design of emergency relief systems. As of the date of this report, the DIERS Users Group was revising the second edition of the CCPS book *Guidelines for Pressure Relief and Effluent Handling Systems* to include technology updates, simplify the text, and provide an advanced software tool for pressure relief design.

---

[21] The EPA Reactive Chemical Hazards web page is http://www.epa.gov/oem/content/learning/chemical_hazards.htm; the OSHA Chemical Reactivity Hazards web page is http://www.osha.gov/SLTC/reactivechemicals/index.html.

6.3.2 Center for Chemical Process Safety

CCPS identifies and addresses process safety needs within the chemical, pharmaceutical, and petroleum industries. CCPS members define, develop, and publish engineering and management best practices to prevent or mitigate catastrophic releases of chemicals.

In 1992, CCPS's Safety and Chemical Engineering Education (SAChE) Committee began providing chemical process safety teaching materials for undergraduate chemical engineering curricula. SAChE products address hazards of chemical reactivity, inherently safer design, risk assessment, simplified relief design, and other process safety topics. Two SAChE products address runaway reactions; one of these provides information on experimental characterization and vent sizing using the Advanced Reactive System Screening Tool (ARSST) (Appendix C discusses the CSB's use of the ARSST).

In 2004, CCPS formed the Reactivity Management Roundtable (RMR) to assimilate, implement, maintain, and update effective practices for managing chemical reactivity. RMR provides reactivity hazard recognition and management resources to companies with limited technical sophistication. RMR is developing web-based tools that use decision trees to identify potential chemical reactivity hazards.

In 2008, SAChE began a Student Certificate of Safety Achievement Program which is available free to student members of U.S. universities' AIChE chapters. Two certificate modules, in addition to the general module, address runaway reactions and chemical reactivity hazards. The program issued about 800 certificates in its first year.

References

Center for Chemical Process Safety (CCPS), 2009. *Inherently Safer Chemical Processes: A Life Cycle Approach,* American Institute of Chemical Engineers (AIChE). Previous editions 1996 & 2004.

CCPS, 2003. *Essential Practices for Managing Chemical Reactivity Hazards*, AIChE.

CCPS, 2001. *Layer of Protection Analysis - Simplified Process Risk Assessment*, AIChE.

CCPS, 1999. *Guidelines for Process Safety in Batch Reaction Systems*, AIChE.

CCPS, 1995a. *Plant Guidelines for Technical Management of Chemical Process Safety (Revised Edition)*, AIChE.

CCPS, 1995b. *Guidelines for Chemical Reactivity Evaluation and Application to Process Design*, AIChE.

CCPS, 1993. *Guidelines for Auditing Process Safety Management Systems*, AIChE.

CCPS, 1992. *Guidelines for Hazard Evaluation Procedures – With Worked Examples*, AIChE.

Fisher, H.G., Forrest, H.S., Grossel, S. S., Huff, J.E., Muller, A.R., Noronha, J.A., Shaw, D.A., and Tilley, B.J.,1992. *Emergency Relief System Design Using DIERS Technology: The Design Institute for Emergency Relief Systems (DIERS) Project Manual*, AIChE.

Health and Safety Executive (HSE), 2007. Chemical Reaction Hazards and the Risk of Thermal Runaway, www.hse.gov.uk/pubns/indg254.htm, HSE.

HSE, 2000. *Designing and Operating Safe Chemical Reaction Processes,* HSE.

Hendershot, D.C., Louvar, J.F., and Kubias, F.O, 1999. "Add Chemical Process Safety to the Chemistry Curriculum," *Chemical Health and Safety*, American Chemical Society (ACS).

Mannan, S., 2005. *Lees' Loss Prevention in the Process Industries, 3rd ed.*, Elsevier Butterworth-Heinemann.

Occupational Safety and Health Administration (OSHA), 2007. *Storage and Handling of Liquefied Petroleum Gases,* 29 CFR 1910.110, OSHA.

Urban, P. G., Ed., 2000. *Bretherick's Handbook of Reactive Chemical Hazards, 6th ed,* Elsevier Science and Technology Books.

The United States Chemical Safety and Hazard Investigation Board (CSB), 2007. *Synthron, LLC: Runaway Chemical Reaction and Vapor Cloud Explosion*, Case Study Number 2006-04-I-NC.

CSB, 2006. *MFG Chemical, Inc.: Toxic Chemical Vapor Cloud Release*, Report Number 2004-09-I-GA.

CSB, 2002a. *Improving Reactive Hazard Management*, Report Number 2001-01-H.

CSB, 2002b. *The Explosion at Concept Sciences: Hazards of Hydroxylamine,* Case Study Number 1999-13-C-PA.

CSB, 2000. *Morton International, Inc.: Chemical Manufacturing Incident*, Report Number 1998-06-I-NJ.

Appendix A

MCMT Process Chemistry

T2 manufactured MCMT in three steps. The first step of the reaction required heating to initiate the reaction. After that, all three steps were exothermic (heat-producing) and required cooling.

In the first step (called metalation), molten metallic sodium was reacted with methylcyclopentadiene (MCPD) in the presence of a diethylene glycol dimethyl ether (diglyme) to form sodium methylcyclopentadiene and hydrogen gas. This was accomplished by adding metallic sodium to a mixture of MCPD dimer[22] and diglyme, then heating the mixture. Heating melted the sodium and split the MCPD dimer molecules into two MCPD molecules that reacted with the sodium. T2 vented the hydrogen gas produced by this step to the atmosphere.

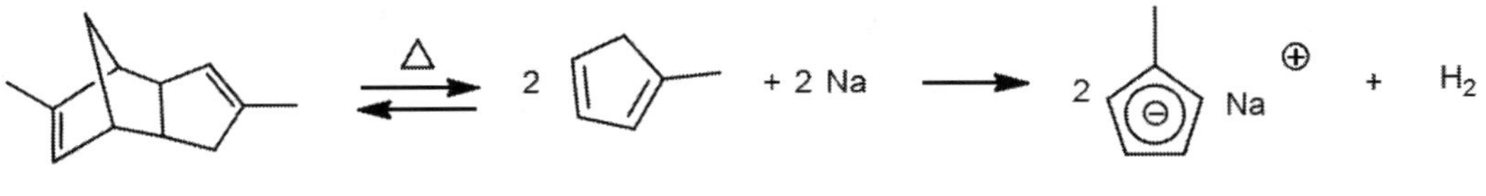

Figure 9. MCMT metalation reaction.

In the second step (called substitution), T2 added manganese chloride (dry powder) to the reactor. The manganese chloride reacted with the sodium methylcyclopentadiene in the reactor to form manganese dimethylcyclopentadiene and sodium chloride.

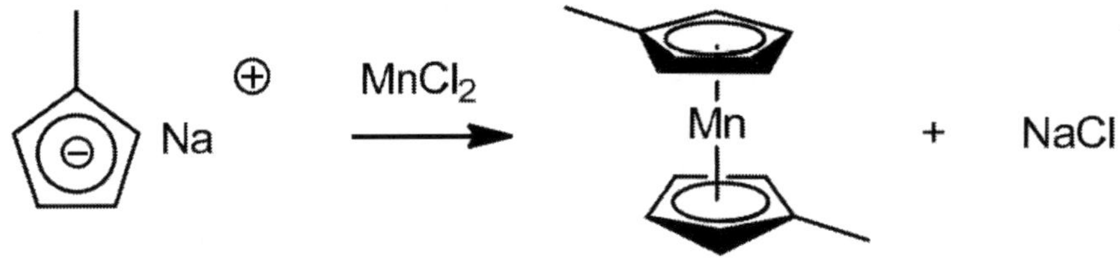

Figure 10. MCMT substitution reaction.

In the final step (called carbonylation), T2 added carbon monoxide gas to the bottom of the reactor, bubbling the gas through the manganese dimethylcyclopentadiene under pressure. In this step, one of the methylcyclopentadiene molecules on each manganese dimethylcyclopentadiene molecule was replaced with three carbon monoxide molecules, forming the MCMT.

---

[22] MCPD dimer is a molecule consisting of two identical MCPD molecules.

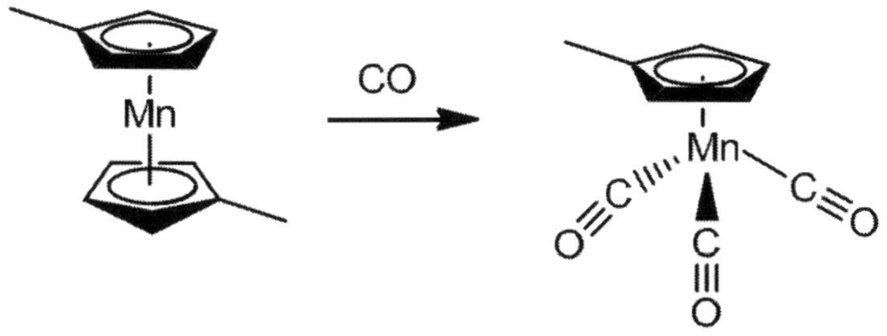

Figure 11. MCMT carbonylation reaction.

Following the carbonylation step, the process operator distilled the mixture to remove the MCMT and the diglyme. The remaining (waste) methylcyclopentadiene and sodium chloride was removed as a solid. The diglyme was recovered and reused in the process.

Appendix B

MCMT Chemical Testing

1.0 Test Description

The CSB performed 12 laboratory tests based on the T2 chemical recipe on the day of the incident, six using the Advanced Reactive System Screening Tool (ARSST) and six using the Vent Sizing Package 2 (VSP2).[23] These systems minimize heat loss through the cell walls to better predict behavior in large vessels, which have much smaller surface-to-volume ratios. All tests were conducted inside a larger containment vessel for safety.

Initial tests were conducted in the ARSST, which uses a small sample size and is open to the containment vessel, to minimize the hazards associated with the potentially destructive chemistry. In these tests, some MCPD and diglyme solvent evaporated because of the open-cell configuration. Two successive exotherms, the second larger, were observed and most pronounced in tests with lower MCPD concentration and excess sodium. The ARSST testing identified conditions required to initiate this second exotherm, but was unable to explore the full runaway potential of the reaction.

Therefore, to more closely match the conditions at T2, the CSB ran a second series of six tests using the larger VSP2 test cell (Figure 12).

---

[23] Both the ARSST and VSP2 are commercially available test apparatus for thermal hazard evaluation testing.

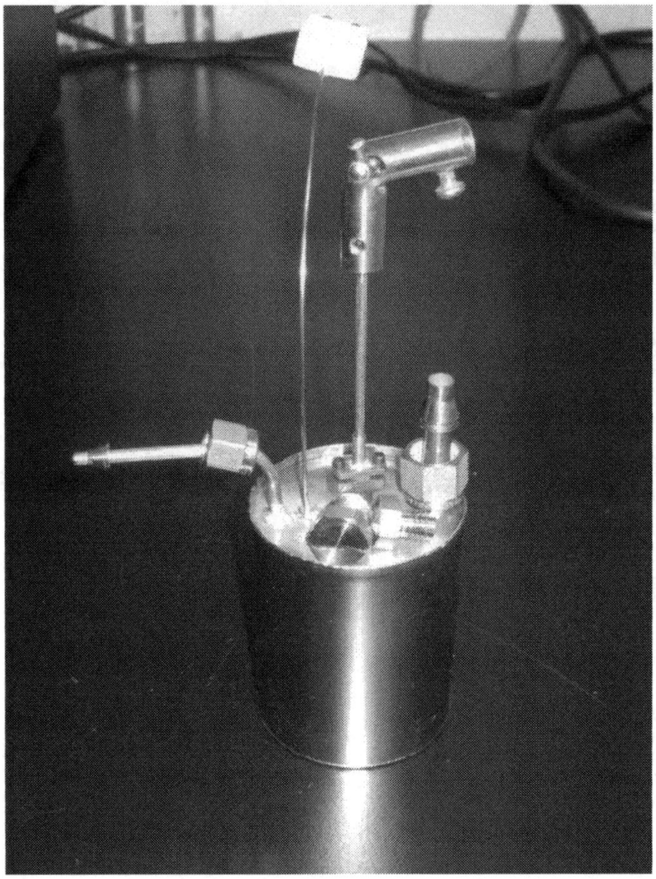

Figure 12. New VSP2 Test Cell.

The VSP2 mimics the T2 facility's physical arrangement more closely and can accommodate more intense stirring to improve mixing of the chemical components. The VSP2 also incorporates baffles in the test cell which provide turbulence and shear in the liquid. These features reduced the molten sodium droplet size and increased contact between the molten sodium and the MCPD, similar to the mixer in the T2 reactor.

In the VSP2 apparatus, a 116 ml test cell is loaded with the raw materials, stir bar, and temperature and pressure measurement devices. The test cell is contained in a protective pressure vessel (containment vessel) that can accommodate high temperature and pressure in the event the test cell ruptures. Tests were either open to containment vessel or sealed in the test cell. A sealed test cell prevents evaporation into the containment vessel, which would have had an undesired cooling effect. This test mimics an event such as a closed vent or undersized relief system.

For the first two VSP2 tests, the test cell was open to the containment vessel and results similar to the ARRST tests were obtained. The CSB conducted the remaining four VSP2 tests in a closed (sealed) test cell. A second and more energetic exothermic reaction was observed when the temperature exceeded 390°F (198.9°C) in two test that used the T2 recipe. The pressure and temperature rise during the second exothermic reaction was about 32,000 psig per minute (2,206 bar per minute) and 2340°F per minute (1300°C per minute) respectively and burst the test cells (Figure 13). The other two tests performed using only sodium and diglyme solvent, suggest that this second more energetic exothermic reaction is a reaction between the sodium and diglyme solvent.

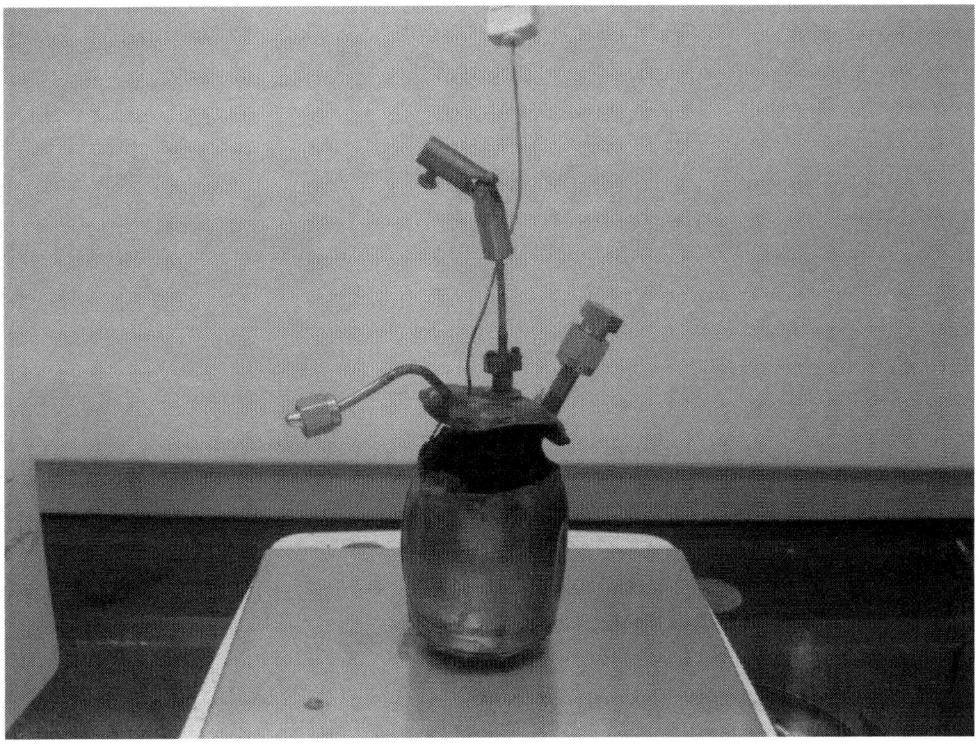

Figure 13. Burst VSP2 Test Cell.

The results of this testing show that the standard T2 chemical recipe is capable of producing the reactor failure observed on December 19, 2007. Using the data obtain from these tests, the CSB determined by calculation failure of the reactor from this second exothermic reaction could not have been prevented by an overpressure relief device. Failure could only be prevented by relieving pressure before the second exothermic reaction and allowing the MCPD and Diglyme solvent to boil and vent. Allowing the MCPD and diglyme to boil and vent removes heat from the mixture and reactants from the reactor. For T2 this would have required a rupture disk set at 75 psig in lieu of the 400 psig disk that was used.

2.0 Test Data

Table 2 presents a summary of the VSP2 test conditions and results.

Table 1. VSP2 Test Results.

VSP2 Run Number	Chemical Components	Test Conditions	Peak Temperature	Peak Pressure	Conclusion
1	Normal recipe	Open to containment	698.4°F 370.2°C	97 psig 6.69 bar	Similar to ARSST
2	Normal recipe	Open to containment	671.4°F 355.2°C	78 psig 5.38 bar	Similar to ARSST
3	Normal recipe	Closed	887.4°F 475.2°C	1512 psig 104.2 bar	Test cell rupture
4	Sodium and Diglyme. No MCPD	Closed.	NA	NA	Leak developed in test cell
5	Sodium and Diglyme. No MCPD	Closed	1281.6°F 694.2°C	1598 psig 110.2 bar	Test cell rupture
6	Normal recipe	Closed	1222.2°F 661.2°C	1788 psig 123.3 bar	Test cell rupture

In VSP2 Run 5, the CSB tested the following chemical recipe:

Table 2. Run 5 Recipe.

Chemical	Actual Charge (g)	Weight Percent of Charge
Sodium	9.412	18.86
Diglyme	37.624	75.38
Mineral Oil[24]	2.877	5.76
Total Sample	49.912	100.00

Patent literature[25] suggests that a sodium-diglyme reaction occurs slightly above T2's normal process temperature. The CSB tested the sodium-diglyme system and determined a strong exothermic reaction can occur between sodium and diglyme at temperatures above 390 °F (198.9°C). T2 conducted the first process step at about 350°F (176.7°C), resulting in a margin of safety of only 40 °F (22.2°C) below the more energetic reaction.

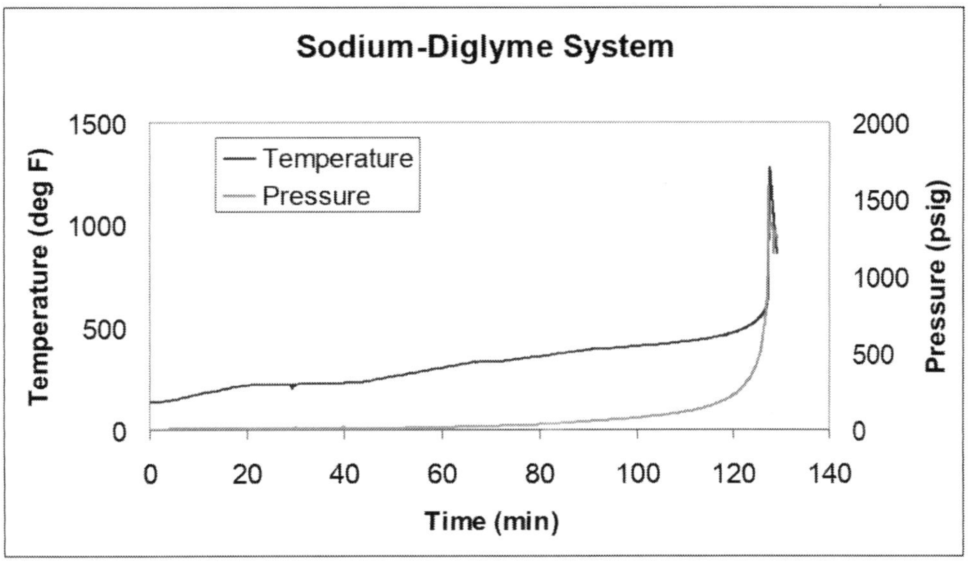

Figure 14. Sodium Diglyme Test Results.

---

[24] Sodium metal arrives packed in mineral oil to prevent oxidation and limit moisture contact with the metal. Some mineral oil is transferred into the test cell with the sodium. The mineral oil does not participate or interfere with the reactions.

[25] US Patent 2,942,040. Manufacture of Sodium Cyclopentadiene from Cyclopentadiene Dimer, June 21, 1960.

In the VSP2 Runs 3 and 6, the CSB tested T2's normal chemical recipe. The recipe used in Run 6 is shown in Table 3 and the results in Figure 15:

Table 3. Run 6 Recipe.

Chemical	**Actual Charge (g)**	**Weight Percent of Charge**
Sodium	5.147	10.59
MCPD Dimer	21.164	43.55
Diglyme (95% purity)	20.690	42.58
Mineral Oil	1.591	3.28
Total Sample	48.592	100.00

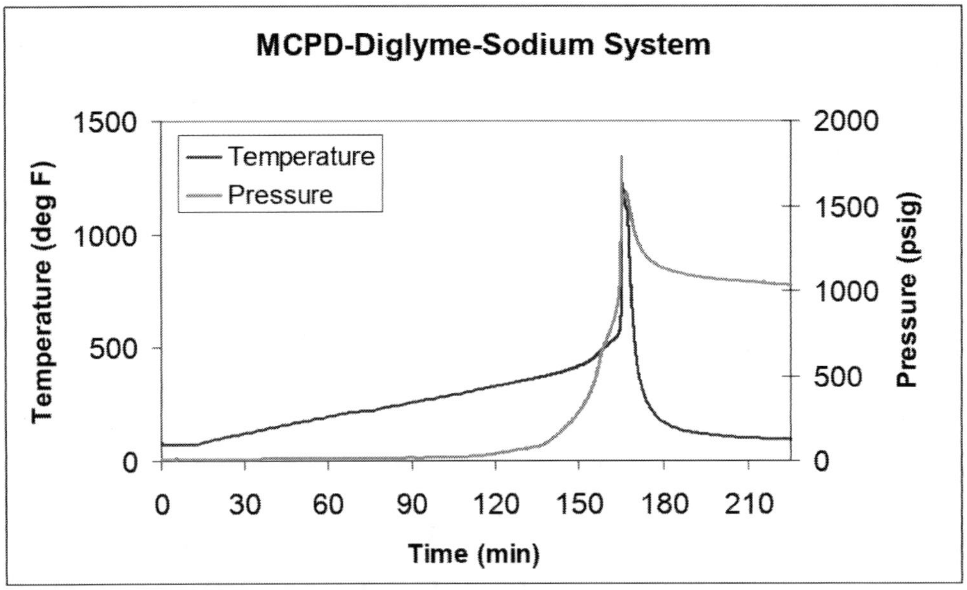

Figure 15. MCPD Diglyme Sodium Test Results.

The peak rate of temperature rise for the final test is 2340°F/minute (1300°C/minute) and the peak rate of pressure rise is 32,000 psi/minute (2206 bar/minute). In the closed system, a strong exothermic reaction occurs using the normal T2 chemical recipe.

An Undergraduate Alarm Documentation & Rationalization Module

Peyton C. Richmond, Chemical Engineering Department, Lamar University, Beaumont, TX

The introduction of Distributed Control Systems (DCS) systems lead to over alarming process plants because alarms became essentially free. However, extraneous process alarms without a clearly defined operator response are a nuisance to process operations and can obfuscate truly critical alarms. In fact, nuisance alarms have frequently been listed as a contributing factor in major industrial accidents. On the other hand, most operating plant personnel are familiar with incidents that could have been prevented by a timely response to a process alarm that was either absent or disabled. Thus, an effective alarm management process must eliminate as many nuisance alarms as possible while still including those alarms necessary to prevent process upsets and incidents.

This presentation describes our experience developing and using an undergraduate PBL module designed to introduce chemical engineering undergraduates to alarm management principles. Groups of 4 to 5 students were assigned to Document & Rationalize (D&R) the process alarms for a reformate stabilizer column. Students were asked to choose one of the following roles: process engineer, process control engineer, superintendent, or operator. They were provided with D&R instructional material, a list of potential concerns about the reformate stabilizer process system, and some example alarm configurations provided by local industry to help them accomplish this module. The module concluded by allowing the students to implement the alarms and test their operation on an actual DCS system.

See more of this Session: Free Forum On Chemical Engineering Education II
See more of this Group/Topical: Education

Preparing for the ABET Accreditation Visit

Gary K. Patterson, Department of Chemical Engineering, Missouri University of Science and Technology, Rolla, MO

Jeffrey J. Siirola, Eastman Research Division, Eastman Chemical Company, Kingsport, TN

Ronald P. Danner, Chemical Engineering, Penn State University, University Park, PA

Edward Rosen, EMR Technology Group, St. Louis, MO

An informal presentation on current expectations for ABET accreditation visits of chemical engineering programs.

See more of this Session: Abet Accreditation Updates and Insights
See more of this Group/Topical: Education

Student Learning in Virtual Laboratories

Milo D. Koretsky, Chemical, Biological, and Environmental Engineering, Oregon State University, Corvallis, OR

Edith Gummer, Education Northwest, Portland, OR

Debra Gilbuena, Chemical, Biological, and Environmental Engineering, Oregon State University, Corvallis, OR

Erick Nefcy, Chemical, Biological, and Environmental Engineering, Oregon State University, Corvallis, OR

Kendra Seniow, Chemical, Biological, and Environmental Engineering, Oregon State University, Corvallis, OR

Eric Gunderson, Chemical, Biological, and Environmental Engineering, Oregon State University, Corvallis, OR

Christine Kelly, Chemical, Biological and Environmental Engineering, Oregon State University, Corvallis, OR

In the context of providing an effective capstone experience in experimental process design, with funding from NSF CCLI (proof-of-concept and Phase 2) and the Intel Faculty Fellows Program, we have developed two virtual process laboratories, the Virtual Chemical Vapor Deposition (CVD) laboratory and the Virtual BioReactor laboratory. In a virtual laboratory, simulations based on mathematical models implemented on a computer can replace the physical laboratory. Since real systems do not deterministically adhere to fundamental models, random and systematic process and measurement variation are added to the output. In our case, students engage in the virtual laboratory in a situated manner, where they attempt to optimize the performance of an industrial process by investigating the effect of the input variables. In the School of Chemical, Biological and Environmental Engineering (CBEE) at Oregon State University (OSU), the virtual laboratory capstone project has been used since Winter 2005, where 270 students in teams of 2-3 have completed the assignment. Through the use of cyber-enabled networks and communications technologies, the laboratories are also remotely available to other institutions. The Virtual CVD laboratory project has been completed remotely at the University of Oregon, UC Berkeley, and the University of Montana, as well as at the high school and community college levels. In the latter cases, the instructional design was significantly modified to align with the knowledge and abilities of high school students. In addition to the software development, instructional design and implementation, we have been developing methods to measure the cognitive processes of students undergoing the virtual laboratories, the social interactions between members of the teams, and student expectations of and metacognitions about their virtual laboratory experience. By simulating the physical operation of the process and metrology equipment, this cyber-enabled learning tool greatly simplifies the physical aspect of performing the experiments and allows a student to experience a different emphasis on the experimental design process than he/she typically encounters in a physical laboratory. This feature allows instructors to focus on aspects of student learning that have been historically elusive.

The study reported investigates student cognitions and metacognitions as they engage in the virtual laboratory project. The mixed methodological perspective of this research targets key project activities to collect data. The primary means of data collection is Protocol Analysis where student groups "think aloud" as they perform the assignment. In addition, samples of written work (Journals and Final Reports) have been analyzed and student survey responses after completion of three different laboratories, two physical laboratories and a virtual laboratory, have been coded. Task analysis of "think-aloud" sessions has verified that students are engaged in the intended, iterative experimental design approach of practicing engineers. In all cases, students engaged in multiple "design" and "analysis" loops. Due to time constraints, this type of iterative experimentation is difficult to achieve in a physical laboratory. Additionally, this laboratory experience was demonstrated to promote higher level cognition in students. Modified Perry's levels were applied to quantify the team's tolerance for ambiguity. Evaluation results elucidating tolerance for ambiguity indicate that by completing this open-ended problem most students evolve past "blind acceptance of authority" and become aware of a "multiplicity of views"; however, while some students continued to climb Perry's levels, eventually becoming comfortable with the idea of "contextual relativism," other students did not. An interesting parallel to these differences is found in the nature of the sociocognitive interactions found in the different student teams; these interactions seem to be able to either promote the desired learning, or they can be detrimental to the intended learning outcomes. Further research into the impact of reflection and sociocognitive interactions is planned to address these issues more fully.

Students' perceptions of the three different laboratory experiences are discussed from the focus of intended (metacognitive questions) and actual learning (cognitive questions). Student perceptions about the laboratory experiences were correlated to student performance in the class, as measured by the students' scores on all graded assessments for the course. Analysis of metacognitive statements of students show enhanced awareness of experimental design, and greater occurrences of critical thinking and higher order cognition in the virtual laboratories.

See more of this Session: Fundamental Research in Education
See more of this Group/Topical: Education

Creation of Reliable Instruments for Assessment of Conceptual Understanding in Heat Transfer and Thermodynamics

Margot A.-S. Vigeant, Chemical Engineering, Bucknell University, Lewisburg, PA

Michael Prince, Chemical Engineering, Bucknell University, Lewisburg, PA

Katharyn Nottis, Education, Bucknell University, Lewisburg, PA

A consistent challenge in rigorous research in engineering education research is the availability of valid and reliable instruments with which to document student gains. This presented a significant obstacle to the course of our NSF-funded work to enhance student conceptual learning in chemical engineering Heat Transfer and Thermodynamics through creation and implementation of inquiry-based activities. Because the goal of our work is to enhance students' conceptual understanding in these areas, a concept inventory is the optimal assessment tool. Concept inventories are multiple choice tests where questions focus on conceptual rather than mathematical understanding and where the "wrong" answers are designed to appeal to commonly held misconceptions. While there were concept inventories in Heat Transfer and Thermodynamics under development, there were none that both 1) addressed the list of concepts under study and 2) had been tested for sufficient reliability and validity.

In this paper, we describe the development and testing of our concept inventories. In Heat Transfer, the concepts of Temperature vs. Energy, Temperature vs. "Feeling" Hot Rate vs. Amount, Radiation. The test was piloted with 202 engineering students at a variety of institutions as a pre- and post- test for relevant courses, and was found to have a split-half reliability of 0.84 and a KR20 of 0.86 for the post-test, both of which indicate a sufficiently high reliability for a research instrument. In the field of Thermodynamics, the concept inventory addresses Entropy, Reversibility, Equilibrium vs. Steady State, Internal Energy vs. Enthalpy, and Reaction Rate vs. Reaction Equilibrium. This test was piloted with 114 students at several institutions, and found to have a split-half reliability of 0.74 and a KR20 of 0.76 on the post- test, which again indicates a sufficiently high reliability for research work.

Future work includes the use of these concept inventories to assess student conceptual gains after use of our inquiry-based activities. We invite instructors in relevant courses to contact us if you wish to be part of the study.

See more of this Session: Fundamental Research in Education
See more of this Group/Topical: Education

Paper Presented at the AIChE Annual Meeting
Nashville, Tennessee
November 10, 2009

Identifying and Repairing Students Misconceptions in Thermal and Transport Science

Ronald L. Miller, Colorado School of Mines
Ruth A. Streveler, Purdue University
Dazhi Yang, Purdue University
Aidsa Y. Santiago Roman, University of Puerto Rico, Mayaguez

Introduction

This paper describes an integration of two research lines combining *identification of* students' misconceptions of difficult engineering concepts with efforts to *repair* some particularly robust misconceptions. Previous studies reported that misconceptions related to heat transfer, fluid mechanics, thermodynamics, and other engineering and science concepts persist among engineering students even after they completed college-level courses in the subjects. [Duit, 2009] Therefore, the first line of our research is focused on two research questions:

- "What important concepts in thermal and transport science are difficult for engineering students to learn?"
- "How an instrument can be developed to identify engineering student misconceptions of these difficult and important concepts?"

The first research question was investigated by conducting a Delphi study [Streveler, et al. 2003] with experienced engineering faculty to identify important and difficult concepts in thermal and transport science. The second research question was investigated by developing the Thermal and Transport Concept Inventory (TTCI). The TTCI is an instrument that measures the conceptual understanding of key ideas in thermodynamics, fluid mechanics, and heat transfer for undergraduate engineering students. Thus, TTCI is also a tool for identifying students' misconceptions in thermal and transport science. Details of misconception identification and TTCI development are provided in the following section.

The second line of our research is focused on how to repair students' misconceptions of difficult engineering concepts once the misconceptions are identified. As indicated in the misconception literature, some of these misconceptions are particularly robust and therefore are particularly difficult to repair using traditional pedagogical strategies. Thus, for this line of the research, we are developing and testing schema training strategies for helping engineering students develop more fundamentally accurate mental models of selected engineering and science concepts.

The schema training strategies were based on the assumption that students learn new concepts by assimilating or encoding new information into an existing schema or category. Assimilation helps students make inferences about and assign attributes to a new concept or phenomenon. However, when students begin learning some particularly challenging engineering concept that are fundamentally different from their commonsense conception, they can make the wrong inference or assign incorrect attributes to the new concept based on their existing incomplete or incorrect schema. For example, the hot or cold sensation we sense when touching an object indicates its temperature (actually it is a measure of how fast energy is transferred into or out of our finger).

To repair such robust misconceptions, Chi and her colleagues proposed an innovative instructional approach involving schema training methods which focus on helping students develop appropriate schemas or conceptual frameworks for learning difficult and challenging engineering concepts. [Chi, 2005] Such methods were effective in helping middle school students and undergraduate psychology students learn difficult science concepts. We are testing Chi's theoretical framework by creating effective schema training protocols and materials that help engineering students create appropriate mental models of fundamentally important dynamic processes and concepts, especially those operating at small length scales.

There is ample evidence in the literature to suggest that students of all ages (including science and engineering students) do not easily understand fundamental small-scale phenomena such as heat transfer, diffusion, fluid mechanics, and electricity. Given the current interest in advances in nanotechnology (e.g. microfluidics, biotechnology, genetic engineering, nanoscale machines), new engineering graduates must have a firm grip of fundamental processes that are characterized by small-scale dynamic systems. Therefore, schema training methods hold promise not only for thermal and transport science but also for other disciplines in engineering.

Discussion

In this section, we discuss the methodology used to identify important and difficult thermal and transport science concepts that are included in the TTCI instrument. We also discuss the methodology and procedures used to generate TTCI items and the results of validity and reliability analyzes from TTCI pilot testing with engineering students. Development of schema training methodologies and materials is discussed followed by a brief summary of schema training results obtained so far.

Identifying important but difficult concepts in thermal science. After considering several methods of identifying important but difficult concepts in thermal and transport science, we choose the Delphi method, which focuses on expert opinion. Our experts consisted of 31 engineering professors ranging from assistant to full professors. Five of the experts were also textbook authors. They were asked to complete a generative round to develop candidate concepts and then 3 rating rounds in which each concept was rated on two scales: importance and difficulty. The non-parametric median and interquartile range were used (rather than mean and standard deviation) because an ordinal scale was used to rate the concepts. The rankings for most concepts stabilized by round 2 (the median for 19 of the 28 concepts changed by a value of 0.5 or less) as suggested by other Delphi studies reported in the literature (Linstone and Turoff, 1996).

The goal of this part of the study was to identify concepts which had high importance (those that were given a high ranking in the "importance" scale) and were also conceptually difficult (those that were given a low ranking on the "conceptual understanding" scale. A total of 12 concepts (from an original list of 28 concepts from the generative round) were identified as meeting the criteria of high importance but low conceptual understanding. These items included key topics in thermal science and transport disciplines such as the 2nd law of thermodynamics including reversible vs. irreversible processes, conservation of fluid momentum, viscous momentum transfer, several energy-related topics (heat, temperature, enthalpy, internal energy), and steady-state vs. equilibrium processes. Two of the concepts (differential vs. integral analysis and system vs. control volume analysis) were deemed mathematical rather than physical concepts and were not included in the TTCI instrument. At the request of several Delphi participants, we also included the ideal gas law and conservation of mass concepts in the TTCI since both are fundamental concepts in fluid mechanics and thermodynamics. Finally, we temporarily set aside the thermal radiation concept that might eventually be included in the instrument. This decision was made so that we could focus development of early versions of the instrument on what were deemed more fundamental heat transfer topics.

TTCI development and results. Based on results of the Delphi study, items were developed for the TTCI assessment instrument. [Streveler, et al., 2009] Each item was developed using a seven-step process recommended by Downing and Haladyna [2006] including:

- drafting open-ended questions about the concept
- collecting student response data orally (think-aloud problem solving sessions) and in written form
- using the responses to convert the open-ended questions to multiple choice items with distractors describing plausible but incorrect answers
- beta testing the drafted items on groups of engineering students
- collecting expert reviews on each item (to establish content validity)
- revising the items based on statistical performance and expert feedback
- collecting additional beta test data.

Updated versions of the TTCI have been created with items from the initial instrument (version 1.0) deleted and new ones based on difficulty and discrimination indices. Statistics for each succeeding version of the instrument has indicated improved reliability. The present version of the instrument (version 3.04) had consistently demonstrated reliabilities of 0.8 and higher for each of the following sub-instruments:

- heat transfer – 12 items containing 18 questions
- fluid flow – 19 items containing 26 questions
- thermodynamics – 17 items containing 24 questions

Version 3.04 is now available as an on-line version (www.thermalinventory.com). Nearly 400 students at more than 20 engineering schools have used at least one of the 3 TTCI instruments.

Schema training development. Chi has argued that students have robust misconceptions because they have no existing mental framework for understanding some complicated science and engineering processes. [Chi, 2005] A particular ontological class of difficult concepts identified by Chi is termed "emergent processes", which are fundamentally different from "sequential processes." Emergent processes occur in systems of constituent elements (e.g. molecules) interacting over time in a random and simultaneous pattern, often in conjunction with equilibration. In contrast, sequential processes occur in systems of interacting agents in a causal and dependent pattern.

Many of the concepts with which engineering students struggle can be identified as emergent processes including heat transfer, diffusion and electricity. [Chi, 2005; Slotta, Chi & Joram, 1995] Emergent processes misconceptions are particularly resistant to traditional instruction because they are made at the ontological level – where students ascribe a fundamental characteristic to the concept that is at odds with the scientifically normative view. [Chi, 2005] In order to help students learn concepts of the emergent process ontology, instruction should help students develop a "schema" or mental framework for that ontology that would make subsequent related concepts easier to understand. [Chi, Roscoe, Slotta & Roy, in press; Slotta & Chi, 2006] Since prior work has demonstrated that even advanced engineering students still hold misconceptions about fundamental concepts in thermal sciences and other scientific subjects [Miller et al., 2006], this study is intended to test whether the schema training framework is effective in helping repair engineering students' misconceptions in heat transfer, diffusion, and microfluidics.

Following the work of Chi and her colleagues, the schema training experiment used a mixed methods design by collecting both qualitative and quantitative data. As shown in Figure 1, the experiment uses both experimental and control groups of students matched for equivalent levels of engineering education. A pre-test in heat transfer concepts is used as a further measure of the "equivalence" of the two groups prior to schema training. The experimental group completes an on-line training module describing the characteristics of sequential and emergent processes and describes why diffusion as an emergent process. The control group completed an on-line training module of approximately equivalent length and complexity (words and figure) that describes the nature of science. Diffusion is described but no mention is made of emergent processes. The purpose of the training module is to help experiment group participants with a "schema" for thinking about diffusion in emergent terms.

Both groups complete the same instruction module on heat transfer principles. Post-test concept questions are answered by both groups followed by a far transfer experiment – microfluidics instruction and concept assessment. Since microfluidics involves ultra-laminar flow, the diffusion behavior of dye molecules and particles such as viruses and bacteria in these flows represents an ideal application of emergent process principles in a context for which undergraduate engineering students are unfamiliar prior to their participation in schema training.

Procedure	Experimental Group	Control Group
Demographic Questions Survey	Demographic Information	
Pre-Test	Heat Transfer concept questions	
Training Module	Sequential and Emergent Processes; diffusion as an example of an emergent process	The Nature of Science; diffusion example with **no** mention of emergent processes
Target Instruction	Heat transfer instruction	Heat transfer instruction
Test for Understanding	Diffusion concept questions	
Post-Test	Heat Transfer concept questions (repeated measure)	
Far Transfer Instruction	Microfluidics	
Test for Understanding	Microfluidics concept questions	

Figure 1. Schema Training Experimental Design

Schema training preliminary results. The first group of students participating in our schema training experiment consisted of 60 juniors and seniors at a large Midwestern research university. Table 1 summarizes the demographic data for these students. Each student completed schema training protocols over a two-day period (2 hours per day) breaking before the target instruction.

Table 1. Demographic Data for Schema Training Participants

Demographic Information Category		Number of Participants	Total
Gender	Female	23	
	Male	37	60
Year of College	Junior	21	
	Senior	39	60
Ethnicity	Caucasian	45	
	Asian/Pacific Islander	15	60
Major	Mechanical engineering	20	
	Chemical engineering	24	
	Other	16	60

Both groups of participants (experimental and control) completed the pre-test for heat transfer at the beginning of the study on the first day and the post-test for heat transfer on the second day. The pre- and post-tests consisted of 18 multiple choice questions which were chosen from the Thermal and Transport Concept Inventory (TTCI) for identifying students misconceptions (Olds, Streveler, Miller & Nelson, 2004).

The overall mean gain (the average of post test scores minus pre test scores) for the experimental group (1.10) was larger than that (.97) of the control group (see Table 2). However, there was no significant difference ($p>.05$) between the two groups in terms of mean gains. The non-significant statistic showed that the schema training approach did not help those engineering students in repairing their misconceptions with concepts in heat transfer.

Table 2. Descriptive Statistics for Mean Gain in Heat Transfer Assessment (18 total questions)

Group	N	Mean Gain	Std
Experimental	30	1.10	1.972
Control	30	.97	2.591

Based on the data from 19 multiple choice diffusion concept questions, the overall mean for the experimental group (15.40) was larger than that (13.87) of the control group (see Table 3). In addition, there was a significant difference ($p<.05$) between the two groups. This result showed that the schema training approach did help those engineering students in the experimental group with their understanding of some diffusion concepts.

Table 3. Descriptive Statistics for Performance on Diffusion Assessment (19 questions)

Group	N	Mean	Std
Experimental	30	15.40	2.673
Control	30	13.87	2.886

Based on the data from 5 multiple choices questions on microfluidics, the overall mean for the experimental group (3.60) was larger than that (2.77) of the control group (see Table 4). In addition, there was a significant difference ($p<.05$) between the two groups. This result showed that the schema training approach did help those engineering students in the experimental group with their understanding of diffusion concepts in a microfluidics system.

Table 4. Descriptive Statistics for Performance on Microfluidics Assessment (5 questions)

Group	N	Mean	Std
Experimental	30	3.60	1.380
Control	30	2.77	1.455

Although preliminary, these data suggest that schema training methods can be designed to help engineering students repair strongly held misconceptions of concepts in which a well-developed emergent schema is required for correct understanding. The absence of a measurable improvement in heat transfer scores for the experimental group is still under investigation but perhaps may be attributed to students taking multiple heat transfer courses which may solidify (sometimes incorrect) cognitive structures of heat transfer processes. For example, a statistically significant improvement in heat transfer scores was found for students taking two or fewer heat transfer courses while no improvement was noted for students who completed three or more courses.

Conclusions

This paper has described two related lines of research involving development and testing of an assessment instrument to identify strongly held student misconceptions in thermal and transport science and development of a clinical method for helping students develop accurate schema for describing and understanding emergent processes that are common in heat transfer, molecular diffusion and molecular momentum transfer. Results show that the assessment instrument, the Thermal and Transport Concept Inventory can reliably identify misconceptions related to 12 important but poorly understood concepts in heat transfer, thermodynamics, and fluid flow.

Preliminary schema training results show that materials informed by relevant psychological theory can be used to help students develop correct mental schema for understanding robust misconceptions involving emergent processes important to thermal and transport science. Additional data collected in this project will provide more detailed guidance about how to transform clinical schema training to classroom contexts.

References Cited

Chi, M.T.H. (2005). Commonsense conceptions of emergent processes: Why some misconceptions are robust. *Journal of the Learning Sciences, 14*(2), 161-199.

Chi, M. T. H., Roscoe, R., Slotta, J., & Roy, M. Domain-general qualitative understanding of emergence transfers to learning of specific science concepts," *Journal of the Learning Sciences*, (in press).

Downing, S.M. and Haladyna, T.M., eds. (2006). *Handbook of Test Development.* Mahwah, NJ: Lawrence Erlbaum Associates.

Duit, R. (2009). Bibliography: Students' and Teachers' Conceptions and Science Education. Kiel, Germany: Institute for Science Education. Available at http://www.ipn.uni-kiel.de/aktuell/stcse/

Linstone, H.A., Turoff, M., eds. (1996). *The Delphi Method: Techniques and Applications.* Reading, MA: Addison-Wesley Publishing Company.

Miller, R. L., Streveler, R. A., Olds, B., Chi, M. M. T. H., Nelson, A., & Geist, M. R. (2006). Misconceptions about rate processes: preliminary evidence for the importance of emergent conceptual schemas in thermal and transport sciences. *Proceedings of 2006 American Society for Engineering Education Annual Conference*, Chicago, Illinois.

Slotta, J. D., Chi, M. T. H., & Joram, E. (1995). Assessing students' misclassifications of physics concepts: An ontological basis for conceptual change. *Cognition and Instruction, 13*, 373-400.

Slotta, J. D., & Chi, M. T. H. (2006). Helping students understand challenging topics in science through ontology training, *Cognition and Instruction, 24*, 261-289.

Streveler, R.A., Olds, B.M., Miller, R.L., and Nelson, M.A. (2003). Using a delphi study to identify the most difficult concepts for students to master in thermal and transport science. *Proceedings of the American Society for Engineering Education Annual Conference* (electronic), Nashville, Tennessee.

Streveler, R.A., Miller, R.L., Santiago Roman, A.I., Nelson, M.A., Geist, M.R., and Olds, B.M. (2009). Using the 'assessment triangle' as a framework for developing concept inventories: a case study using the thermal and transport concept inventory. Submitted to the Journal of Engineering Education.

Textbooks Are So Twentieth Century

Carl R. Lund, Chem & Bio. Engr, SUNY-Buffalo, Buffalo, NY

National statistics show that lecturing predominates as the instructional method in engineering education even though research shows other methods to be more effective and even while professional and governmental bodies proclaim the need for changes in engineering education. One premise underlying the project described here is that many faculty would adopt more effective teaching methods if doing so took no more of their time than teaching by lecturing and was equally "comfortable." Textbooks in use today do not provide the resources necessary for meeting these requirements. While many, many teaching resources are available, instructors must find them, figure out how to use them, modify them to fit their course, and implement them in their class. These activities are time-consuming and may still leave the instructor uncertain as to how to work the activity into their class (which in turn causes discomfort or anxiety when they teach).

The present project addresses this situation by developing a new kind of teaching resource that goes beyond the present-day textbook. Using the acronym TExT (Toolkit for Exceptional Teaching), it provides three important components. First, it provides all the materials found in a current-day textbook, including written presentation of the technical subject matter, illustrative example problems and unsolved problems that can be used as homework assignments. Second, it provides video files that essentially present "lectures" for the entire course. These are intended for use by the students prior to coming to class. Two important aspects of these videos are (a) that the professor need not feel guilty for not lecturing in class because the students still get a lecture, and (b) that by assigning students to watch the videos prior to class, class time can then be put to more effective use. The third important component of the TExT is a complete set of in-class learning activities for the teacher to choose from and use in class.

The learning activities are varied in type and style. They include activities such as object lessons, "one-minute" papers, explore and learn activities (sometimes using provided simulators), three slide presentations, group or individual problem solving, brainstorming sessions, contests, debates and panel discussions. Each activity includes a list of necessary resources, a detailed lesson plan, slides and other visuals, and tips on conducting the activity. The instructor does not need to locate the activity online or elsewhere, it is already presented using nomenclature consistent with all other parts of the TExT, and there isn't any additional pre-requisite knowledge that will need to be presented in order to use it.

A TExT has been developed for teaching undergraduate kinetics and reaction engineering. The use of the TExT will be described, and examples of the different kinds of learning activities will be presented. Initial assessment results related to student learning when using the TExT will also be presented. However, the primary reason for creating this TExT is to test a few hypotheses related to how to effect a change in the methods used in engineering education. The following are among the hypotheses that will be tested using this TExT: that an instructor would be able to implement active learning via the resources supplied with the TExT without investing significantly more time than would be invested in teaching via the traditional lecture, that by providing lesson plans and teaching instructions for each learning activity, an instructor without an education background or prior experience involving active learning, will be able to implement active learning effectively (i. e. so that student learning improves relative to the situation where the same instructor used traditional lectures) and that engineering instructors who currently teach using conventional lectures will be willing to change to active learning via a TExT (assuming the first two points are successfully demonstrated). The next phase of the project will entail using the TExT to test these hypotheses.

See more of this Session: Communication and Collaboration Using New Technology
See more of this Group/Topical: Education

Student Electronic Portfolios for Professional Development Using Google Apps

Jessica Anderson Kuczenski, Chemical and Biomolecular Engineering, University of Notre Dame, Notre Dame, IN

Joshua A. Enszer, Chemical and Biomolecular Engineering, University of Notre Dame, Notre Dame, IN

Mark J. McCready, Chemical and Biomolecular Engineering, University of Notre Dame, Notre Dame, IN

Jay B. Brockman, College of Engineering, University of Notre Dame, Notre Dame, IN

Xiaojing Duan, Office of Informational Technology, University of Notre Dame, Notre Dame, IN

Paul Turner, Office of Informational Technology, University of Notre Dame, Notre Dame, IN

The primary goal of our undergraduate program is to produce engineers who are one step ahead of their peers, who have begun to prepare themselves for more than just their entry-level jobs. In order to accomplish this, we seek improvements to the educational process outside the classroom and ways to encourage students to have a more direct role in their own personal development. For this purpose, we present here the Notre Dame Electronic Portfolio (NDeP) project, which is designed to help us meet this goal.

An electronic portfolio is a collection of electronic files that is created and maintained by a user as evidence of professional development. In particular, we hope to achieve four fundamental objectives with our NDeP project: 1) engage students in their learning and allow students to play a more direct role in their personal development, 2) enable students to link together diverse parts of their learning, including the formal and informal curriculum, and allow for reflection and awareness of learning strategies and needs, 3) encourage students to create a professional portfolio of their achievements, both from the classroom and beyond, with the ability to publicize their achievements as evidence towards future employers, etc. and 4) offer the academic department (in this case the Chemical and Biomolecular Engineering department) a meaningful mechanism for accessing and organizing evidence of student learning.

To achieve the goals for this project, we chose to implement a Google Apps ePortfolio using the entire Google Apps suite. This includes gDocs, gSpreadsheets, Picasa, Google Video, Blogger, and the like for evidence collection. For student presentation, reflection, and electronic documentation, we are planning to use gSites.

The e-portfolio also serves as a tool to make communication between students and advisers more efficient, and to help students formulate short- and long-term plans for their own career development. As a portion of the academic advising experience, NDeP includes a set of surveys designed to help students consider their academic coursework, in which we also have students assess their own progress in the program. The survey addresses the six main tools every engineer in our program should have: analytical and problem solving skills, comprehension of engineering concepts, understanding of technical problems, communication skills, creativity and independence, and organizational skills. For each tool, students are asked to consider what deliverable proof exists that demonstrates their current proficiency. These artifacts can be electronically archived on their own NDeP site. The surveys also have the ability to be compiled for use in ABET accreditation and other department assessments of student learning, interests, and achievements.

Therefore, we present here the current progress and future goals of the NDeP project, including information from the pilot group of students and plans to make the process more interactive for both students and faculty.

See more of this Session: Communication and Collaboration Using New Technology

See more of this Group/Topical: Education

Keeping up with the Student by Introducing New Technology Into Chemical Engineering Courses

Noelle K. Comolli, Chemical Engineering, Villanova Univesity, Villanova, PA

William J. Kelly, Chemical Engineering, Villanova Univesity, Villanova, PA

Our current students probably spend more time on their laptops, blackberries and IPod's than reading their textbooks. In order to keep up with the students we need to make our engineering classes more interactive and take advantage of technologies such as tablet PC's, "clickers", WebCT and Facebook. Presented here are some ideas used by the Villanova Chemical Engineering Department for integrating technology into our courses. Some of the courses developed using these technologies include both traditional chemical engineering core courses (Heat Transfer Operations) as well as chemical engineering elective courses (Biomaterials, and Special Topics in Drug Delivery).

For all of the courses, a simple method to include technology started with keeping all course files on WebCT. This allows students constant access to lecture notes, homework solutions and their grades in the course. This also allows the student to see where their grade is relative to the class average for all assignments. WebCT can also be used to create practice quizzes for students to test their knowledge of the course material.

Since students are issued laptops at Villanova University as part of their tuition, it is easy for us to take advantage of this and make all notes "paperless". Most of the faculty has access to tablet PC's where we can write on the screen during lectures. This allows us to send out to the students the incomplete or "starter" set of notes that are on the screen. These notes include some key words, definitions, equations and images. This allows the students to time to add their own notes either by hand or on their laptops, and gives them more time to listen to teacher. To make the learning even more interactive, a set of "clickers" were purchased from Turning Point Technologies Inc. These "clickers" and corresponding software allow the teacher to poll the class and have the results appear graphically on the screen. The "clickers" have proven a fun and effective way to see if the students are following the theory that was just introduced in the lecture. The "clickers" are also extremely effective and keeping students interested in longer class periods.

One of the newest methods of keeping up with the students is to introduce "on-line" office hours. Many have done this using AOL's instant messenger. Almost all students are now on Facebook, so taking advantage of this technology, online office hours can be hosted here as well. A course page can easily be made that only current students can be allowed to join. In this page anyone can post questions and answers. It can also be extremely effective method of connecting students who do not live on campus, or who have questions late at night.

These ideas for integrating technology have been successful for our courses and we look forward to sharing these and hearing more ideas from colleagues at AIChe.

See more of this Session: Communication and Collaboration Using New Technology
See more of this Group/Topical: Education

Using Class Wikis as Reflective Learning Tools

David L. Silverstein, University of Kentucky, Paducah, KY

At the end of a lecture there is a tendency for a student to rapidly move as far away from the classroom and the topics discussed as possible. This apparent instinct to flee and thereby free their mental energies from class work discourages reflection on the lecture subject by the student. This tends to reduce retention and absorption of the concepts presented during the class.

One effective means of encouraging reflection as a mode of learning is portfolio development. They foster active learning and give the student the perspective needed to ensure that they are progressively learning over time. The downside of portfolios, whether paper-borne or electronic, is that they take significant student time to prepare and manage. McGourty notes that "full portfolios are very labor intensive and are sometimes abandoned due to the work required in assessment, despite the richness of results." In this project, the reflective element of portfolios was extracted and applied in a collaborative, online setting using an electronic wiki.

A wiki is a web-based document that can be edited by multiple users. For this project, students in a material and energy balance course were assigned the weekly task of maintaining a wiki page on the current textbook chapter by entering what they perceived as the most important items learned during class. This was similar to other active learning activities suggested in the literature, but in this case the student contributions were collaborative and archival. Students were encouraged to be complete and accurate with the promise that their entries would be available during an exam. Other wiki pages the students developed included a set of suggestions in preparing for the first exam for future students enrolled in the course.

Instructor assessment on learning indicated that once students understood the expectations of the reflective wiki entries they more rapidly demonstrated understanding of the key points made during previous lectures. The students in this small course section rapidly formed the sort of positive interdependence relationship that Johnson describes as a result of collaborative learning. Students that the instructor suspected at the start of the course would be unable to complete the course were successful early in the course and were able to demonstrate learning sufficient to pass the course with a 'C' or better. Student assessment of the process concluded that reflection was an important mechanism for learning.

See more of this Session: Communication and Collaboration Using New Technology
See more of this Group/Topical: Education

RAPID GENERATION OF VIDEOTAPED LECTURES FOR A COURSE OF "INTRODUCTION TO MODELING AND COMPUTATION", USING TABLET PC AND THE CAMTASIA RECORDER.

Mordechai Shacham, Ben-Gurion University of the Negev, Beer-Sheva, Israel

Introduction

The recent outbreak of the swine flu epidemic brought into focus the issue of preserving "Academic Continuity". Academic continuity, as defined by the Sloan Consortium (a nonprofit group that encourages technology use in education, www.academiccontinuity.org), is: "Academic continuity is the process of maintaining continuity of learning in a crisis situation caused by a natural disaster, human-induced ("man-made") disaster, or other precipitating factor".

Young (2009) has recently reviewed several instances where universities and colleges had to be closed because of emergency situations for several weeks and there were attempts to continue the teaching online. Switching from classroom teaching to online teaching in emergency conditions and in a very short notice was extremely difficult for most instructors and students. Following these experiences, institutions of higher learning started to offer workshops to prepare the instructors for the move to online teaching in a short notice. Other institutions have developed "quick start" faculty guides to continuing instruction during an emergency (Young, 2009)

In Israel there were several instances, in the last few years, when institutions of higher learning had to be closed because of emergency situations caused by wars. In the summer of 2006 all the universities and colleges in north part of Israel were closed for about one month. In January of 2009 the institutes in the south part were closed for three weeks. Thus, it became eminent that a viable plan is needed to enable immediate switch to online teaching whenever the need arises. Our preferred approach for the solution of this problem is to attempt to modify of the teaching technique used by the faculty and the learning habits of the students to ensure that new courses are developed and existing courses are modified so that lectures can be delivered in the classroom or online on the same level of efficiency. Keeping the cost of the development of the online courses at minimal level (both in terms of money and instructor's time) is also of at most importance.

Fortunately, we already had a model for a new teaching technique which can meet the above requirements. Ben-Yaakov, 2008, an electrical engineering faculty in Ben-Gurion University of the Negev (BGU), developed several online courses using this new technique. The development of online lectures, using Ben-Yaakov's approach, involves the use Tablet PC and the Camtasia recorder (http://www.softwarecasa.com/) during regular lectures that are delivered in class. The recording is done by the instructor. The recorded lecture becomes available on-line within a day or two after some editing.

The committee for the advancement of e-learning at BGU together with the Dean of the faculty of engineering sciences initiated a pilot study in which tablet Pcs were purchased for five engineering faculty members with the commitment that they teach at least one course in the spring semester of the 2008/9 academic year using Ben-Yaakov's technique. Eventually, seven courses were developed in the framework of this pilot study. In the following sections the use of this technique for teaching the "Introduction to Modeling and Computation for Chemical

Engineers" (IMCCE) will be described in some detail. The results of student and instructor evaluation of the IMCCE course will also be presented.

The "Introduction to Modeling and Computation for Chemical Engineers" course

The IMCCE course was developed in the last ten years with the main objective of replacing the traditional computer programming course with a course centered on Computer Based Problem Solving (CBPS). The course was described in detail by Shacham (2005). Only its main features will be briefly reviewed here.

The principal objectives of the course are: 1. Enabling the students to formulate engineering problems as mathematical models belonging to one of the following categories: complex consecutive calculations, linear and nonlinear algebraic equations, ordinary differential equations and multiple linear and polynomial regressions and 2. Enabling them to solve the resultant mathematical models using Polymath (copyrighted by M. Shacham, M. B. Cutlip and M. Elly, http://www.polymath-software.com), Excel (trademark of Microsoft Corporation, http://www.microsoft.com) and MATLAB (trademark of The Math Works, Inc., http://www.mathworks.com).

We have found that the combination of these three packages fulfills a major portion of the computational needs of undergraduate and graduate chemical engineering students. Polymath is used in over 120 universities worldwide for problem solving and it is also included in several textbooks of mass and energy balances, chemical reaction engineering and numerical problem solving (Cutlip and Shacham, 2008). Excel is a computational tool that has a very wide range of capabilities and it is extensively used by practicing engineers. MATLAB, with its wide range of capabilities, is used in some undergraduate and graduate engineering courses and can also be considered as a programming language. The three packages are learned in a sequence, where the concepts are build up gradually, making the learning of programming with MATLAB considerably easier.

To enhance the learning effectiveness the following advanced features are included in the course: *Real life problems* – the study of each subject starts by presenting a real life problem, which requires numerical solution . *Multi-stage problems* - the problems require computer solution in several stages. Preliminary stages are solvable with the easy to use software package Polymath, while advanced stages require the use of Excel or MATLAB programming. *Programming by modification* - solved examples are provided. The students are encouraged to modify and extend the solved examples to solve their assignments. *Self – grading* – students check and grade their own internet-based homework assignments. The self-test program provides immediate feedback on errors made (Shacham, 1998).

The course is divided into six chapters according to the mathematical/numerical method that has to be used for solving the problems: 1. Complex Consecutive Calculations, 2. Iterative Solution of a Nonlinear Equation, 3. Matrix Operations and Solution of Systems of Linear Equations, 4. Multiple-Linear, Polynomial and Nonlinear Regression, 5. Introduction to Solution of Systems of Nonlinear Algebraic Equations and 6. Introduction to solution of Systems of Ordinary Differential Equations

The course is given to freshmen chemical engineering students who have already taken an introductory "Material and Energy Balance" course or the two courses are run in parallel.

Delivering and Recording Lectures using the Tablet PC and the Camtasia Recorder

Using the new technique, the Tablet PC serves three purposes: 1. It is used for presenting pre-prepared power point slides; 2. It replaces the blackboard (or whiteboard) by enabling the display of material written or marked on the computer screen using a special pen; and 3. The computer screen display and the voice of the instructor are continuously recorded enabling the creation of the online lecture.

In the IMCCE course we prepared power point slides on which only part of the material was put in advance. The students are encouraged to bring to the class hard copy of the incomplete slides. The missing parts are completed in class. The instructor writes on the tablet PC screen while explaining the material, and the students complete the missing information on the hard copies. This method gives enough time to absorb the more difficult material and to ask questions if needed.

Consider for example the "Linear Recycle" process that is defined in Figure 1. The process that includes a mixer, a reactor and two separators can be modeled using basic matrix operations such as matrix multiplications, matrix additions and solution of a system of linear algebraic equations. The first step in solving this problem is defining the matrices that represent the various process units. The slide intended for definition of the pertinent matrices and operations, for four of the unit operations is shown in Figure 2. The details added during the lecture are shown in Figure 3. Observe that the preliminary information, that requires time-consuming preparation but easy to understand (the sketch of the process flow-sheet and process units and the operational data) is prepared in advance. The new material, which is the conversion of the process data and process operations into matrices and vectors and their operations, is demonstrated and written in detail (Figure 3). The option of writing and putting markings on the pre-prepared slides is especially beneficial, in this course and similar ones, for explanations related to computer programs. The programs must be prepared and put on the slides in advance to make sure that no typographical errors are introduced when copying the programs manually. Adding comments in writing when explaining the structure and purpose of the various commands ensures that the explanation will be available at the time when the student has to write his own program.

Instructors who teach more theoretical courses where most of the writing includes development and explanation of equations may prefer not to prepare slides in advance but to do all the writing on the Tablet PC during the lecture. In such cases the advantage of the use of the Tablet PC is that, unlike the blackboard which must be erased after filled up the its "board" can be rolled back to other important material that was developed, say, an hour ago.

Before starting the lecture the instructor should initiate the recording using the Camtasia recorder. This is a critical stage and requires some training, as if the recording is not started correctly the complete hour of recording is lost. It is recommended that a technical person to be present in the first few lectures given by an inexperienced instructor, to make sure that the technical set up of the recording is correct.

Because of the need to write on the tablet PC and to use its internal microphone for voice recording it is preferable to deliver the lecture while sitting in front of the desk where the Tablet PC is situated. This requirement may frighten instructors who are used to stand and walk while talking and writing on the blackboard. Also, it seems that it is more difficult to control the class while sitting. However, from personal experience, it is fairly easy to get used to lecturing while sitting.

As for disciplinary problems during the lecture, the experience shows that there are practically none. The reason for that is, probably, that students who tend to be noisy or create other distractions during lectures, have the choice of not attending the class lecture but watching the online version, instead.

Student Survey

Toward the end of the semester student survey forms were put on the web sites of all the courses that participated in the Tablet PC project. The survey was intended for getting feedback regarding the use of the Tablet PC and the online lectures and did not include other aspects of the course. The results concerning the IMCCE course will be reported here.

Out of the 65 students who attended the course 61 filled out the evaluation forms. The first section of the survey contained 8 statements and the students were asked to select from 5 possible answers: Strongly agree, agree, neutral, disagree and strongly disagree. Summary of the results for this part is shown in Table 1. The first two statements are related to the improvement of the quality of the class presentation using the "writing on the screen" and "previous material reviewing" capabilities of the tablet PC. More than 70% of the students "agreed" or "strongly agreed" that the utilization of these capabilities improves the quality of the presentation. There was similar but somewhat lower level (59%) of agreement regarding the statement that the "use of the tablet PC improves my understanding of the course material.

Since the introduction of the PowerPoint™ into classroom presentations there were severe criticisms of its extensive use. Felder and Brent (2005) in their paper entitled "Death by PowerPoint" mention the following undesired effects: 1) The attendance in the class session may drop as students have access to copies of the presentation in the course web site. 2) The class may become mind-numbing for students who have studied the material, and the pace may be too fast for students who did not. 3) The students may not appreciate the knowledge of the instructor as he/she only "repeats what is already written on the slides". The first part of the survey results shows that the use of the Tablet PC with the proposed technique essentially eliminated the shortcomings of the extensive use of pre-prepared slides in the classroom.

Very few (less than 6%) of the students encountered technical difficulties while attempting to watch the online lectures. There was considerable disagreement among the students regarding the need to attend the lectures in class (see Statement 5). More than one third (36.5%) "agreed" or "strongly agreed" that watching the recorded lectures, only, would be sufficient for them, the same number of students was "neutral" and only 27% felt that they needed class lectures as well. The great majority of the students (91%) agreed that the recorded lectures helped them to complete material that they missed in class. Watching the online lectures improved the understanding of the course material for ~75% of the students. Only 54% thought that it is desirable that the lectures in all the courses be recorded for online delivery using the Tablet PC, 24.6% were "neutral" regarding to this statement and 21% disagreed.

In Table 2 the student responses regarding the extent of the use of the online lectures are summarized. The largest section of the class (41%) watched all the online lectures in order to understand the course material better, about 20% watched only the lectured that they missed in class, 18% watched only the parts covering difficult material which was not understood in the lecture in class and only 20% did not watch any of the online lectures.

Conclusions

In this article the development of an online IMCCE course using a Tablet PC and Camtasia recorder and its use in class were described in detail. It was shown that the development of such a course, using the technology presented, requires minimal training of the instructor, a few hours of technical help and minimal investment. Development and online delivery of such a course provides the training necessary for the instructor and the students to ensure smooth transition from on campus learning to online, distance learning in emergency situations. The online course can also be used, on its own, should the need arise (teaching small number of students in remote locations, for example).

The use of the proposed technique for course development and delivery has several pedagogical advantages, as well:

1. Most students agree that the use of the Tablet PC with the "writing on the screen" and "reviewing previous content" capabilities improves the quality of the presentation and enhances the student understanding of the course material.
2. Most students agree that reviewing the recorded, online lectures helps them to understand the course material better and most of them make use of the online lectures at various extents. More than one third of the students make use of all the recorded lectures.
3. The availability of the online lectures provides the students the freedom to choose whether to attend the live lectures or to watch the online versions instead. This ensures that only students who are really interested in the live lectures show up in class. This creates a more pleasant atmosphere in class with minimal distractions and disciplinary problems.

As for the question whether the online version of the course can completely replace classroom learning this study provides a partial answer. It seems that the freedom provided to choose between classroom and online learning is appreciated very much by the students. Removing the classroom option completely may offend many of the students. This attitude may change in the future. However, it is preferably to let the change of attitude to happen naturally without trying to force it.

Development and preparation of the course, for the first time, using the new technique may take a fair amount of extra effort on the instructor's behalf. Even if PowerPoint files of lecture material are already available, careful consideration must be given to what material that should remain on the slides and what details must be added during the lecture. One hour of lecture material has to be divided to smaller sections that are clearly identified, so that the students can reach the section they are interested, in the recorded version, easily. The lecture has to be prepared carefully, in very much detail, to ensure that no re-recording of parts of the lecture will be required.

This new instructional technique seems very promising and there is no doubt that considerable growth in its use in engineering and exact science education will be witnessed in the near future.

References

1. Ben-Yaakov, S., Engineering Education Improvement - E_learning, Webcast (in Hebrew) 2008. Link: http://www.ee.bgu.ac.il/~pel/presentation/Engineering%20Education%20Improvement/Engr_Edu_video.html
2. Cutlip, M. B. and Shacham, M. *Problem Solving In Chemical and Biochemical Engineering with Polymath, Excel and MATLAB.* Prentice-Hall, Upper Saddle River, New-Jersey, 2008.
3. Felder, R. M. and Brent, R., "Death by PowerPoint", *Chemical Engineering Education*, 39 (1), 28-29 (2005)
4. Shacham, M., "Computer Based Exams in Undergraduate Engineering Courses", *Comput. Appl. Eng. Educ*, 6(3), 201-209(1998)
5. Shacham, M., "An Introductory Course of Modeling and Computation for Chemical Engineers", *Comput. Appl. Eng. Educ.* 13, 137 – 145 (2005)
6. Young, J. R., "In Case of Emergency, Break Tradition-Teach Online", *The Chronicle of Higher Education, Technology*, Aug. 17, 2009

Table 1. Summary of student responses to the statements they were provided concerning the use of the Tablet PC and the recorded lectures

No.	Statement	Strongly Agree	Agree	Neutral	Disagree	Strongly Disagree
1	Emphasizing points and writing explanations on the slides using the Tablet PC improves the quality of the presentation	34.4%	36.0%	16.4%	6.6%	6.6%
2	The use of the Tablet PC enables reviewing previous content. This option improves the quality of the presentation	41.0%	29.5%	13.1%	9.8%	6.6%
3	The use of the Tablet PC improves my understanding of the course material	33.9%	25.4%	23.7%	10.2%	6.8%
4	Technical difficulties were encountered while attempting to watch the recorded lectures	2.0%	3.9%	3.9%	11.8%	78.4%
5	Watching the recorded lectures is sufficient for me. It is the same as attending the lectures in class	17.3%	19.2%	36.6%	11.5%	15.4%
6	When I missed a lecture the recorded lecture enabled me to complete the material I missed	62.2%	28.9%	6.7%	2.2%	0.0%
7	Watching the recorded lectures improved my understanding of the course material	49.0%	25.5%	13.5%	6.0%	6.0%
8	It desirable that all the instructors record their lectures for online delivery using Tablet PC	39.3%	14.8%	24.6%	4.9%	16.4%

Table 2. Summary of student responses regarding the extent of the use of the recorded lectures

No.	Statement	Agree
1	I did not use any of the online lectures for studying	19.7%
2	I studied from an online lecture only when I missed a live lecture	19.7%
3	I studied using the online lecture only the parts which I did not understand in class	18.0%
4	I used the online lectures for briefly reviewing the material before recitation sessions.	1.6%
5	I watched all the online lectures in order to understand the course material better	41.0%

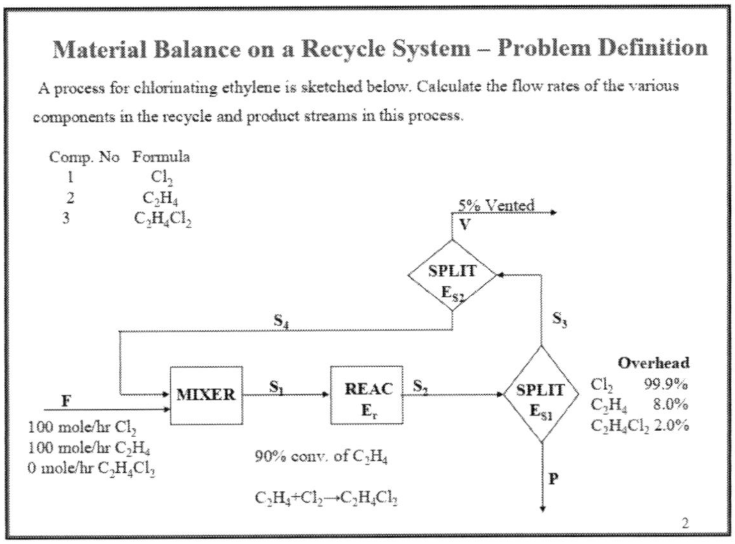

Figure 1. Linear recycle problem definition.

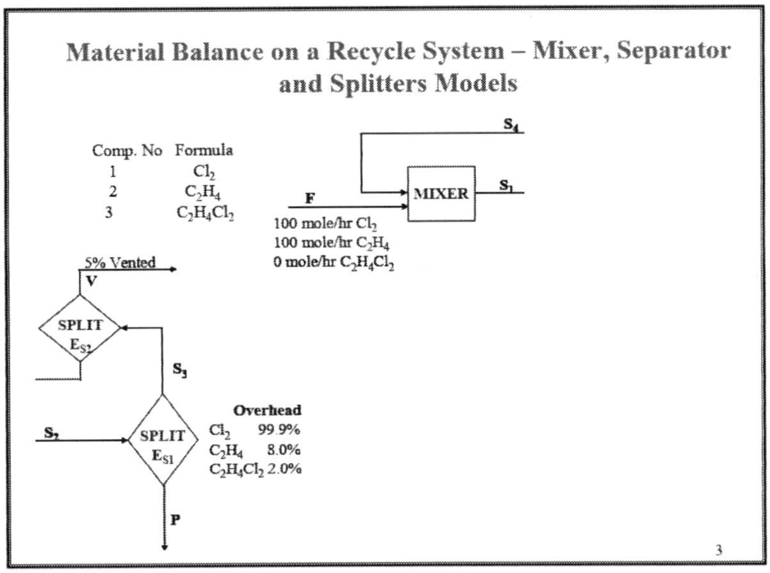

Figure 2. Incomplete unit operation's models definition slide

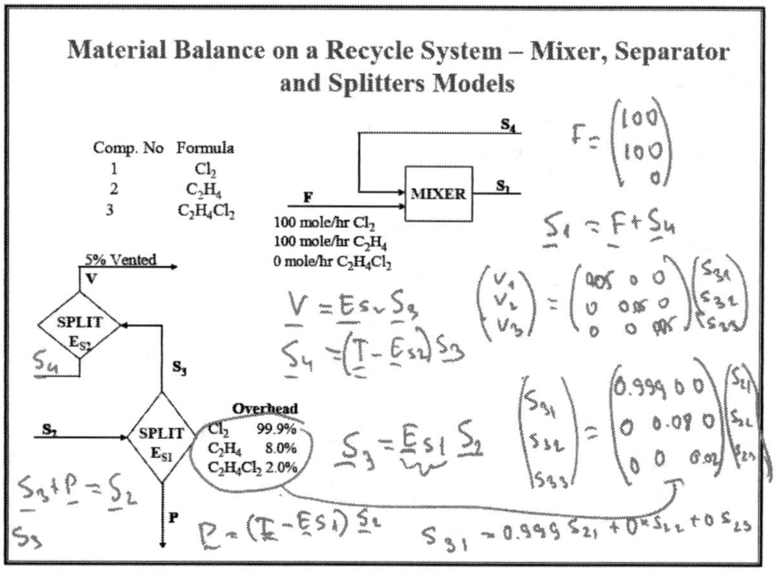

Figure 3. Completed unit operation's models definition slide

Understanding Student Learning in Remote and Hands-On Laboratory Experiences

Jim Henry, Chemical Engineering, University of Tennessee at Chattanooga, Chattanooga, TN

Marina Miletic, Department of Chemical and Biomolecular Engineering, University of Illinois Urbana-Champaign, Urbana, IL

David DiBiasio, Chemical Engineering Department, Worcester Polytechnic Institute, Worcester, MA

Background------------- Our interest is in understanding student learning and the advantages and disadvantages of remotely-operated laboratory experiences compared to hands-on. Ideally, remote experiments can be conducted at any time from any place. They are particularly useful for students at universities where resources are severely limited and there is no access to any significant experimental equipment. In such contexts, knowing how to optimize learning in a remote lab experiment is critical. And, knowing the limitations compared to hands-on is important. Remote experiments are also a way to introduce a lab-like experience in any university where large enrollments prevent practical access to laboratory equipment. At WPI distillation is taught in the sophomore year throughout a project-based spiral curriculum. All projects are team-based. We introduce basic concepts early in the sophomore year then revisit distillation throughout the year with successively more complex assignments and projects. This includes at least two lab experiences. The first experiment uses a batch column operated at total reflux to introduce students to multistage distillation including efficiency and energy balances at total reflux. A follow-on course explores pressure swing distillation without a lab component. Near the end of the year, the batch column operated at a constant external reflux ratio is used again in a lab project. This project engages students in process dynamics and challenges them to compare differences between theory and reality using the Rayleigh analysis. Large enrollments forced us to teams of 8-10 people (not optimal) or to run experiments during times when safety and lab monitoring because serious issues. Simultaneously with struggling to deal with this problem, we began collaboration with UTC involving remotely accessible experiments. In fact, the UTC distillation column is physically nearly identical to the WPI column but much more flexible in its operation. In 2008 we ran the first trial comparison of remote lab compared to hands-on labs. ---------------Methodology------------------We have completed two cycles of a pilot study that would inform a subsequent larger, more rigorous investigation. In 2008 we recruited 7 volunteer WPI teams from the cohort enrolled in the final sophomore year course to do the remote-only experiment using the UTC column. We also had 7 different WPI teams run the identical experiment locally using the WPI system. In 2009, 7 remote teams were assigned and 8 local teams were assigned, hence the bias introduced through self-selection was eliminated. Remote-only and hands-on-only teams conducted identical experiments with identical assignments and project reporting requirements. Evaluation had three components with direct and indirect assessments, giving us some degree of triangulation. Both cohorts completed surveys managed by a third partner: UIUC. The online survey included closed-end, Likert-scaled responses and open-ended questions. The open-ended questions included student attitudes but also probed team processing for both local and remote teams. We wanted to learn more about how each team handled roles, logistics, scheduling, data collection, and analysis to see if there were meaningful differences between an online team process and a hands-on team process. Course instructors compared final reports from remote and hands-on teams. A pre/post in-class quiz compared individual learning for students in the remote cohort compared to hands-on cohort in a quantitative manner. Qualitative measures include evaluating student attitudes about the experience and assessing students' improvement in ABET outcomes such as: (b) - design, analysis and interpretation of data; (d) - functioning on multidisciplinary teams; (g) - effective remote communication; (h) - have broad education necessary to understand the impact of engineering solutions in a global, economic, and environmental societal context; and (k) - using techniques, skills, and modern engineering tools (such as computers and web interfaces) necessary for engineering practice.

See more of this Session: Communication and Collaboration Using New Technology
See more of this Group/Topical: Education

Collaborative Education Across the Curriculum Using Cogeneration System

F. Carl Knopf, Chemical Eng., Louisiana State University, Baton Rouge, LA

Kerry.M. Dooley, Chemical Engineering, Louisiana State University, Baton Rouge, LA

Mohammed Shafi Syed, Chemical Engineering, Louisiana State University, Baton Rouge, LA

Combined Heat and Power (CHP) or cogeneration, is an efficient way of generating power and thermal energy from a single fuel source, which greatly increases the facility's operational efficiency and decreases the energy costs. Cogeneration, is critical to the economic viability of many industries, including fuels, chemicals, plastics, paper, and ferrous and nonferrous metals. CHP systems provide an excellent platform for both ChE and ME students to study thermodynamics, certain unit operations and process design principles, optimization, data reconciliation, and emissions control. Even better, many Universities have ready (albeit remote) access to such a system on their own campus. At LSU, we are opening our newer (commissioned 2005, 20 MW, $20 MM) cogeneration system to the world of engineering education through real-time data delivery using formats which can communicate with virtually any control/data acquisition system and then broadcast compressed data via the net. Experimental We purchased OSISoft PI® software to interface to and collect data from various control systems (JCI Metasys, ABB Symphony) of the 20 MW LSU Cogeneration Facility. Extracted data are then archived by the PI Server and linked to the web using a web server running PI's RtWebParts software. RtWebParts allows users to create, customize and deploy continuously updated views (e.g., Fig.1 showing the turbine section of the LSU Cogeneration System) of PI and non-PI data using a web browser. RtWebParts provides data in time series display, calculated data and contextual data. Students can access the data from the web as displays of individual units of the cogeneration system, or trends of important tags which they can create themselves. Users can view these in Microsoft Excel. Therefore access to real-time data from our cogeneration system can be broadcast to students worldwide. Some of the learning modules based on this technology are listed below. These modules have already been tested in 6 different classes within Chemical Engineering - Junior and Senior labs, Numerical Methods, Simulation and Optimization, Unit Operations and Process Design. 1. Data Reconciliation / Gross Error Detection - Students use real-time data from the cogen system to illustrate the difficulties in closing material and energy balances. 2. Energy Dispatching / On-line Optimization - Students fit data from the cogeneration system and then develop a plan to reduce energy costs. 3. Interface Construction for Process Control - Students are provided with a list of all available tags and are asked to create an operator-friendly interface using the tags important in control of the process.

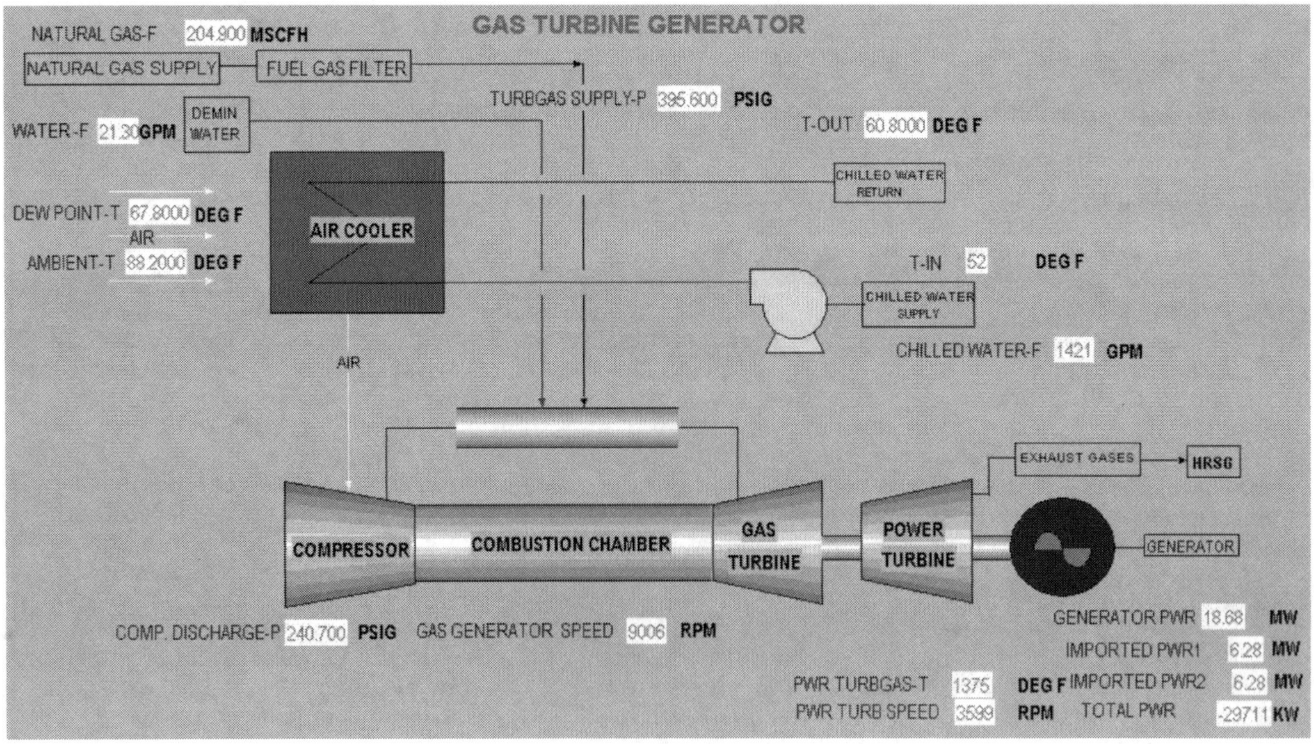

See more of this Session: Communication and Collaboration Using New Technology
See more of this Group/Topical: Education

A Simple Optical Experiment On Polymer Diffusion for Undergraduates That Incorporates Web-Cam Capture, Data Digitization, and Multi-Variable Regression

Vinay K. Gupta, Chemical & Biomedical Engineering, University of South Florida, Tampa, FL

Bijith Mankidy, Chemical & Biomedical Engineering, University of South Florida, Tampa, FL

Cecil Coutinho, Chemical & Biomedical Engineering, University of South Florida, Tampa, FL

Diffusion is studied due to its fundamental importance in various fields of biomedicine, chemical technology and nanotechnology. Measuring diffusivity of polymers is necessary in the development and design of different processes involving polymer diffusion. A laboratory experiment to emphasize concepts of macromolecular diffusion for undergraduate students can, therefore, be valuable. Sophisticated equipment that have evolved in the past for measuring diffusivity are not suitable for undergraduate instructional laboratories because of cost, complexity and difficulty in demonstrating diffusion concepts in a straightforward manner. A simple, cost effective and visually instructive optical method to measure polymer diffusion coefficients will be presented that is based on Weiner's method. Refraction of laser light is employed to analyze the relationship of size, shape and solution environment on diffusion coefficients of common linear and globular polymers such as poly(ethylene glycol), poly(vinyl alcohol), chitosan and bovine serum albumin. We have also added features such as digital capture of data using a webcam, data digitization, and multi variable regression that provides opportunities for students to learn tools that can be valuable in their academic and professional career.

See more of this Session: Communication and Collaboration Using New Technology
See more of this Group/Topical: Education

NSF CBET Overview and Other NSF Programs

John McGrath, Division of Chemical, Bioengineering, Environmental and Transport Systems, National Science Foundation, Arlington, VA

An overview of CBET programs

See more of this Session: NSF Workshop 1
See more of this Group/Topical: Education

Highlights of CBET Cluster On Chemical, Biochemical & Biotechnology Systems

Rosemarie D. Wesson, Division of Chemical, Bioengineering, Environmental and Transport Systems, National Science Foundation, Arlington, VA

Cluster on Chemical, Biochemical & Biotechnology Systems

See more of this Session: NSF Workshop 1
See more of this Group/Topical: Education

Highlights of CBET Cluster On Transport and Thermal Fluids Phenomena

Henning Winter, Division of Chemical, Bioengineering, Environmental and Transport Systems, National Science Foundation, Arlington, VA

Cbet Cluster on Transport and Thermal Fluids Phenomena

See more of this Session: NSF Workshop 1
See more of this Group/Topical: Education

Highlights of CBET Cluster On Environmental Engineering & Sustainability

Gregory Rorrer, Chemical, Bioengineering, Environmental and Transport Systems, National Science Foundation, Arlington, VA

Cbet Cluster on Environmental Engineering & Sustainability.

See more of this Session: NSF Workshop 1
See more of this Group/Topical: Education

Highlights of CBET Cluster On Biomedical Engineering and Engineering Healthcare

Alexander Simonian, Division of Chemical, Bioengineering, Environmental and Transport Systems, National Science Foundation, Arlington, VA

An overview of the CBET cluster on Biomedical Engineering and Engineering Healthcare

See more of this Session: NSF Workshop 1
See more of this Group/Topical: Education

Interactive Question and Answer Session with NSF Program Directors

Robert M. Wellek, Division of Chemical, Bioengineering, Environmental and Transport Systems, The National Science Foundation, Arlington, VA

A discussion with NSF program directors

See more of this Session: NSF Workshop 1
See more of this Group/Topical: Education

Proposal Writing Tutorial

John R. Regalbuto, U. Illinois at Chicago, Chicago, IL

James M. Lee, Washington State University, Pullman, WA

A tutorial for proposal writing

See more of this Session: NSF Workshop 2
See more of this Group/Topical: Education

Interactive Breakout Panels

Robert M. Wellek, Division of Chemical, Bioengineering, Environmental and Transport Systems, The National Science Foundation, Arlington, VA

Breakout panels coordinated by NSF Program Directors

See more of this Session: NSF Workshop 2
See more of this Group/Topical: Education

Organizing a Multidisciplinary Student Design Project through An Innovative Partnership with Local Industry

Jeffrey R. Seay, Department of Chemical and Materials Engineering, University of Kentucky, Paducah, KY

David L. Silverstein, University of Kentucky, Paducah, KY

Jimmy L. Smart, Chemical & Materials Engineering, University of Kentucky, Paducah, KY

The design of a chemical process is often the culmination of the undergraduate education in chemical engineering. There are significant benefits to be realized by introducing students to the design process earlier in the curriculum. However, these efforts are often hampered by the fact that sophomore and junior level students haven't yet had the coursework required for a typical capstone type design project.

This contribution will describe an innovative project conceived through a partnership with local industry that involves a multidisciplinary team of students from the sophomore, junior and senior classes working together on an actual design project provided by a local industrial partner. By grouping students of different levels of experience, the underclass students are able to contribute at an appropriate level, while still being involved in the overall design project as organized by the senior class.

Several challenges must be addressed when working on projects provided by industry, namely, confidentiality, intellectual property ownership and ability to present and/or publish findings. The strategy for addressing each of these challenges will be presented. Despite these challenges, the potential rewards of such a partnership are tremendous. By working on an actual project provided by local industry, the students get a real sense of the challenges faced by practicing engineers. These challenges include underspecified problem statements, lack of process and property data and open-ended design goals. This contribution will present some of the student experience with this unconventional approach to a non-capstone design project. Finally, a path forward for maintaining the partnership with industry and plans for additional outcome assessment will be presented.

See more of this Session: Teaching Design Outside the Capstone Course
See more of this Group/Topical: Education

Development of Classroom Experiments to Introduce Decision Making Under Uncertainty to Students in a Design I Course

Dimitrios V. Papavassiliou, Chemical, Biological, and Materials Engineering, University of Oklahoma, Norman, OK

Chintamani Jog, Economics, University of Oklahoma, Norman, OK

Phuong Le, Chemical, Biological and Materials Engineering, University of Oklahoma, Norman, OK

Margaret Freeman, Chemical, Biological and Materials Engineering, University of Oklahoma, Norman, OK

Georgia Kosmopoulou, Economics, University of Oklahoma, Norman, OK

Chemical Engineering graduates will face challenges at the work place that even their peers who graduated a mere decade ago were not expected to face. One such major challenge is the management and operation of companies and plants under conditions of uncertainty and the need to make decisions in competitive situations. Modern developments in Economics offer the tools to address this challenge with rather well developed concepts, such as decision theory and financial risk analysis. In this work we will describe our efforts to integrate modern economics concepts and techniques in the teaching of Engineering Design. In addition, we will discuss the development of illustrative classroom experiments and case studies that will showcase such concepts.

The Chemical Engineering curriculum at the University of Oklahoma includes three courses in Design: Design Lab (2 credit hours), Design I (3 credit hours) and Design II (3 credit hours). The capstone projects are part of Design II, while Design I is the course that introduces Engineering Economics to students, as well as concepts such as profitability and cost analysis. The majority of the students who enroll in Design I have no background in Economics, since introductory Economics courses are part of the general education electives (as are several other courses in the social sciences). The type of Engineering Economics taught to them focused on cost analysis and profitability estimation with limited references to uncertainty and decision making under uncertainty. This is typical of Engineering Economics courses offered in programs around the country. For the past several years, we have included material on risk analysis and decision making in the Design I class. Lately, through collaboration between Chemical Engineering and the Department of Economics, we have developed classroom experiments that demonstrate concepts such as game theory, the winner's curse, and the utility function that can be used to quantify risk and facilitate decision making under uncertainty. In this presentation, we will discuss the development of these experiments and the educational objectives of each experiment. We will also demonstrate the basic components of these experiments and we will discuss the mechanics of conducting these experiments in the classroom.

See more of this Session: Teaching Design Outside the Capstone Course
See more of this Group/Topical: Education

Design and Experimental Investigation in the Mass and Energy Balance Course

Galen J. Suppes, University of Missouri-Columbia, Columbia, MO

Bryan Sawyer, University of Missouri-Columbia, Columbia, MO

A semester long project was implemented in the sophomore mass and energy balance course including the steps of contemplation of the underlying theory, simulation, experimental experimental investigation, modeling, and evaluation, and product design. The objective was to qualitatively and quantitatively describe the energy and mass balances in a circuit comprised of an AA battery connected in series with a resistor. The design aspect included using data and understanding gained from the experimental investigation of the temperature profiles of this system to design a battery of the same chemistry to power a 10W light bulb for 1 hour. The scale-up included directly scaling up the masses of anode and cathode materials as well as consideration of constraints on how battery efficiency was linked to the mass transfer and cross sectional area for mass transfer within the battery.

See more of this Session: Teaching Design Outside the Capstone Course
See more of this Group/Topical: Education

The Converging-Diverging Model of Design and Dixon's Taxonomy: Useful Tools for Introducing Students to Design

Kevin Dahm, Chemical Engineering, Rowan University, Glassboro, NJ

Rowan University's Engineering curriculum includes the Sophomore Engineering Clinic, a team-taught two-course sequence intended to teach engineering design and technical communication. Students complete open-ended design projects in lab, supervised by Engineering faculty. The concurrently learn principles of technical communication from Communications faculty; specifically, technical writing in the fall and public speaking in the spring. Most of the graded deliverables in the courses are reports and presentations about the design projects, and are graded jointly by Engineering and Communications faculty. This course provides students with their first exposure to open-ended design challenges. In recent years, the Engineering faculty noted widespread shortcomings in student approaches to their design problems. Most notably, students routinely made significant design decisions without a sound quantitative basis.

This presentation will focus on changes made to the course in an attempt to improve student mastery of design practice. Specifically it will describe two published design models- Dym's Converging-Diverging model for design and Dixon's Taxonomy of students of knowledge- and will outline how these models were integrated into the course. The presentation will also provide an assessment of how the changes led to improvement in student performance, both in the Sophomore Clinic and in the capstone design course.

See more of this Session: Teaching Design Outside the Capstone Course
See more of this Group/Topical: Education

Using Simulators to Teach Process Safety

Andrew I. Biaglow, Russell P. Lachance, Robert G. Bozic, Chemical Engineering Program, Department of Chemistry and Life Science, United States Military Academy, West Point, NY

Donald C. Glaser, Simulation Solutions, Inc., Shrewsbury, NJ

Abstract

This work describes our efforts to enhance process safety and unit operations education using simulator technology. Simulators provide an excellent opportunity for students to experiment with and learn about a process, something that is not feasible in an actual plant. The specific software tools that were used included simulations of various unit operations, including a heat exchanger, cooling tower, and distillation column. Students were presented with various scenarios in which they were asked to execute a cold start-up or to troubleshoot and repair a fault. As student expertise in the simulators increased, they were exposed to more complicated processes, including an FCC unit and an atmospheric crude unit. The more complicated processes were accompanied by a "virtual reality" outside operator view in which the students could locate and monitor the various control instruments in the process. This presentation will focus on our classroom methods and some of the preliminary feedback from the students.

Introduction

The use of realistic process simulators can augment many levels of the undergraduate chemical engineering curriculum. Such simulators allow students to operate equipment and processes that are typically too expensive for experimentation or for training exercises. While simulators cannot replace actual hands-on experience, they add a new dimension to such experiences by allowing more complicated processes to be studied.

The operation of a chemical plant has many interacting dynamic systems, the operation of which is very different from what can be gleaned from textbook problems. Steady-state CAD simulators such as ChemCAD or Aspen can add another level of understanding particularly when performing perform dynamics simulations. However, another set of simulation tools exists for the performance of real-time simulations.[1-4] This type of experience is very different in the sense that the students are placed into the situation and must think on their feet, without the benefit of time for reconsideration or development of multiple courses of action.

As an example, consider the teaching problem of describing to someone how to drive an automobile. The driver must be aware of speed and acceleration, direction and steering, braking systems, and clutch (if used). The driver must also be aware of a constantly changing environment, with traffic signals and other motorists, as well as legal restrictions on allowed maneuvers. However, this level of awareness is insufficient. The motorist must also learn to anticipate events while driving, and to formulate and execute responses based on those anticipations in a nearly instantaneous fashion. This anticipatory response is impossible to learn from textbook scenarios and must be learned by experience. Chemical plants behave in a very similar manner. During a disturbance, a complex array of events occurs, causing operators and plant engineers to have to use a similar type of anticipatory behavior. Again, for safe operation of a plant, this behavior must be learned. Recognizing this, many companies now provide operator and plant engineer training courses using simulators.[1,2,4] In this paper, we discuss the use of some of these same methods in an undergraduate chemical engineering curriculum.

Process Description

A very good example of a chemical engineering process simulator is the fluid catalytic cracking unit, or FCCU, provided by Simulation Solutions, Inc. The simulator contains a process model that is accessed through a number of control screens. The overview of the process is shown in Figure 1. Several other detail control screens are available in subsequent screens and can be accessed quickly during the simulation. An example of a detail screen is shown in Figure 2, which is a detail view of the regenerator section.

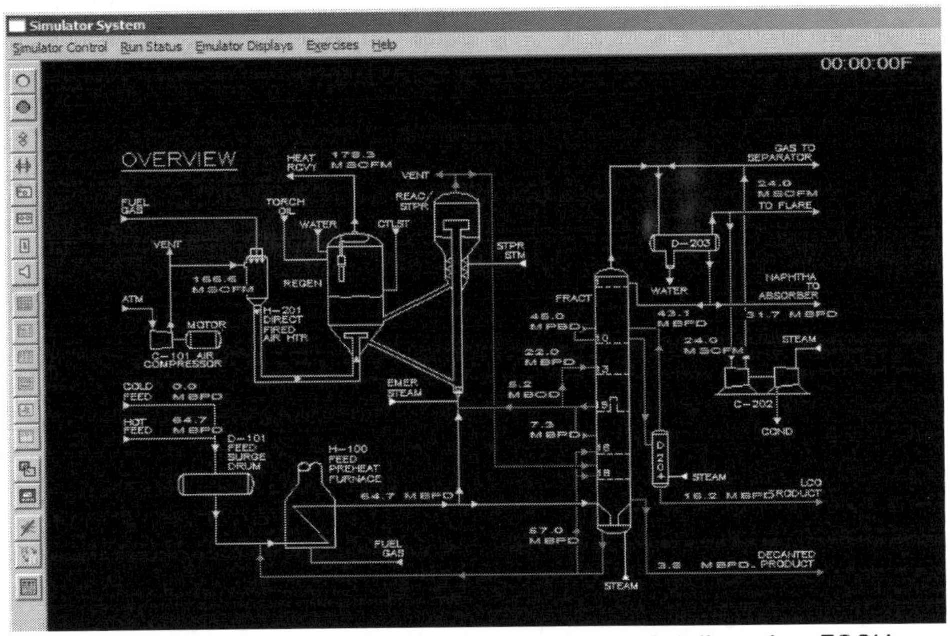

Figure 1. Overview screenshot of the Simulation Solutions, Inc. FCCU.

The FCCU process shown in Figure 1 is comprised of a reactor section on the left side of the figure and a separations section on the right. The reactor section includes a fluidized bed reactor, catalyst separator, and catalyst regenerator. Catalyst is added to the system and continuously circulates between these three units. Feed enters at the bottom of the reactor and cracks to smaller molecular weight products, after which it leaves the top of the reactor and is sent to the fractionator section for purification. During the contacting time, a significant amount of coke forms on the catalyst. The coked catalyst is sent to the regenerator where the coke is burned off, after which the catalyst is returned to the reactor. The heat produced by the combustion of the coke is carried back into the reactor with the hot catalyst. When the system is at or near steady state, controlling the catalyst flow provides thermal control of the reactor. During startup operations, an auxiliary torch located inside the regenerator can be brought online to add additional heat. Product separation and feed recovery are accomplished in the fractionator section. One of the interesting features of the fractionator column is that heat removal is required at both the reboiler and condenser.

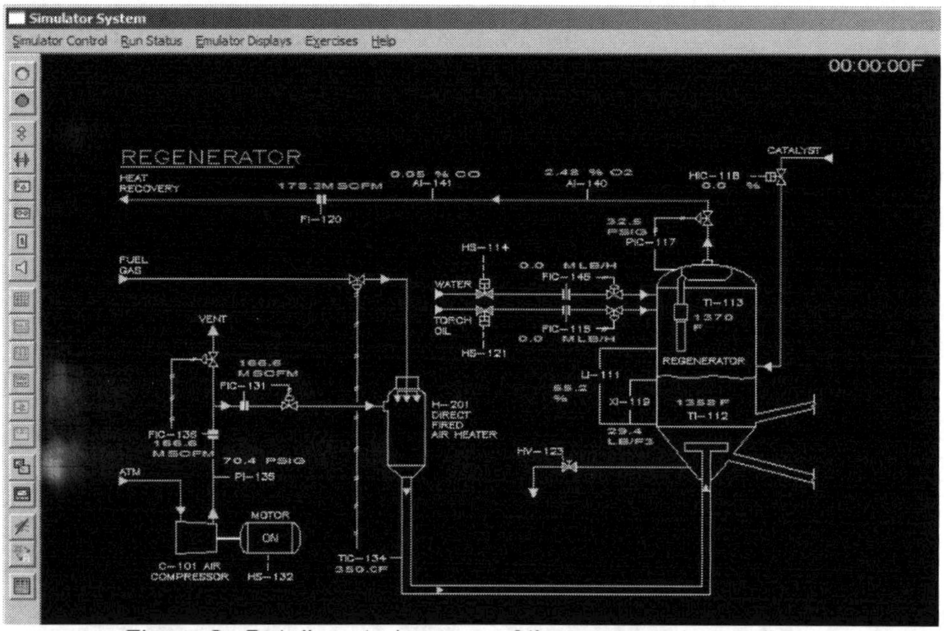

Figure 2. Detail control screen of the regenerator section.

The detail screen in Figure 2 shows some of the controllers and analyzers that are available to the operator. Each major subsection of the process has a separate control screen, including the regenerator shown in Figure 2, as well as the reactor, reactor-regenerator, fractionator overhead, pumparounds, and bottoms. Control actions are accomplished with a mouse click on a process variable. For example, clicking on the setpoint for FIC-131 above, namely, the number 166.6, accesses the controller window, shown below in Figure 3. Clicking the Detail button in the controller window accesses the controller detail screen, also shown in Figure 3, where the controller parameters can be adjusted as desired for tuning.

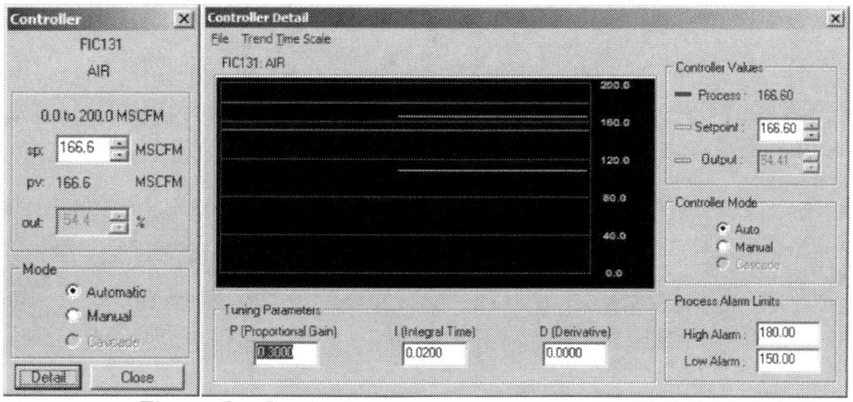

Figure 3. Controller access screens for FIC131.

One of the exciting features of the Simulation Solutions FCCU is the "virtual reality" outside operator view. The outside operator overview for the FCCU is shown below in Figure 4 (a). The operator can enter any of three sections of the plant simply by clicking on the desired section. For example, clicking on the "Furnace & Overhead" section results in the screen shot shown in Figure 4 (b). After entering the plant, the operator can proceed to monitor analyzers and sensors, or to activate outside switches and valves. The outside and control room views can be configured to operate on separate computers on the network. The outside operator and control room trainees can then communicate with handheld radios, or even by text messaging or chat rooms, etc. This makes for interesting training possibilities during startup and troubleshooting operations.

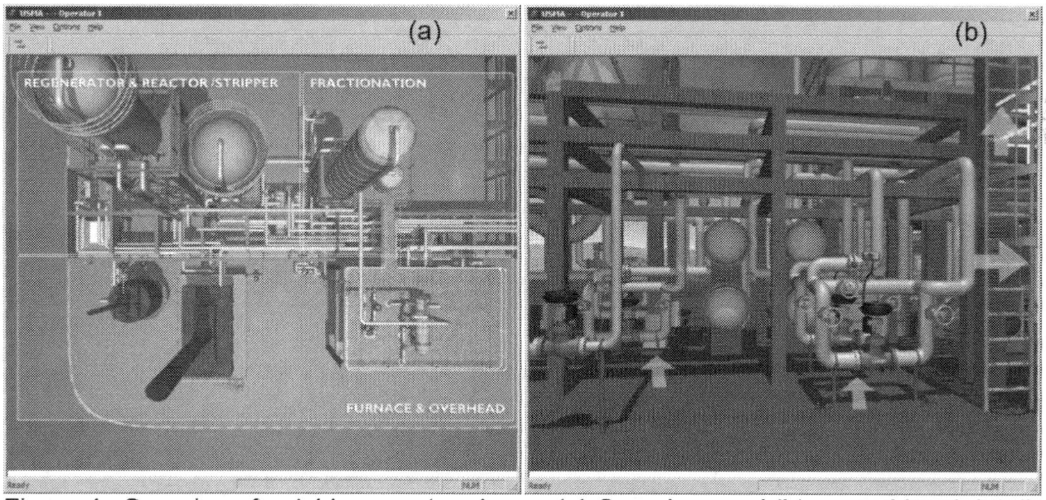

Figure 4. Samples of outside operator views. (a) Overview, and (b) ground-level detail.

The FCCU process outlined above is complicated and requires a significant amount of time for students to acclimate to the simulator environment. Simulation Solutions provides a number of individual process units, such as reactors, distillation columns, heat exchangers, etc., that can be used to introduce controller basics in a more simplified setting. These smaller units make excellent laboratory experiments in a unit operations course. Even after training on the smaller units, for large processes such as the FCCU, it is a good idea to allow for some warm-up exercises before proceeding into actual control scenarios. Sample Exercise 1 below is an example.

Sample Exercise 1 – Orientation and Introduction to Control Groups

In this exercise, students were asked to familiarize themselves with the piping and instrumentation diagrams (P&IDs) as a way to understand the complement and layout of the plant equipment, controls, instrumentation, and safety systems. Afterwards, they performed a simulated walkdown to the units using the virtual reality outside operator's representation of the plant. The students could then learn how to relate the control room to the plant. They were asked to record and map the location of the control group elements in the outside operator view for future reference. In this way, students can be introduced to topics such as controller grouping and advanced trending, while preparing for actual operations involving remote switches and valves. The control groups are easily accessed through the advanced trending window in the control room screens. The advanced trending window, shown in Figure 5a below, is available with a single button click on the ▤ button in the control room view. Using the advanced trending window, students can identify the controllers and sensors in each control group. The preferred mapping method used by most of the students was to develop a grid system for the outside operator overview screen. This is shown in Figure 5a, with grids labeled vertically from A through H and horizontally from 1 through 10. By exploring the outside operator views, students can then locate the control elements. In this example, the locations are indicated with a dark circle and a number corresponding to the respective element, ❶, ❷, etc., as shown in Figure 5b. Locations can then be easily referenced by their coordinates. It is interesting that this method was developed independently by almost all of the students in the class.

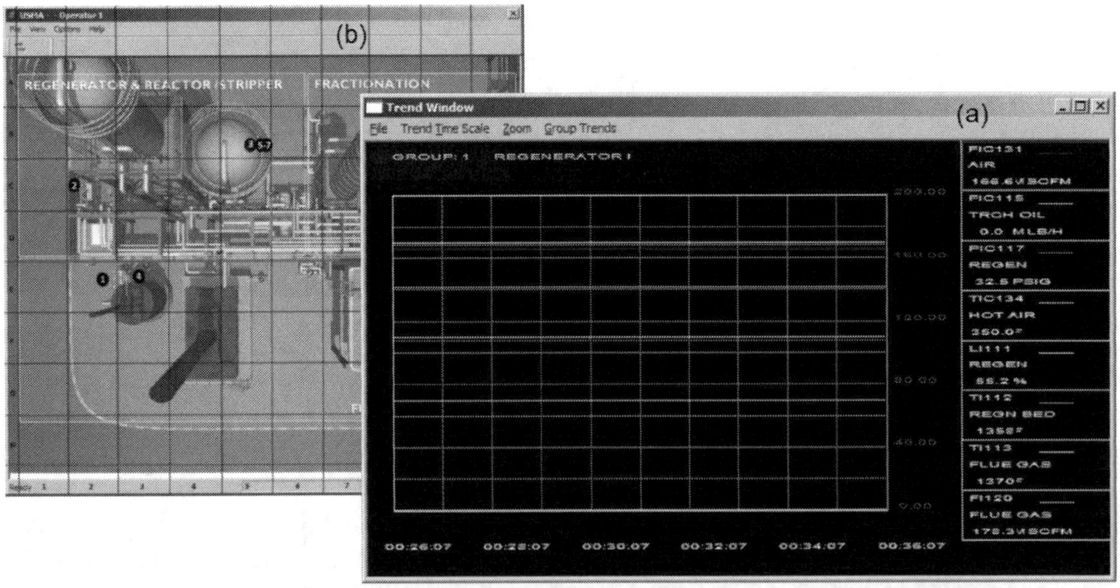

Figure 5. Controllers and sensors in Control Group 1 for the regenerator. (a) Advanced trending window, and (b) outside operator locations.

Sample Exercise 2 – Development of "What-If" Exercises

In this exercise, students are asked to identify key process variables in various subsystems, and to investigate the effects of changing these variables. By learning to make predictions about how disturbances propagate through the process, students become more familiarized with the dynamic nature of the process and the sensitivity of key control loops. This provides practice with the system and experience needed for later exercises where students need to quickly anticipate system behavior. A portion of a sample worksheet is shown in Figure 6a. In the left-hand column, students list a variable to change, such as the flow rate measured by FIC-136. Along the top row, the student identifies key process variables that will change in response to changes in FIC-136. Students are also given a key for recording predicted changes, that includes increases, decreases, increase followed by decrease, etc.

(a)

"What if" Exercises

Directions: Determine the direction of variable change for the specified event variables and conditions.

⇧ INCREASE	⇩ DECREASE	↱ INC→DEC	↳ DEC→INC	⇨ NO CHANGE

Variable→ / Event ↓						

(b)

Variable→ / Event ↓	FIC131 (PV)	FIC131 (CO)	PI135	TIC134 (PV)	TIC134 (CO)	TI112
SP FIC136 ⇩	⇨	⇨	⇨	⇨	⇨	⇨
SP FIC136 ⇧	⇨	⇨	⇨	⇨	⇨	⇨
SP FIC136 ⇧⇧	↳	↱	↳	⇨	↳	⇨

Figure 6. (a) Blank, and (b) completed "What if" exercises for C-101.
SP – setpoint; PV – process variable; CO – controller output.

Figure 6b shows how this worksheet might be completed. As an example, we consider the controllers for the air compressor C-101 (C-101 is shown in Figures 1 and 2). Referring to Figure 2, we see that FIC-136 controls an air vent from the compressor. This vent serves to protect the compressor. If the flow rate of air into the process (measured on FIC-131) drops below the setpoint on this controller for whatever reason, the vent line is opened, maintaining the compressor flow rate and preventing surges or premature shutdown. The first row indicates that no changes are expected in response to a drop in setpoint on FIC-136, since the flow rate on FIC-131 is still above FIC-136's setpoint, and there is no reason to open the vent. If the setpoint on FIC-136 is raised, but only raised slightly, we see very slight or no changes as long as the new setpoint is not changed too much and is still below the setpoint on FIC-131. If, however, we raise the setpoint further, a disturbance is created on the controller output, opening the vent momentarily, and causing a flow of air out of the vent. This in turn causes a drop in the flow through FIC131, which compensates by opening slightly. If the setpoint on FIC-136 is opened further yet so that it exceeds the setpoint on FIC-131, then the system responds by opening the vent to maintain the flow rate into the process.

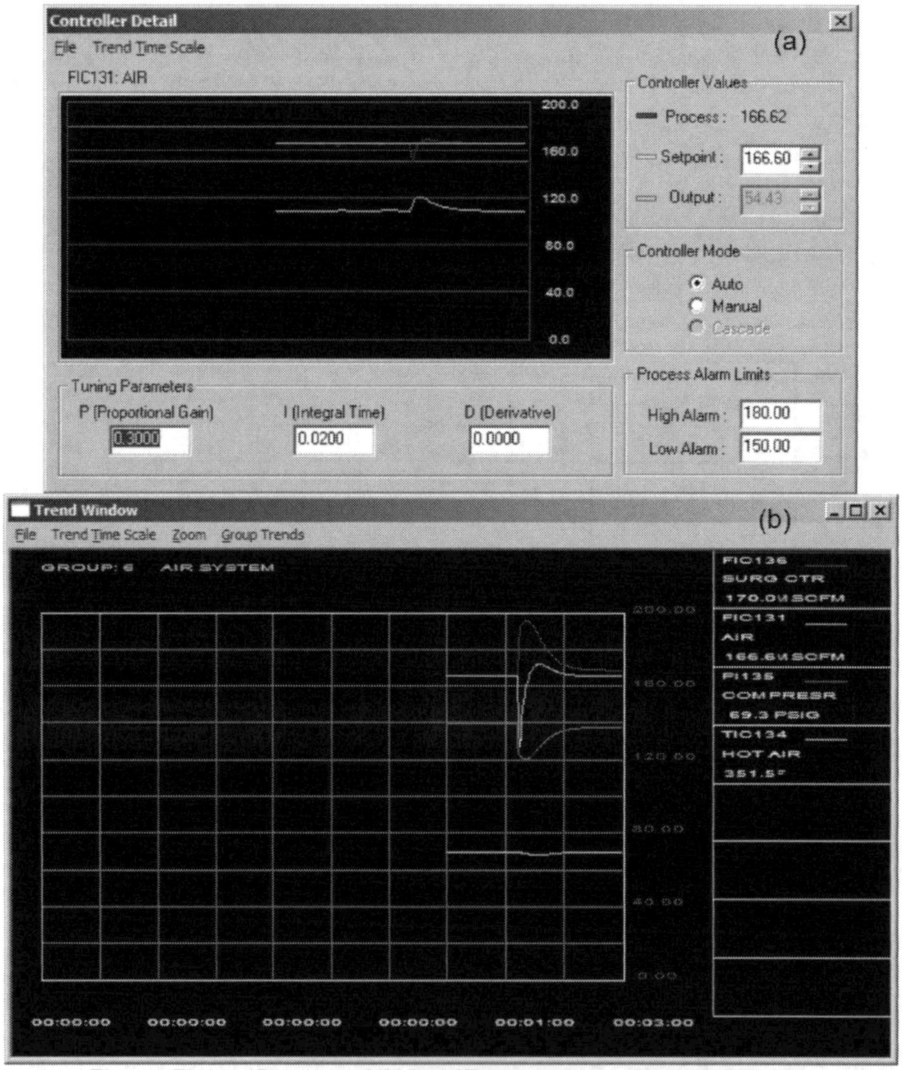

Figure 7. Verification of "What if" exercises using (a) FIC131 controller detail and (b) advanced trending.

Student verification of the results would proceed by inspection of the advanced trending and controller detail screens, as shown below in Figure 7a and b. Students are also encouraged to explore each of the different sections of the process to see where the disturbance might be creating additional responses. Students should also be encouraged to explore different types of disturbances with controllers in manual mode. The provides the students with an appreciation of why that part of the plant is automated from a safety and production standpoint

Additional Exercises

There are a wide range of exercises that can be explored with the Simulation Solutions software. An interesting feature of the software is that student computers can be set up to interact with the instructor computer, which acts as a server. In this mode, the instructor can issue "disturbances" to the student systems while the simulation is running. An example activity might be for the students to troubleshoot the problem, identify the cause, and provide a remedial action. A very different type of experience is to run the "cold-startup," in which the students bring the process online with all controllers, flows, etc. set to an initial off condition. This type of exercise requires a deep understanding of the overall process, as well as the interplay between subsystems.

Immersive Design Experience – Second and Third Order Benefits of Using Simulators

While the focus of our simulator use was to enhance student knowledge of process safety, we witnessed several second and third order benefits of using process simulators in a comprehensive manner in our process design course. First, student teams were asked to execute exercises similar to those described above to familiarize themselves with the simulators and the operation of process equipment. Then the groups had to review the P&IDs for the FCCU plant, as well as an atmospheric crude unit (ACU) to ensure they understood the relationships among the diagrams, the control screens and the 3D outside operator views. Interestingly, our student teams were able to identify several errors in the P&IDs that we brought to the attention of Simulations Solutions, Inc. Once all teams had a working knowledge of the simulator, the outside operator view and the P&IDs, the teams were then tasked to re-design a particular section of the process using parameters extracted from the simulation. This month-long immersive exercise involving these simulations greatly enhanced our student knowledge of the FCCU and ACU processes. This knowledge and familiarity with the plant could normally only be gained by working at an actual facility. However, the simulators provided an excellent opportunity for the students to work essentially as plant engineers before diving into the design project. We also incorporated a comprehensive cold-start exercise as a final exam, which did an excellent job of presenting a realistic series of problems requiring solutions that integrated knowledge from across our chemical engineering curriculum.

Student Feedback

Table 1 below shows the responses students gave to a survey conducted after they had been working with the simulators for approximately 6 weeks. As discussed above, the simulators were incorporated into the unit operations and design courses in an effort to familiarize the students with plant operations and professional practice. At the end of the experience, students were asked to respond to each question using a 1-5 Likert scale.

Part of the exercises that the students were tasked with was to re-create the simulation in a flow process simulator environment. In this case, the students used ChemCAD. The survey results shown in Table 1 are somewhat mixed here. While Q1 suggests student confidence in process design increased, Q4 shows that student confidence with ChemCAD did not. Interestingly, the low responses to Q4 provide a measure of the sensitivity of the survey, by indicating how low the students were willing to respond, which in turn provides somewhat of a baseline for the survey.

Students felt that their knowledge of equipment theory (Q3), as well as their understanding of process controllers (Q5), had increased as a result of the exercises. Other areas, such as knowledge of process economics (Q2) and mechanical design of equipment (Q10), showed somewhat strong responses in the mid-three range. The response to Q9, ability to design and conduct experiments, was surprisingly low. This suggests that the students do not consider the simulations to be equivalent to actual hands-on pilot plant operations in the lab.

Questions 6, 7, and 8 are associated with the students' perceptions of process safety. Q6 indicates that the students felt their knowledge of process safety improved. Knowledge of plant operator personnel, Q7, and knowledge of plant operations, Q8, also showed strong favorable responses. The high scores on these questions are particularly interesting since the students had already participated in field trips to actual plants and pilot operations in the unit operations lab prior to the simulator exercises.

Table 1. Results of the student surveys.

Question	Response
1. Use of simulators has enhanced my understanding of process design.	3.85 ± 0.99 (13)
2. Use of simulators has improved my understanding of process economics.	3.62 ± 1.33 (13)
3. Use of simulators provided additional insight into theory of equipment design or operation.	4.31 ± 0.48 (13)
4. Use of simulators provided additional insight into use of ChemCAD.	2.54 ± 0.88 (13)
5. Use of simulators has improved my understanding of process controllers.	4.08 ± 0.86 (13)
6. Use of simulators improved my knowledge of process safety.	3.15 ± 1.28 (13)
7. Use of simulators improved my knowledge of the job performed by process operator personnel.	4.23 ± 0.73 (13)
8. The outside operator views in the simulator enhanced my knowledge of plant operations.	3.77 ± 0.83 (13)
9. The use of simulators has enhanced my ability to design and conduct experiments.	2.92 ± 1.04 (13)
10. Use of simulators has improved my understanding of mechanical design of chemical engineering equipment.	3.54 ± 1.05 (13)

Additional Observations

The plant simulator provides simultaneous control room and outside operator views. During the student troubleshooting exercises, the students frequently observed that there was no strong correlation between the equipment that field operator had within his or her view to either control room graphics or control groups. In the actual plant, the same person does not have access to a simultaneous view of the control room and the field. In the simulator, this capability is provided. This is shown in Figures 2 and 4b. In Figure 4b, the outside operator must be instructed as to where to find the equipment.

Furthermore, the control room operator, outside operator, and plant engineers viewing the P&IDs are each looking at different layouts of the same thing, and must learn to coordinate movements, especially for emergency situations. During the role playing exercises, the students became adept at this. The students also learned to map pathways to specific equipment, which could be vital to field operator safety during likes, fires, etc. The absence of this type of practice and mastery can result in serious plant safety issues.[5]

In the future, we plan to implement different types of computer networking strategies to enhance student communication skills. In this context, the control room and outside operator views will run on separate computers at separate locations. In this type of scenario, the students can communicate by text messaging or radio, adding a further dimension of reality to the exercises. This type of communication skills is essential to plant safety and is rarely practiced in the college curriculum using lifelike situations.

References

1. Virtual Reality 3D Simulators, Simulation Solutions, Inc., http://www.simulation-solutions.com

2. Dynamic Process Simulation, Technical Simulation Consultants, Ltd., http://www.tscsimulation.co.uk

3. Douglas Cooper and Danielle Dougherty, "Enhancing Process Control Education with the Control Station Training Simulator," *Computer Applications in Engineering Education*, 7 (1999) 203-212.

4. Vinod Patel, Jeffrey Feng, Surajit Dasgupta and Jack Kramer, "Use of Dynamic Simulation in the Design of Ethylene Plants," *20th Ethylene Producers' Conference, #34 Rotating Equipment Paper Session (T2008)*, April 7, 2008.

5. Jason Laberge, Peter Bullemer, and Stephen Whitlow, "Communication and Coordination Failures in the Process Industries," Proceedings of the Human Factors and Ergonomics Society 52nd Annual Meeting (2008) 498-502.

ACCESS AND UTILIZATION OF THE DIPPR PHYSICAL PROPERTY DATABASE IN CH. E. EDUCATION

Michael B. Cutlip, University of Connecticut, Storrs, CT

Mordechai Shacham, Ben-Gurion University of the Negev, Beer-Sheva, Israel

Michael Elly, Intel Corp., Qiryat Gat, Israel

Introduction

Physical property data are essential for problem solving in chemical engineering and other related fields such as environmental and biochemical engineering. Physical property data may be required for large number of compounds over wide ranges of temperatures, pressures and compositions in chemical process design and optimization as well as environmental impact assessment.

In the not very distant past, property data was often very difficult to find. This can be demonstrated with reference to the "semi-batch steam distillation" problem that will be discussed in the next section. This problem was originally suggested by Prenosil in 1976. At that time he had to collect the data required for solving the steam distillation problem from various sources.

He provided the details of these sources, as follows: "The constants of the Antoine equations for the vapor pressures of *n*-octane and *n*-decane were taken from The Handbook of Chemistry and Physics (1970) and for water they were obtained by non-linear regression from steam tables (Perry, 1963). The *Cp* values of the liquid components were also found in Perry (1963) and their temperature dependence was neglected. The *Cp* data for the vapor components were calculated as functions of temperature from the equation $Cp = A + BT + CT^2$ with the constants taken from Balshiser et al. (1972). The latent heats of evaporation for the organic material came from Cox and Pilcher (1970) and that of water from Landolt-Börnstein (1960). The Handbook of Chemistry and Physics (1970) also served as a source of critical data. The molar volume at 95 °C was only found for *n*-octane in Timmermans (1965). The molar volume of *n*-decane was obtained by extrapolation from the data tabulated at lower temperatures and our own measurements up to a temperature of 70 °C."

Finding the necessary data in that era represented a "small research project." There was usually a need to make some assumptions (such as neglecting the temperature dependence of the *Cp* values), or to extrapolate the data with no basis to evaluate the validity of the assumptions and the reliability of the extrapolated values. Also converting the data and correlations collected from difference sources to the same consistent set of units represented a challenge and often may have introduced errors into the mathematical model of the process.

Currently, most of the needed properties can be found in available databases. The database we have been using is the DIPPR® database, (http://dippr.byu.edu/, Rowley et al., 2006). The DIPPR database developers have invested considerable efforts to provide complete sets of constant pure component properties. Additionally temperature dependent correlations for the compounds are provided, even if some of the properties must be predicted. Estimations for the uncertainty of the measured or predicted values are also supplied. The completeness of the datasets as well as the availability of the uncertainty estimations has great value in both chemical engineering education and industrial practice. The provision of only experimental data, as several data bases do, puts the burden on the user to predict the property data which is essential for solving particular problems.

Unfortunately, the use of physical property databases in engineering education and practice is not as widespread as might be desired. Firstly, property databases are rarely mentioned in current textbooks, so that even the instructors are unaware of their potential educational benefits. Students who use property data provided by the instructor or from limited lists appearing in the appendices of the textbooks do not acquire the habit of obtaining all property data from one reliable source. Consequently they may not use databases in their graduate research or engineering practice. Secondly, many of the current property databases do not efficiently enable the transfer of property and correlation data from the databases to computer codes, and the transfer can be a cumbersome, complex and error prone process. The name of the property, its units, value, uncertainty and additional information are usually stored in separate cells in the database. This typically requires considerable copying, pasting and manually manipulation so that important information can be introduced and documented in the computer code. For temperature dependent properties, the name of the property, its units, the coefficients of the correlations, the range of validity of the correlation, the uncertainty of the calculated values and some additional information are all stored separately. Here, also, after copying the data, manual introduction into the respective equations and the computer code is required. Shacham et al. (2009) have shown that careless rounding of coefficients while copying from the database to various software packages may cause significant errors.

In order to simplify the transfer of the data from the database to computer code and thus minimize the probability of the introduction of errors during the process, we have developed a "DIPPR to Code" interface. This interface enables searching the database for a particular compound, marking the desired properties and obtaining as output the necessary data and correlations in a format that can be copied and pasted directly into a computer code. The formats that are currently supported are for Polymath Software™ (a product of Polymath Software, http://www.polymath-software.com) and MATLAB® (a product of The Math Works, Inc. http://www.mathworks.com). The Polymath package is a user-friendly, easy-to-learn package which is widely used for problem solving in engineering. MATLAB can be considered as an advanced computer language that is widely used in graduate engineering and scientific education and research. Property correlation equations coded in MATLAB format can be introduced into programs written in various computer languages with very minimal changes.

The Polymath package is also capable of exporting equations to Excel® (a product of Microsoft Corporation, http://www.microsoft.com) by provision of automatic conversion of the DIPPR correlation equations to Excel formulas. This is very important as practically all engineers use Excel for their computational tasks. Thus the "DIPPR to Code" interface enables exporting DIPPR data and equations to many of the widely used software packages.

The use of the DIPPR database, the interface program and their potential benefits will be demonstrated by modeling and simulating a semi-batch steam distillation process.

Semi-Batch Steam Distillation of a Binary Organic Mixture – Problem Definition

A typical example of a process model that includes a large amount of constant and temperature dependent data is steam distillation of an organic mixture. Such an example is provided by Prenosil (1976). This illustrative example involves semi-batch steam distillation of an *n*-octane (comp. 1) and *n*-decane (comp. 2) mixture. A simplified version of this problem appears in the textbook of Ingham et al. (2007) and this version will be presented here.

A schematic plot of the steam distillation apparatus is shown in Figure 1. The organic mixture is charged into the still initially and steam is bubbled through continuously until the desired degree of separation has been reached. There are two different periods in the operation of the still: the heating period, until the boiling point temperature of the organic mixture is reached, and the distillation period. A brief description of the mathematical models for the two periods follows.

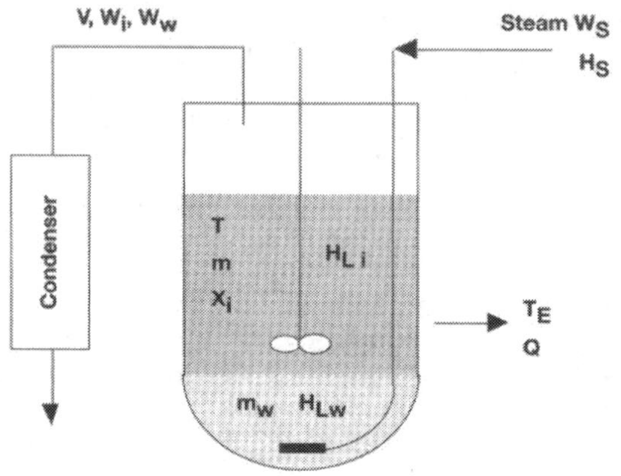

Figure 1 – Schematic plot of steam distillation

Heating Period

A simple mass balance on the water phase yields

$$\frac{dm_W}{dt} = W_S \tag{1}$$

where W_S is the steam flow-rate and m_W is the mass of water in the still. It is assumed that al the steam condenses in the distillation vessel and that the organic phase masses remain constant during the heating period.

An energy balance on the still provides the equation for the change of the temperature (T)

$$\frac{dT}{dt} = \frac{W_S(H_S - H_{Lw}) + Q}{m_w c_{pLw} + m(x_1 c_{pL1} + x_2 c_{pL2})} \tag{2}$$

where H_S is the enthalpy of the steam, H_{Lw} is the enthalpy of liquid water, Q is the rate of heat transfer to the surrounding, c_{pLw} is the molar specific heat of the water, m is the mass of the organic phase in the still, x_1 and x_2 are the mole fractions and c_{pL1} and c_{pL2} are the molar specific heats of *n*-octane and *n*-decane, respectively. The heat transfer rate to the surrounding is calculated from Eq. 3

$$Q = UA(T_E - T) \tag{3}$$

where U is the overall heat transfer coefficient, A is the contact area with the surrounding and T_E is the ambient temperature.

Assuming ideal liquid behavior, Rault's law can be used to calculate the vapor mole fraction of the components in the organic phase. Thus

$$y_1 = \frac{x_1 P_1}{P} \qquad y_2 = \frac{x_2 P_2}{P} \tag{4}$$

where P is the total pressure and P_1 and P_2 are the vapor pressures of the organic compounds. The mole fraction of the water which is immiscible in the organic phase is given by $y_W = P_W / P$. The heating period continues until the sum of vapor pressures of the organic compounds and the water is equal to the total pressure. Thus, the following "bubble point" equation has to be satisfied.

$$f(T) = 1 - (y_1 + y_2 + y_W) = 0 \tag{5}$$

Distillation Period

During the distillation period, there is output of water vapor from the still. Thus Eq. 1 must be modified to

$$\frac{dm_W}{dt} = W_S - Vy_W \qquad (6)$$

where V is the outlet vapor flow rate. Material balances on the two organic compounds yield two additional differential equations

$$\frac{d(mx_1)}{dt} = -Vy_1 \qquad \frac{d(mx_2)}{dt} = -Vy_2 \qquad (7)$$

The organic mass in the still at any time is given by: $m = mx_1 + mx_2$. The temperature in the still changes in a manner so that the bubble point equation (Eq. 5) is satisfied. The energy balance at a particular temperature yields the momentary vapor flow rate

$$V = \frac{W_S(H_S - H_{Lw}) + Q}{H_V - [y_w h_{Lw} + (y_1 h_{L1} + y_2 h_{L2})]} \qquad (8)$$

where H_V is the molar enthalpy of the vapor phase, h_{Lw}, h_{L1} and h_{L2} are the liquid phase molar enthalpies of water, *n*-octane and *n*-decane respectively. Material balances on the water and organic phases in the still can provide the amount and the mole fractions of the various components in the distillate.

A typical assignment in using this semi-batch distillation apparatus may involve the calculation of the distillation time when a pre-specified separation can be reached. Various initial organic phase compositions can be investigated.

Solution Algorithm for the Semi-Batch Steam Distillation Problem

Two separate models can be used for solving this problem. In the solution of the "heating period model," the special challenge is to determine the time when the function of Eq. 5 approaches close enough to zero. For one or a few sets of data, this can be done by some kind of "trial and error" procedure. However, for larger sets of input data it is preferable to find the most appropriate time by solving a "two-point boundary value problem" as demonstrated, for example, in Prob. 6.4 of the Cutlip and Shacham (2008) textbook. When the boiling point is first achieved, the temperature and mole fraction values must be passed to the "distillation period" model, as initial values.

A specific challenge in the solution for the distillation period is the need to follow the bubble point temperature curve (Eq. 5) during the time of the integration. This requires solving a nonlinear algebraic equation for each and every integration step. (This is basically a differential-algebraic or DAE problem.) A simple method that we recommend using for this purpose is the "controlled integration" technique of Shacham and Brauner (1997). Using this method Eq. 5 is rewritten

$$\varepsilon = 1 - (y_1 + y_2 + y_W) \qquad (5A)$$

and a "proportional controller" is added to the system of equations for changing the temperature:

$$\frac{dT}{dt} = K_c \varepsilon \qquad (9)$$

The proportional gain, K_c, is selected large enough so as to keep ε below a pre-specified tolerance. Setting K_c at a large positive value will often lead to a "stiff" system of differential equations which has to be solved using specific stiff algorithms.

Thus, the semi-batch steam distillation problem can be considered a Multiple-Model, Multiple-Algorithm (MMMA) problem. There are two different models for the heating and the distillation periods and two different integration algorithms (non-stiff for the heating period, stiff for the distillation period) are used. The solution of MMMA problems has been discussed in detail recently by Cutlip et al. (2009).

Physical Property Needs and Sources

In addition to the mathematical models, pure component physical property data is required for all three compounds involved in the steam distillation process. Most of the data is needed in the form of equations (correlations) that enable calculation of a particular pure component property as function of the temperature. Correlations are needed for liquid vapor pressure, liquid heat capacity and enthalpy, ideal gas heat capacity and enthalpy. For calculating the ideal gas enthalpy, the heat of vaporization at a reference temperature is also needed. The volume of the liquid in the still need to be determined (for preventing overflow of liquid at any stage of the distillation, for example), and thus correlations for liquid densities are required.

The information that is provided by the interface program in the Polymath format is demonstrated in Table 1 for the needed properties of *n*-octane that are required for the steam distillation problem. This table shows the exact information as provided by the "DIPPR to Code" interface when the Polymath format is specified.

Table 1. Physical properties of n-octane as obtained from the DIPPR database by the "DIPPR to Code" interface for use in Polymath

No.	Equation/ # Comment
1	# Liquid Vapor Pressure of n-OCTANE (C8H18)
2	# Error: < 1%; Data Type: Experimental; Source Type: Unevaluated; Source Detail: Primary; Error Source: Staff
3	# Min_T=216.38, Max_T=568.7, Min_Val=2.1083, Max_Val=2467300
4	VP_C8H18 = exp(96.084 - 7900.2 / T - 11.003 * ln(T) + 7.1802E-06 * T ^ 2) # Pa
5	# Liquid heat Capacity (at 1 atm below normal boiling point, saturation pressure at and above) of n-OCTANE (C8H
6	# Error: < 1%; Data Type: Smoothed; Source Type: Evaluated; Source Detail: Evaluated; Error Source: Staff
7	# Min_T=216.38, Max_T=460, Min_Val=229340, Max_Val=341890
8	LCP_C8H18 = 0 - 186.63 * T + 0.95891 * T ^ 2 + 224830 # J/kmol*K
9	# Ideal Gas Heat Capacity of n-OCTANE (C8H18)
10	# Error: < 1%; Data Type: Predicted; Source Type: Evaluated; Source Detail: Evaluated; Error Source: Staff
11	# Min_T=200, Max_T=1500, Min_Val=145290, Max_Val=497640
12	#HIG_C8H18 = 135540*T + 443100 *1635.6* (coth(1635.6 / T)) - 305400 *746.4* (tanh(746.4 / T))+Hcon # J/kmol
13	ICP_C8H18 = 135540 + 443100 * (1635.6 / T / sinh(1635.6 / T)) ^ 2 + 305400 * (746.4 / T / cosh(746.4 / T)) ^ 2 # J
14	# Liquid Density (at 1 atm below normal boiling point, saturation pressure at and above) of n-OCTANE (C8H18)
15	# Error: < 1%; Data Type: Smoothed; Source Type: Evaluated; Source Detail: Evaluated; Error Source: Staff
16	# Min_T=216.38, Max_T=568.7, Min_Val=6.705, Max_Val=2.05
17	LDN_C8H18 = 0.5266 / 0.25693 ^ (1 + (1 - T / 568.7) ^ 0.28571) # kmol/m^3
18	# Heat of Vaporization of n-OCTANE (C8H18)
19	# Error: < 1%; Data Type: Experimental; Source Type: Unevaluated; Error Source: Staff
20	# Min_T=216.38, Max_T=568.7, Min_Val=45898000, Max_Val=0, Error:< 1%
21	HVP_C8H18_273 = 4.2902E+07 # J/kmol (T=273.15)

The code that is generated by the interface program includes correlation equations, definition of constant values and comments (text that starts with the "#" sign and ends with the end of the line). The row numbers shown in Table 1 are not part of the output generated by the interface program; they were added as references for the explanations that follow. Lines 1-17 can be generated at once by selecting the desired "property equations" and the "basic" log level. The "log level" determines the amount of information that is to be included as comments. The data for the heat of vaporization at 273.15 K (lines 18 to 21) have to be generated separately as it is not an equation but rather a point value of the property.

Lines 1 through 4 contain the information related to the vapor pressure of *n*-octane. In line 1 the full name of the property, the full name of the compound and its formula are shown (as a comment). In line 2, information concerning the reliability of the vapor pressure correlation, as determined by the DIPPR staff is shown. The uncertainty (error) in the predicted vapor pressure is < 1% and the correlation is based on experimental data. There are additional details given in this line regarding the reliability of the data source. In line 3, the range of validity of the correlation equation is shown. "Min_T" is the lower temperature limit of the range of validity, "Max_T" is the upper temperature limit, "Min_Val" is the property (vapor pressure, in this case) value at the lower temperature limit and "Max_Val" is the property value at the upper temperature limit.

In line 4, the property correlation is shown. The variable into which the calculated value of the vapor pressure is entered made up from the symbol of the property (VP in this case) and the chemical formula of the compound involved. The Riedel equation is used to model the change of the vapor pressure with the temperature. The units of the property (vapor pressure in this case) are included in a "comment" in the same line with the equation. The units of the temperature are not given explicitly as the only units used are K.

Similar information is provided for liquid heat capacity (lines 5 through 8) and liquid density (lines 14 through 17). The point value of the heat of vaporization of n-Octane is provided in lines 18 through 21. The variable name in which the point property is stored (see line 21) includes the property symbol (HVP), the formula of the compound (C8H18) and the integer digits of the temperature (273). The comment provides the units of the property and the exact temperature value.

The ideal gas heat capacity (lines 9 – 13) deserves special discussion. DIPPR provides the coefficients for the Aly and Lee (1981) equation for this property. The integrated form of this equation, for calculating ideal gas enthalpy is considerably different than the heat capacity equation. For the benefit of the users, whenever the ideal gas heat capacity is requested by the user, the equation for ideal gas enthalpy is also included, as a comment (see line 12). Note that the enthalpy equation includes an integration constant *Hcon*. This constant depends on the standard state selected for the enthalpy calculation. For example, selecting as standard state pure liquid component at 0 °C (273.15 K) yields Hcon as ideal gas enthalpy at 273.15 K subtracted from the heat of vaporization of the component at the same temperature. The equations for calculating *Hcon* for *n*-octane are shown in Table 2. This set of equations yields: *Hcon* = -4.928E+08 J/kmol.

Table 2. Calculation of Hcon for n-Octane (reference state for enthalpy: pure liquid at 273.15 K)

No.	Equation/ # Comment
1	T0 = 273.15 #K
2	HIG_C8H18 = 135540*T0 + 443100 *1635.6* (coth(1635.6 / T0)) - 305400 *746.4* (tanh(746.4 / T0)) # J/kmol
3	HVP_C8H18_273 = 4.2902E+07 # J/kmol (T=273.15)
4	Hcon_C8H18= HVP_C8H18_273-HIG_C8H18 # J/kmol

Heating Period Model for the Steam Distillation Example

The Polymath program for modeling the heating period of the semi-batch distiller is shown in Table 3. In lines 1 and 2 the principal equations of the model (Eqs. 1 and 2) presented. The numerical data for the problem, as provided by Ingham et al. (2007) are included in lines 3 through 12. The auxiliary equations (3), (4) and (5A) are in lines 13 through 17. All the physical property equations necessary are shown in lines 18 through 26. These include the Riedel equations for calculating the vapor pressure of the three compounds, the equations for calculating the heat capacity of the individual compounds as well as the heat capacity of the liquid phase (in line 23). The enthalpy of the steam and the enthalpy of the water (required in Eq. 1) are calculated in lines 25 and 26 respectively. In lines 27 through 30, the initial values of the dependent variables T and m_W and the initial and final values of the time: t (s) are shown.

Table 3. Polymath program for simulating the "heating period" in the steam distiller

No.	Equation/ # Comment
1	d(Temp)/d(t) = (MS * (HS - HL_H2O) +Q)/ CpL # Still temperature from heat balance
2	d(MW)/d(t) = MS # Mass of water in the still (mol) from mass balance
3	TK = Temp + 273.15 # Absolute temperature (K)
4	T0=273.15
5	TSK = 99.2 + 273.15 # Steam temperature (K)
6	P = 9.839E+04 # Ambient Pressure (Pa)
7	MS = 3.85e-5 # Steam flow rate (kmol/s)
8	U = 1.05 # Heat transfer coeff (J/s-K)
9	Ta = 25 # Ambient temperature (deg C)
10	x1=0.725 #Mole Fraction of n-octane
11	x2=1-x1 #Mole Fraction of n-decane
12	M=0.015 # Initial amount of organics, kmol
13	Q = U* (Ta - Temp) # Heat removed from the still (J/s)
14	Y1= VP_C8H18* x1 / P # n-octane vapor mole fraction
15	Y2 = VP_C10H22* x2 / P # n-decane vapor mole fraction
16	YW =VP_H2O / P # Water vapor mole fraction
17	err = 1 - (Y1 + Y2 + YW) # Error used for bubble point determination
18	VP_C8H18 = exp(96.084 - 7900.2 / TK - 11.003 * ln(TK) + 7.1802E-06 * TK ^ 2) # Pa
19	VP_C10H22 = exp(112.73 - 9749.6 / TK - 13.245 * ln(TK) + 7.1266E-06 * TK ^ 2) # Pa
20	VP_H2O = exp(73.649 - 7258.2 / TK - 7.3037 * ln(TK) + 4.1653E-06 * TK ^ 2) # Pa
21	LCP_C8H18 = (0 - 186.63 * TK + 0.95891 * TK ^ 2 + 224830) # J/kmol*K
22	LCP_C10H22 = (0 - 197.91 * TK + 1.0737 * TK^ 2 + 278620) # J/kmol*K
23	LCP_H2O = (276370 - 2090.1 * TK + 8.125 * TK ^ 2 - 0.014116 * TK ^ 3 + 9.3701E-06 * TK ^ 4) # J/kmol*K
24	CpL = MW * LCP_H2O + M * (x1 * LCP_C8H18 + x2 * LCP_C10H22) # Heat capacity of the liquid J/kmol*K
25	HS = (33363*TSK + 26790 *2610.5* (coth(2610.5 / TSK)) ^ 2 + 8896 * 1169*(tanh(1169 / TSK)) -4.471E+07)# Steam Enthalpy (J/kmol)
26	HL_H2O = (276370*(TK-T0) - 2090.1 * (TK^2-T0^2)/2 + 8.125 * (TK ^ 3-T0^3)/3 - 0.014116 * (TK ^ 4-T0^4)/4 + 9.3701E-06 * (TK ^ 5-T0^5)/5)# J/kmol)
27	Temp(0) = 25 # Temperature in the still (deg. C)
28	MW(0) = 0 # Mass of water in the still (kmol)
29	t(0) = 0
30	t(f) = 181.7 # s

Note that Polymath automatically reorders the equations, as needed, before starting the computations. Thus initially the equations can be arranged so as to provide clear and concise documentation of the model if desired. The comments related to the range of applicability and the uncertainty of the physical property correlations, that were shown in Table 1, have been removed for brevity; however, it is recommended that these comments be retained in the final program to provide more detailed documentation.

The completion time of the heating period (in line 30 of Table 3), that brings $f(T)$ of Eq. 5 close to zero, was found by trial and error to be: $t_f = 181.7$ s. At this point the temperature in the still is $T = 90.65$ °C and the mass of water in the still is $m_W = 0.07$ kmol. These values are transferred to the distillation period model as initial values.

Distillation Period Model for the Steam Distillation Example

Part of the Polymath program for modeling the distillation period of the semi-batch distiller is shown in Table 4. Only the lines that were not included in Table 3 are shown in this table.

Table 4. Polymath program for simulating the "distillation period" in the steam distiller

No.	Equation/ # Comment
1	d(Temp)/d(t) = 1000 * err # Still temperature by controlled integration
2	d(Mx1)/d(t) = -V * Y1 # Mass of n-octane in the still (kmol) from mass balance
3	d(Mx2)/d(t) = -V * Y2 # Mass of n-decane in the still (kmol) from mass balance
4	d(MW)/d(t) = MS - V * YW # Mass of water in the still (kmol) from mass balance
-	lines 3-9 from Table 3
5	x01 = 0.725 # Initial n-octane organic liquid mole fraction
6	x02 = 0.275 # Initial n-decane organic liquid mole fraction
7	M0 = 0.015 # Initial amount of organics, kmol
8	M = Mx1 + Mx2 # Organic mass in the still (kmol)
9	x1 = Mx1 / M # n-octane organic liquid mole fraction
10	x2 = Mx2 / M # n-decane organic liquid mole fraction
-	lines 13-17 from Table 3
11	V =(MS*(HS-HL_H2O)+ Q) / (HV - (HL_H2O * YW + (Y1 * HL_C8H18 + Y2 * HL_C10H22))) # Vapor flow rate (kmol/s)
12	M1dist = M0 * x01 - Mx1 # Mass of n-octane in the distillate (kmol)
13	M2dist = M0 * x02 - Mx2 # Mass of n-decane in the distillate (kmol)
14	MWdist = MS * t - MW # Mass of water in the distillate (kmol)
15	Mdist = M1dist + M2dist # Distilled organic phase (kmol)
16	x1dist = If (Mdist > 0) Then (M1dist / Mdist) Else (0) # n-octane distillate mole fraction
17	x2dist = If (Mdist > 0) Then (M2dist / Mdist) Else (0) # n-decane distillate mole fraction
-	lines 18-26 from Table 3
18	HL_C8H18 = (224830*(TK-T0) - 186.63 *(TK^2-T0^2)/2 + 0.95891 * (TK ^ 3-T0^3)/3) # J/kmol
19	HL_C10H22 = (278620*(TK-T0) - 197.91 * (TK^2-T0^2)/2 + 1.0737 *(TK ^ 3-T0^3)/3)# J/kmol
20	HL= x1*HL_C8H18+x2*HL_C10H22 # Enthalpy of the organic liquid phase (J/kmol)
21	HV1 = (135540*TK + 443100 *1635.6* (coth(1635.6 / TK)) - 305400 *746.4* (tanh(746.4 / TK))-4.928E+08) # n-octane vapor enthalpy (J/kmol)
22	HV2= (167200*TK+ 535300 * 1614.1*(coth(1614.1 / TK)) - 378200 *742* (tanh(742 / TK)) -5.791E+08) # n-decane vapor enthalpy (J/kmol)
23	HVW = (33363*TK + 26790 *2610.5* (coth(2610.5 / TK)) ^ 2 + 8896 * 1169*(tanh(1169 / TK)) -4.471E+07) # Water vapor enthalpy (J/kmol)
24	HV = YW * HVW + Y1 * HV1 + Y2 * HV2 # Vapor phase enthalpy (J/kmol)
25	Temp(0) = 90.64995 # Temperature in the still (deg. C)
26	Mx1(0) = 0.010875 # Mass of n-octane in the still (kmol)
27	Mx2(0) = 0.004125 # Mass of n-decane in the still (kmol)
28	MW(0) = 0.006995 # Mass of water in the still (kmol)
29	t(0) = 181.7
30	t(f) =2000

The lines that remained the same and are included also in this program are clearly marked in Table 4. In line 1, the equation for changing the temperature using controlled integration (Eq. 9) is shown. The value of the proportional gain K_c (K_c = 1000) was selected so that to keep the deviation of Eq. (5A) below the value of 10^{-5} ($\varepsilon < 10^{-5}$). In lines 2 and 3, the change in the mass of the organic compounds is calculated using Eq. (7), and in line 3, the change of the mass of the water in the still is obtained from Eq. (6). The vapor flow rate is calculated using Eq. (8) in line 11 and the masses and mol fractions of the various compounds in the distillate are calculated from material balance equations in lines 12 through 17. Additional physical properties (liquid and vapor phase enthalpies) are calculated in lines 18 through 24. The initial values of the dependent variables and t (obtained as the final values of the heating period model) are shown in lines 25 through 28. The integration is continued until the amount of *n*-octane in the still gets below 10^{-4} kmol (t_f= 2000 s, line 30 in Table 4).

Results and Discussion

The calculated temperature profile in the steam distiller is shown in Figure 2. There is a rapid increase within the first three minutes during which the temperature increases from the initial value of 25 °C to the boiling temperature of 90.65 °C. During the distillation period there is a more gradual increase of the temperature because of the depletion of the more volatile *n*-octane in the still. After 30 min of distillation the temperature reaches 96 °C.

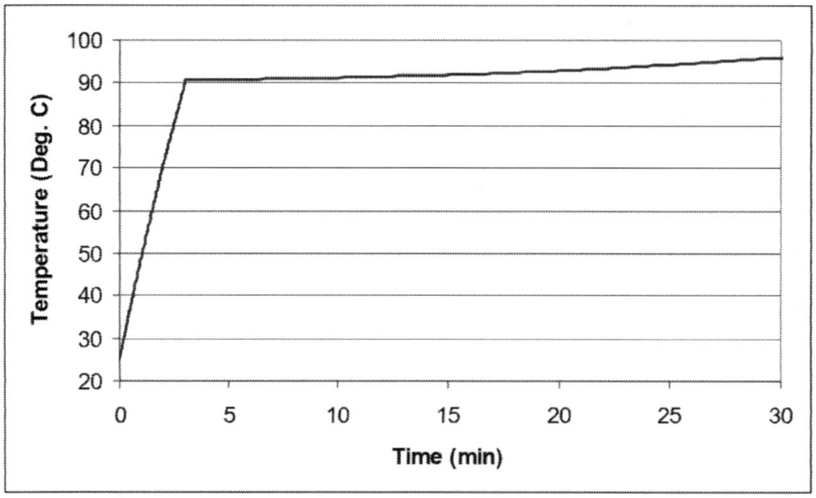

Figure 2. Temperature change during semi-batch steam distillation

The variation of the concentration of the organic compounds in the still (liquid and vapor phases) is shown in Figure 3. Starting at the beginning of the distillation period, there is a steady decrease in the liquid mole fraction of the more volatile component and steady increase in the mole fraction of the less volatile component. After 30 minutes, the concentration of n-octane reaches the value of x_1 = 0.085 while the concentration of *n*-decane reaches: x_2 = 0.915. During the heating period there is a rapid increase in the concentration n-octane in the vapor phase however, after distillation starts the vapor phase concentration follows the same trend as the liquid phase.

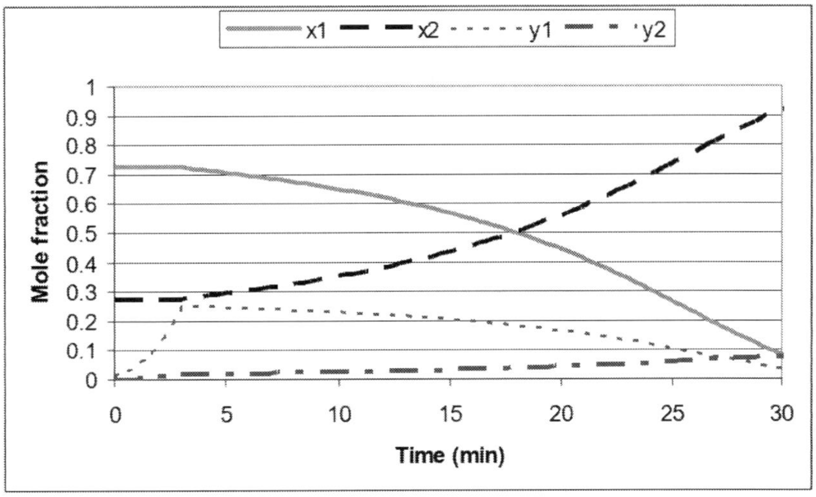

Figure 3. Change of organic phase composition during semi-batch steam distillation

Figure 4 shows the concentration of the organic compounds in the distillate. The mole fraction of the *n*-octane at the start of the distillation is x_{1dist} = 93% and after 30 minutes it is reduced to x_{1dist} = 82%. The distillation can be stopped when the desired level of separation is reached. For example if the desired n-octane concentration is x_{1dist} = 90%, the distillation should be stopped after 18 min.

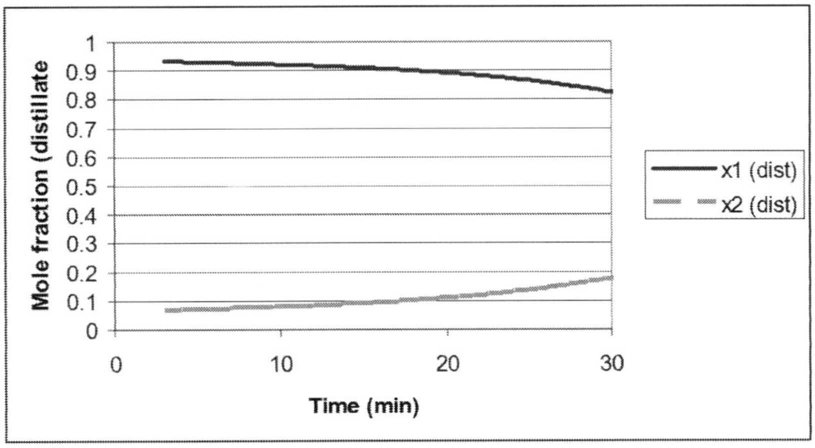

Figure 4. Change of organic phase distillate composition during semi-batch steam distillation

We have solved this problem previously with Aspen Properties® (a product of AspenTech, http://www.aspentech.com/) by utilizing the Aspen Properties Add-In for Excel (Shacham and Cutlip, 2004). The results using the two different property packages (Aspen Properties and DIPPR) are essentially the same.

Exporting the Model to Excel and MATLAB

The Polymath models presented in Tables 3 and 4 can be exported from Polymath to Excel by a single key-press. The spreadsheet, which is automatically generated by Polymath, includes the Excel formulas that are equivalent to the Polymath equations. The variable names and Polymath equations are also exported to Excel as text, to provide a clear documentation for the Excel worksheet. Intrinsic functions and logical conditions are also automatically converted to Excel formulas. An ODE solver Add-Inn (provided as a separate utility along with Polymath) is available for solving the ordinary differential equations within Excel.

The Polymath models can also be exported to MATLAB in the form of MATLAB functions. Library functions for solving ODE's (such as ode45.m) can be used to solve the problem within MATLAB.

The Excel and MATLAB export options can not be demonstrated here because of lack of space; however, these options are explained in detail in chapters 4 and 5 of the Cutlip and Shacham (2008) textbook.

Conclusions

This work demonstrates the many advantages of the use of a database, such as DIPPR, in computer-based problem solving. These advantages include:

1. Minimization of the effort for searching or predicting properties as all the necessary data are available in one source.
2. Presentation of the reliability and uncertainty of the data as all this information is available in the database.
3. Provision of the range of applicability so that correlations can be used to assure accurate results.
4. Reduction of conversion factor errors as all of the data are available in consistent units.

The use of the DIPPR data base will introduce students to an important aspect of problem solving – that good results in detailed calculations are achieved with accurate data from physical property databases. This experience will carry over database usage to their professional engineering careers.

For the instructor, the availability of database access to students can minimize the effort in making sure that needed property data are available for homework and design project assignments.

The new "DIPPR to Code" interface will enable users to have convenient export of DIPPR data and property equations to several widely used software packages, thus minimizing the required effort and reducing error introduction during this process.

References

1. Aly, F. A.; Lee, L. L. Self-Consistent Equations for Calculating the Ideal Gas Heat Capacity Enthalpy, and Entropy. *Fluid Phase Equilib.* **6**, 169(1981)
2. Balshiser, R. E., Samuels M. R. and J. D. Eliassen, *Chemical Engineering Thermodynamics*, Prentice-Hall, Englewood Cliffs, N.J., 1972.
3. Cox, J. D. and G. Pilcher, *Thermochemistry of Organic and Organomerallic Compounds*, Academic Press, New York, 1970.
4. Cutlip, M. B. and Shacham, M. *Problem Solving In Chemical and Biochemical Engineering with Polymath, Excel and MATLAB.* Prentice-Hall, Upper Saddle River, New-Jersey, 2008.

5. Cutlip, M. B., Brauner, N. and M. Shacham, " Biokinetic Modeling of Imperfect Mixing in a Chemostat – an Example of Multiscale Modeling", *Chemical Engineering Education* , Vol. 43. No. 3, 243-248 (summer 2009)
6. *Handbook of Chemistry and Physics*, 51st edn., Chemical Rubber Co., Cleveland, Ohio, 1970-71.
7. Ingham, J, Dunn, I. J., Heinzle, E. and Prenosil, J. E., *Chemical Engineering Dynamics: An Introduction to Modelling and Computer Simulation*,3rd Ed, Wiley-VCH; 2007
8. Landolt—Börnstein Tabellen, 6th edn., Springer-Verlag, Berlin, 1960.
9. Perry, J. H., *Chemical Engineers' Handbook*, 4th edn., McGraw-Hill, New York, 1963.
10. Prenosil, J. E., "Multicomponent Steam Distillation: Comparison between Digital Simulation and Experiment", The Chemcal Engineering Journal, 12, 59-68 (1976).
11. Rowley, R.L.; Wilding, W.V.; Oscarson, J.L.; Yang, Y.; Zundel, N.A., DIPPR Data Compilation of Pure Chemical Properties, Design Institute for Physical Properties, (http//dippr.byu.edu), Brigham Young University, Provo, Utah, 2006.
12. Shacham, M. and N. Brauner, "What To Do if Relative Volatilities Cannot Be Assumed To Be Constant? - Differential-Algebraic Equation Systems in Undergraduate Education", *Chem. Eng. Educ.*, 31(2), 86-93(1997)
13. Shacham, M and M. B. Cutlip, "Reactor Design within Excel Enabled by Rigorous Physical Properties and Advanced Numerical Computation Package", presented at Aspen Word 2004, Orlando, Florida, Oct. 10-15, 2004.
14. Shacham, M., M. B. Cutlip and M. Elly., "Common Errors in Numerical Problem Solving – How can They be Detected and Prevented", Chemical Engineering Progress (Accepted for Publication, 2009 , also available at http://cee.che.ufl.edu/)
15. Timmermans, J., Physico-chemical Constants of Pure Organic Compounds, Elsevier, Amsterdam, 1965.

Appendix (Selected Screen Displays of the Polymath "DIPPR to Code" Interface)

Figure A-1. Compound Search Screen

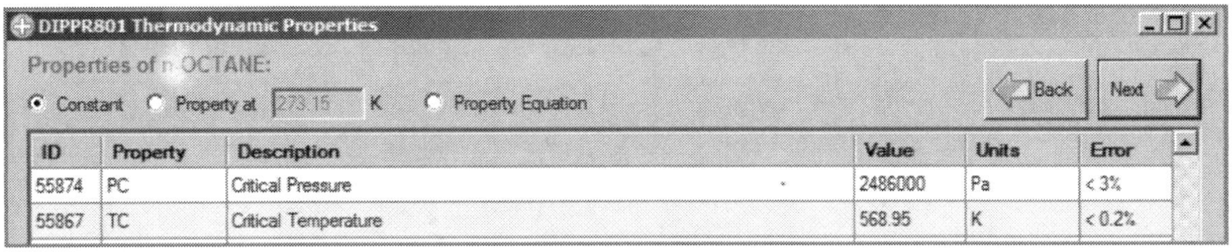

Figure A-2. List of Properties Screen

Figure A-3. Example of the Property Equation Polymath Output for n-OCTANE

Reverse Osmosis Desalination Research in the Undergraduate Laboratory

Han Gu, Chemical & Biomolecular Engineering, University of California, Los Angeles, Los Angeles, CA

Alex Bartman, Chemical and Biomolecular Engineering, University of California Los Angeles, Los Angles, CA

Panagiotis Christofides, Department of Chemical and Biomolecular Engineering and Department of Electrical Engineering, University of California, Los Angeles, Los Angeles, CA

Yoram Cohen, Chemical & Biomolecular Engineering, University of California, Los Angeles, Los Angeles, CA

In many parts of world, desalination of saline waters is becoming a necessity for augmenting dwindling fresh water supplies. Reverse Osmosis (RO) is, at present, the leading technology for desalting saline waters, both from brackish water sources and seawater. Despite of its growing importance, RO water purification technologies remain largely underrepresented in the Chemical Engineering undergraduate curriculum. This presentation will describe a series of experiments designed and implemented for use in chemical engineering undergraduate laboratory courses. Two separate approaches will be described. The first involves a commercial mini Reverse Osmosis (RO) desalination system, consisting of a single RO membrane element, that was adapted to demonstrate the basic concepts of RO desalination. Using this system students carry out experiments with the objective of characterizing RO membrane performance, including membrane permeability, intrinsic salt rejection, and permeate flux. Based on experimental results, students developed mass transfer correlations to estimate the degree of concentration polarization in the membrane modules and develop process-based models for targeted water production. In the second approach, teams of upper division undergraduate students carry out research term projects using a novel automated pilot-scale RO desalination platform developed by the UCLA Water Technology Research Center. Team projects were designed to enable large data sets to be compiled using the pilot-scale system. A web-based data sharing capability was developed to enable students to access and contribute to the overall experimental database. Based on the experimental data, students develop models to describe the performance of RO membranes, evaluate specific energy consumption and propose approaches to energy minimization. Given the modularity of the M3 research platform different membrane modules can be used in the system that can be utilized for desalting of both brackish and seawater. Examples will be given of the type of experiments, data and data analysis that are possible, and samples of student project results and analysis will be described.

See more of this Session: Teaching Design Outside the Capstone Course
See more of this Group/Topical: Education

AIChE Academic Salary Survey

Geoffrey L. Price
Department of Chemical Engineering
University of Tulsa
Tulsa, OK

Abstract

Each US Chemical Engineering department listed at www.che.utexas.edu/che-faculty/ has been contacted via e-mail through the chair/head/director of the department with a request to return a salary survey for their department. Responses will be averaged nationally and by region and presented at the Department Head's Forum. Results will also be compared to results from previous surveys.

ABET Update

Jeffrey J. Siirola, Eastman Research Division, Eastman Chemical Company, Kingsport, TN

An update on ABET

See more of this Session: Department Heads Forum
See more of this Group/Topical: Education

ChE Core Course Content and Pedagogies - Survey Across Departments

David L. Silverstein, University of Kentucky, Paducah, KY

Pedagogy across departments.

See more of this Session: Department Heads Forum
See more of this Group/Topical: Education

Federal Research Funding: Impact of a New Administration and Changed Global Economic Climate

Maria K. Burka, National Science Foundation, Arlington, VA

John McGrath, Division of Chemical, Bioengineering, Environmental and Transport Systems, National Science Foundation, Arlington, VA

Impact of federal funding for this new administration

See more of this Session: Department Heads Forum
See more of this Group/Topical: Education

Insights Into Engineering a Successful Academic Career

Christopher Roberts, Department of Chemical Engineering, Auburn University, Auburn, AL

Presentation from Dept. Chair.

See more of this Session: New Faculty Forum
See more of this Group/Topical: Education

Tips for Busy New Professors

Phillip C. Wankat, Department of Chemical Engineering, Purdue University, West Lafayette, IN

This is a presentation by the author.

See more of this Session: New Faculty Forum
See more of this Group/Topical: Education

Presentations by NSF CBET Program Managers

National Science Foundation, NSF, Arlington, VA

Presentation

See more of this Session: New Faculty Forum
See more of this Group/Topical: Education

Break-out Session with NSF Program Managers

National Science Foundation, NSF, Arlington, VA

Presentation

See more of this Session: New Faculty Forum
See more of this Group/Topical: Education

Expect the Unexpected - Introducing Uncertainty in the Chemical Engineering Laboratory

Marvi A. Matos, Chemical Engineering, University of Washington, Seattle, WA

Danilo Pozzo, Chemical Engineering, University of Washington, Seattle, WA

The evolution of Chemical Engineering imposes a unique challenge to the design of adequate modules in the laboratories. The constant design and manufacture of new chemical products is characterized by fast-paced product and process development. This requires that engineers and managers work under highly uncertain scenarios. We have designed and implemented a novel experimental module in the Chemical Engineering laboratory that intentionally exposes the students to high uncertainty. The module also includes contemporary topics in colloids, complex fluids and biotechnology, thus also giving more emphasis to molecular interactions. The objective of this quarter-long project is the extraction of lysozyme from egg whites. Students use Aqueous Bi-Phasic Systems (ABS) in three separate experimental phases that range from benchtop to pilot scale experiments. During the first phase, students optimize the separation of lysozyme from ovalbumin using bench scale extractions. In the second phase, students scale the separation to the pilot scale using a reciprocating liquid-liquid extraction column. In the third phase, students optimize the process and perform the final extraction of lysozyme from egg whites. Students in each phase review scientific literature and reports from the other groups. This emphasizes the value of adequate critical reviews of literature and reports from peers. Our goal is to help students develop "soft" skills such as composure, critical thinking, creativity, group work, task division, time management and forward thinking. We are intentionally exposing the students to high degrees of uncertainty which serves as a tool to prepare the students for a rapidly changing industrial world.

See more of this Session: Chemical Engineering Modules
See more of this Group/Topical: Education

Enriching the Undergraduate Chemical Engineering Curriculum through Tracks: A Case Study of the Biopharmaceutical Engineering Track at the University of Kentucky

Thomas Dziubla, Department of Chemical and Materials Engineering, University of Kentucky, Lexington, KY

Brad Anderson, Department of Pharmaceutical Sciences, University of Kentucky, Lexington, KY

J. Zach Hilt, Chemical and Materials Engineering, University of Kentucky, Lexington, KY

Kimberly W. Anderson, Chemical & Materials Engineering, University of Kentucky, Lexington, KY

D. Bhattacharyya, Chemical and Materials Engineering, University of Kentucky, Lexington, KY

Tate Tsang, Department of Chemical and Materials Engineering, University of Kentucky, Lexington, KY

The advantage to the chemical engineering degree is the wide diversity of potential career paths available to the graduating student. It is the solid education in the engineering fundamentals that provides this high flexibility. However, as the job market changes and competition for positions increase, it is important that students who are particularly interested in a specific field obtain skills and knowledge that would make them more competitive in the global market. One method to achieve this goal is to develop modular track programs within the chemical engineering curriculum that enriches the core curriculum.

As an example of this education philosophy, the College of Engineering and College of Pharmacy at the University of Kentucky collaborated to form a Biopharmaceutical Engineering Track, which offers students an exposure to both upstream (Drug Research and Development) and downstream processes (e.g., Good manufacturing Practices) The goal of this program is to provide students with an understanding of the multiple aspects of the industry that will better prepare them for either a job in manufacturing & processing or to enter graduate school. In addition to taking core courses in the College of Pharmacy (e.g., Basic Pharmaceutics and Good Manufacturing Practices), additional electives (e.g., Drug Delivery) are cross taught between the colleges. This provides a true multidisciplinary take on special topics integrating the pharmaceutical aspects with engineering fundamentals such as mass transfer and process control.

In this talk, we shall provide overview of the curriculum, examples of student experiences and future goals of the track program.

See more of this Session: Chemical Engineering Modules
See more of this Group/Topical: Education

PyRosetta: An Interactive Platform for Teaching Protein Structure Prediction and Design

Jeffrey J. Gray, Chemical & Biomolecular Engineering, Johns Hopkins University, Baltimore, MD

Sidhartha Chaudhury, Program in Molecular Biophysics, Johns Hopkins University, Baltimore, MD

Sergey Lyskov, Chemical & Biomolecular Engineering, Johns Hopkins University, Baltimore, MD

Structures of proteins and protein complexes help explain biomolecular function, and computational methods provide an inexpensive way to predict unknown structures, manipulate behavior, or design new proteins or functions. However, current computational protein structure prediction and design packages are complex and difficult to learn. To make these approaches broadly accessible to biomolecular engineers with varied backgrounds, we have developed PyRosetta, a Python-based interactive platform for accessing the objects and algorithms within the Rosetta protein structure prediction suite. Rosetta, developed by a consortium of laboratories in the Rosetta Commons, has an unmatched variety of functionalities and is one of the most accurate protein structure prediction and design approaches (Das & Baker Ann Rev Biochem 2008; Gray Curr Op Struct Biol 2006). In PyRosetta, users can measure and manipulate protein conformations, calculate energies in low- and high-resolution representations, fold proteins from sequence, model variable regions of proteins (loops), dock proteins or small molecules, and design protein sequences. Furthermore, with access to the primary Rosetta optimization objects, users can build custom protocols for operations tailored to particular biomolecular applications. Since the Python-based program can be run within the visualization software PyMol, search algorithms can be viewed on-screen in real time. In this talk, we will detail how PyRosetta was used in an undergraduate elective course to teach both the fundamentals and the practical application of protein structure prediction and design.

See more of this Session: Chemical Engineering Modules
See more of this Group/Topical: Education

Introduction of Biological Applications in Process Dynamics and Control Courses

Laurent Simon, Otto H. York Department of Chemical, Biological and Pharmaceutical Engineering, New Jersey Institute of Technology, Newark, NJ

As part of the curriculum, undergraduate chemical engineering students are required to take a course in Process Dynamics and Control (PDC) in their senior years. Among the topics covered are: i) Introduction to dynamics and derivation of state models for chemical processes; ii) Laplace transforms for solving linearized state model equations, iii) Transfer functions and dynamic response of open-loop systems and iv) Design of control systems and controller tuning. Illustrations and examples provided to help anchor core concepts and facilitate learning are often based on familiar processes such as chemical reactors and heat exchangers. As the need to develop and commercialize bio-based products becomes more prevalent, ChE departments throughout the nation have begun to shift their focus from a curriculum centered on the knowledge of chemical plant operations to a program that includes biological applications. The transition will help ensure that ChE graduates secure employment in industries where knowledge in bioprocess engineering is mandatory. Pharmaceutical companies, specializing in the production of therapeutics using biological processes, are the first beneficiaries of this trend. A qualified labor force, with competencies in process analysis, design and control will assure that the target quantity and quality of the end-products are met and in line with federal and state regulations. Among educators, there is no consensus as to the ratio of chemical and bio content of a syllabus that would foster the needed skills. However, it is clear that the instructors, themselves, need to have access to adequate resources to supplement their teaching. This contribution focuses on an approach that prepares faculty to effectively incorporate biological applications into PDC. A series of problems that deal with process control applications in the biotechnology industries are designed to respond to the demand.

See more of this Session: Chemical Engineering Modules
See more of this Group/Topical: Education

Integrating Dcs and MPC Into Problem-Based Undergraduate Chemical Engineering Education

Peyton C. Richmond, Dan F. Smith Department of Chemical Engineering, Lamar University, Beaumont, TX

Daniel H. Chen, Dan F. Smith Department of Chemical Engineering, Lamar University, Beaumont, TX

Kuyen Li, Lamar University, Beaumont, TX

This paper describes a Lamar University Chemical Engineering Department initiative to incorporate modern manufacturing tools including an industrial Distributed Control System (DCS), into a problem-based learning (PBL) educational program. A Honeywell Experion DCS system including simulated Honeywell C200 controllers is stimulated using process values obtained from AspenTech's HYSYS dynamic process simulation cases. The loop is closed between the DCS and dynamic process simulation using Microsoft's automation technology and OPC data transfer protocol. This environment allows us to incorporate realistic modern manufacturing tools for learning Advanced Regulatory Control, Model Predictive Control (MPC), and DCS elements such as data storage and data analysis.

As part of this initiative, we plan to adapt a senior undergraduate second control class (CHEN 4332, Process Control II) which covers elements of Distributed Control System, Advanced PID Control, and MPC and a senior Advanced Analysis class (CHEN 4350) in which students learn data screening and validation in the chemical process industries. Students in the advanced analysis class use historized data of process streams around a de-ethanizer to practice and learn data screening techniques. They then validate the screened data using process simulation tools. Students in the second process control class tune traditional PID and advanced regulatory control structures and are introduced to industrial MPC using software for multivariable constraint control developed at Lamar University. The Lamar MPC software includes three separate programs: (1) Model Identification (2) Controller Configuration (3) Controller Simulation consistent with the industrial steps used to implement this advanced control technology. In future implementations of this lab the students will connect their controller to the DCS system to control a dynamically simulated chemical process.

Tests were used as evaluation instruments to assess student understanding of the data analysis and MPC modules. Then, surveys were performed to assess student satisfaction.

See more of this Session: Chemical Engineering Modules
See more of this Group/Topical: Education

Development of Problem Sets for Undergraduates on Pharmaceutical Particulate Systems

C. Stewart Slater, Mariano J. Savelski,
Christopher A. DelVecchio, Adrian J. Kosteleski, Sarah A. Wilson
Department of Chemical Engineering
Rowan University
Glassboro, NJ, USA

Educational problem sets have been developed on structured organic particulate systems (SOPS) used in pharmaceutical technology. The sets present topics such as pharmaceutical particle properties and powder flow and can be integrated into K-12 and college level curriculum. The materials educate students in specific areas of pharmaceutical particulate processing, such as tableting and mixing, and engineering topics, such as material and energy balances. The problems are designed to be incorporated into existing courses by presenting an alternative class assignment while still covering the same course objective. Initial efforts have focused on introductory chemical engineering courses at the freshman and sophomore level. Some materials for K-12 have also been proposed. The educational development is part of the Rowan outreach efforts associated with the NSF-sponsored ERC on SOPS at Rutgers University, New Jersey Institute of Technology, Purdue University and the University of Puerto Rico-Mayaguez.

INTRODUCTION

Rowan University is an educational outreach partner institution of the NSF Center for Structured Organic Particulate Systems (SOPS) (http://ercforsops.rutgers.edu/). The Center's research focus is on pharmaceutical processing which include: manufacturing science; composites structuring and characterization; and particle formation and functionalization. The areas of research can then be applied to the three development programs, which describe each of three test beds being created at the Center. Development Program I concentrates on the continuous manufacturing of pharmaceutical tablets. Utilizing a continuous process rather than a batch process could lead to increased tablet uniformity and stability, reduced production and labor costs and simplified scale up from experimental testing to full scale manufacturing [1]. Development Programs II and III focus on novel methods for drug delivery. Development Program II hopes to stabilize API nanoparticles in edible substrates [1]. The use of nanoparticles increases the surface area of the particles, resulting in higher bioavailability. Finally, Development Program III will use a system called drop-on-demand to layer API's on an edible substrate [1]. The system could be portable and compact for use in third world countries and military applications.

Rowan University partnered with the ERC-SOPS Center in 2008 to provide outreach and training components to support the educational mission of the Center. During our first two years on the project, Rowan has worked to translate SOPS concepts into educational modules for undergraduate use [2]. This paper describes the development of problems sets that can be utilized in introductory chemical engineering classes. These problems were developed by Rowan students in our Engineering Clinic program in consultation with Rutgers doctoral students, Frank Romanski and William Engisch (coincidently who were Rowan alums). In addition, one of our students, Sarah Wilson spent a summer REU at Rutgers and was able to use this experience to help further the development of the curricular materials. Students also attended the annual meeting of ERC-SOPS at Rutgers to present a poster paper on the project and solicit feedback from the faculty and students at other ERC schools. Through this active engagement, we were able to develop problem sets that represent an area of interest in pharmaceutical particulate systems that ties to Center research topics and also meets educational objectives of basic science and engineering courses.

CURRICULAR MATERIALS

Problems sets were developed to "link" to various research advances of the Center (Figure 1). This was accomplished by "coding" the problems to research thrusts of the Center. Each problem had a target audience in both grade level and subject matter. This way an index of all problems developed could be organized and updated easily. For example Problem TB1.04-*Pharmaceutical Granulation Production* relates to Test Bed 1, Problem 4. Problem TA2.03-*Why Powders are Unique* relates to Thrust Area 2, Problem 3. The majority of the problems were developed in relationship to Test Bed 1, since for chemical engineers, this is much more easily assimilated into curricular materials for courses such as material and engineering balances and unit operations. Problems were aligned to student grade levels by modifying technical content. Over 30 problems have been developed with fully-executed solutions. In the majority of cases these included multiple parts with varying levels of difficulty. Supplemental materials, e.g., journal papers, vendor information, pharma web sites, are also referred to in many problems which therefore require the students to do additional reading. Examples of some completed problem sets are provided. The complete set of problems are available in the education section of the Center web site (http://ercforsops.rutgers.edu/).

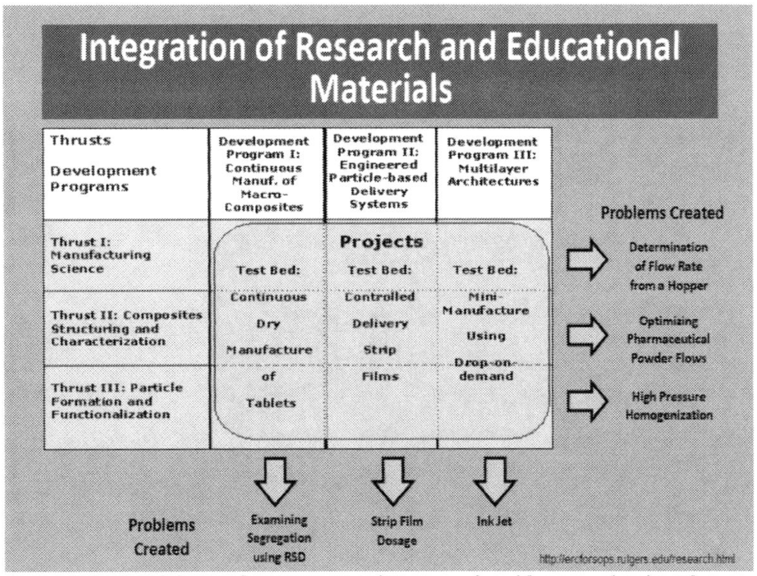

Figure 1. Linkage of Center research areas and problem sets developed.

Example Problem - Problem TB3.01 – Drop on Demand (Introductory Engineering)

Problem Statement

Research is being conducted at the NSF Center for Structured Organic Particulate Systems (C-SOPS) to introduce a novel technology to the pharmaceutical industry. The new process utilizes drop on demand (DoD) technology to create a new form of drug administration. The Center hopes to create a device similar to an ink jet printer that will be able to "print" drug doses that are tailored specifically to the patient [1]. By knowing the concentration of the active pharmaceutical ingredient (API) cartridge, the technician will be able to use the printer to administer the exact dose necessary for the patient, rather than a dosage predetermined by tablet size. The cartridge used in the printer will contain an API that has been dissolved in a volatile liquid. When the drop makes contact with an edible substrate, the liquid will quickly evaporate leaving the API crystallized on the substrate. The patient would then consume the crystals and substrate. The benefits of the system would be that the printer is compact and portable. This would allow the system to be used in third world countries and military applications.

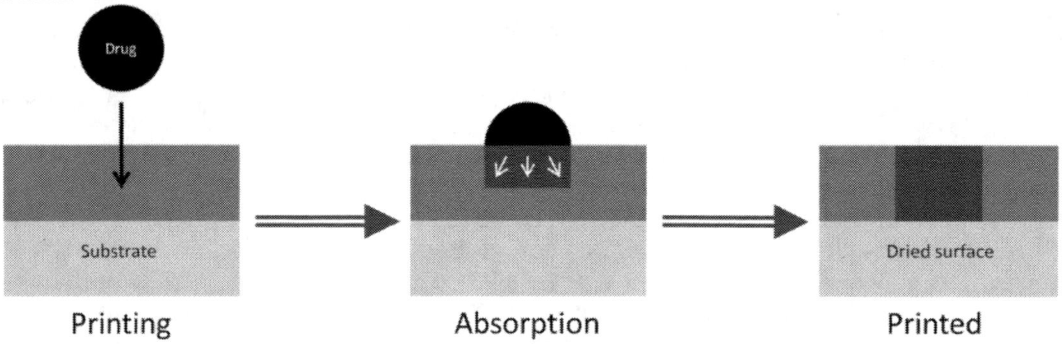

Illustration of the drug being "printed" onto the dry surface for drug delivery.

A brand new ink jet printer has been sent to a camp in Iraq to allow doctors to quickly administer ibuprofen for aches and pains. The cartridge contains 0.5 liters of 1 molar ibuprofen dissolved in hexane [3].

a) If each drop is 1 mL of liquid, determine how many drops would be needed to administer a 350 mg dose of ibuprofen.
b) The doctors have found that it is best to administer 2.5 milligrams of API per pound of body weight to experience the most effective pain relief. Determine how many drops of the solution need to be used to administer ibuprofen to:

i) 10 year old boy weighing 100 pounds
ii) Solider weighing 245 pounds

c) The density of ibuprofen is 0.7 g/cm^3. Determine the volume of ibuprofen required to treat a 300 pound soldier. How many drops of solution were required?
Molecular Weight of ibuprofen: 206.3 g/mole

Problem Solution

Using the given information from the problem dimensional analysis can determine the number of drops for each specific weight.

a)

$$350\text{ mg} \times \frac{1\text{ g}}{1000\text{ mg}} \times \frac{1\text{ mole}}{206.3\text{ g}} \times \frac{1\text{ liter}}{1\text{mole}} \times \frac{1000\text{ ml}}{1\text{ liter}} \times \frac{1\text{ drop}}{1\text{ ml}} = 1.7\text{ drops}$$

b-i)

$$100\text{ pounds} \times \frac{2.5\text{ mg}}{1\text{ pound}} = 250\text{ mg}$$

$$250\text{ mg} \times \frac{1\text{ g}}{1000\text{ mg}} \times \frac{1\text{ mole}}{206.3\text{ g}} \times \frac{1\text{ liter}}{1\text{mole}} \times \frac{1000\text{ ml}}{1\text{ liter}} \times \frac{1\text{ drop}}{1\text{ ml}} = 1.21\text{ drops}$$

b-ii)

$$250 \text{ pounds} \times \frac{2.5 \text{ mg}}{1 \text{ pound}} = 500 \text{ mg}$$
$$500 \text{ mg} \times \frac{1 \text{ g}}{1000 \text{ mg}} \times \frac{1 \text{ mole}}{206.3 \text{ g}} \times \frac{1 \text{ liter}}{1\text{mole}} \times \frac{1000 \text{ ml}}{1 \text{ liter}} \times \frac{1 \text{ drop}}{1 \text{ ml}} = 2.42 \text{ drops}$$

c)

$$300 \text{ pounds} \times \frac{2.5 \text{ mg}}{1 \text{ pound}} = 750 \text{ mg}$$
$$750 \text{ mg} \times \frac{1 \text{ g}}{1000 \text{ mg}} \times \frac{1 \text{ mole}}{206.3 \text{ g}} \times \frac{1 \text{ liter}}{1 \text{ mole}} \times \frac{1000 \text{ ml}}{1 \text{ liter}} \times \frac{1 \text{ drop}}{1 \text{ ml}} = 3.64 \text{ drops}$$

Example Problem - Problem TB3.03 - Ink Jet Drug Printer (Introductory Engineering)

Problem Statement

A printer cartridge contains tiny colored particles suspended in a quick-drying liquid. The particles are so small and light that they are suspended in the ink. When the ink is printed on paper, the liquid is allowed to dry leaving the solid particles behind. Research advances are being made at the NSF Center for Structured Organic Particulate Systems to investigate this concept applied to drug delivery using a compact medicine "printer" to which different cartridges containing drugs can be loaded ("Thrust Area 3." Research. Rutgers C-SOPS. 29 Feb 2008, http://ercforsops.rutgers.edu/research.html.). Technology like this "prints" solid dose drug forms onto biodegradable and edible sheets of paper called a substrate. The substrate will be able to be ingested by the patient and function as a carrier for the drug.

This drop-on-demand (DoD) method of delivering drugs has untapped potential as it could be used quickly and efficiently for the military, third world countries, or any other emergency situation. The biodegradable substrates will absorb, protect, and maintain the drug in stable form. Some substrates will also have the capability to accept more than one drug at a time which will permit multiple drugs to be given to the patient in one dose. The following is the diagram of how the printer will apply the drug:

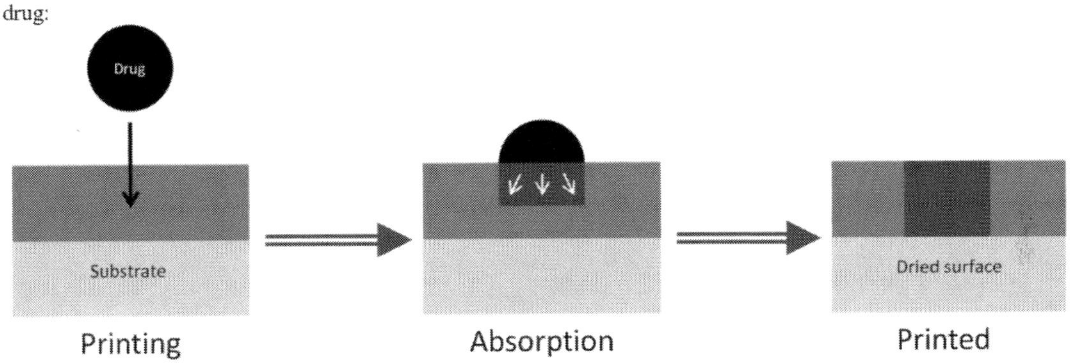

Steps is the ink jet "printing" the drug onto the dry surface ready for delivery

Consider the following situation: An analytical chemical lab has received a revolutionary "ink-jet particle printer". Their summer intern was calibrating the machine when he accidentally printed three squares of the product *Lafta*. This drug is valuable to the company. The squares must be salvaged and their concentrations discovered. The squares were each printed using a different concentration cartridge in the printer. The printer was set to give 75 mg dosages of *Lafta* using 2.5 mm thickness substrate capable of absorbing 45% of its own volume of printable drug. Here is some additional information about the squares

Square #1: Diagonal measured 0.707 inches
Square #2: Diagonal measured 0.400 inches
Square #3: Volume is 25% larger than square #2.

What is the concentration of *Lafta* in each of the squares?
What was the original concentration of *Lafta* in the three cartridges? Report all answers in mg/ml.

Problem Solution

To find the concentration of each square, the volume must be found using the appropriate equations; then the concentration can be found by using the given dosage and dividing it into the volume.

Square # 1

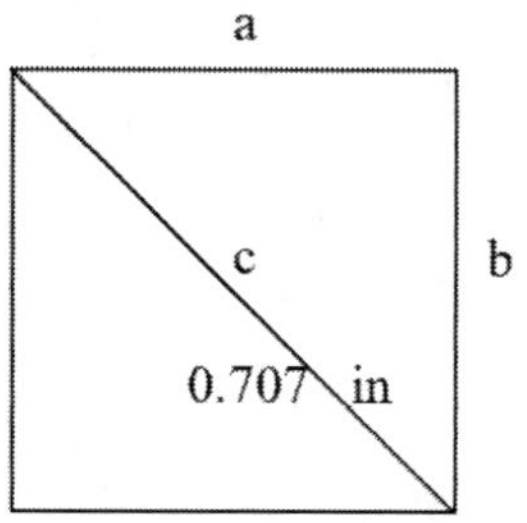

$$a^2 + b^2 = c^2$$
$$a = b$$
$$2a^2 = (0.707)^2$$
$$a = 0.500 \text{ in}$$

$$0.500 \text{ in} \times \frac{0.0254 \text{ m}}{1 \text{ in}} \times \frac{100 \text{ cm}}{1 \text{ m}} = 1.27 \text{ cm}$$

$$2.5 \text{ mm} \times \frac{1 \text{ cm}}{10 \text{ mm}} = 0.25 \text{ cm}$$

$$V_{square} = bhl$$
$$V = (1.27 \text{ cm})(1.27 \text{ cm})(0.25 \text{ cm})$$
$$V = 0.40 \text{ cm}^3 = 0.40 \text{ ml}$$

$$\frac{75 \text{ mg}}{0.040 \text{ ml}} = 1900 \frac{\text{mg}}{\text{ml}} \Rightarrow \text{Total concentration}$$

Substrate absorbs 45%

$$(V)(0.45) = \text{Vol drug in substrate}$$
$$(0.40 \text{ cm}^3)(0.45) = 0.18 \text{ cm}^3$$
$$0.18 \text{ cm}^3 = 0.18 \text{ ml}$$

$$\frac{75 \text{ mg}}{0.18 \text{ ml}} = 420 \frac{\text{mg}}{\text{ml}} \Rightarrow \text{Original concentration}$$

Square # 2

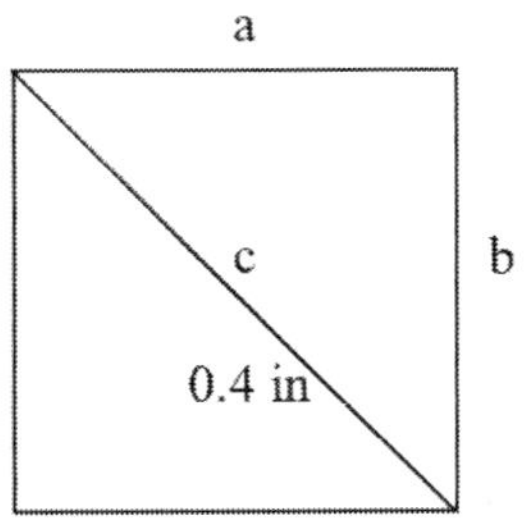

$$a^2 + b^2 = c^2$$
$$a = b$$
$$2a^2 = (0.400)^2$$
$$a = 0.283 \text{ in}$$

$$0.283 \text{ in} \times \frac{0.0254 \text{ m}}{1 \text{ in}} \times \frac{100 \text{ cm}}{1 \text{ m}} = 0.719 \text{ cm}$$

$$2.5 \text{ mm} \times \frac{1 \text{ cm}}{10 \text{ mm}} = 0.25 \text{ cm}$$

$$V_{square} = bhl$$
$$V = (0.719 \text{ cm})(0.719 \text{ cm})(0.25 \text{ cm})$$
$$V = 0.13 \text{ cm}^3 = 0.13 \text{ ml}$$

$$\frac{75 \text{ mg}}{0.13 \text{ ml}} = 580 \frac{\text{mg}}{\text{ml}} \Rightarrow \text{Total concentration}$$

Substrate absorbs 45%

$$(V)(0.45) = \text{Vol drug in substrate}$$
$$(0.13 \text{ cm}^3)(0.45) = 0.059 \text{ cm}^3$$
$$0.059 \text{ cm}^3 = 0.059 \text{ ml}$$

$$\frac{75 \text{ mg}}{0.059 \text{ ml}} = 1300 \frac{\text{mg}}{\text{ml}} \Rightarrow \text{Original concentration}$$

Square # 3

$$V_{SQ3} = 1.25 V_{SQ2}$$
$$V_{SQ3} = 1.25 \times 0.13 \text{ cm}^3$$
$$V_{SQ3} = 0.16 \text{ cm}^3 = 0.16 \text{ ml}$$
$$\frac{75 \text{ mg}}{0.16 \text{ ml}} = 470 \frac{\text{mg}}{\text{ml}} \Rightarrow \text{Total concentration}$$

Substrate absorbs 45%

$$(V)(0.45) = \text{Vol drug in substrate}$$
$$(0.16 \text{ cm}^3)(0.45) = 0.072 \text{ cm}^3$$
$$0.072 \text{ cm}^3 = 0.072 \text{ ml}$$

$$\frac{75 \text{ mg}}{0.072 \text{ ml}} = 1000 \frac{\text{mg}}{\text{ml}} \Rightarrow \text{Original concentration}$$

SUMMARY

Basic problems on structured organic particulate systems relevant to pharmaceutical technology have been developed for integration into the engineering curricula. Educational topics include fundamental units of measure and conversion factors, material and energy balances, fluid mechanics, and applied mathematics. These educational materials will increase student interest in research and development in the pharmaceutical industry by introducing students to basic pharmaceutical engineering concepts within the framework of existing courses.

ACKNOWLEDGEMENTS

The authors acknowledge the assistance of Frank Romanski and Bill Engisch of Rutgers University in reviewing the course materials. This project has been supported by National Science Foundation Engineering Research Center grant # 0540855.

REFERENCES

1. "Research - Welcome to ERC for SOPS." Engineering Research Center for Structured Organic Particulate Systems, http://ercforsops.org/ last accessed August 6, 2009

2. S. Farrell, Z. O. Gephardt, R. Krchnavek, M. J. Savelski, C. S. Slater, E. Batten, J. Carroll, C. DelVecchio, A. Kosteleski, K. Ross, T. Sokal, K. Whitaker, S. Wilson "Educational Outreach Materials for ERC Structured Organic Particulate Systems: Project Overview," Paper 1626, *Proceedings of the 2009 American Society for Engineering Education Annual Conference*, Austin, TX, June 2009.

3. http://www.ibuprofen-foundation.com/what-ibuprofen/documents/TIIbuprofen.pdf, last accessed August 6, 2009

Hydrogen Education Curriculum at Michigan Technological University

Jason M. Keith, Department of Chemical Engineering, Michigan Technological University, Houghton, MI

Daniel Crowl, Department of Chemical Engineering, Michigan Technological University, Houghton, MI

David Caspary, Department of Chemical Engineering, Michigan Technological University, Houghton, MI

Jeffrey Allen, Department of Mechanical Engineering - Engineering Mechanics, Michigan Technological University, Houghton, MI

Dennis Meng, Department of Mechanical Engineering - Engineering Mechanics, Michigan Technological University, Houghton, MI

Jeff Naber, Department of Mechanical Engineering - Engineering Mechanics, Michigan Technological University, Houghton, MI

Abhijit Mukherjee, Department of Mechanical Engineering - Engineering Mechanics, Michigan Technological University, Houghton, MI

Jay Meldrum, Keweenaw Research Center, Michigan Technological University, Houghton, MI

Barry Solomon, Department of Social Sciences, Michigan Technological University, Houghton, MI

The chemical engineering aspects of energy conservation and sustainability can have important societal benefits. One emerging area that requires additional study to determine if it is fully feasible is hydrogen and fuel cells. As such, it is imperative to educate students so that they can be informed of the advantages and disadvantages as well as the challenges and opportunities of hydrogen technology.

This presentation will describe a Department of Energy funded project on curriculum development in hydrogen and fuel cells at Michigan Technological University. There are four major emphases of this project.

1. Project work in the Alternative Fuels Group Enterprise, in which students from any discipline and class standing can enroll in as electives or towards the Enterprise Minor.

2. Development of courses specifically related to hydrogen technology for undergraduate and graduate students

- Fundamentals of Hydrogen as an Energy Carrier

- Hydrogen Measurements Laboratory

- Fuel Cell Fundamentals

3. Development of hydrogen technology related course material for core courses in chemical engineering, mechanical engineering, and electrical engineering.

4. Development of hydrogen technology related supplementary material for the text Elementary Principles of Chemical Processes (authors R. M. Felder and R. W. Rousseau) and for the text Transport Processes and Separation Process Principles (author C. J. Geankoplis).

These topics and projects are designed in a module format so that they can be easily incorporated as stand-alone modules in individual courses in the engineering curriculum.

See more of this Session: Chemical Engineering Modules
See more of this Group/Topical: Education

A New Desktop Experimental Module (DEMo) for Heat Transfer

Adrienne R. Minerick, Dave C. Swalm School of Chemical Engineering, Mississippi State University, Mississippi State, MS

A new Desktop Experiment Module (DEMo) has been developed for a chemical engineering junior level heat transfer course. The DEMo system can illustrate conduction of various materials, thermal energy generation, thermal contact resistance, heat dissipation from fins, and convection across surfaces of varying geometry. This DEMo learning tool is versatile, fairly inexpensive, and portable such that it can be positioned on student desks throughout a classroom.

Advantages of this approach are that students can closely examine and fiddle with the system to see cause and effect better than with instructor led demonstrations. These learning tools can also serve as vibrant hands-on experiments with high school students.

See more of this Session: Chemical Engineering Modules
See more of this Group/Topical: Education

The Documentation Cycle: A Tool within a Learning Cycle

Paula Arce-Trigatti, Economics, Finance and Marketing, Tennessee Technological University, Cookeville, TN

Jennifer Anne Pascal, Chemical Engineering, Tennessee Tech University, Cookeville, TN

Jeffery W. Thompson, Chemical Engineering, Tennessee Technological University, Cookeville, TN

Pedro Arce, Chemical Engineering, Tennessee Technological University, Cookeville, TN

On the educator-front, it is suggested by the abundance of literature related to general note-taking procedures that educators are perhaps overly concerned with providing students a singular technique for learning and ignoring the substantial impact a cycle of documentation might have on learning efficacy. While note-taking is just one tool in the arsenal of options students may wish to use in their learning paths, a macro-level approach to learning may prove to be more successful. In this presentation, we will propose a documentation cycle that details and outlines a procedure that students may repeatedly employ in their classroom endeavors in order to more completely grasp the material at hand. The cycle begins with the students' initial interaction with the material in the form of "dirty notes," followed by a cycle of activities ranging from the students' analysis of the material, asking questions, seeking answers, obtaining feedback, and beginning the data collection process once again. The results from such a series of events are the "clean notes," accompanied by the greater aftereffect, the potential mastery of the material by the students. The tangible documents (for example, in the form of a personal class binder, [1]) obtained as a direct result of the cycle provides the students with a set of materials which are then accessible throughout their careers.

[1] Rawlings, R., S. Allen, P. Arce. "The Personalized Class Binder (PCB): A Powerful Tool for Enhancing Active and Collaborative Learning Environments." ASEE Southeastern Conference, 2005.

See more of this Session: Strategies for Overcoming Course Compartmentalization and Aiding in Retention of Knowledge
See more of this Group/Topical: Education

Integrating Course Content through the Use of Concept Maps

John R. Schlup, Department of Chemical Engineering, Kansas State University, Manhattan, KS

In many engineering classes, it can be difficult for students to develop a perspective for how course content fits into the broader picture of their discipline. This is particularly true in larger service courses that cross several fields of study and that may be required as a part of the student's broader education. The introductory materials science and engineering courses at Kansas State University fit this description. They are larger service classes required by students in five different departments. Two common frustrations expressed by students are that they do not always see how the various topics tie together and that they do not see how the course fits into their discipline. Concept maps have been employed recently within one of these classes to facilitate student integration of specific course content. The implementation of and early experiences with this approach will be discussed. Preliminary data showing the impact of concept maps on student performance will be provided.

See more of this Session: Strategies for Overcoming Course Compartmentalization and Aiding in Retention of Knowledge
See more of this Group/Topical: Education

Organizing Thinking and Problem Solving Stategies That Cross Chemical Engineering Topics Areas and Can Guide Students to Success in Tasks Across the Curriculum

Richard L. Long, New Mexico State University, Las Cruces, NM

A central task for chemical engineers is to formulate and solve problems and commuicate the results found. Part of this task is acquisition of information and discrimination between reliable and unrelaiabe sources. Orgnaizing student thinking at the beginning of the curriculum can be a real asset to equalizing opportunities for students who are committed to success.

See more of this Session: Strategies for Overcoming Course Compartmentalization and Aiding in Retention of Knowledge
See more of this Group/Topical: Education

Interactive Problem-Based Learning for Enhanced Interdisciplinarity and Retention

Marcel A. Liauw, Industrial Chemistry, ITMC, RWTH Achen, Aachen, Germany

Dina Treu, RWTH Aachen, Institute of Technical and Macromolecular Chemistry, Aachen, Germany

Two strategies are presented for an enhanced education in chemical engineering:

1) Multi-layered chemical engineering / chemistry problems are presented via appropriate media like games or internet portals. Individual students or student teams have to solve trouble-shooting tasks in a virtual chemical factory. Clues required for success come from a wide variety of areas, from reaction engineering over chemistry to materials sciences. The interactivity is said to enhance the longevity of the acquired skills, and hence knowledge retention. The preparation of problems is as educational for the students as is the attempt to solve them.

2) With the support of an actual chemical company, a well-defined task has to be solved by a "Virtual Team", 5-10 real students at various universities with different specializations. Work on the problem is carried out via electronic communication, the final period is hosted by the industry partner at their site. A joint presentation of the results is given by the students before the company's experts.

Examples and experiences are given from the European project IMPULSE and others.

See more of this Session: Strategies for Overcoming Course Compartmentalization and Aiding in Retention of Knowledge
See more of this Group/Topical: Education

Enigma of Teaching the Second Law of Thermodynamics: AN Investigation of DIFFERENT Methodologies IN Search for Finding Most Effective Pedagogic MODEL

M. P. Sharma, Chemical and Petroleum Engineering, University of Wyoming, Laramie, WY

Based on surveys conducted with thermodynamics students, it is well known that Thermodynamics poses enigmatic and difficult learning experience for students. And, therefore this a challenging and interesting subject for teachers and instructors to teach in such a way that results in deeper learning and understanding. In my view no one has expressed this enigmatic and mysterious nature of this subject better than Arnold Sommerfeld (a noted physicist):

"Thermodynamics is a funny subject. The first time you go through the subject, you don't understand it at all. The second time you go through it, you think you understand it except for one or two small points. The third time you go through it, you know you don't understand it, but by that time you are so used to the subject, it doesn't bother you anymore."

Sommerfeld's comment applies more to the learning and teaching of the "second law of thermodynamics" than any other topic of this subject. I participated in a National Science Foundation's supported research project entitled "Delphi Study to Identify the Most Difficult Concepts for Students to Master in Thermal and Transport Sciences" which was a part of "Developing an Outcomes Assessment Instrument for Identifying Engineering Student Misconceptions in Thermal and Transport Sciences". Out of 28 different "most difficult to understand concepts" that were selected (through a rigorous process of surveying the instructing professors in US), the concept of "Entropy and Second Law of Thermodynamics" was rated to be in the "most misunderstood concept" category and at the same time it was rated as "most important concept" by the same group of teaching professors. This project was conducted jointly by Colorado School of Mines and University of Colorado. As a teacher of Thermodynamics, the author's responsibility and keen interest has been in researching, developing and implementing several methods, tools, and strategies for improving the effectiveness of teaching/learning of every aspect of the "engineering thermodynamics" course, but particularly dealing with the difficulties of teaching the understanding and the skill of using (in problem solving) the Second Law and Entropy. This paper will present details and outcomes of this project. Different methods of teaching the "second law and entropy" that exist in the literature (textbooks of thermodynamics) were researched and studied in this project. Those methods were then classified based on common characteristics based criteria. Special features, strength, and shortcomings of these methods are then analyzed focusing on why students tend to develop misconceptions and confusions with some of these models of teaching compared with some others. Then, a new model of teaching the "second law and entropy" topic is presented that claims to be more effective in developing understanding and skills of this topic for students.

See more of this Session: Strategies for Overcoming Course Compartmentalization and Aiding in Retention of Knowledge
See more of this Group/Topical: Education

LABORATORY ON SYSTEMS IDENTIFICATION, A KEY TOOL IN THERMODYNAMICS THEORY

I. - INTRODUCTION

Thermodynamics is an essential tool and plays a significant role in the training of chemical engineers, since the use of its laws and fundamental concepts contributes to a better understanding of the energetic needs involved in diverse type of processes (chemical, environmental, mechanical transformations, etc.). The main advantage of Thermodynamics consists of the simplicity of its concepts and the universality of its applications in different fields from a discipline. As Albert Einstein wrote down: "*A theory is the more impressive the greater the simplicity of its premises, the more different kinds of things it relates, and the more extended its area of applicability. Therefore the deep impression that classical thermodynamics made upon me. It is the only physical theory of universal content which I am convinced will never be overthrown, within the framework of applicability of its basic concepts.*" (1)

In chemical engineering, as well as in mechanical, environmental or food engineering, Thermodynamics allows evaluating, among others things:

- The feasibility of a process and its energetic requirements,
- The maximum possible conversion of a reaction, and
- Under which specific conditions phase equilibrium can be reached.

The Thermodynamics course helps students apply problem-solving techniques when dealing with the laws of Thermodynamics, in which it is needed the handling of information contained in tables (thermodynamics properties tables) or generated by computer programs, and to predict the operation of a system.

Despite its importance, the persistence of some misunderstandings about Thermodynamics concepts as well as a lack of integration between theory and practice makes students fail when interpreting, analyzing, or producing accurate outcomes. The difference between energy and heat, or energy and work are not clear, and what is the systems investigated is not easy topics, for examples. Hence, they are more likely to find this subject useless and tedious.

In order to modify this perception, it was developed a laboratory practice called ***Identification of thermodynamic systems,*** focused on core concepts, in order that the students can recognize, in practice concepts such as:

- What portion of the universe is to be studied?
- What type of transformations happens in the system, and
- What equations are required to quantify the involved power changes in the transformations of the system?

This practice can be implemented as integral part of a completely theoretical subject at an undergraduate level or as a first session of laboratory in an experimental course, without great investments and with resources currently found at a laboratory.

II. METHODOLOGY

The main purpose of this laboratory practice is to achieve what the Chinese proverb says: "*Tell me and I forget. Show me and I remember. Involve me and I understand*"· providing the students significant experiences enables them to participate actively in their own learning process, acquiring understanding and constructing meaning by introducing a familiar context.

In order to be able to implement this laboratory practice, the following aspects are required:

a) Use the practice and the laboratory itself, as tools of active learning.

b) Interact, preferably, experimental equipment or experimental assemblies with which students are familiar.

c) Identify what the students have assimilated during the development of the practice and

d) Give permanent feedback so that misconceptions and voids can be identified.

III. METHODOLOGICAL IMPLEMENTATION

The aim pursued is to implement an inductive methodology through observations in which students become aware, in a simple and forceful way, that to observe is much more than to see. In order to obtain so, the experiment of the description of an ignited candle was implemented. Here students were given a candle and were asked to observe it, after that it was ignited and they were requested to describe what it happened. Later, the description of the experiment that appears in O'Connor, David, Haenisch, MacNab and McClellan (2) was read. After this, the true lab practice starts, but first, students were asked to answer the following questions:

- What is a thermodynamic system?
- What is a limit of a system?
- What difference does exist between energy and work?
- What difference does exist between energy and heat?
- Which is the difference between function of state and function of trajectory?
- What are intensive and extensive variables?

Once the answers were picked up, the students went to the laboratory where some assemblies with their respective systems were located.

IV. DEVELOPMENT OF LABORATORY PRACTICES

In the laboratory there were eight systems, to be observed by the students for 10 minutes each, from which they could answer the following questions:

- Which is the system?
- Which are the boundaries of the system?
- Is it an open or a closed system?
- How many degrees of freedom does the system have?

• What kind of process is it being carried out?
• If the system interchanges heat, in which parts of the system is it being held and what is the respective sign?
• To establish, in case of a positive answer to the previous question, how the first law of thermodynamics is in each case. If possible, to quantify flows.
• To determine if the involved variables are intensive or extensive.

The assemblies implemented were:

- A thermo with a liquid in its interior.
- A system of boiling of a liquid and condensation of a vapor to constant pressure.
- A system of boiling of a liquid and condensation of a vapor to constant volume.
- A drop of liquid contained in a refractometer.
- A calorimeter.
- Three beakers of the same size, whose water was warmed up in different ways. One by means of a Bunsen burner, another one by means of an electric hot plate and another one by means of microwaves. (3)
- A system with two phases (phenol water).
- An absorption or distillation column.

The pictures of the system of boiling and condensation of a liquid to constant pressure and constant volume are show in the figure 1.

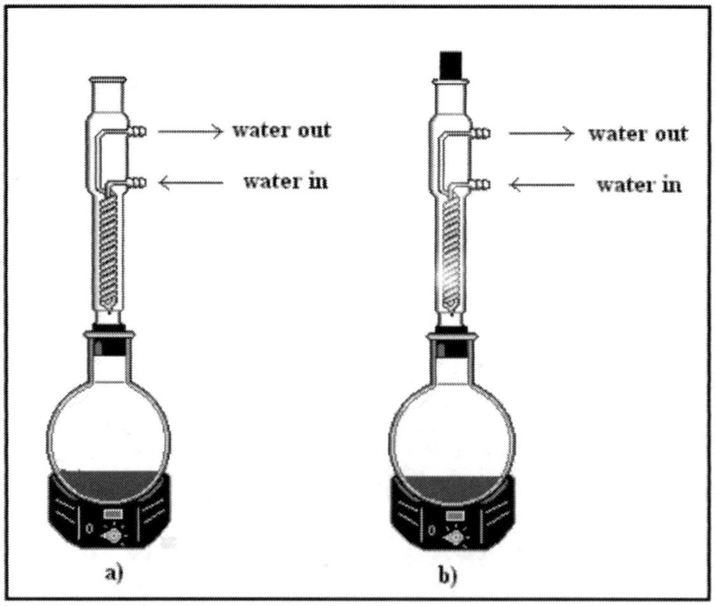

Figure 1. Picture of the systems a) Constant Pressure and b) Constant volume.

A picture of the three different way of heating a liquid is shown in the figure 2.

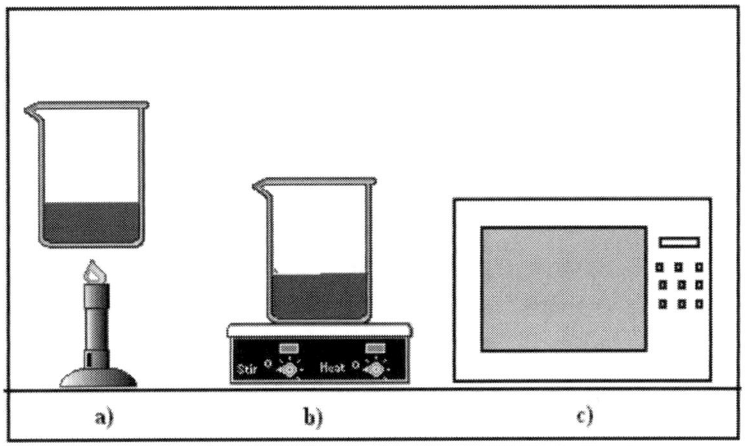

Figure 2. Picture of the systems with three different methods of heating a liquid. a) Bunsen Burner, b) Electric hot plate, and c) Microwave

RESULTS AND CONCLUSIONS

The results obtained during the implementation of the practice were positive because it was possible to establish actually and by means of simple assemblies what the system of interest was, and the type of transformation that happened (isobaric, isothermal, adiabatic, etc.). The time of observation never assigned could be obtained and the students consumed less time solving a problem of the first law in open systems that responding to the questions posed in this practice. Therefore the number of experimental assemblies that is suggested is 5. The evaluations of the subject realized by the students were satisfactory.

Regarding to the first activity related to the description of an ignited candle practice, the results allowed to conclude that the students realized that observing is an activity that consumes time and requires a greater intellectual effort.

About the observations registered by the students the following facts can be reported:
- 10% wrote the initial characteristics of the system studied (candle).
- Only a 5% wrote how the experiment started
- 100% identified the occurrence of a chemical reaction
- 60% noticed the fusion process of the candle.

Depending on the chemistry courses studied, the students were able to distinguish concepts such as oxidation, reduction and energy of reactions. But in a general way, 10 % wrote down misconceptions about the concept of oxidation.

After reading the description that the authors of the book showed, all the students showed that they perceived some changes; however, they could not describe them by the little provided time to register their answers. In the part of identification of systems,

90% of the students did not have any problem at identifying the system, although it became difficult to them to distinguish in the case of the systems of boiling and condensation if the processes were carried out under isobaric conditions or of constant volume. These two systems in all the cases were those that consumed more time, although they recognized where the system received and yielded heat, always there was difficulties with the form in which the first law in those cases must be expressed.

The system most commented was the one where there was three ways to warm up a liquid, and contributed to distinguish the differences between the trajectory and state functions. In this system also there were difficulties at determining the thermodynamic efficiency, although it was clear the involved forms of energy in each case.

The evaluations of the subject realized by the students were satisfactory and in tables 1, 2 and 3 some results obtained in the final evaluation of the course realized by the students are registered in an experimental course of thermodynamics, during three different academic periods:

In all these academic periods this practice was realized and there were used tests based on scientific topics written by renowned scientific investigators (4).

TABLE No 1. Results of Final Evaluation in 2004 – II

		TEACHER	DEPT.	FACULTY
G1	TEACHER	4.5	4.4	4.4
G2	METHOD	4.4	4.1	4.1
G3	COURSE	4.4	4.2	4.1
G4	EQUIPMENT	4.3	4.0	3.9
G5	HOMEWORK	4.5	4.2	4.1
Q25	SATISFACTION	4.8	4.1	4.1
Q26	SATISFACTION	4.3	4.0	3.9

Total average: 4.52

TABLE No 2 Results Final total evaluation: 2005 - I

		TEACHER	DEPT.	FACULTY
G1	TEACHER	4.8	4.5	4.4
G2	METHOD	4.8	4.1	4.1
G3	COURSE	4.6	4.2	4.1
G4	EQUIPMENT	4.6	4.0	3.9
G5	HOMEWORK	4.6	4.2	4.1
Q25	SATISFACTION	4.7	4.1	4.1
Q26	SATISFACTION	4.6	4.0	3.9

Total average: 4.73

TABLE No 3 Results of Final total evaluation 2005 – II

		TEACHER	DEPT.	FACULTY
G1	TEACHER	4.6	4.3	4.3
G2	METHOD	4.7	4.0	4.0
G3	COURSE	4.8	4.2	4.1
G4	EQUIPMENT	4.3	3.9	4.0
G5	HOMEWORK	4.5	4.0	4.0
Q25	SATISFACTION	4.6	3.8	3.9
Q26	SATISFACTION	4.2	3.7	3.8

Total average: 4.61

In these surveys the students were asked about:

- the professor
- the pedagogic method
- the relevancy of the course in their training process
- the equipment used in the laboratory
- the assignments and forms of evaluation

The last two rows correspond to the degree of satisfaction with the teacher and with the course in general. In these tables, the last two columns mention how this course was compared with others of the department and of the faculty respectively. It is necessary to emphasize that this practice obtained one of the highest notes in the evaluations of the course in the different semesters. Similar outcomes were obtained where this practice was implemented as part of a theoretical course in mechanical, environmental and food engineering. In general the averages of these evaluations were 4.0.

The same practice was realized as part of a course of physicochemical to postgraduate level in Master Engineering with Alternative Energies emphasis. In this course, professionals of diverse types of engineering, many of which did not deal thermodynamic or physicochemical in their pre-degree. The main differences consisted on the assemblies had to be with equipments of analysis Orsat, engines diesel, mixtures of ethanol, petrol and the simulation of the nuclear plant that is available in the electronic address: http://www.ida.liu.se/~her/npp/demo.html. The boiling and condensate vapor was used in postgraduate level too

In all cases, under and postgraduate level, this practices was useful for reinforce the energy balance between a condensate vapor and a refrigerant liquid. The student said that the different between latent and sensible heat were clear. The same happened for the reaction heat as energy source for a boiling liquid. In the case of a refractometer the systems was trouble for some students, especially when this equipment had a control system temperature.

In the case of heterogeneous mixture systems, the degree of freedom was useful to show, in the practice the equilibrium conditions, if pressure and temperature are constant what variable is changing; and with the help of a phase diagrams the lever rule.

Finally, the course where this practice was implemented, as a whole, had a population of 30 undergraduate students and 8 postgraduate level students.

BIBLIOGRAPHY

1) Sandler, Stanley. *Chemical and Engineering Thermodynamics 2a E*d. Solution Manual. John Wiley and Sons.1989.
2) O'Connor, David, Haenisch, MacNab and McClellan. *Chemistry and experimental Science.* Ed Reverte. 1987.
3) American Chemical Society, ACS. *Introduction to green Chemistry*. The need to conserve energy. 2002.
4) Camargo G, Tirano J. Asimov Reading in thermo chemistry laboratory. Memories of 236th Congress of American Chemical Society. Philadelphia 2008

Longitudinal Contact with Individual Students as a Route of Encouraging Self-Determination in Chemical Engineers

Paul Blowers, Chemical and Environmental Engineering, The University of Arizona, Tucson, AZ

It is difficult to have contact with individual students over a sustained period of time due to constrained schedules and competing time demands. However, frequent contact with individuals over time allows advisors to build student strengths in self-determination while tailoring advice directly to changing interests.

The talk will highlight advising opportunities from outreach, through retention, continuing to graduation, and post graduate interactions that fit within student progress towards their individual careers. Appropriate advising content for a technically rigorous chemical engineering program will be used as examples of how to motivate students towards exploring options and making decisions that open new doors to professional development. The issue explored is that self-determination comes from inside the student and that confidence in personal evaluation is fostered through directed activities throughout students' time with us.

Objectives that attendees should be able to meet will be:

• to have the ability to identify points of contact where longitudinal contact can be encouraged

• to have the knowledge of how to remove impediments to longitudinal advising through reconstruction of advising duties in larger programs with many staff members, or through directed contact with students in smaller departments

• To have the ability to link student possibilities to exploratory activities that lead to self-determination

Students in chemical engineering are generally very strong in academic abilities, but often have not been exposed to a breadth of activities and possibilities that enable them to construct their own paths. Sustained advising contact has led to strong statistically-based success of students who have progressed through our program with approximately 30% of our graduates entering top ten chemical engineering graduate programs, 100% student placement of graduates almost every year, and a host of university-based, state, and national awards going to our students.

See more of this Session: Strategies for Overcoming Course Compartmentalization and Aiding in Retention of Knowledge
See more of this Group/Topical: Education

Engaging Early Engineering Students: An NSF STEP Project to Increase Student Retention

Michigan State University College of Engineering,
the Department of Chemical Engineering & Materials Science,
and the College of Natural Science's Division of Science and Mathematics Education
Daina Briedis, Neeraj Buch, Nat Ehrlich, Denise Fleming, Tim Hinds, Colleen McDonough,
Jon Sticklen, Mark Urban-Lurain, and Tom Wolff.

Retention of early engineering students has been identified as a nation-wide concern that will affect the strength of the future engineering workforce and, hence, the role of the United States as a dominant world player in engineering and technology[4]. This requires that we increase the number of B.S. graduates in these disciplines. We must be particularly earnest about this given the current global fiscal downturn and the technological advances needed to stimulate the national and world economies. This paper describes one of a synergistic assembly of projects being undertaken at our institution to address the issue of retention in undergraduate engineering programs.

Over the last decade, enrollments have declined substantially in the College of Engineering at Michigan State University (MSU), although recent upswings have been observed in certain disciplines. The overall decline at MSU is exaggerated above and precedes the national decline largely due to the decline of the manufacturing industry in our state. Thus, our state is in a process of "reinventing" its economy, which will require an increased number of engineers graduated from our universities as a critical key to this process.

Increasing the number of undergraduate engineers can be accomplished by recruitment, retention, or a combination of both. This paper will address the preliminary phases of work being done in our college to address the *retention* of early engineering students. Recruitment efforts are outside the scope of this work and are being effectively addressed by other college efforts. The project described in this abstract is part of a larger, integrated retention effort that has been recently funded by a five-year NSF STEP (STEM **T**alent **E**xpansion **P**rogram) grant. The scope of the project transcends several colleges and two institutions, and includes a multidisciplinary research team of thirteen project investigators and personnel. The project goals are to increase student retention locally by 10 percentage points and to provide a transferable model for increasing retention at other large state institutions. Specifically this paper describes the background and preliminary work that has been implemented for a "Connector Faculty" mentoring program.

Background

Research-active universities award the majority of undergraduate degrees in the United States. Therefore, it is important that these large institutions implement effective means of retaining and educating students at all levels and in all disciplines. Our NSF project focuses mainly on the retention of early engineering students—those in the first one and one-half years of their degree programs.

Causes of attrition from STEM majors have been studied extensively and are fairly well known. The noted National Survey of Student Engagement (NSSE)[5] and a related report by Kuh[2] encourage institutions to critically examine practices that could engage or alienate students. Five benchmarks of effective education practice are used in the NSSE study, and two deal directly with various aspects of the work described in this paper—student-faculty interactions and a supportive campus environment. These studies also place special emphasis on beginning students and emphasize the importance of establishing positive interactions with students as soon as they step over the threshold of the university campus. Young students in particular find it most difficult to discover who they should approach for which kinds of help.

Ohland *et al.*[7], in their analysis of National Survey of Student Engagement (NSSE)[5] data, discuss the fact that persistence in engineering may not be the major cause of a deficiency in the number of engineers being prepared in the United States. However, they do underscore the 40% departure rate of first-year students. While some movement is to be expected, particularly after the freshmen year, Ohland *et al.* also emphasize that benefits of retention programs and the culture change that they bring to engineering faculty have great potential for attracting *more* students to engineering, and have particular value for women and students of color.

The often-cited and perennially seminal work of Seymour and Hewitt[9] describes these early "leavers" as typically falling into one of two categories--those who leave because of inadequate academic qualifications and those who leave because they find the academic culture unsupportive or not engaging. MSU's College of Engineering NSF project presents a suite of four integrated programs that address both categories of leavers and their transition from high school into engineering undergraduate programs, thereby increasing retention. These components are shown on Figure 1. They are

- A peer-assisted learning program for key core courses taken by early engineering students; this is similar to a nationally known program called "Supplemental Instruction." It has been given an acronym PAL (Peer Assisted Learning) on our campus.
- Content cross-linkages among key technical core courses taken by early engineering students. This is an effort to provide relevance to the basic mathematics and science courses required for admission to the college.
- Formative course diagnostics and linked capability-building exercises that students can use to strengthen skills in the same prerequisite courses.
- "Connector Faculty:" Increased faculty connections and mentoring—a friendly face—for early engineering students.

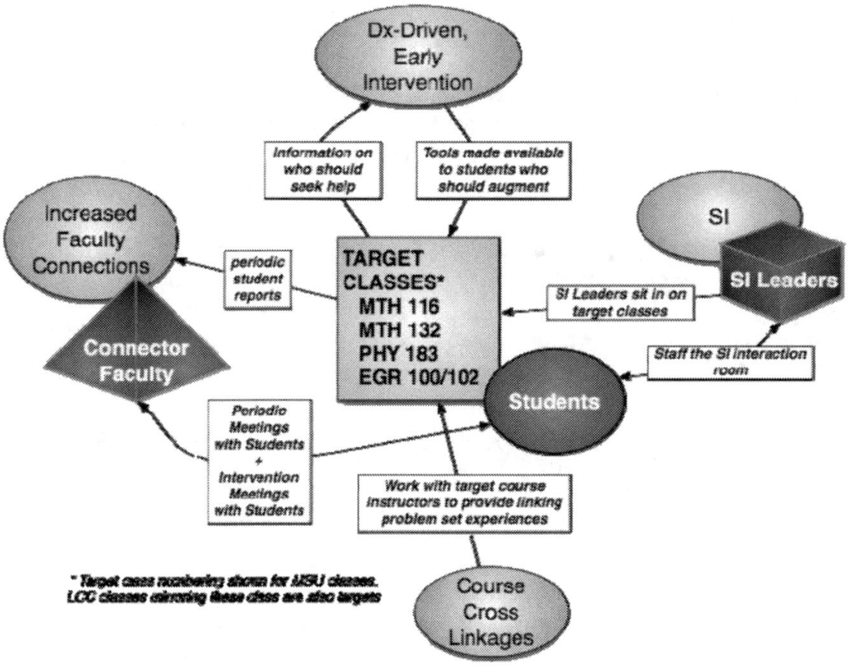

Figure 1. Integrated structure of project components

While such techniques have been used at other institutions and in other STEM disciplines, their use as a package addresses many of the benchmarks of effective educational practice described by Kuh[2]. The table below illustrates how the four project components integrate with each other to address the major concerns of student migration out of engineering. It is clear that Connector Faculty (CF) have a major role in the successful outcome of this project.

Table 1. How Project Components Address Retention

Project Component	Academic difficulty	Disconnected coursework	Isolated as individual student	Lack of faculty concern
Connector Faculty	✓	✓	✓	✓
Peer Assisted Learning	✓	✓	✓	
Course Cross-Linkages		✓	✓	
Diagnostic-Based Intervention	✓			

What is 'Connecting'?

Students have clear needs in their help-seeking behavior with faculty and advisors. These include the expected issues of advice on academic and career alternatives, accurate information on curriculum, help in understanding course material, and practical advice on personal matters that affect academics[9]. Relevant to the work described in this paper is the need that students have to have someone take an interest not only in their academic progress, but in them as a person[1,10]. Vogt[12] has shown that a faculty member's manner toward a student is critical to a student's perception of self-efficacy, a major determinant of student achievement. Even small attitudinal changes toward students can make a huge difference. Tinto[11] points out that "students are more likely to persist when they find themselves in settings that hold high expectations for their learning, provide needed academic and social support, and actively involve them with other students and faculty in learning." We view this as a significant opportunity for adding value to the first-year student's experience.

Yet, there is also the fundamental conflict of the faculty's role as an academic or career advisor and their role as gatekeepers of solid talent for their professions. For this reason, the Connector Faculty (CF) in this project are volunteers who are neither directly instructors nor academic advisors of the students. They have no evaluative role, and their purpose is three-fold: (a) to serve as a role model for the engineering profession; (b) to project the core value of the college that engineering faculty really care about the early engineering students--to be the friendly "face" of the faculty and of the profession to their students; and (c) to provide a "value-added" experience that benefits the attitude and learning of the first-year students.

Our college data reflect national trends in that our most significant attrition is from the early engineering students. Our "leavers" fit into the same two categories as described in the literature[9]: those who perform well academically, but choose to leave; and students who fall just short of the academic admissions threshold. The target of the CF portion of this work is to address both groups of students through better interactions with faculty mentors. While the goal is not an "at all costs" rescue of each and every potential "leaver," our desire is to keep those students whose true desire is to study engineering. This will be done for all students, not just those deemed at risk.

Given that the enrollment of our first-year students is currently around 900 students and that we have about 160 tenure stream faculty who are involved in undergraduate education, the key to the success of the CF component of our project is to establish a cadre of engineering faculty members who will be willing to serve in a significant way. This represents a significant culture change at our college that is addressed in the following section.

Developing a Climate for Connecting

Bringing change to engineering colleges is not always straightforward and requires negotiation of the traditional steps to climate change—denial, resistance, willingness, and commitment.[8] At MSU, the climate for renewed emphasis on undergraduate program at MSU has undergone recent changes. Our college's administration has brought significant and renewed emphasis to the historic importance of excellence in our undergraduate engineering programs. With major impetus provided by the dean, the college has recently revitalized its freshman engineering offerings freshman courses expanding on the former ROSES program (Residential Option for Science and Engineering) to add two new courses: EGR 100, Introduction to Engineering Design, and EGR 102, Introduction to Engineering Modeling. Both courses are required in all but two engineering programs. They are taught by a group of disciplinary faculty and are managed by instructors dedicated to these courses.

In addition, the university has also approved and financially supported the significant expansion of a residential program in engineering that received its first residents in Fall, 2009.

Thus, the confluence of these activities and projects—the new freshmen courses, the residential program, and the five-year NSF grant--are mutually supportive and catalytic and provide the best possible environment in which to establish connections between young students and engineering faculty.

MSU students may declare a major as they matriculate into the university, but they are not "admitted" to the College of Engineering until they satisfy several admission requirements: declaration of a degree-granting engineering major; at least12 credits in residence; completion of six core courses (math, science, EGR 100, EGR 102); and achievement of a minimum grade point average (2.8 to 3.0 depending on the major). Before being admitted to a college, students may change majors freely. After admission to a college, an application must be submitted to the new major. The former group includes the early engineering students we are working to retain.

After significant groundwork and preparation, the CF project was piloted in the spring, 2009 semester. The students in EGR 100 served as the target student population, because it represented a smaller group of first-year students all within the confines of one course. The course enrollment in Spring, 2009 was 195 students of whom 34 were women. The recruitment of CFs was done on a volunteer basis. The second major effort was accomplished in Fall, 2009 with almost 600 students being 'connected' to 69 faculty. Of these 69 faculty, 27 were returning "veterans" from the spring effort.

The Connector Faculty Effort to Date

The first step in our approach to implementing the climate needed to engage faculty volunteers with first-year students was to inform faculty and college administrators of the "opportunities" that would present themselves as a result of the funding of our NSF grant.

We publicized the NSF grant to the college, the institution, and the community. This was accomplished through web postings, press releases, and announcements at faculty meetings. Important and influential subsets of the engineering faculty were addressed directly and separately by project team members. This included the college's administrative group (chairpersons and associate and assistant deans), where the college dean indicated that department faculty were going to be involved in various parts of the projects. Although the dean did not establish a requirement, the expectation for faculty involvement was clearly set. While the faculty reward system has not been changed at this point, we view strong administrative support as an important first step in raising faculty awareness and in changing faculty expectations about the need to support early engineering students. This type of administrative support is clearly critical in the success of this expansive of a project.

Several of the project team members also gave a presentation and led discussion at a monthly college noontime "Brown Bag" Symposium. The topic of the Brown Bag was announced to the entire college, and about twenty faculty members attended this event. The presenters were quick to emphasize the positive influence that CFs could have on the perceptions that students have of engineering careers and of the college. They also pointed out that the faculty involvement was fluid enough that they could, to some degree, tailor their student interactions to suit their schedules and their personalities. The promise of staff support and a small supplies stipend were also offered as enticement. Members of the project team also led discussions at faculty meetings in individual departments. After this early contact, several faculty members stepped up as volunteers for the CF program.

<u>Orientation</u>

The first major Connector Faculty event was an orientation session for the Spring, 2009 volunteer faculty. This event had to occur quickly in the semester so that faculty volunteers would be prepared for a quick start in the program. Except for advising a small number of honors students, engineering faculty typically did not interact with students until the sophomore year. In order to support positive interactions between the CFs and their students, the faculty had to be made aware of who these young students were and how best to make them feel connected.

To help with this aspect of the project, the project team sought assistance from our institution's Counseling Center's Director. Because of her exceptional expertise in developing counseling programs that support academic success, student retention, and faculty development, she was uniquely qualified to lead the orientation and training. The fact that she was a voice of expertise from outside the College of Engineering also carried significant weight with the audience.

The one and one-half hour orientation session covered the following topics:

- Review of the most recent NSSE findings on student engagement[5]—a source of external validation for the project
    - Emphasis on the reluctance of young students to approach faculty either in or outside of class
    - Discussion of "student-faculty interactions" as a benchmark of effective educational practice
- Review of our institution's Cooperative Institution Research Program (CIRP) data[6]
    - Only 26% of our high school students asked a teacher for advice after class
    - Over half the students were frequently bored in their high school classes
    - 69.2% spent less than six hours per week studying or doing homework
- The viewing of connecting with students as an opportunity, not a burden
- Interactive discussion of what to do when the students showed up at the faculty's office
- Role-playing and scenarios that could be typical of a first-year engineering students comments
    - I'm afraid I won't get the grades to get into engineering.
    - I used to get As in high school, but now I am flunking chemistry!

    - I really get nervous for my tests, and I feel as if I don't show what I really know.
    - Engineering is too much work. My roommate never has any homework!
- Active listening skills
- Suggestions for informal meetings with students
- Resources and on-campus referrals
- Recognizing a troubled student
- Information on the structure and course content of EGR 100, the class to whose students the faculty members were 'connecting' (emphasized in Fall, 2009)

The sessions were highly interactive and generated an enthusiasm for the project because of the synergistic interactions among the faculty attending the sessions and the excellent session facilitation by the Director of the Counseling Center. As early participants in the project, this group of faculty was also encouraged to contribute to the design and evolution of the CF program.

The orientation was also used to emphasize the close collaboration that must exist between faculty and the advisers. It is not the purpose of the CFs to take the role of the advisers. The project leaders are particularly sensitive to the need to coordinate CF efforts to academic advising.

A similar orientation was conducted early in the Fall, 2009 semester. The topics were adjusted based on survey feedback and also to include more information about the EGR 100 course.

<u>Student-Faculty Interactions</u>

The next major landmark in the CF project was the first formal meeting between the faculty and their students. The project coordinator assigned three to seven students per faculty member in the pilot (Spring 2009) of this program and as many as nine students per faculty member in the Fall, 2009 in order to accommodate the larger enrollments. Several faculty volunteers, including the NSF project team, were assigned smaller numbers of students. Typically, untenured tenure-stream faculty did not participate as CF or were assigned smaller numbers of students. Students were first matched to faculty according to declared major, but this was not possible for all students. Several women faculty members also requested that they be assigned only women students, but gender was used only as a secondary criterion for assignment.

As suggested by CIRP and NSSE data, getting students over the threshold of faculty offices may be one of the most difficult steps. Students often view faculty as having threatening personas, so enhancing the approachability of faculty is a major factor in establishing continuing relationships. To do this in a non-threatening way, the Connector Faculty met with the target student group (EGR 100) during the lecture section of the class, i.e., the faculty attended the EGR 100 class solely for the purpose of meeting with their students. Students had been briefed about and prepared for this meeting; the idea that these EGR 100 students would have the opportunity to be linked with and mentored by an engineering faculty member had already been introduced to them in classroom presentations early in the semester.

On the day of the meeting, all available Connector Faculty volunteers gathered in the large EGR 100 classroom and were introduced to the entire student group. After brief introductions, each faculty member met with his or her student group. Since the classroom is in a residence hall (the site of our residential program), a cafeteria and coffee shop were readily available for informal first meetings.

In order to improve on several aspects of connecting students with faculty, a more formal first communication between students and faculty was implemented in Fall, 2009. One of the students' first assignments in EGR 100 was to interview their connector faculty to gather information for an essay. This provides a formal means of encouraging students to take the first step across the faculty member's office threshold. Because the students are required to interview their CF, e-mail contact has already been well established. This assignment is ongoing at the time of this writing. Details of the outcome will be provided in the oral presentation.

Beyond the first formal interactions, CFs conducted additional meetings with their students including meetings during office hours, discussions over coffee, lunch and dinner get-togethers, and an organized trip to the MSU Cyclotron.

Project Evaluation

A strong project evaluation plan involving the MSU Institute for Public Policy and Social Research (IPPSR) is in place. One purpose of the evaluation aspect is to document our project activities, processes, products, and progress. A major component is also to improve the five-year project in a formative manner.

Survey #1: In this role, our IPPSR collaborator first distributed a survey to all faculty in engineering, mathematics, and physics to learn more about their awareness of the project and their willingness to participate. [Mathematics and physics faculty members are involved in another aspect of the NSF project and do not serve as connector faculty.] This survey was conducted in November, 2008. Response rates from the College of Engineering were high with 50 engineering faculty responding to the web-based survey. Responses from mathematics and physics were lower and are not cited here.

We were most interested in the initial attitude of the respondents as to whether the program would meet the goal of increasing retention of students by 10% over five years. We asked for a response on a 7-point Likert scale with end points defined as Very Likely and Very Unlikely to succeed. The results for engineering faculty are shown in Table 1. The overall estimated probability of success based on engineering, physics, and math faculty attitudes is 71%.

Table 1. Survey responses of faculty attitudes on project success.

Very Likely			Uncertain			Very Unlikely
9%	32%	26%	23%	4%	2%	4%

Some comments provided by those who rated the program in one of the first two positive categories included the following:

> "When students feel more support and like they are part of a community, they are less likely to leave."
>
> "Early engineering students have not previously been the focus of much attention from engineering faculty, and a program that changes that should have some significant impact."
>
> "There is a significant level of enthusiasm and sound upper administrative support."

Survey #2: The Spring 2009 CF volunteers were surveyed at mid-semester to obtain feedback on the orientation session and to gather early data on student meetings. Of the 33 participating connector faculty, 23 responded; 14 rated the orientation as excellent or good. At this mid-semester point, on average, CFs had already met with about 1/3 of their assigned students.

Nineteen connector faculty also met for a mid-semester debriefing/feedback meeting. The discussion provided constructive comments for formative and summative application.

Survey #3: A final survey was administered to the CFs at the end of the semester. This time 27 of the 33 responded. Sixty-seven percent *agreed* and an additional 19% *agreed strongly* that the CF program will aid in achieving our goals of increased student retention (no one disagreed strongly). With regards to student meetings, 1/3 of the faculty had already met with more than one student, 1/3 had met with one, and 1/3 had met with no students.

Survey #4: Students were also surveyed at the end of the spring semester with a 21% response rate. Forty-one percent saw the CF program as helpful and a good use of time; 13% judged it helpful but too time-consuming. Another 41% thought the program was not helpful and didn't spend much time with it. Five percent did not think the program was helpful and spent too much time on it.

Students also rated the topics discussed at their meetings with faculty. Students used six topics to judge the "usefulness" of the program (with an α better than 0.05). Topics which appeared to be most useful to students (correlated to best satisfaction with the program and their CF) included:

- Student's study habits
- MSU in general
- Student's hobbies/interests
- Engineering program requirements
- Faculty member's work
- Initiation of next meeting by faculty member

Students provided a wide assortment of comments about the CF program which could be summarized in the following list:

- A "waste of time"
- Amazing! Take advantage of it!

- Good chance to get to know faculty
- Just talk to them; they're people too!
- Helped get an internship
- Make meetings more frequent
- Make meetings less frequent

The variety here is not surprising!

Project Improvement Plans for Year 2

Based on analysis of the data and observations, plans for Year 2 include:

- Establishing a better means of communicating with the students; they do not answer e-mail
- Enabling CF access to their student records
- Integrating better into EGR 100
- Working with FIRST SEMESTER students (the pilot was with 2nd-semester students, who have already established "habits")
- Recruiting a larger pool of CF volunteers

At the time of this writing, several of these plans have already been implemented. As mentioned earlier in this paper, we have recruited 69 faculty volunteers to serve almost 600 students. The instructor of EGR 100 is a more active, integrated member of the project team and has supported improved linkages between the course and the CF project, such as the writing assignment requiring students to interview their CFs.

Conclusions

It appears that at least a core group of faculty, probably most already with significant commitment to excellence in undergraduate education, has already contributed significant effort to making the CF program successful. The sustainability of the project will depend on obtaining buy-in from the faculty at large. It is clear that a supportive administration at the college and department levels, faculty who understand that students are not the only ones who benefit from engagement, and effective project leadership and coordination will be vital to the sustainability of this project.

"Student success is the product of thousands of small gestures extended on a daily basis by caring, supportive educators sprinkled throughout the institution" George Kuh[3]

This material is based upon work supported by the National Science Foundation under award 0757020 (DUE). Any opinions, findings, and conclusions or recommendations expressed in this material are those of the author(s) and do not necessarily reflect the views of the National Science Foundation (NSF).

Bibliography

1. Bransford, J.D., A.L. Brown et al, eds. (1999). How people learn: Brain, Mind, experience and school. Washington, D.C., National Academy Press.

2. Kuh, G.D. (2008). High-impact educational practices: What they are, who has access to them, and why they matter. Washington D.C.: Association of American Colleges and Universities.

3. Kuh, G.D. (2007). Student Success in College: Lessons from High Performing Colleges and Universities. Presented at the University of Maine System, March 20, 2007.

4. National Science Board (2006). Science and Engineering Indicators 2006 (Two volumes), National Science Foundation, Vol. 1, NSB 06-01; Vol. 2 NSB 06-01A.

5. National Survey of Student Engagement (2008). Promoting Engagement for All Students: The Imperative to Look Within.

6. http://www.gseis.ucla.edu/heri/cirp.php [Cooperative Institutional Research Program (CIRP)]

7. Ohland, M.W., S. D. Sheppard, G. Lichtenstein, O. Eris, D. Chachra, and R. A. Layton, (2008). "Persistence, Engagement, and Migration in Engineering Programs," J. Eng. Educ., Vol. 97(3), 259.

8. Rogers, Gloria. Institute for the Development of Excellence in Assessment Leadership. Workshop presented July 31-August 4, 2006, Baltimore, MD.

9. Seymour, Elaine and Nancy M. Hewitt (1997). Talking about Leaving: Why Undergraduates Leave the Sciences. Boulder, CO, Westview Press.

10. Stipek, D.J. (1996). Motivation and instruction. Handbook of educational psychology. D.D. Berliner and R.C. Calfee, eds. New York.

11. Tinto, V. (2003). Establishing conditions for student success. Improving completion rates among disadvantaged students. Pp. 1-10. L. Thomas, M. Cooper, and J. Quinn, eds. Sterling, VA, Trentham Books Limited.

12. Vogt, C. (2008). "Faculty as a Critical Juncture in Student Retention and Performance in Engineering Programs," J. Eng. Educ., Vol 97(1), 27.

Graduate Student's Perceptions of Academia

Laura Dykes, University of Maryland, Baltimore County, Baltimore, MD

The student/ advisor relationship is a primary source of the first hand knowledge doctoral students acquire about academic careers. Of particular interest is the role of advisors in the decision making process of underrepresented minority students. A study to analyze the advisor's influence is being conducted on students from Maryland's Alliance for Graduate Education and the Professoriate (AGEP) which is comprised of three schools: University of Maryland-College Park, University of Maryland-Baltimore County and the University of Maryland-Baltimore.

A 31-item Likert response survey was developed to assess student attitudes towards doctoral advisors, AGEP mentors and careers in academia. The survey was administered at two AGEP events and resulted in 102 respondents, 44 of which were from STEM majors. The survey results indicate significant Spearman's rho correlations between positive attitudes towards academic careers and plans to pursue an academic career. The results did not show significant correlations between doctoral advisors or AGEP mentors and positive attitudes towards academia. This paper discusses the survey construction, results and plans for future work.

See more of this Session: Retention, Course Development and Curriculum
See more of this Group/Topical: Education

Analytical Chemistry for Chemical Engineers

Darrell L. Sparks, Chemical Engineering, Mississippi State University, Mississippi State University, MS

Tracy Benson, Chemical Engineering, Lamar University, Beaumont, TX

Ashli Brown, Biochemistry and Molecular Biology, Mississippi State University, Mississippi State University, MS

William E. Holmes, Mississippi State Chemical Laboratory, Mississippi State University, Mississippi State, MS

Rafael Hernandez, Chemical Engineering, Mississippi State University, Mississippi State University, MS

A common challenge facing chemical engineering graduate students is a lack of basic knowledge in analytical chemistry techniques. To help students surmount this challenge, a new elective course at Mississippi State University entitled BioAnalytical Research Methods was developed. This course was a collaborative effort between the Swalm School of Chemical Engineering, Department of Biochemistry and Molecular Biology, and the Mississippi State Chemical Laboratory. The topics covered in the course included chromatography, mass spectrometry, spectroscopy, and proteomics. The class combined theory of analytical techniques with hands-on experience to aid students in understanding the applicability of common analytical methods. Students were also encouraged to teach others about research methods they had personally used in their research. This paper will discuss the first iteration of the new course in terms of material covered, laboratory experiments, grading components, student response, and curriculum modifications that will be implemented for the next iteration of the class.

See more of this Session: Retention, Course Development and Curriculum
See more of this Group/Topical: Education

Ideas to Consider for NEW Chemical Engineering Educators: Junior and Senior Level Courses

Jason M. Keith, Department of Chemical Engineering, Michigan Technological University, Houghton, MI

David L. Silverstein, University of Kentucky, Paducah, KY

Donald P. Visco, Chemical Engineering, Tennessee Technological University, Cookeville, TN

So, you are going to teach a core chemical engineering course next term that you have not taught before. It's time to come up with some new ideas to revolutionize that core course in ways that will amaze students and maximize learning, right? Or perhaps the maxim about "an hour in the library is worth a month in the laboratory" might be meaningful in the context of teaching. This paper summarizes the authors' selection of the most effective, innovative approaches reported recently in the literature or discussed at previous conferences for upper-division core courses in chemical engineering, as presented at the 2007 ASEE Summer School for Chemical Engineering Faculty. The challenges associated with particular courses and solutions successfully applied to address those challenges will also be described. Courses covered in this paper include solution thermodynamics, heat and mass transfer, kinetics, and process control.

See more of this Session: Retention, Course Development and Curriculum
See more of this Group/Topical: Education

Impact of Rose-Hulman Institute of Tecnology's Freshman Engineering Course in Chemical Engineering On Retention

Scott McClellan, Chemical Engineering, Rose-Hulman Institute of Technology, Terre Haute, IN

Sharon Sauer, Chemical Engineering, Rose-Hulman Institute of Technology, Terre Haute, IN

The transition from the freshman to sophomore years is often considered to be the critical stage in the retention of engineering students. Our students' first exposure to the chemical engineering curriculum is during the spring quarter of the freshman year in our Introduction to Design course. This recently revised course exposes the students to a wide range of chemical engineering applications in a team-focused, problem-based learning environment, with one of the goals being the retention of freshmen in the chemical engineering curriculum. In this presentation, we will examine the impact of this course on freshman retention as well as the effect the changes have had on the students' perception of chemical engineering as a profession.

See more of this Session: Retention, Course Development and Curriculum
See more of this Group/Topical: Education

Development of New Curriculum in Pharmaceutical Bioprocessing: a Novel Educational Area in Pharmaceutical Engineering

Piero M. Armenante, Otto H. York Department of Chemical, Biological, and Pharmaceutical Engineering, New Jersey Institute of Technology, Newark, NJ

Laurent Simon, Otto H. York Department of Chemical, Biological and Pharmaceutical Engineering, New Jersey Institute of Technology, Newark, NJ

The Otto H. York Department of Chemical Engineering at NJIT, recently renamed the Otto H. York Department of Chemical, Biological and Pharmaceutical Engineering to clearly signal that it is expanding in areas that it considers strategic, has been a pioneer in the development of educational programs for students and professionals who plan or are already working in the pharmaceutical industry. While significant activities both on the educational and the research fronts have taken place in the department in the now "traditional" pharmaceutical engineering area, much less emphasis has been placed so far on the biological engineering component, although significant biological expertise already exist in the department and this is clearly an area of development for the entire pharmaceutical industry.

With this contribution, we would like to present a case study on the development of a novel MS program in Pharmaceutical Bioprocessing, which can be conceived as a new area of growth for the pharmaceutical industry and an area in need of new educational programs. The driving force for this development is in the pharmaceutical industry itself, which is rapidly and massively moving into the biotech area, since it sees this as the new growth area for the industry and the only major alternative to the declining prospects for the development of future block-busters using small-molecule pharmaceuticals. There is the potential for a strong student demand in this area, also fueled by the need of the industry to re-tool some of its workforce.

In this presentation, we show how existing pharmaceutical and chemical engineering programs can be partially modified and readapted to address the new needs of the pharmaceutical industry, as far as the area of processing engineering applied to bioprocesses is concerned.

See more of this Session: Retention, Course Development and Curriculum
See more of this Group/Topical: Education

Structured Organic Particulate Systems in the Chemical Engineering Curriculum

Stephanie Farrell, Chemical Engineering, Rowan University, Glassboro, NJ

Z. Otero Gephardt, Chemical Engineering, Rowan University, Glassboro, NJ

Mariano Savelski, Chemical Engineering, Rowan University, Glassboro, NJ

C. Stewart Slater, Chemical Engineering, Rowan University, Glassboro, NJ

Eileen Batten, Chemical Engineering, Rowan University, Glassboro, NJ

Christopher DelVecchio, Chemical Engineering, Rowan University, Glassboro, NJ

Michael Glasspool, Chemical Engineering, Rowan University, Glassboro, NJ

Adrian Kosteletski, Chemical Engineering, Rowan University, Glassboro, NJ

Danielle Lussier, Chemical Engineering, Rowan University, Glassboro, NJ

Katherine Ross, Chemical Engineering, Rowan University, Glassboro, NJ

Tatsiana Sokal, Chemical Engineering, Rowan University, Glassboro, NJ

Sarah Wilson, Chemical Engineering, Rowan University, Glassboro, NJ

Kathryn Whitaker, Chemical Engineering, Rowan University, Glassboro, NJ

Educational modules on structured organic particulate systems (SOPS) used in pharmaceutical technology have been developed for the Chemical Engineering curriculum. Modules present topics such as drug delivery, pharmaceutical particulate system properties, and powder flow through problem sets, laboratory experiments, and highly visual and cost effective demonstrations. The materials and activities educate students in specific areas of pharmaceutical particulate processing, such as tableting and mixing, and engineering topics, such as material and energy balances. The problems are designed to be incorporated into existing ChE courses by presenting an alternative class assignment or activity related to the core course topics. Initial efforts have focused on introductory chemical engineering courses at the freshman and sophomore level, mass transfer, and senior/graduate electives. The educational development is part of the Rowan outreach efforts associated with the NSF-sponsored ERC on SOPS at Rutgers University, New Jersey Institute of Technology, Purdue University and the University of Puerto Rico-Mayaguez.

See more of this Session: Retention, Course Development and Curriculum
See more of this Group/Topical: Education

An Integrated Approach to Chemical Engineering Undergraduate Curriculum Reform

Lâle Yurttas, Chemical Engineering, Texas A&M University, College Station, TX

Larissa Pchenitchnaia, Chemical Engineering, Texas A&M University, College Station, TX

Jeffrey Froyd, Center for Teaching Excellence, Texas A&M University, College Station, TX

Mahmoud El-Halwagi, Chemical Engineering, Texas A&M University, College Station, TX

Charles Glover, Chemical Engineering, Texas A&M University, College Station, TX

Irvin W. Osborne-Lee, Chemical Engineering, Prairie View A&M University, Prairie View, TX

Patrick L. Mills, Dept of Chemical & Natural Gas Engineering, Texas A&M University-Kingsville, Kingsville, TX

Sponsored by the National Science Foundation (NSF), three chemical engineering departments within the Texas A&M University System have made significant progress in their efforts to renew their curricula to address pressures of multi-disciplinary technological developments and the growing breadth of abilities and knowledge areas expected for competitive chemical engineering graduates. Three broad strategies to accomplish project goals have been implemented: (1) curriculum content reform and development; (2) student assessment activities, and (3) faculty development initiatives.

The three strategies are being implemented through six key mechanisms:

(i) Identifying and organizing curriculum development activities around four course strings to improve integration of learning outcomes and activities.

(ii) Developing interlinked curriculum components to organize and reinforce core ideas in chemical engineering curricula.

(iii) Using service learning in required chemical engineering courses.

(iv) Integrating comprehensive assessment plans and processes throughout the chemical engineering curriculum and using the data to make and evaluate changes.

(v) Offering faculty development activities to offer knowledge and development opportunities for chemical engineering faculty members.

(vi) Sharing our experiences with audiences beyond the Texas A&M University System.

In the course of the project, faculty incorporated innovative pedagogies, such as teams, cooperative learning, and project based learning. Faculty in a community-centered effort have developed outcomes and expectations for incoming students in the courses that comprise the four course strings. Web-based modules that span courses and application areas have been developed and are being integrated into core courses. Consistent conversations about continuous improvement and assessment plan engaged faculty in a collective effort for sustained change. This paper will discuss the outcomes and experiences of the three-year, NSF-sponsored project to reform chemical engineering undergraduate curricula.

See more of this Session: Retention, Course Development and Curriculum
See more of this Group/Topical: Education

AUTHOR INDEX

AUTHOR INDEX

AUTHOR INDEX

AUTHOR INDEX

AUTHOR INDEX

AUTHOR INDEX

AUTHOR INDEX

AUTHOR INDEX

AUTHOR INDEX